UNITEXT for Physics

T0172040

UNITEXT for Physics series, formerly UNITEXT Collana di Fisica e Astronomia, publishes textbooks and monographs in Physics and Astronomy, mainly in English language, characterized of a didactic style and comprehensiveness. The books published in UNITEXT for Physics series are addressed to graduate and advanced graduate students, but also to scientists and researchers as important resources for their education, knowledge and teaching.

More information about this series at http://www.springer.com/series/13351

Gleb Arutyunov

Elements of Classical and Quantum Integrable Systems

 Springer

Gleb Arutyunov
II. Institute for Theoretical Physics
University of Hamburg
Hamburg, Germany

ISSN 2198-7882   ISSN 2198-7890   (electronic)
UNITEXT for Physics
ISBN 978-3-030-24200-8   ISBN 978-3-030-24198-8   (eBook)
https://doi.org/10.1007/978-3-030-24198-8

This Springer imprint is published by the registered company Springer Nature Switzerland AG
The registered company address is: Gewerbestrasse 11, 6330 Cham, Switzerland

Preface

The theory of integrable models constitutes a basis and at the same time a theoretical laboratory for the study of dynamical systems in general. This theory is most advanced in making quantitative statements about the behaviour of classical and quantum dynamical systems that fall in the category of being integrable. The development of the theory has a remarkable history which started from the 1853 paper by Joseph Liouville. The twenty-first century has already been marked by finding new important applications of integrability that go beyond the traditional area of condensed matter physics. One of these applications is in the context of the gauge-string duality conjecture, also known as the AdS/CFT correspondence. Unravelling the integrable nature of the quantum four-dimensional supersymmetric $N = 4$ gauge theory and its string theory counterpart has led to significant progress both in the spectral problem and also for determination of the gauge theory correlation functions. The application of methods and ideas of integrable systems to the quantum field theory on the two-dimensional string worldsheet allowed to obtain spectacular results in the four-dimensional gauge theory avoiding the evaluation of zillions of Feynman diagrams. Perhaps, this is for the first time that string theory reveals, through its integrable structure, its wonderful elegance and simplicity in comparison to the standard field-theoretic approach based on perturbation theory. Still, an underlying structure behind the gauge field and string integrability is largely unconventional and requires further theoretical explorations, which is partly our motivation to discuss the phenomenon of integrability in this book.

The book arose from a number of special courses taught by the author for undergraduate and graduate students, its aim is to acquaint the reader with some elements of the modern theory of integrable systems. The choice of the material is largely based on the author's taste and his personal involvement in the subject, as well as natural restrictions put on the size. In this respect, this book is not meant as a competitor to the existing excellent textbooks and reviews on the subject, but as a valuable addition where some of the topics are treated and emphasised perhaps more than usual. For instance, we decided to restrict the discussion to finite-dimensional integrability leaving out the theory of classical and quantum integrable field theories. On the other hand, once the Bethe Ansatz approach to the spectrum of quantum-mechanical models is understood through the framework of the Factorised Scattering Theory, it takes little effort to adapt it for the field-theoretic case. Perhaps, one feature that makes this book rather unique is a large amount of detailed computations to uncover all possible technical issues. Also, our discussion of the Poisson reduction in the context of the classical and quantum Ruijsenaars-Schneider models is more monographic rather than of text-book style and in this respect, it might be useful for experts working in this area.

The book consists of six chapters. We start from the Liouville theorem in Chap. 1, which allows us to introduce the concept of integrability. We further discuss symmetries and conservation laws and also a number of important tools, such as the Lax pair and the classical r-matrix. We also exhibit a variety of integrable models, some of them are also explicitly solved in the spirit of the Liouville theorem. The material of this chapter belongs to textbooks and can be easily adapted in general courses on integrable models. Chapter 2 deals with symmetries and the group-theoretical origin of integrable systems. We introduce the Hamiltonian and Poisson reduction techniques and give a detailed discussion of a number of important phase spaces with symmetries such as the cotangent bundle of a Lie group and the Heisenberg double. The integrable models of Calogero-Moser-Sutherland and Ruijsenaars-Schneider types are discussed from this reduction point of view. These models are prototypical for finite-dimensional integrability and we study them throughout the book not only to exemplify various integrability concepts but also because of their own intrinsic interest. In particular, in Chap. 3 we show how their exact spectrum can be determined in terms of orthogonal polynomials in the case of the periodic potentials. Chapter 4 is devoted to the Factorised Scattering Theory that arises upon combining the standard quantum-mechanical scattering theory with the existence of a sufficiently large family of commuting operators. The central role here is played by the Bethe wave function and the two-body scattering matrix. In Chap. 5 we discuss one of the main techniques of quantum integrable models—the Bethe Ansatz. We start from the coordinate Bethe Ansatz that arises upon imposing the periodic boundary conditions of the Bethe wave function. Diagonalisation of the matrix Bethe-Yang equations is then done by means of the Algebraic Bethe Ansatz which allows us to introduce the spin chain realisation of highest weight representations of the symmetric group. Finally, Chap. 6 is about thermodynamics of integrable models, where it is explained how to derive the basic thermodynamic quantities by exploiting the integrable structure of the corresponding model. In total, the book provides a rather broad spectrum of integrability topics ranging from the area of geometry to statistical physics. For additional material the reader is invited to consult the bibliography at the end of each chapter.

It is my distinct pleasure to thank Sergey Frolov for enjoyable long-term collaboration on integrable models. Some of the results presented in the book would not be possible without this collaboration. I greatly appreciate the help of László Fehér, Sylvain Lacroix, Jules Lamers, Hannes Malcha, Enrico Olivucci, Alessandro Sfondrini and Stijn van Tongeren, who have read various parts of the book at different stages of its creation and gave me many valuable comments. I am especially thankful to Rob Klabbers, who carefully read the whole manuscript and proposed a number of important improvements. Finally, I am deeply indebted to my wife Anna and my daughter Sophia for their constant support, encouragement and for keeping me in the winning mood through the whole project. This book is dedicated to them.

Halstenbek, Germany   Gleb Arutyunov
April 2019

Note on Notation

The summation symbol is used whenever an indication of the summation range is necessary and sometimes also for aesthetic reasons. Otherwise, we resort to Einstein's convention for summation of indices. In some cases, to lighten the notation, we omit the range of a summation index which should be clear from the context. We reserve the letter N for the dimension of the configuration space and for the number of particles. Sometimes we want to indicate the physical dimension of a certain quantity and we use for this purpose the brackets $[]$, for instance, $[a] = [\hbar]$ means that the dimension of the quantity a coincides with the dimension of the Planck constant. The space of $N \times N$ matrices with complex coefficients is denoted as $\mathrm{Mat}_N(\mathbb{C})$, the notation $\mathrm{GL}_N(\mathbb{C})$ stands for the general linear group over $\mathbb{C}$.

The same variable ℓ is used to denote the length parameter of the hyperbolic or trigonometric CMS and RS models, and also as an element of the dual space of a Lie algebra. It is, of course, clear from the context which meaning it should be attached to. The same concerns μ, either it is the mass parameter $\mu = mc$ of the RS models or the moment map $\mu : \mathscr{P} \to \mathfrak{g}^*$.

Contents

Chapter 1
Liouville Integrability

When, however, one attempts to formulate a precise definition of integrability, many possibilities appear, each with a certain intrinsic theoretic interest.

George D. Birkhoff
Dynamical systems, AMS, 1927

In fact, the theorem of Liouville ...covers all the problems of dynamics which have been integrated to the present day.

Vladimir I. Arnold
Mathematical methods of classical mechanics

The Liouville theorem in classical mechanics states the conditions under which the equations of motion of a dynamical system can always be solved by means of a well-established mathematical procedure. As such, this theorem naturally provides a definition of an integrable system. After a brief reminder on classical mechanics, we present a modern formulation of the Liouville theorem due to Arnold, discuss the symmetry origin of conservation laws and give a number of representative examples of integrable models. Also, we introduce the main tools for exhibiting and studying classical integrability such as the Lax pair and classical r-matrix.

1.1 Liouville Integrable Models

To create the necessary background for the discussion of the Liouville theorem and Liouville integrable models, we need some general facts from classical mechanics, Poisson and symplectic geometry [1, 2].

© Springer Nature Switzerland AG 2019
G. Arutyunov, *Elements of Classical and Quantum Integrable Systems*,
UNITEXT for Physics, https://doi.org/10.1007/978-3-030-24198-8_1

1.1.1 Dynamical Systems of Classical Mechanics

We start with recalling the two ways dynamical systems are described in classical mechanics. The first description is known as the lagrangian formalism and is equivalent to the "principle of stationary action". Consider a point particle with mass m which moves in an N-dimensional space with coordinates $q = (q^1, \ldots, q^N)$ and a potential $V(q)$. Newton's equations which describe the particle's trajectory are

$$m\ddot{q}^i = -\frac{\partial V}{\partial q^i}. \tag{1.1}$$

These equations can be obtained by extremising the following action functional

$$S[q] = \int_{t_1}^{t_2} dt\, L(q, \dot{q}, t) = \int_{t_1}^{t_2} dt\, \left(\frac{m\dot{q}^2}{2} - V(q)\right). \tag{1.2}$$

According to the principle of stationary action, the actual trajectories of a dynamical system (particle) are the ones that extremise S.

In general, we consider the *lagrangian* L as an arbitrary function of q, $\dot{q}$ and time t. The equations of motion are obtained by extremising the corresponding action

$$\frac{\delta S}{\delta q^i} = \frac{\partial L}{\partial q^i} - \frac{d}{dt}\left(\frac{\partial L}{\partial \dot{q}^i}\right) = 0$$

and they are called the *Euler-Lagrange equations*. An assumption that L does not involve higher order time derivatives implies that the corresponding dynamical system is fully determined by specifying initial coordinates and velocities. Indeed, for a system with N degrees of freedom there are N Euler-Lagrange equations of second order. Thus, the general solution will depend on $2N$ integration constants, which are determined by specifying e.g. the initial coordinates and velocities.

Note that adding to the lagrangian a time derivative of a function which depends on coordinates and time only: $L \to L + \frac{d}{dt}\Lambda(q, t)$ will not influence the Euler-Lagrange equations. Indeed, the variation $\delta S'$ of the new action S' will be

$$\delta S' = \delta S + \int_{t_1}^{t_2} dt\, \frac{d}{dt}\delta\Lambda(q, t) = \delta S + \frac{\partial\Lambda}{\partial q^i}\delta q^i\Big|_{t=t_1}^{t=t_2},$$

where δS is the variation of the original action S. Since in deriving the equations of motion the variations of coordinates are assumed to vanish at the initial and final moments of motion, we get that $\delta S' = \delta S$ and, as a result, the Euler-Lagrange equations remain unchanged.

If L does not explicitly depend on t, then

$$\frac{dL}{dt} = \frac{\partial L}{\partial \dot{q}^i}\ddot{q}^i + \frac{\partial L}{\partial q^i}\dot{q}^i.$$

Substituting here $\frac{\partial L}{\partial q^i}$ from the Euler-Lagrange equations, we get

$$\frac{dL}{dt} = \frac{\partial L}{\partial \dot{q}^i}\ddot{q}^i + \frac{d}{dt}\left(\frac{\partial L}{\partial \dot{q}^i}\right)\dot{q}^i = \frac{d}{dt}\left(\frac{\partial L}{\partial \dot{q}^i}\dot{q}^i\right).$$

Therefore, we find

$$\frac{d}{dt}\left(\frac{\partial L}{\partial \dot{q}^i}\dot{q}^i - L\right) = 0,\tag{1.3}$$

as a consequence of the equations of motion. Thus, the quantity

$$H \equiv \frac{\partial L}{\partial \dot{q}^i}\dot{q}^i - L\tag{1.4}$$

is conserved under the time evolution of our dynamical system. For our particular example,

$$H = m\dot{q}^2 - L = \frac{m\dot{q}^2}{2} + V(q) = T + V \equiv E,$$

where T is the kinetic energy, $\dot{q}^2 \equiv \dot{q}^i\dot{q}^i$. Thus, H is nothing else but the energy E of the system; the energy is conserved due to the equations of motion. In general, dynamical quantities which remain unchanged under the time evolution are called *conservation laws* or *integrals of motion*. Conservation of energy is one of the main examples of conservation laws.

Introduce the quantity called the *canonical* momentum

$$p_i = \frac{\partial L}{\partial \dot{q}^i}, \qquad p = (p_1, \ldots, p_N).$$

Obviously, for a particle $p_i = m\dot{q}^i$. If $V = 0$, then

$$\dot{p}_i = \frac{d}{dt}\left(\frac{\partial L}{\partial \dot{q}^i}\right) = 0$$

by the Euler-Lagrange equations. Thus, in the case of vanishing potential, the particle momentum is an integral of motion. This is another example of a conservation law.

Let us now we recall the second description of dynamical systems, which exploits the notion of the hamiltonian. The energy of a system expressed via canonical coordinates and momenta is called the *hamiltonian*:

$$H(p, q) = \frac{p^2}{2m} + V(q).$$

where $p^2 \equiv p_i p_i$. Given the hamiltonian, Newton's equations can be rewritten as

$$\dot{q}^j = \frac{\partial H}{\partial p_j} , \qquad \dot{p}_j = -\frac{\partial H}{\partial q^j} . \tag{1.5}$$

These are equations of motion in the hamiltonian form or Hamilton's equations. These equations can also be obtained by means of the variational principle. The corresponding action has the form, *cf.* (1.2) and (1.4),

$$S[p, q] = \int_{t_1}^{t_2} \left(p_i \dot{q}^i - H(p, q) \right) \mathrm{d}t . \tag{1.6}$$

Varying this action with respect to p and q, *considered as independent variables*, one obtains the hamiltonian equations.

Hamilton's equations can be represented in the form of a single equation. Introduce two $2N$-dimensional vectors

$$x = \begin{pmatrix} q \\ p \end{pmatrix} , \qquad \nabla H = \begin{pmatrix} \frac{\partial H}{\partial q^j} \\ \frac{\partial H}{\partial p_j} \end{pmatrix}$$

and $2N \times 2N$ matrix J:

$$J = \begin{pmatrix} 0 & -\mathbb{1} \\ \mathbb{1} & 0 \end{pmatrix} , \tag{1.7}$$

where $\mathbb{1}$ is the $N \times N$ unit matrix. Then (1.5) are concisely written as

$$\dot{x} = -J \cdot \nabla H , \qquad \text{or} \qquad J \cdot \dot{x} = \nabla H . \tag{1.8}$$

The vector $x = (x^1, \ldots, x^{2N})$ defines a state of a dynamical system in classical mechanics. The set of all states forms the *phase space* $\mathcal{P} = \{x\}$ of the system which in the present case is the $2N$-dimensional space with the euclidean metric $(x, y) = \sum_{i=1}^{2N} x^i y^i$. Solving Hamilton's equations with given initial conditions (p_0, q_0) representing a point in the phase space, we obtain a phase space curve

$$p \equiv p(t; p_0, q_0) , \qquad q \equiv q(t; p_0, q_0)$$

passing through this point. As follows from the uniqueness theorem for ordinary differential equations, there is one and only one phase space curve through every phase space point.[1]

Let $\mathcal{F}(\mathcal{P})$ be the space of smooth real-valued functions on $\mathcal{P}$. It carries the structure of an algebra with respect to the pointwise multiplication and its elements are called *observables*. Using the matrix J, one can define on $\mathcal{F}(\mathcal{P})$ the following *Poisson bracket*

---

[1] The phase curve may consist of a single point. Such a point is called an *equilibrium position*.

$$\{f, g\}(x) = J^{ij}\partial_i f \partial_j g = \sum_{i=1}^{N}\left(\frac{\partial f}{\partial p_i}\frac{\partial g}{\partial q^i} - \frac{\partial f}{\partial q^i}\frac{\partial g}{\partial p_i}\right)$$

for any $f, g \in \mathcal{F}(\mathcal{P})$. The Poisson bracket is a map $\mathcal{F}(\mathcal{P}) \times \mathcal{F}(\mathcal{P}) \to \mathcal{F}(\mathcal{P})$ which has the following properties

(1) Linearity $\{f + \alpha h, g\} = \{f, g\} + \alpha\{h, g\}$;
(2) Skew-symmetry $\{f, g\} = -\{g, f\}$;
(3) Jacobi identity $\{f, \{g, h\}\} + \{g, \{h, f\}\} + \{h, \{f, g\}\} = 0$;
(4) Leibniz rule $\{f, gh\} = \{f, g\}h + g\{f, h\}$

for arbitrary functions $f, g, h \in \mathcal{F}(\mathcal{P})$ and $\alpha \in \mathbb{R}$. The first three properties imply that the Poisson bracket introduces on $\mathcal{F}(\mathcal{P})$ the structure of an infinite-dimensional Lie algebra, while the Leibniz rule expresses the compatibility of the bracket with multiplication in $\mathcal{F}(\mathcal{P})$. Due to this rule, the bracket is fully determined by its values on the coordinate functions x^i for which $\{x^i, x^j\} = J^{ij}$ or, explicitly,

$$\{q^i, q^j\} = 0, \quad \{p_i, p_j\} = 0, \quad \{p_i, q^j\} = \delta_i^j. \tag{1.9}$$

Using the Poisson bracket, Hamilton's equations for the coordinate functions can be rephrased in the following concise form

$$\dot{x}^j = \{H, x^j\}.$$

As a consequence, evolution of any function f on the phase space is governed by the equation

$$\dot{f} = \{H, f\}.$$

Due to the skew-symmetry property of the Poisson bracket, this form of Hamilton's equations makes the conservation law for H obvious.

Poisson and symplectic manifolds. The properties $(1) - (4)$ provide a general definition of the Poisson bracket for an arbitrary smooth manifold $\mathcal{P}$. Any Poisson bracket is described by a skew-symmetric tensor J on $\mathcal{P}$ satisfying the Jacoby identity. In local coordinates this identity takes the form

$$\sum_{(i,l,m)} J^{ik}\partial_k J^{lm} = 0,$$

where the sum is over the cyclic permutation of indices. A manifold endowed with a Poisson bracket is called *Poisson*.

For later we will need the notion of a *Poisson map*. For Poisson manifolds $\mathcal{M}$ and $\mathcal{N}$, a smooth map $\varphi : \mathcal{M} \to \mathcal{N}$ is called Poisson, if for any $f, h \in \mathcal{F}(\mathcal{N})$

$$\{f, h\}_{\mathcal{N}}(\varphi(x)) = \{\varphi^* f, \varphi^* h\}_{\mathcal{M}}(x), \tag{1.10}$$

where $\varphi^* f(x) = f(\varphi(x))$ and $\varphi^* h(x) = h(\varphi(x))$, $x \in \mathcal{M}$, are pullbacks of f and h. Here $\{ , \}_{\mathcal{M}}$ and $\{ , \}_{\mathcal{N}}$ stand for the Poisson brackets on the respective manifolds.

In general, the rank r of the matrix J is less than or equal to the dimension $\dim \mathcal{P}$ of a manifold and it might change from point to point. In the case when $r = \dim \mathcal{P}$ at every point, the matrix J is invertible and the corresponding Poisson bracket is called non-degenerate. This is only possible if $\dim \mathcal{P}$ is even. Indeed, since $J^t = -J$, one has

$$\det J = \det(-J) = (-1)^{\dim \mathcal{P}} \det J ,$$

so that $(-1)^{\dim \mathcal{P}} = 1$ since $\det J \neq 0$.

A manifold $\mathcal{P}$ supplied with a non-degenerate Poisson bracket is called *symplectic*. The inverse of J with entries ω_{ij}, where $J^{ik}\omega_{kj} = \delta^i_j$, defines a skew-symmetric bilinear differential 2-form ω on $\mathcal{P}$

$$\omega = -\tfrac{1}{2}\omega_{ij}(x)\, dx^i \wedge dx^j .$$

The Jacobi identity for J implies that this form is closed, i.e. $d\omega = 0$. A closed non-degenerate 2-form is called *symplectic*.

An example of a symplectic manifold is the space $\mathbb{R}^{2N}$ with the bracket (1.9). The corresponding symplectic form is

$$\omega = dp_i \wedge dq^i = d(p_i dq^i) .$$

The 1-form $\alpha = p_i dq^i$ is called the *canonical 1-form*.

Given a Poisson manifold, to any function $f \in \mathcal{F}(\mathcal{P})$ one can associate a vector field ξ_f defined as

$$\xi_f = \{f, \cdot \} . \tag{1.11}$$

This field is called the *hamiltonian vector field* generated by f, and f is the generating or *hamiltonian function* of ξ_f. In local coordinates x^i we have

$$\xi_f = J^{ij}\partial_i f \partial_j . \tag{1.12}$$

If we let $\xi_f = \xi^j_f \partial_j$, then the relation above gives

$$\xi^j_f = J^{ij}\partial_i f , \qquad \partial_j f = \omega_{ij}\xi^i_f . \tag{1.13}$$

The Jacobi identity for the Poisson bracket implies

$$\xi_{\{f,g\}} = [\xi_f, \xi_g] . \tag{1.14}$$

Hence, the map $f \to \xi_f$ is a homomorphism $\mathcal{F}(\mathcal{P}) \to \mathfrak{X}(\mathcal{P})$, where $\mathfrak{X}(\mathcal{P})$ is the Lie algebra of vector fields on $\mathcal{P}$. If $\mathcal{P}$ is symplectic, the definition (1.11) of the hamiltonian vector field can be formulated with the help of the interior product i_ξ

$$i_{\xi_f}\omega + df = 0,\tag{1.15}$$

while the one-to-one correspondence between the Poisson bracket and the symplectic form ω can be expressed as

$$\omega(\xi_f, \xi_h) = \{f, h\} = \xi_f h = -\xi_h f.\tag{1.16}$$

A function C is called a *central* or *Casimir function* if it Poisson-commutes with any element of $\mathcal{F}(\mathcal{P})$, that is

$$\{C, f\} = 0, \quad \forall f \in \mathcal{F}(\mathcal{P}).$$

Casimir functions form a ring. If C is a Casimir function then it is annihilated by any hamiltonian vector field ξ_f, i.e. the latter lies everywhere tangent to the level set of the function C. On the other hand, the hamiltonian vector field ξ_C vanishes as the one-form dC belongs to the kernel of J: $JdC = 0$. Thus, the existence of non-constant Casimir functions means that $r \neq \dim \mathcal{P}$, i.e. the Poisson bracket is degenerate.

Let $\{C_i\}$, $i = 1, \ldots, m$, be a complete set of independent Casimir functions. Consider a level set $\mathcal{P}_c = \{x \in \mathcal{P} : C_i(x) = c_i\}$, where c_i are constants. Any hamiltonian vector field is tangent to $\mathcal{P}_c$

$$\xi_f C_i = \{f, C_i\} = 0, \quad \forall f \in \mathcal{F}(\mathcal{P}).$$

The same is true for the commutator of any two hamiltonian vector fields. Thus, by the Frobenius theorem,[2] the level set $\mathcal{P}_c$ is an integral submanifold in $\mathcal{P}$. On $\mathcal{P}_c$ one can naturally define a 2-form ω

$$\omega_x(\xi_f, \xi_g) = \{f, g\}(x), \quad x \in \mathcal{P}_c,\tag{1.17}$$

where ω_x is the value of ω at x. The differential of ω can be computed with the help of the formula

$$3d\omega(\xi_f, \xi_g, \xi_h) = \xi_f \omega(\xi_g, \xi_h) + \xi_g \omega(\xi_h, \xi_f) + \xi_h \omega(\xi_f, \xi_g)$$
$$- \omega([\xi_f, \xi_g], \xi_h) - \omega([\xi_h, \xi_f], \xi_g) - \omega([\xi_g, \xi_h], \xi_f).$$

Using (1.14), definition (1.17) and the Jacobi identity, we get $d\omega = 0$. Since the hamiltonian vector of any Casimir function vanishes, the form ω is non-degenerate and, therefore, it is symplectic, i.e. $\mathcal{P}_c$ is a symplectic manifold. Thus, the hamiltonian vector fields foliate $\mathcal{P}$ into integral even-dimensional sub-manifolds called *symplectic*

---

[2]The Frobenius theorem and its proof are discussed in Appendix A.

leaves, each of which inherits a symplectic form from the original Poisson bracket on $\mathscr{P}$.

Canonical transformations. Consider a smooth coordinate transformation $x \to x' = x'(x)$. In terms of these new coordinates Hamilton's equations (1.8) take the form

$$\frac{dx'^i}{dt} = \frac{\partial x'^i}{\partial x^k}\frac{dx^k}{dt} = \frac{\partial x'^i}{\partial x^k}J^{km}(x)\nabla_m^x H = \frac{\partial x'^i}{\partial x^k}\frac{\partial x'^j}{\partial x^m}J^{km}(x)\nabla_j H' \equiv J'^{ij}(x')\nabla_j H',$$

where

$$J'^{ij}(x') = \frac{\partial x'^i}{\partial x^k}\frac{\partial x'^j}{\partial x^m}J^{km}(x), \qquad (1.18)$$

that is under coordinate transformations J transforms as a contravariant antisymmetric tensor field. Here $H'(x') = H(x(x'))$. Evidently, the equations for x' are of the hamiltonian form with the new hamiltonian $H'(x')$ if and only if

$$\frac{\partial x'^i}{\partial x^k}\frac{\partial x'^j}{\partial x^m}J^{km}(x) = J^{ij}(x'). \qquad (1.19)$$

Diffeomorphisms of the phase space which satisfy this condition are called *canonical*. In other words, canonical transformations do not change the form of the Poisson (tensor) bracket. An infinitesimal diffeomorphism $x'^k = x^k + \xi^k$ is generated by a vector field ξ. Under such a diffeomorphism the form of an arbitrary contravariant tensor J varies according to (1.18),

$$\left(\mathscr{L}_\xi J\right)^{ij} \equiv J^{ij}(x) - J'^{ij}(x) = \xi^k \partial_k J^{ij} - \partial_k \xi^i J^{kj} - \partial_k \xi^j J^{ik}. \qquad (1.20)$$

Here $\mathscr{L}_\xi$ is the Lie derivative of J along the vector field ξ. It is now obvious that infinitesimal canonical transformations correspond to those ξ for which $\mathscr{L}_\xi J = 0$.

If a manifold $\mathscr{P}$ is symplectic, then canonical transformations preserve the corresponding symplectic form, that is

$$\mathscr{L}_\xi \omega = 0. \qquad (1.21)$$

For this reason, these transformations are also called *symplectic* or *symplectomorphisms*.

An important class of canonical transformations is comprised by the hamiltonian vector fields. Consider a diffeomorphism generated by a hamiltonian vector field ξ_f. From the definition (1.20) of the Lie derivative we deduce that

$$\left(\mathscr{L}_{\xi_f} J\right)^{ij} = J^{km}\partial_m f \partial_k J^{ij} - \partial_k(J^{im}\partial_m f)J^{kj} - \partial_k(J^{jm}\partial_m f)J^{ik}$$

$$= -\partial_m f \sum_{(i,j,m)} J^{ik}\partial_k J^{jm} = 0,$$

where the sum over the cyclic permutation of indices i, j, k vanishes due to the Jacobi identity. The same result follows immediately from the Cartan formula

$$\mathcal{L}_{\xi_f}\omega = d(i_{\xi_f}\omega) + i_{\xi_f}(d\omega) = -d^2 f = 0,$$

since ω is closed. Hence, any hamiltonian vector field generates a canonical transformation. If a Poisson manifold is not symplectic, then hamiltonian vector fields generate symplectomorphisms of the corresponding symplectic leaves.

Generally, a hamiltonian system is characterised by a triple $(\mathscr{P}, \{\ ,\ \}, H)$: a phase space $\mathscr{P}$, a Poisson structure $\{\ ,\ \}$ and a hamiltonian function H. For any function f on the phase space, evolution equation is

$$\frac{df}{dt} = \{H, f\}.$$

Since $\{H, H\} = 0$, the hamiltonian is automatically conserved. Therefore, the motion of the system takes place on the submanifold of the phase space defined by the equation $H = E$ where E is a fixed constant.

1.1.2 Liouville Theorem and Complete Integrability

Among a large variety of physically relevant dynamical systems, those which admit an exact solution turn out to be rather rare. Remarkably, however, for a special class of systems solutions of the corresponding Hamilton's equations can always be found by quadratures, i.e. by solving a finite number of algebraic equations and computing a finite number of definite integrals. Dynamical systems falling in this class are generally known as *Liouville integrable systems* because they satisfy the assumptions of the famous Liouville theorem. In essence, for a dynamical system with a $2N$-dimensional phase space $\mathscr{P}$ this theorem states that if there exist N independent functions $f_i \in \mathscr{F}(\mathscr{P})$ including the hamiltonian H, which Poisson commute, $\{f_i, f_j\} = 0$, then the corresponding equations of motion can be solved by quadratures. Since $\{H, f_i\} = 0$, the functions f_i do not depend on time, i.e. they are integrals of motion. In general, two functions on a phase space that Poisson commute are said to be in involution. Thus, Hamilton's equations of any dynamical system that admits an involutive family of integrals of motion which is equal to half the dimension of its phase space in number can be solved, at least in principle, by means of well-established mathematical operations. For this reason Liouville integrable systems are also called *completely integrable systems*.

To demonstrate the concept of Liouville integrability in a simple setting, consider the example of the one-dimensional harmonic oscillator. The hamiltonian is (mass $m = 1$)

$$H = \frac{1}{2}p^2 + \frac{\omega^2}{2}q^2$$

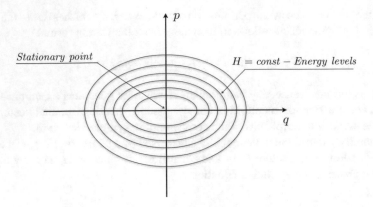

Fig. 1.1 Phase space trajectories of the harmonic oscillator

and the Poisson bracket is given by $\{p, q\} = 1$. The energy is conserved, therefore, the phase space is fibered into ellipses $H = E$. Perform a change of variables

$$p = \rho \cos(\theta), \qquad q = \frac{\rho}{\omega} \sin(\theta).$$

Then for the Poisson bracket one gets $\{\rho, \theta\} = \frac{\omega}{\rho}$. The hamiltonian becomes $H = \frac{1}{2}\rho^2$, so ρ is an integral of motion. The variable θ evolves according to

$$\dot{\theta} = \{H, \theta\} = \rho\{\rho, \theta\} = \omega \quad \rightarrow \quad \theta(t) = \omega t + \theta_0.$$

Thus, the phase space trajectories are ellipses with fixed values of ρ (Fig. 1.1).

The generalisation to the n-dimensional harmonic oscillator is straightforward. The corresponding hamiltonian is[3]

$$H = \sum_{i=1}^{N} \left(\frac{1}{2}p_i^2 + \frac{\omega_i^2}{2}q_i^2 \right),$$

while commuting integrals are

$$f_i(p, q) \equiv \frac{1}{2}p_i^2 + \frac{\omega_i^2}{2}q_i^2, \quad i = 1, \ldots, N.$$

Define the common level set

$$\mathscr{P}_c = \{x \in \mathscr{P} : \; f_i(p, q) = c_i, \; i = 1, \ldots, N\},$$

---

[3]To uniformise notations, for the rest of this section we label coordinates by using lower indices.

where c_i are constants. This set is a manifold isomorphic to an N-dimensional real torus $\mathbb{T}^N$. These tori foliate the phase space and can be parametrised by N angle variables θ_i that evolve linearly in time with frequencies ω_i.

Consider the equation

$$k_1\omega_1 + \cdots k_N\omega_N = 0, \tag{1.22}$$

where $k = (k_1, \ldots k_N)$ is a vector with integer components. If (1.22) has at least one non-zero solution solution, the frequency set $(\omega_1, \ldots, \omega_N)$ is called *resonant*, otherwise it is *non-resonant*. For a non-resonant set of frequencies every trajectory is dense on the torus $\mathbb{T}^N$ and the corresponding motion is called *conditionally periodic*. Evidently, if all the frequencies are commensurable (rationally comparable), that is for any ω_i and ω_j there exist integers m and n such that

$$\omega_i\, m = \omega_j\, n\,,$$

then the motion is periodic.

The multi-dimensional harmonic oscillator is a beautiful example of a Liouville integrable system as any such system exhibits a very similar structure of its phase space flows, the latter are described by the Liouville theorem. The modern version of this theorem and the corresponding proof is due to Arnold [2].

Arnold-Liouville theorem. Let $\mathscr{P}$ be a $2N$-dimensional symplectic manifold. Suppose there exist N functions $f_i \in \mathscr{F}(\mathscr{P})$ that are pairwise in involution with respect to the corresponding Poisson bracket

$$\{f_i, f_j\} = 0\,, \quad \forall i, j = 1, \ldots, N\,.$$

Consider a common level set $\mathscr{P}_c$ of these functions,

$$\mathscr{P}_c = \{x \in \mathscr{P} :\ f_i(x) = c_i,\ i = 1, \ldots, N\}\,, \tag{1.23}$$

where c_i are constants. Assume that functions f_i are independent on $\mathscr{P}_c$, which means that the 1-forms df_i are linearly independent at each point of $\mathscr{P}_c$. Then

(1) $\mathscr{P}_c$ is a smooth manifold invariant under the hamiltonian flow with $H = H(f_i)$.
(2) If $\mathscr{P}_c$ is compact and connected then it is diffeomorphic to the N-dimensional torus

$$\mathbb{T}^N = \{(\varphi_1, \ldots, \varphi_N)\ \text{mod}\ 2\pi\}\,.$$

(3) The motion on $\mathscr{P}_c$ under H is conditionally periodic, that is,

$$\frac{d\varphi_i}{dt} = \omega_i(c)\,.$$

(4) The equations of motion can be integrated by quadratures.

We sketch the proof of the Arnold-Liouville theorem referring the reader for the full treatment to [2]. Consider the hamiltonian vector fields ξ_i corresponding to the functions f_i. Since $\xi_i f_j = 0$, these vector fields are tangent to $\mathscr{P}_c$ and their linear independence implies that they span the tangent space of $\mathscr{P}_c$ at any point. Taking into account that the vector fields are in involution $[\xi_i, \xi_j] = 0$, we conclude on the base of the Frobenius theorem, see Appendix A, that $\mathscr{P}_c$ is a maximal integral submanifold for the distribution spanned by ξ_i. Clearly, the manifold $\mathscr{P}_c$ is invariant under the hamiltonian flow triggered by any $H = H(f_i)$. Varying the constants c_i, we obtain a foliation of almost all[4] $\mathscr{P}$ into invariant submanifolds, see Fig. 1.2.

The main part of the proof consists in showing that whenever $\mathscr{P}_c$ is compact and connected, it is a torus but not, for instance, a sphere. Let $g_i^{t_i}$, $t_i \in \mathbb{R}$, be a one-parametric group of diffeomorphisms of $\mathscr{P}$ corresponding to the hamiltonian vector field ξ_i. The one-parametric groups corresponding to different vector fields commute because the vector fields commute. As a result, one can define the following action of the abelian group $\mathbb{R}^N = \{t_1, \ldots, t_N\}$ on $\mathscr{P}_c$:

$$g^t(x) = g_1^{t_1} \cdots g_N^{t_N}(x). \tag{1.24}$$

Since $\mathscr{P}_c$ is an integral manifold for the distribution spanned by ξ_i, this action is transitive and, therefore, $\mathscr{P}_c$ is a homogeneous space. Thus, $\mathscr{P}_c$ is diffeomorphic to the quotient $\mathbb{R}^N / \Gamma$, where Γ is the isotropy subgroup of $\mathbb{R}^N$, i.e. a set of all points $t \in \mathbb{R}^N$ for which $g^t(x) = x$. The fact that the fields ξ_i are independent at any point of $\mathscr{P}_c$ implies that the action (1.24) is locally free and, therefore, Γ must be a discrete subgroup of $\mathbb{R}^N$. By assumption $\mathscr{P}_c$ is compact and, therefore, Γ should be nothing else[5] but an integral lattice $\mathbb{Z}^N$, so that $\mathscr{P}_c$ is diffeomorphic to $\mathbb{R}^N / \mathbb{Z}^N = \mathbb{T}^N$. By the standard construction of a homogeneous space as a coset, the vector fields ξ_i are mapped by this diffeomorphism to the translation-invariant vector fields on $\mathbb{T}^N$. The angle variables $\{\varphi_i \bmod 2\pi\}$ parametrising the torus provide a coordinate system on $\mathscr{P}_c$ and they can be linearly expressed via $t_1, \ldots, t_N$. The uniform motion on the torus $\mathbb{T}^N$ happens according to the law $\varphi_i = \varphi_i^0 + \omega_i t$ and is conditionally periodic. The numbers $\omega_i = \omega_i(c)$ are called frequencies.

Further, we note that the Arnold-Liouville theorem can be extended to the case when $\mathscr{P}_c$ is not necessarily compact [3]. With an additional assumption that the hamiltonian vector fields ξ_i are complete[6] on $\mathscr{P}_c$, it is possible to show that each connected component of $\mathscr{P}_c$ is diffeomorphic to $\mathbb{T}^k \times \mathbb{R}^{N-k}$.

Action-Angle variables. The variables $f_i, \varphi_j, i, j = 1, \ldots, N$ featuring in the Arnold-Liouville theorem are not in general canonical coordinates on $\mathscr{P}$. However, such coordinates can be constructed. First we note that in a small neighbourhood of $\mathscr{P}_c$ the symplectic manifold $\mathscr{P}$ is diffeomorphic to the direct product $D \times \mathbb{T}^N$, where D is a small domain in $\mathbb{R}^N$. It turns out that in $D \times \mathbb{T}^N$ there exist coordinates I_i, θ_j, where $I_i \in D$, $\theta_j \in \mathbb{T}^N$ such that in these variables all f_i depend only on I_j and the

---

[4]There could be values of c_i for which the equations $f_i = c_i$ cease to be independent.

[5]All discrete subgroups of $\mathbb{R}^n$ correspond to integral lattices $\mathbb{Z}^k$, $k \leqslant N$.

[6]A vector field is complete if any of its flow curves exists for all values of time.

Fig. 1.2 Foliation of a phase space by invariant tori. Each torus coincides with a level set $\mathscr{P}_c$. All trajectories on a given torus have the same frequencies $\omega_i(c)$, so one may speak of the "frequency set of a torus"

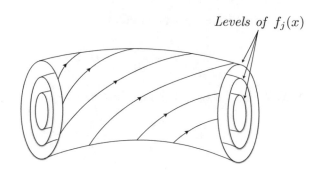

Levels of $f_j(x)$

symplectic structure has the canonical form $\omega = dI_i \wedge d\theta_i$. An explicit construction of the canonical variables I_i, θ_j proceeds as follows.

For simplicity we consider a Liouville integrable system with the phase space $\mathbb{R}^{2N}$. According to the Liouville theorem, the motion occurs on a N-dimensional torus $\mathbb{T}^N$ being a common level of N commuting integrals. Let $\gamma_j, 1 \leqslant j \leqslant N$, be the fundamental cycles of this torus depending continuously on the level $\{c_j\}$. Consider a set of equations $f_j(p, q) = c_j$ and solve it for p_j: $p_j = p_j(c, q)$. Introduce the so-called action variables[7]

$$ I_j(c) = \frac{1}{2\pi} \oint_{\gamma_j} p_i(q, c) dq_i = \frac{1}{2\pi} \oint_{\gamma_j} \alpha, \tag{1.25} $$

where $\alpha = p_i dq_i$ is the canonical 1-form. Since c_j are time-independent as they are values of the integrals of motion, the variables $I_j = I_j(c)$ are also time-independent. Moreover, assuming that I_i are independent functions of c_j, the map $c_j \rightarrow I_j(c)$ given by (1.25) has an inverse. The angle variables θ_j are constructed by requiring that the transformation

$$ (p_j, q_j) \rightarrow (I_j, \theta_j) \tag{1.26} $$

is canonical. To construct this canonical transformation, we will use the following generating function depending on the "old" coordinates q and the "new" momenta I

$$ S(I, q) = \int_{q_0}^{q} p_i(\tilde{q}, I) d\tilde{q}_i , $$

where an integration path lies on $\mathscr{P}_c$. We have

$$ p_j = \frac{\partial S}{\partial q_j} \quad \rightarrow \quad p_j = p_j(I, q). \tag{1.27} $$

---

[7]The physical dimension of I_j coincide with the dimension of the action that is the same as the dimension of angular momentum.

The angle variables are introduced as

$$\theta_j = \frac{\partial S}{\partial I_j} \quad \rightarrow \quad \theta_j = \theta_j(I, q). \tag{1.28}$$

Thus, for the differential of S we then have

$$dS = \frac{\partial S}{\partial q_j} dq_j + \frac{\partial S}{\partial I_j} dI_j = p_j dq_j + \theta_j dI_j.$$

Acting on this relation with d and taking into account that $d^2 S = 0$, we get

$$\omega = dp_j \wedge dq_j = dI_j \wedge d\theta_j,$$

which shows that I_i, θ_j are canonical variables.

A subtle point here concerns a dependence of S on the integration path. Consider a closed path: from q_0 to q and further from q to q_0. If this path is contractable, then by Stokes' theorem

$$\Delta S = \oint_{q_0}^{q_0} \alpha = \int d\alpha = \int \omega = 0.$$

Here the vanishing of the integral of ω is due to the fact that ω vanishes on $\mathcal{P}_c$

$$\omega(\xi_i, \xi_j) = \{f_i, f_j\} = 0.$$

If an integration path encloses a non-trivial cycle γ, the generation function undergoes a shift by an integral of α over this cycle

$$\Delta_\gamma S = \int_\gamma \alpha$$

that depends on I_j only. As a result, going over the cycle the variables θ_j undergo a jump

$$\Delta_\gamma \theta_j = \frac{\partial}{\partial I_j} \int_\gamma p_i(q, I) dq_i,$$

i.e. θ_j are multi-valued functions on $\mathcal{P}_c$. In particular, $\Delta_{\gamma_i} \theta_j = 2\pi \delta_{ij}$. This shows that θ_j are independent angle coordinates on the cycles. The same conclusion can be also drawn from the following consideration

$$\oint_{\gamma_j} d\theta_i = \oint_{\gamma_j} d\frac{\partial S}{\partial I_i} = \frac{\partial}{\partial I_i}\left(\oint_{\gamma_j} dS\right) = \frac{\partial}{\partial I_i}\left(\oint_{\gamma_j} \frac{\partial S}{\partial q_k} dq_k\right) = \frac{\partial}{\partial I_i}\left(\oint_{\gamma_j} p_k dq_k\right) = 2\pi \delta_{ij},$$

as on $\gamma_j \in \mathbb{T}^N$ the variables I_j are constants and the function $S(I, q)$ depends on q only.

In the variables I, θ the Hamiltonian is a function of I. Then equations of motion become

$$\dot{I}_j = -\frac{\partial H}{\partial \theta_j} = 0\,, \quad \dot{\theta}_j = \frac{\partial H}{\partial I_j} \equiv \omega_j(I)$$

and they are trivially solved, $I_j(t) = I_j^0$, $\theta_j(t) = \theta_j^0 + \omega_j(I^0)t$. On the way of constructing the angle coordinates θ_j, algebraic operations were used to find p_j from $f_j(p, q) = c_j$ and a computation of a definite integral was implicitly done to obtain $S(I, q)$. Finally, the inverse of (1.26) was constructed by solving equations (1.28) for $q_j = q_j(I, \theta)$, which is also an algebraic operation. This way of solving a Liouville integrable system is behind the term "quadrature".

Note that even in the one-dimensional case the action-angle variables are not uniquely defined. The action variable is defined up to an additive constant and the angle variable can be shifted by an arbitrary function h of I: $I \to I + \text{const}$, $\theta \to \theta + h(I)$.

Example of action-angle variables. We illustrate the construction of the action-angle variables by using the harmonic oscillator as an example. We have

$$E = \frac{1}{2}(p^2 + \omega^2 q^2) \quad \to \quad p(E, q) = \pm\sqrt{2E - \omega^2 q^2}$$

and, therefore,

$$I = \frac{1}{2\pi} \oint_E dq \sqrt{2E - \omega^2 q^2} = \frac{2}{2\pi} \int_{-\frac{\sqrt{2E}}{\omega}}^{\frac{\sqrt{2E}}{\omega}} dq \sqrt{2E - \omega^2 q^2} = \frac{E}{\omega}\,.$$

The generating function of the canonical transformation reads

$$S(I, q) = \int^q dx \sqrt{2I\omega - \omega^2 x^2}\,,$$

while for the angle variable we obtain

$$\theta(I, q) = \frac{\partial S}{\partial I} = \omega \int^q \frac{dx}{\sqrt{2I\omega - \omega^2 x^2}} = \arctan \frac{\omega q}{\sqrt{2I\omega - \omega^2 q^2}}\,. \quad (1.29)$$

The change of θ for the period of motion, which is the same as an integral over the cycle of constant energy, is

$$\frac{1}{2\pi} \oint_E d\theta = \frac{1}{\pi}\omega \int_{-\sqrt{\frac{2I}{\omega}}}^{\sqrt{\frac{2I}{\omega}}} \frac{dx}{\sqrt{2I\omega - \omega^2 x^2}} = 1\,. \quad (1.30)$$

Inverting (1.29) with respect to q, we get

$$q = \sqrt{\frac{2I}{\omega}} \sin \theta \,.$$

We can verify that the transformation to the action-angle variables is indeed canonical

$$dp \wedge dq = \left(\frac{\omega dI}{\sqrt{2I\omega - \omega^2 q^2}} - \frac{\omega^2 q dq}{\sqrt{2I\omega - \omega^2 q^2}} \right) \wedge dq$$

$$= \frac{\omega}{\sqrt{2I\omega - \omega^2 q^2}} dI \wedge \sqrt{\frac{2I}{\omega}} \, d(\sin \theta) = dI \wedge d\theta \,.$$

1.1.3 Systems with Closed Trajectories

It may happen that a dynamical system defined on a symplectic manifold $\mathscr{P}$ of dimension $2N$ exhibits more than N integrals of motion. In this case the maximal number of independent pairwise commuting integrals f_i can not exceed N. Indeed, the hamiltonian vector fields of $f_i, i = 1, \ldots, k$, span a subspace $V \subset T\mathscr{P}$ of the tangent bundle $T\mathscr{P}$ at any given point $x \in \mathscr{P}$. Assuming f_i in involution, one gets $\omega|_V = 0$ and, therefore, $V \subset V^\perp$, where $V^\perp$ is a skew-orthogonal complement of V in $T\mathscr{P}$. The last observation implies that $k = \dim V \leqslant \dim V^\perp$. On the other hand, $\dim V + \dim V^\perp = 2N$, as ω is non-degenerate. Hence, $k \leqslant 2N - k$, so that $k \leqslant N$.

The Liouville integrable systems of phase space dimension $2N$ are characterised by the requirement to have N globally defined, pairwise Poisson-commuting integrals of motion $f_i(p, q)$. Singling out the level set (1.23), we obtain (in the compact case) the N-dimensional torus. In general, frequencies ω_i characterising the motion on the Liouville torus are not commensurable and, as a result, the corresponding trajectories are not closed.

A special situation arises when at least two frequencies become rationally comparable. The corresponding motion is called *degenerate*. Here we will be interested in the situation of the completely degenerate motion, i.e. when all N frequencies ω_j are comparable. In this case the classical trajectory is a closed curve and the number of globally defined integrals of motion increases to $2N - 1$.[8] A dynamical system admitting such a large number of integrals of motion is called *superintegrable*. Certainly, these $2N - 1$ integrals cannot all Poisson-commute, because, as we remarked earlier, the maximal number of pair-wise commuting integrals cannot exceed N. Below we give a typical example of such a degenerate system.

Consider a two-dimensional harmonic oscillator with the hamiltonian

$$H = \tfrac{1}{2}(p_1^2 + p_2^2) + \tfrac{1}{2}(\omega_1^2 q_1^2 + \omega_2^2 q_2^2) \,.$$

---

[8]For the corresponding quantum-mechanical system classical degenerate motion implies degeneracy of energy levels.

There are two independent and commuting integrals

$$f_1 = \tfrac{1}{2}p_1^2 + \tfrac{1}{2}\omega_1^2 q_1^2\,, \qquad f_2 = \tfrac{1}{2}p_2^2 + \tfrac{1}{2}\omega_2^2 q_2^2\,.$$

Obviously, $H = f_1 + f_2$, and we can replace f_1, f_2 by H and the other integral $K_1 = f_1 - f_2$. If the ratio ω_1/ω_2 is irrational, oscillator trajectories are everywhere dense on the Liouville torus given by a common level set of these integrals. However, if

$$\frac{\omega_1}{\omega_2} = \frac{r}{s}\,, \tag{1.31}$$

where r, s are (positive) coprime integers, there appear two additional real and globally defined integrals

$$K_2 = \bar{a}_1^s a_2^r + a_1^s \bar{a}_2^r\,, \qquad K_3 = i(\bar{a}_1^s a_2^r - a_1^s \bar{a}_2^r)\,,$$

where

$$a_i = \frac{1}{\sqrt{2\omega_i}}(p_i - i\omega_i q_i)\,, \qquad \bar{a}_i = \frac{1}{\sqrt{2\omega_i}}(p_i + i\omega_i q_i)\,.$$

From the equations of motion $\dot{q}_i = p_i$ and $\dot{p}_i = -\omega_i^2 q_i$, one finds the corresponding evolution equations for the amplitudes $\dot{a}_i = -i\omega_i a_i$, $\dot{\bar{a}}_i = i\omega_i \bar{a}_i$. With the help of these equations it is readily seen that $\dot{K}_i = 0$, i.e. K_i commute with H. Concerning the structure of K_2 and K_3, they are homogenous functions of degree $r + s$ over both the coordinates and momenta. Further, K_i and H are functionally dependent:

$$K_2^2 + K_3^2 - 4\left(\frac{H + K_1}{2\omega_1}\right)^s \left(\frac{H - K_1}{2\omega_2}\right)^r = 0\,, \tag{1.32}$$

so that there are three independent integrals of motion for our dynamical system with four-dimensional phase space. Thus, for the case (1.31) all oscillator trajectories are closed; on the screen of an oscilloscope they are visualised as *Lissajous figures*.

Concerning the Poisson relations of K_i, we find

$$\{K_1, K_2\} = (s\omega_1 + r\omega_2)K_3\,, \qquad \{K_1, K_3\} = -(s\omega_1 + r\omega_2)K_2\,, \tag{1.33}$$
$$\{K_2, K_3\} = \frac{rs}{2^{r+s-2}\omega_1^s \omega_2^r}\left(\omega_1(H + K_1)^s (H - K_1)^{r-1} - \omega_2(H + K_1)^{s-1}(H - K_1)^r\right).$$

The hamiltonian is central with respect to K_i. For generic positive integers r, s the bracket is non-linear. However, in the special case of equal frequencies, which without loss of generality can be taken equal to one, we have $r = s = 1$ and the formulas (1.33) yield the linear Poisson relations

$$\{K_i, K_j\} = 2\,\epsilon_{ijk}K_k\,, \tag{1.34}$$

isomorphic to the defining relations of the $\mathfrak{su}(2)$ Lie algebra. Later we will encounter further examples of superintegrable systems.

Non-commutative integrability. Given a symplectic manifold $\mathscr{P}$ and the hamiltonian, in general, linearly independent integrals of motion form an infinite-dimensional Lie algebra. Even if there exists a maximal integral submanifold spanned by the corresponding hamiltonian vector fields, the latter can be immersed in $\mathscr{P}$ in a very intricate way, in particular, it might not be closed.[9] There is, however, a special situation where a non-abelian algebra of integrals still implies the existence of the Liouville torus. This situation is described by the theorem on non-commutative integrability, various variants of which have been established in [4, 5], see also [6]. Below we formulate this theorem, for the proof the reader is invited to consult the original literature.

Let $\mathscr{P}$ be a symplectic manifold of dimension $2N$ and let the hamiltonian H has a set of $k \geqslant N$ integrals $f_1, \ldots, f_k$, independent in a neighbourhood of a connected level set

$$\mathscr{P}_c = \{x \in \mathscr{P} : \ f_i(x) = c_i, \ \ i = 1, \ldots, k\}.$$

Let $\{f_i, f_j\} = \Psi_{ij}(f)$ be a matrix of the rank $\Psi(c') = 2(k - N)$ for all c' close to c. Then

(1) In the neighbourhood of $\mathscr{P}_c$ the trajectories of the hamiltonian vector field of H lie on the isotropic surfaces $\mathscr{P}_{c'}$ of dimension $2N - k$.
(2) If $\mathscr{P}_{c'}$ are compact and connected then they are diffeomorphic to $2N - k$-dimensional tori and the phase flow on these tori is conditionally periodic.

An example of the theorem on non-commutative integrability at work is provided by Euler's top, which dynamics will be discussed later.

1.2 Symmetries and Conservation Laws

As we have seen, the very possibility of having an integrable model with equations of motion solvable by the Liouville theorem or by other means, relies on the existence of a sufficiently large number of conservation laws. In some cases these conservation laws derive their origin from variational type symmetries that a dynamical model might have. Variational symmetries are the symmetries of the model action and they are shared by the corresponding equations of motion. Noether's theorem relates these symmetries to integrals of motion and gives an explicit formula for the latter. For a non-abelian symmetry group, these integrals form a non-commutative algebra with respect to the Poisson brackets. Apart from explaining the origin of degenerate motion, this algebra can serve as an algebra of constraints for the reduction techniques illuminated in Chap. 2. In this section we discuss Noether's theorem in a finite-dimensional mechanical context.

---

[9]A closed manifold is a compact manifold without boundary.

1.2.1 Noether's Theorem

Consider a mechanical system with N degrees of freedom governed by a lagrangian $L \equiv L(q, \dot{q})$. Assume that there exists an infinitesimal transformation $q \to q + \delta q$ such that the corresponding variation of the lagrangian takes the form[10] of the total time derivative of some function $\delta \Lambda$:

$$\delta L = \frac{d \delta \Lambda}{dt} .$$

This transformation δq is called a variational symmetry. Variational symmetries give rise to conservation laws, as described by Noether's theorem.

Noether's theorem. To any s-parametric variational symmetry of L correspond s quantities $J_n, n = 1, \ldots, s$, which are conserved on solutions of the Euler-Lagrange equations.

The proof goes as follows. Let an infinitesimal transformation $q \to q' = q + \delta q$ be a variational symmetry. Here

$$\delta q^i = \sum_{1 \leqslant n \leqslant s} Q_n^i \epsilon^n ,$$

where ϵ^n are constant, that is time-independent parameters. Then

$$\delta L = \frac{\partial L}{\partial q^i} \delta q^i + \frac{\partial L}{\partial \dot{q}^i} \delta \dot{q}^i = \frac{\partial L}{\partial q^i} \delta q^i + \frac{\partial L}{\partial \dot{q}^i} \frac{d}{dt} \delta q^i = \frac{d \delta \Lambda}{dt} .$$

The last equality can be cast in the form

$$\frac{d}{dt} \left(\frac{\partial L}{\partial \dot{q}^i} \delta q^i - \delta \Lambda \right) = \left(\frac{d}{dt} \left(\frac{\partial L}{\partial \dot{q}^i} \right) - \frac{\partial L}{\partial q^i} \right) \delta q^i .$$

On solutions of the Euler-Lagrange equations the right hand side of this expression vanishes and, therefore, the quantity

$$J = \frac{\partial L}{\partial \dot{q}^i} \delta q^i - \delta \Lambda = p_i \delta q^i - \delta \Lambda$$

is time-independent. Since $J = J_n \epsilon^n$ and $\delta \Lambda = \Lambda_n \epsilon^n$, where

$$J_n = p_i Q_n^i - \Lambda_n , \tag{1.35}$$

the quantities J_n, called Noether's charges, are separately conserved

---

[10]Without use of equations of motion.

$$\frac{dJ_n}{dt} = 0 \, .$$

Some well-known applications of Noether's theorem include examples given below.

(1) *Momentum conservation.* Momentum conservation is related to the freedom of arbitrarily choosing the origin of the coordinate system. Consider a lagrangian

$$L = \frac{m}{2} \eta_{ij} \dot{q}^i \dot{q}^j \, , \tag{1.36}$$

where η_{ij} is some constant metric, which will be used to raise and low the indices. Consider a displacement

$$q'^i = q^i + \epsilon^i \;\; \Rightarrow \;\; \delta q^i = \epsilon^i \, ,$$
$$\dot{q}'^i = \dot{q}^i \;\; \Rightarrow \;\; \delta \dot{q}^i = 0 \, .$$

Obviously, under this transformation the lagrangian remains invariant and we can take $\Lambda = 0$. Thus,

$$J = p_i \delta q^i = p_i \epsilon^i \, ,$$

Since the ϵ^i are arbitrary, all the components p_i are conserved.

(2) *Angular momentum conservation.* Consider again the lagrangian (1.36) and make a transformation

$$q'^i = q^i + \epsilon^{ij} q_j \;\; \Rightarrow \;\; \delta q^i = \epsilon^{ij} q_j \, .$$

Then,

$$\delta L = m \dot{q}_i \epsilon^{ij} \dot{q}_j \, .$$

Clearly, if ϵ^{ij} is anti-symmetric, the variation of the lagrangian vanishes. Again, we can take $\Lambda = 0$ and obtain

$$J = p_i \delta q^i = p_i \epsilon^{ij} q_j \, ,$$

Since ϵ^{ij} is an arbitrary anti-symmetric tensor, we find the conservation of angular momentum components

$$J_{ij} = p_i q_j - p_j q_i \, .$$

(3) *Particle in a constant gravitational field.* The lagrangian of a particle in the gravitational field with the gravitational acceleration constant g is

$$L = \frac{m}{2} \dot{z}^2 - mgz \, .$$

Consider a shift $z \rightarrow z + a$, i.e. $\delta z = a$. We get $\delta L = -mga = \frac{d}{dt}(-mgat)$ so that $\delta \Lambda = -mgat$. Thus, the quantity

$$J = m\dot{z}\delta z - \Lambda = m\dot{z}a + mgat$$

is conserved. This is a conservation law of the initial velocity $\dot{z} + gt = \text{const} \equiv b$. Integrating the last equation we find a trajectory of a particle in the gravitational field $z(t) = z_0 + bt - \frac{gt^2}{2}$, where z_0 is an integration constant.

(4) *Conservation of energy.* Energy conservation is related to the freedom of arbitrarily choosing the origin of time (performing an experiment today or in a hundred years will lead to the same result provided the same initial conditions are fulfilled).

We derive now the conservation law of energy in the framework of Noether's theorem. Suppose we make an infinitesimal time displacement $\delta t = \epsilon$. The response of the lagrangian is

$$\delta L = \frac{dL}{dt}\epsilon .$$

On the other hand,

$$\delta L = \frac{\partial L}{\partial q^i}\delta q^i + \frac{\partial L}{\partial \dot{q}^i}\delta \dot{q}^i + \frac{\partial L}{\partial t}\delta t = \frac{d}{dt}\left(\frac{\partial L}{\partial \dot{q}^i}\right)\delta q^i + \frac{\partial L}{\partial \dot{q}^i}\delta \dot{q}^i ,$$

where we have used the Euler-Lagrange equations and assumed that L does not explicitly depends on time. Obviously, $\delta q^i = \dot{q}^i \epsilon$ and $\delta \dot{q}^i = \ddot{q}^i \epsilon$, so that

$$\delta L = \frac{d}{dt}\left(\frac{\partial L}{\partial \dot{q}^i}\right)\dot{q}^i \epsilon + \frac{\partial L}{\partial \dot{q}^i}\ddot{q}^i \epsilon = \frac{dL}{dt}\epsilon .$$

Cancelling ϵ, we recover the conservation law for the energy

$$\frac{dH}{dt} = 0, \quad H = p_i \dot{q}^i - L .$$

For all the symmetry transformations we have considered so far the integration measure dt in the action did not transform (even in the last example $dt \rightarrow d(t + \epsilon) = dt$).

Notice that in all examples the action of symmetry transformations can be translated to the space of initial data, the latter can be identified with the phase space of a model. For instance, the general solution of the equations of motion in the first and second examples is

$$q^i(t) = a^i t + b^i ,$$

and the symmetry transformations act as $a^i \rightarrow a^i + \epsilon^{ij} a_j$ and $b^i \rightarrow b^i + \epsilon^{ij} b_j + \epsilon^i$. In the last example dealing with conservation of energy, the initial coordinates are universally shifted under time translations $b^i \rightarrow b^i + \epsilon a^i$.

1.2.2 *Variational Symmetries and Equations of Motion*

Here we show that under variational symmetries solutions of the equations of motion
transform into the other solutions. More precisely, we prove that under these symmetries the lagrangian derivatives

$$\mathcal{E}_i \equiv \frac{\partial L}{\partial q^i} - \frac{d}{dt} \frac{\partial L}{\partial \dot{q}^i}$$

transform in a covariant way.

Given a variational symmetry, the lagrangian fulfils the following relation

$$L(q', \dot{q}', t') \equiv L(q, \dot{q}, t) \frac{dt}{dt'} + \frac{d\Lambda}{dt'}, \tag{1.37}$$

where the symmetry transformations involve both independent coordinates and the
time variable

$$q^i = q^i(q', t'), \quad t = t(q', t'). \tag{1.38}$$

In the full generality we assume that L can have an explicit dependence on time,
while $\Lambda \equiv \Lambda(q', t')$ is some function of coordinates and time only. By definition,

$$\dot{q}^i = \frac{dq^i}{dt}, \quad \dot{q}'^i = \frac{dq'^i}{dt'}.$$

Let us denote the lagrangian derivative of the left hand side of (1.37) with respect to
the transformed variables as $\mathcal{E}'_i \equiv \mathcal{E}_i(q', \dot{q}', \ddot{q}', t')$. Then we have

$$\begin{aligned}
\mathcal{E}'_i &= \left(\frac{\partial}{\partial q'^i} - \frac{d}{dt'} \frac{\partial}{\partial \dot{q}'^i} \right) \left(L(q, \dot{q}, t) \frac{dt}{dt'} \right) \\
&= \frac{dt}{dt'} \left(\frac{\partial q^j}{\partial q'^i} \frac{\partial L}{\partial q^j} + \frac{\partial \dot{q}^j}{\partial q'^i} \frac{\partial L}{\partial \dot{q}^j} + \frac{\partial t}{\partial q'^i} \frac{\partial L}{\partial t} \right) + L \frac{\partial}{\partial q'^i} \left(\frac{dt}{dt'} \right) \\
&\quad - \frac{d}{dt'} \left(\frac{dt}{dt'} \frac{\partial \dot{q}^j}{\partial \dot{q}'^i} \frac{\partial L}{\partial \dot{q}^j} + L \frac{\partial}{\partial \dot{q}'^i} \left(\frac{dt}{dt'} \right) \right).
\end{aligned} \tag{1.39}$$

In particular, the function $\frac{d\Lambda}{dt'}$ decoupled as it is the total time derivative. We get

$$\frac{dt}{dt'} = \frac{\partial t}{\partial q'^j} \dot{q}'^j + \frac{\partial t}{\partial t'}. \tag{1.40}$$

With this formula at hand we compute

$$\frac{\partial}{\partial \dot{q}^{\prime i}}\left(\frac{dt}{dt'}\right) = \frac{\partial t}{\partial q^{\prime i}}, \qquad \frac{\partial}{\partial q^{\prime i}}\left(\frac{dt}{dt'}\right) = \frac{\partial^2 t}{\partial q^{\prime i} \partial q^{\prime j}}\dot{q}^{\prime j} + \frac{\partial^2 t}{\partial q^{\prime i} \partial t'}.$$

It follows from these formulae that terms in (1.39) proportional to L cancel. Indeed, combining these terms, we have[11]

$$\frac{\partial}{\partial q^{\prime i}}\left(\frac{dt}{dt'}\right) - \frac{d}{dt'}\left(\frac{\partial}{\partial \dot{q}^{\prime i}}\left(\frac{dt}{dt'}\right)\right) = \frac{\partial^2 t}{\partial q^{\prime i} \partial q^{\prime j}}\dot{q}^{\prime j} + \frac{\partial^2 t}{\partial q^{\prime i} \partial t'} - \frac{d}{dt'}\left(\frac{\partial t}{\partial q^{\prime i}}\right) = 0.$$

Thus, we are left with

$$\mathscr{E}_i' = \frac{dt}{dt'}\left(\frac{\partial q^j}{\partial q^{\prime i}}\frac{\partial L}{\partial q^j} + \frac{\partial \dot{q}^j}{\partial q^{\prime i}}\frac{\partial L}{\partial \dot{q}^j} + \frac{\partial t}{\partial q^{\prime i}}\frac{\partial L}{\partial t}\right) - \frac{d}{dt'}\left(\frac{dt}{dt'}\frac{\partial \dot{q}^j}{\partial q^{\prime i}}\frac{\partial L}{\partial \dot{q}^j}\right) - \frac{dL}{dt'}\frac{\partial t}{\partial q^{\prime i}}.$$

We proceed by isolating the terms

$$\begin{aligned}\frac{\partial t}{\partial q^{\prime i}}\left(\frac{dt}{dt'}\frac{\partial L}{\partial t} - \frac{dL}{dt'}\right) &= \frac{\partial t}{\partial q^{\prime i}}\left(\frac{dt}{dt'}\frac{\partial L}{\partial t} - \frac{dq^j}{dt'}\frac{\partial L}{\partial q^j} - \frac{d\dot{q}^j}{dt'}\frac{\partial L}{\partial \dot{q}^j} - \frac{dt}{dt'}\frac{\partial L}{\partial t}\right) \\ &= -\frac{\partial t}{\partial q^{\prime i}}\left(\frac{dq^j}{dt'}\frac{\partial L}{\partial q^j} + \frac{d\dot{q}^j}{dt'}\frac{\partial L}{\partial \dot{q}^j}\right).\end{aligned}$$

Taking the last formula into account, we arrive at

$$\mathscr{E}_i' = \left(\frac{dt}{dt'}\frac{\partial q^j}{\partial q^{\prime i}} - \frac{\partial t}{\partial q^{\prime i}}\frac{dq^j}{dt'}\right)\frac{\partial L}{\partial q^j} + \frac{dt}{dt'}\frac{\partial \dot{q}^j}{\partial q^{\prime i}}\frac{\partial L}{\partial \dot{q}^j} - \frac{d}{dt'}\left(\frac{dt}{dt'}\frac{\partial \dot{q}^j}{\partial q^{\prime i}}\frac{\partial L}{\partial \dot{q}^j}\right) - \frac{\partial t}{\partial q^{\prime i}}\frac{d\dot{q}^j}{dt'}\frac{\partial L}{\partial \dot{q}^j}.$$

Further we have

$$\dot{q}^j = \frac{dq^j}{dt} = \frac{\frac{\partial q^j}{\partial q^{\prime k}}dq^{\prime k} + \frac{\partial q^j}{\partial t'}dt'}{\frac{\partial t}{\partial q^{\prime k}}dq^{\prime k} + \frac{\partial t}{\partial t'}dt'}.$$

Dividing the numerator and denominator by the differential dt', we get

$$\dot{q}^j = \frac{\frac{\partial q^j}{\partial q^{\prime k}}\dot{q}^{\prime k} + \frac{\partial q^j}{\partial t'}}{\frac{\partial t}{\partial q^{\prime k}}\dot{q}^{\prime k} + \frac{\partial t}{\partial t'}}. \tag{1.41}$$

This formula allows one to determine

$$\frac{\partial \dot{q}^j}{\partial \dot{q}^{\prime i}} = \frac{\frac{\partial q^j}{\partial q^{\prime i}}}{\frac{\partial t}{\partial q^{\prime k}}\dot{q}^{\prime k} + \frac{\partial t}{\partial t'}} - \frac{\frac{\partial q^j}{\partial q^{\prime k}}\dot{q}^{\prime k} + \frac{\partial q^j}{\partial t'}}{(\frac{\partial t}{\partial q^{\prime k}}\dot{q}^{\prime k} + \frac{\partial t}{\partial t'})^2}\frac{\partial t}{\partial q^{\prime i}}.$$

---

[11] The expression below obviously vanishes, as a result of applying the lagrangian derivative to a particular function $\frac{d\Lambda}{dt'}$, where $\Lambda = t$.

Multiplying this expression with (1.40), we find that

$$\frac{dt}{dt'}\frac{\partial \dot{q}^j}{\partial \dot{q}^{\prime i}} = \frac{\partial q^j}{\partial q^{\prime i}} - \frac{\frac{\partial q^j}{\partial q^{\prime k}}\dot{q}^{\prime k} + \frac{\partial q^j}{\partial t'}}{\frac{\partial t}{\partial q^{\prime k}}\dot{q}^{\prime k} + \frac{\partial t}{\partial t'}}\frac{\partial t}{\partial q^{\prime i}} = \frac{\partial q^j}{\partial q^{\prime i}} - \frac{\partial t}{\partial q^{\prime i}}\dot{q}^j\,,$$

where (1.41) was used. Further, using the last formula, we note that

$$\frac{dt}{dt'}\frac{\partial \dot{q}^j}{\partial \dot{q}^{\prime i}}\frac{d}{dt'}\frac{\partial L}{\partial \dot{q}^j} = \Big(\frac{\partial q^j}{\partial q^{\prime i}} - \frac{\partial t}{\partial q^{\prime i}}\dot{q}^j\Big)\frac{dt}{dt'}\frac{d}{dt}\frac{\partial L}{\partial \dot{q}^j} = \Big(\frac{dt}{dt'}\frac{\partial q^j}{\partial q^{\prime i}} - \frac{\partial t}{\partial q^{\prime i}}\frac{dq^j}{dt'}\Big)\frac{d}{dt}\frac{\partial L}{\partial \dot{q}^j}\,.$$

Next, with the help of this relation, we reduce the expression for $\mathscr{E}_i'$ to the form

$$\mathscr{E}_i' = \Big(\frac{dt}{dt'}\frac{\partial q^j}{\partial q^{\prime i}} - \frac{\partial t}{\partial q^{\prime i}}\frac{dq^j}{dt'}\Big)\Big(\frac{\partial L}{\partial q^j} - \frac{d}{dt}\frac{\partial L}{\partial \dot{q}^j}\Big) + \Big(\frac{dt}{dt'}\frac{\partial \dot{q}^j}{\partial q^{\prime i}} - \frac{d}{dt'}\Big(\frac{dt}{dt'}\frac{\partial \dot{q}^j}{\partial \dot{q}^{\prime i}}\Big) - \frac{\partial t}{\partial q^{\prime i}}\frac{d\dot{q}^j}{dt'}\Big)\frac{\partial L}{\partial \dot{q}^j}$$

$$= \Big(\frac{dt}{dt'}\frac{\partial q^j}{\partial q^{\prime i}} - \frac{\partial t}{\partial q^{\prime i}}\frac{dq^j}{dt'}\Big)\mathscr{E}_j + \Big(\frac{dt}{dt'}\frac{\partial \dot{q}^j}{\partial q^{\prime i}} - \frac{d}{dt'}\Big(\frac{\partial q^j}{\partial q^{\prime i}} - \frac{\partial t}{\partial q^{\prime i}}\dot{q}^j\Big) - \frac{\partial t}{\partial q^{\prime i}}\frac{d\dot{q}^j}{dt'}\Big)\frac{\partial L}{\partial \dot{q}^j}\,.$$

To compute the coefficient in front of $\frac{\partial L}{\partial \dot{q}^j}$, we first consider

$$\frac{dt}{dt'}\frac{\partial \dot{q}^j}{\partial q^{\prime i}} = \frac{\partial^2 q^j}{\partial q^{\prime i}\partial q^{\prime k}}\dot{q}^{\prime k} + \frac{\partial^2 q^j}{\partial q^{\prime i}\partial t'} - \frac{\frac{\partial q^j}{\partial q^{\prime k}}\dot{q}^{\prime k} + \frac{\partial q^j}{\partial t'}}{\frac{\partial t}{\partial q^{\prime k}}\dot{q}^{\prime k} + \frac{\partial t}{\partial t'}}\Big(\frac{\partial^2 t}{\partial q^{\prime k}\partial q^{\prime i}}\dot{q}^{\prime k} + \frac{\partial^2 t}{\partial q^{\prime i}\partial t'}\Big)$$

$$= \frac{\partial^2 q^j}{\partial q^{\prime i}\partial q^{\prime k}}\dot{q}^{\prime k} + \frac{\partial^2 q^j}{\partial q^{\prime i}\partial t'} - \dot{q}^j\Big(\frac{\partial^2 t}{\partial q^{\prime k}\partial q^{\prime i}}\dot{q}^{\prime k} + \frac{\partial^2 t}{\partial q^{\prime i}\partial t'}\Big)\,,$$

where we again made use of the formulae (1.40) and (1.41). Finally, using the expression above, we compute

$$\frac{dt}{dt'}\frac{\partial \dot{q}^j}{\partial q^{\prime i}} - \frac{d}{dt'}\Big(\frac{\partial q^j}{\partial q^{\prime i}} - \frac{\partial t}{\partial q^{\prime i}}\dot{q}^j\Big) - \frac{\partial t}{\partial q^{\prime i}}\frac{d\dot{q}^j}{dt'} = \frac{dt}{dt'}\frac{\partial \dot{q}^j}{\partial q^{\prime i}} - \frac{d}{dt'}\Big(\frac{\partial q^j}{\partial q^{\prime i}}\Big) + \frac{d}{dt'}\Big(\frac{\partial t}{\partial q^{\prime i}}\Big)\dot{q}^j$$

$$= \Big[\frac{\partial^2 q^j}{\partial q^{\prime i}\partial q^{\prime k}}\dot{q}^{\prime k} + \frac{\partial^2 q^j}{\partial q^{\prime i}\partial t'} - \frac{d}{dt'}\Big(\frac{\partial q^j}{\partial q^{\prime i}}\Big)\Big] + \Big[\frac{d}{dt'}\Big(\frac{\partial t}{\partial q^{\prime i}}\Big)\dot{q}^j - \dot{q}^j\Big(\frac{\partial^2 t}{\partial q^{\prime k}\partial q^{\prime i}}\dot{q}^{\prime k} + \frac{\partial^2 t}{\partial q^{\prime i}\partial t'}\Big)\Big] = 0\,,$$

where the terms within the square brackets cancel each other. In this way it is shown [7, 8] that

$$\mathscr{E}_i(q',\dot{q}',\ddot{q}',t') = \Big(\frac{dt}{dt'}\frac{\partial q^j}{\partial q^{\prime i}} - \frac{\partial t}{\partial q^{\prime i}}\frac{dq^j}{dt'}\Big)\mathscr{E}_j(q,\dot{q},\ddot{q},t)\,.$$

Thus, under variational symmetries the lagrangian derivatives transform covariantly and, therefore, solutions of the equations of motion transform into the other solutions of the same equations.

It should be stressed that the inverse statement is not true – there could be symmetries of the equations of motion that are not shared by the action. As a simple and illustrative example of this situation, consider the lagrangian $L = \frac{m\dot{q}^2}{2}$ which implies the equation of motion $\ddot{q} = 0$. Consider the transformations [7]

$$q' = \frac{q}{1 - \epsilon q}, \quad t' = \frac{t}{1 - \epsilon q},$$

where ϵ is an arbitrary parameter. One finds

$$\dot{q}' = \frac{\dot{q}}{1 - \epsilon q + \epsilon t \dot{q}}, \quad \ddot{q}' = \frac{dt}{dt'}\dot{q}' = \left(\frac{1 - \epsilon q}{1 - \epsilon q + \epsilon t \dot{q}}\right)^3 \ddot{q}.$$

Thus, the transformation we are dealing with is a continuous one-parametric symmetry of the equation of motion. On the other hand,

$$L(q', \dot{q}', t')\frac{dt'}{dt} - L(q, \dot{q}, t) = \frac{m\dot{q}'^2}{2}\frac{dt'}{dt} - \frac{m\dot{q}^2}{2} = \frac{m\dot{q}^2}{2}\left(\frac{1}{(1 - \epsilon q + \epsilon t \dot{q})(1 - \epsilon q)^2} - 1\right).$$

The right hand side here is not a linear function of $\dot{q}$ and for this reason cannot originate from $\frac{d\Lambda(q,t)}{dt}$ for some function $\Lambda(q, t)$. Thus, this symmetry is not variational.

Hamiltonian symmetries. In the hamiltonian formalism the variables p and q are independent and Hamilton's equations follow from the action principle for

$$S = \int_{t_1}^{t_2} \left(p_i \dot{q}^i - H(p, q)\right) dt. \tag{1.42}$$

Then the integrals of motion for H generate symmetries of (1.42). Indeed, suppose $I = I(p, q)$ is an integral of motion, $\dot{I} = \{H, I\} = 0$. The hamiltonian vector field ξ_I of I yields the corresponding transformations of the dynamical variables

$$\begin{aligned} \delta p_i &= \xi_I p_i = \{I, p_i\}, \\ \delta q^i &= \xi_I q^i = \{I, q^i\}, \end{aligned} \tag{1.43}$$

These transformations are the symmetries of the hamiltonian as

$$\delta H = \xi_I H = \{I, H\} = 0.$$

On the other hand, since ξ_I is a hamiltonian vector field, it generates a symplectic transformation that preserves the symplectic form ω:

$$\mathcal{L}_{\xi_I}\omega = d i_{\xi_I}\omega + i_{\xi_I} d\omega = d i_{\xi_I}\omega = -d^2 I = 0.$$

As to the canonical 1-form α, with the help of the Cartan formula we find

$$\mathcal{L}_{\xi_I}\alpha = di_{\xi_I}\alpha + i_{\xi_I}d\alpha = d\alpha(\xi_I) + i_{\xi_I}\omega = d\alpha(\xi_I) - dI = d(\alpha(\xi_I) - I).$$

Thus, the canonical 1-form changes by the differential of the function $\alpha(\xi_I) - I$. This implies that under the variation generated by ξ_I the action (1.43) will be changed by the boundary term

$$\delta_I S = \int_{t_1}^{t_2} \frac{d(\alpha(\xi_I) - I)}{dt} dt = (\alpha(\xi_I) - I)\Big|_{t=t_1}^{t=t_2} = \left(p_j \frac{\partial I}{\partial p_j} - I\right)\Big|_{t=t_1}^{t=t_2}.$$

Thus, the hamiltonian action S is invariant up to this boundary term and non-vanishing of the latter can be the only source of violation of the invariance of S. On the other hand, changing S by the boundary term does not influence equations of motion.

The hamiltonian symmetries are symplectomorphisms which commute with the hamiltonian. From the point of view of Noether's theorem, they constitute a special class of symmetry transformations that do not involve non-trivial transformations of time. Those integrals I which generate symmetry transformations involving time do not commute with the hamiltonian, rather one has

$$0 = \frac{dI}{dt} = \{H, I\} + \frac{\partial I}{\partial t}, \tag{1.44}$$

as follows from the conservation of I. In this book we deal exclusively with hamiltonian symmetries. In Sect. 2.1.2 we reformulate Noether's theorem in the context of hamiltonian symmetries on the geometric language of the moment map.

1.3 Some Examples of Integrable Models

Here we describe some finite-dimensional integrable models. Following the tradition, we start with historically one of the first examples of integrable systems, namely, with the Kepler two-body problem. We discuss this dynamical system in much detail with a two-fold purpose: on the one hand, to illustrate an application of the Liouville theorem to a physically relevant model, and, on the other hand, to give an impression to which extent one may hope to find a solution of an integrable model in exact terms. Then, we proceed with multi-body systems, such as the Neumann model, delta-interacting Bose gas, Calogero-Sutherland-Moser and Ruijsenaars-Schneider models, which we touch upon only in a descriptive manner. We end up with a brief introduction into the theory of rigid bodies and focus on the representative example of the integrable Euler's top, for which we also construct a solution. By now there is a very long list of known finite-dimensional integrable systems which we do not attempt to reproduce; the models we discuss will be enough to demonstrate the main concepts of integrability. Later, when we come to the discussion of quantum systems, we will also introduce spin chains, as another interesting and physically relevant class of finite-dimensional integrable systems.

1.3.1 Kepler Problem

Consider $\mathbb{R}^3$ supplied with the euclidean scalar product $\langle \cdot, \cdot \rangle$.[12] In the center of mass frame (with the reduced mass set to one) Newton's equations we want to solve are

$$\frac{d^2 q_i}{dt^2} = -\frac{\partial V(r)}{\partial q_i}, \quad i = 1, 2, 3. \tag{1.45}$$

Here r is the length of the vector $q = (q_i) \in \mathbb{R}^3$. A potential $V(r)$ generates a central force and is invariant under rotations. For the original Kepler problem $V(r) = -\frac{k}{r}$, $k > 0$. Equation (1.45) are of the hamiltonian form for

$$H = \frac{1}{2} \sum_{i=1}^{3} p_i^2 + V(r)$$

with respect to the canonical Poisson structure

$$\{p_i, q_j\} = \delta_{ij}, \quad \{q_i, q_j\} = 0 = \{p_i, p_j\}.$$

Since the potential $V(r)$ is rotationally invariant, Noether's theorem implies the conservation of the vector of angular momentum $J = q \times p$. The components $J_i = \epsilon_{ijk} q_j p_k$ exhibit the well-known Poisson relations

$$\{J_i, J_j\} = -\epsilon_{ijk} J_k.$$

The dimension of the phase space is 6 and one can immediately find three independent commuting integrals, namely,

$$H, \quad J_3, \quad J^2 \equiv \langle J, J \rangle = J_1^2 + J_2^2 + J_3^2, \tag{1.46}$$

which puts the Kepler problem in the class of Liouville integrable systems.

Due to rotational symmetry, it is advantageous to use a spherical coordinate system

$$q_1 = r \sin \theta \cos \phi, \quad q_2 = r \sin \theta \sin \phi, \quad q_3 = r \cos \theta.$$

As we will see, the use of spherical coordinates will allow us to separate the variables in the Kepler problem. Meanwhile, comparing the expression for the canonical 1-form α in Cartesian and spherical coordinates

$$\alpha = p_i dq_i = p_r dr + p_\theta d\theta + p_\phi d\phi,$$

we obtain the relation between the canonical momenta associated to two different sets of canonical coordinates

---

[12]We will not distinguish between upper and lower indices.

$$p_1 = \frac{1}{r}\left(rp_r \cos\phi \sin\theta + p_\theta \cos\theta \cos\phi - p_\phi \frac{\sin\phi}{\sin\theta}\right),$$

$$p_2 = \frac{1}{r}\left(rp_r \sin\phi \sin\theta + p_\theta \cos\theta \sin\phi + p_\phi \frac{\cos\phi}{\sin\theta}\right),$$

$$p_3 = p_r \cos\theta - \frac{1}{r}p_\theta \sin\theta.$$

In spherical coordinates the conserved quantities (1.46) take the form

$$H = \frac{1}{2}\left(p_r^2 + \frac{1}{r^2}p_\theta^2 + \frac{1}{r^2 \sin^2\theta}p_\phi^2\right) + V(r),$$

$$J^2 = p_\theta^2 + \frac{1}{\sin^2\theta}p_\phi^2,$$

$$J_3 = p_\phi.$$

To better understand the physical picture, we note that the motion happens in the plane orthogonal to the vector J. This follows from the fact that $\langle J, q \rangle = \langle q \times p, q \rangle = 0$ and $\langle J, p \rangle = \langle q \times p, p \rangle = 0$, meaning that the vectors q and p lie in the plane orthogonal to J. Thus, without loss of generality, we can choose a coordinate system such that J has only one non-trivial component J_3: $J = (0, 0, J_3)$. Then the motion happens in the plane (x_1, x_2), which can be accounted for by taking in the above formulae $\theta = \frac{\pi}{2}$. With this arrangement we get

$$\dot\phi = \{H, \phi\} = \left\{\frac{p_\phi^2}{2r^2}, \phi\right\} = \frac{p_\phi}{r^2}.$$

This yields the following expression for the integral of motion $p_\phi = J_3 = J$

$$p_\phi = r^2 \dot\phi.$$

This is nothing else but the conservation law of angular momentum, as originally discovered by Kepler through his observations of the motion of Mars. The quantity $p_\phi = J$ has a simple geometric meaning. Kepler introduced the *sectorial velocity* v_s:

$$v_s = \lim_{\Delta t \to 0} \frac{\Delta A}{\Delta t},$$

where ΔA is the area of an infinitesimal sector swept by the radius-vector q for time Δt:

$$\Delta A = \frac{1}{2}r \cdot r\dot\phi \Delta t + \mathcal{O}(\Delta t^2) \approx \frac{1}{2}r^2 \dot\phi \Delta t.$$

This equation expresses the second law by Kepler: *in equal times the radius vector sweeps out equal areas, hence the sectorial velocity is constant.* This is one of the formulations of the conservation law of angular momentum.

Solution by the Liouville theorem. We can now see how the solution can be found by using the general approach based on the Liouville theorem. The expressions for the momenta on the surface of constant energy $H = E$ and $J = J_3$ are

$$p_r = \sqrt{2(E - V(r)) - \frac{J^2}{r^2}}, \quad p_\phi = J_3 = J.$$

As promised, in the spherical coordinate system the dynamical variables got separated: on a surface of constant integrals of motion p_r is a function of its canonical coordinate and p_ϕ is trivially constant. The generating function of the canonical transformation from the Liouville theorem is therefore

$$S = \int^r \sqrt{2(E - V(r)) - \frac{J^2}{r^2}} + \int^\phi J d\phi.$$

The associated angle variables are then

$$\psi_E = \frac{\partial S}{\partial E}, \quad \psi_J = \frac{\partial S}{\partial J}.$$

In the new coordinates (E, ψ_E), (J, ψ_J) equations of motion read

$$\dot{\psi}_E = 1, \quad \dot{\psi}_J = 0.$$

Integrating the first equation, one gets

$$\psi_E = t - t_0,$$

that is

$$t - t_0 = \int^r \frac{dr}{\sqrt{2\left(E - V(r) - \frac{J^2}{2r^2}\right)}}$$

This equation can be used to determine the time evolution of $r = r(t)$. The equation for ψ_J yields

$$\psi_J = -\int^r \frac{J\,dr}{r^2\sqrt{2\left(E - V(r) - \frac{J^2}{2r^2}\right)}} + \phi = 0,$$

so that

$$\phi(r) = \int^r \frac{J\,dr}{r^2\sqrt{2\left(E - V(r) - \frac{J^2}{2r^2}\right)}} \tag{1.47}$$

gives a parametric description of a trajectory.

Various kinds of trajectories are possible depending on values of E and J and on the exact form of the central potential $V(r)$. Here we will be interested in the bounded motion corresponding to the situation when r has two turning points, $r_{\min}$ and $r_{\max}$, called *pericentum and apocentrum*, respectively.[13] In general, the turning points, where $\dot{r} = 0$, are solutions of the following equation

$$E - V(r) - \frac{J^2}{2r^2} = 0. \tag{1.48}$$

At the turning points $\dot{r} = 0$ but $\dot{\phi} \neq 0$. This shows that r oscillates between $r_{\min}$ and $r_{\max}$, while ϕ changes monotonically. The angle between neighbouring apocenter and pericenter is given by

$$\Delta \phi = \int_{r_{\min}}^{r_{\max}} \frac{J\, dr}{r^2 \sqrt{2\left(E - V(r) - \frac{J^2}{2r^2}\right)}}.$$

Evidently, a generic orbit is not closed. It will be closed only if $\Delta \phi = 2\pi \frac{m}{n}, m, n \in \mathbb{N}$, otherwise it is everywhere dense in the annulus defined by $r_{\min}$ and $r_{\max}$. Depending on E and J, the annulus can degenerate into a circle which implies a closed trajectory for these values of the integrals of motion.

Determination of a central potential for which *all* bounded orbits are closed is known as the *Bertrand problem*. It turns out that there are only two potentials for which all bounded orbits are closed, namely,

$$V(r) = ar^2, \quad a > 0,$$
$$V(r) = -\frac{k}{r}, \quad k > 0.$$

We will continue with the second potential, the one of the original Kepler problem describing the two-body gravitational interaction.

Kepler laws. For the Kepler potential $V(r) = -\frac{k}{r}$ and values of energy in the interval $-\frac{k^2}{2J^2} \leqslant E \leqslant 0$, Eq. (1.48) has two solutions, $r_{\min}$ and $r_{\max}$, given by

$$r_{\min} = \frac{-k + \sqrt{2EJ^2 + k^2}}{2E},$$
$$r_{\max} = \frac{-k - \sqrt{2EJ^2 + k^2}}{2E}.$$

Equation (1.47) turns into

---

[13] If the earth is the center then $r_{\min}$ and $r_{\max}$ are called perigee and apogee, if the sun—perihelion and aphelion, if the moon—perilune and apolune.

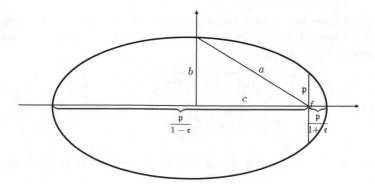

Fig. 1.3 Keplerian ellipse: semi-axes a, b, parameter p and eccentricity e

$$\phi(r) = \int^r \frac{J\,dr}{r^2\sqrt{2(E + \frac{k}{r} - \frac{J^2}{2r^2})}}$$

and can be integrated by elementary means giving the following equation for the orbit

$$\phi(r) = \arccos \frac{\frac{J}{r} - \frac{k}{J}}{\sqrt{2E + \frac{k^2}{J^2}}}.$$

Here the integration constant is set to zero which corresponds to the choice of the reference point for the angle ϕ at the pericenter: $\phi(r_{\min}) = 0$. Defining the following quantities

$$\mathfrak{p} = \frac{J^2}{k}, \qquad \mathfrak{e} = \sqrt{1 + \frac{2EJ^2}{k^2}},$$

we can write the equation for the orbit in the form

$$r = \frac{\mathfrak{p}}{1 + \mathfrak{e}\cos\phi}.$$

This is the so-called *focal equation of a conic section*. When $\mathfrak{e} < 1$, i.e. $E < 0$, the conic section is an ellipse. The number $\mathfrak{p}$ is called the *parameter* of the ellipse and $\mathfrak{e}$ the *eccentricity*. The motion is bounded for $E < 0$.

We can notice three distinguished points

$$\phi = 0: \quad r = \frac{\mathfrak{p}}{1 + \mathfrak{e}},$$

$$\phi = \frac{\pi}{2}: \quad r = \mathfrak{p},$$

$$\phi = \pi: \quad r = \frac{\mathfrak{p}}{1 - \mathfrak{e}},$$

which give us certain geometric intuition about various quantities appearing in the description of elliptic orbits, see Fig. 1.3. In particular, the major semi-axis a is determined as

$$2a = \frac{\mathfrak{p}}{1-\mathfrak{e}} + \frac{\mathfrak{p}}{1+\mathfrak{e}} = \frac{2\mathfrak{p}}{1-\mathfrak{e}^2} \, .$$

We also have

$$c = a - \frac{\mathfrak{p}}{1+\mathfrak{e}} = \frac{\mathfrak{e}\mathfrak{p}}{1-\mathfrak{e}^2} \, .$$

From the last two equations the eccentricity can be determined via the major semi-axes as

$$\mathfrak{e} = \frac{c}{a} = \frac{\sqrt{a^2 - b^2}}{a} = \sqrt{1 - \frac{b^2}{a^2}} \, .$$

We can now formulate the Kepler laws:

(1) The first law: planets describe ellipses with the Sun at one focus.
(2) The second law: the sectorial velocity is constant.
(3) The third law: the period of revolution around an elliptical orbit depends only on the size of the major semi-axes. The squares of the revolution periods of two planets on different elliptical orbits have the same ratio as the cubes of their major semi-axes.

The third law follows from the following considerations. Let T be a revolutionary period and A be the area swept out by the radius vector over the period. An ellipse with the semi-axes a and b encompasses the area

$$A = \pi a b = \pi a^2 \sqrt{1 - \mathfrak{e}^2} = \pi \frac{\mathfrak{p}^2}{(1-\mathfrak{e}^2)^2} \sqrt{1 - \mathfrak{e}^2} = \pi \frac{\mathfrak{p}^2}{(1-\mathfrak{e}^2)^{\frac{3}{2}}} = \frac{\pi k J}{(\sqrt{2|E|})^3} \, ,$$

where we have taken into account that

$$a = \frac{\mathfrak{p}}{1-\mathfrak{e}^2} = \frac{k}{2|E|} \, .$$

On the other hand, since the sectorial velocity v_s is constant, we have

$$\int_0^T v_c = \int_0^T dt \frac{dA}{dt} = A \, , \quad \rightarrow \quad v_s T = \frac{J}{2} T = A \, ,$$

that is,

$$T = \frac{2A}{J} = \frac{2\pi k}{(\sqrt{2|E|})^3} = \frac{2\pi}{\sqrt{k}} a^{3/2} \, .$$

It is interesting that the total energy depends only on the major semi-axis a and it is the same for the whole set of elliptical orbits from a circle of radius a to a line segment of length $2a$. The value of the second semi-axis depends on the angular momentum.

Laplace-Runge-Lenz vector and Liouville torus. The phase space for the motion in a central potential is six-dimensional. There are four conserved quantities: three components of the angular momentum J and the energy E. They ensure that the motion happens on a two-dimensional submanifold. In the case of bounded motion, this manifold is the Liouville torus. Thus, there should be two associated frequencies and if the latter are not commensurable, the orbits are not closed and fill the torus densely. It turns out that the Kepler problem with its specific potential (with any sign of k) admits one more non-trivial conserved quantity that is absent for a generic central potential: *the Laplace-Runge-Lenz vector* $R \in \mathbb{R}^3$ (for definiteness we assume that $k > 0$ and denote $v = \dot{q} = p$, since $m = 1$):

$$R = v \times J - k\frac{q}{r} .$$

Conservation of this vector can be verified in a straightforward manner. Indeed, we have

$$\dot{R} = \dot{v} \times J - k\frac{v}{r} + k\frac{q \langle v, q \rangle}{r^3} = \dot{v} \times (q \times v) - k\frac{v}{r} + k\frac{q \langle v, q \rangle}{r^3} .$$

On the other hand,

$$\dot{v} = -\frac{\partial V}{\partial r}\frac{q}{r} = -k\frac{q}{r^3}$$

and, therefore,

$$\dot{R} = -k\frac{q \times (q \times v)}{r^3} - k\frac{v}{r} + k\frac{q \langle v, q \rangle}{r^3} .$$

Now, taking into account that $q \times (q \times v) = q \langle v, q \rangle - r^2 v$, we get $\dot{R} = 0$.

The Laplace-Runge-Lenz vector has the following properties. First, $\langle R, J \rangle = 0$. Second,

$$R^2 = \langle v \times J, v \times J \rangle - 2k\frac{\langle v \times J, q \rangle}{r} + k^2 = v^2 J^2 - 2k\frac{\langle J, q \times v \rangle}{r} + k^2$$

$$= 2(E - V)J^2 - 2k\frac{J^2}{r} + k^2 = 2EJ^2 + k^2 .$$

Thus, from the three components of R only one is independent. Together with J and E, we have five (not all Poisson pair-wise commuting) integrals of motion that reduce the dimension of the Liouville torus to one and make the corresponding motion periodic (in the bounded case). The Kepler system is therefore an example of a maximally superintegrable model with completely degenerate motion. A non-abelian symmetry responsible for this degeneration can be made explicit by considering the Poisson relations

$$\{J_i, R_j\} = -\epsilon_{ijk} R_k, \quad \{R_i, R_j\} = 2H\epsilon_{ijk} J_k. \tag{1.49}$$

Defining $K_i^\pm = \frac{1}{2}(J_i \pm \frac{i}{\sqrt{2H}} R_i)$, one obtains

$$\{K_i^+, K_j^+\} = -\epsilon_{ijk} K_k^+, \quad \{K_i^-, K_j^-\} = -\epsilon_{ijk} K_k^-, \quad \{K_i^+, K_j^-\} = 0. \tag{1.50}$$

Thus, $K^\pm$ generate two independent copies of the angular momentum algebra, each of them is isomorphic to the Lie algebra $\mathfrak{su}(2)$. Together they are combined into the $\mathfrak{so}(4) \simeq \mathfrak{su}(2) \oplus \mathfrak{su}(2)$ symmetry.

Infinitesimal transformations of the dynamical variables generated by the R_i component of the Laplace-Runge-Lenz vector are

$$\delta_i q_j = \{R_i, q_j\} = \epsilon_{ijk} J_k + q_i p_j - \delta_{ij} \langle p, q \rangle,$$
$$\delta_i p_j = \{R_i, p_j\} = -\delta_{ij} p^2 + p_i p_j + \frac{k}{r}\left(\delta_{ij} - \frac{q_i q_j}{r^2}\right).$$

Consequently, the canonical 1-form $\alpha = \langle p, dq \rangle$ undergoes a change by the following exact form

$$\delta_i \langle p, dq \rangle = \delta_i p_j dq_j + p_j d\delta_i q_j = d\left(p^2 q_i - p_i (p, q) + \frac{k}{r} q_i\right).$$

One can see that the right hand side here coincides with $\alpha(\xi_i) - R_i = p_k \frac{\partial R_i}{\partial p_k} - R_i$, where ξ_i is the hamiltonian vector field corresponding to R_i, in complete agreement with our general discussion of hamiltonian symmetries in Sect. 1.2.2.

Solving Kepler's equation. The last step in solving Kepler's problem is to determine the evolution laws along the already established orbit. This is most easily done in terms of the so-called *eccentric anomaly u*, rather than in terms of the true anomaly ϕ. Looking at Fig. 1.4, it is easy to see that

$$r \cos\phi + a\mathfrak{e} = a \cos u,$$

where $a\mathfrak{e}$ is half the distance between two foci. Expressing the product $r \cos\phi$ from the equation $r = \mathfrak{p}/(1 + \mathfrak{e} \cos\phi)$, we get

$$r = a(1 - \mathfrak{e} \cos u).$$

Now we have two expressions for the radius r

$$\frac{\mathfrak{p}}{1 + \mathfrak{e} \cos\phi} = r = a(1 - \mathfrak{e} \cos u),$$

from which we can find the relation between the true and eccentric anomalies, namely,[14]

---

[14] It is convenient to express $\cos\phi$ and $\cos u$ with the tangent of the corresponding half-argument, *cf.* formulae on Fig. 1.4.

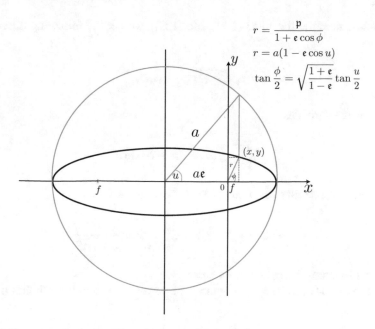

$$r = \frac{p}{1 + e \cos \phi}$$

$$r = a(1 - e \cos u)$$

$$\tan \frac{\phi}{2} = \sqrt{\frac{1 + e}{1 - e}} \tan \frac{u}{2}$$

Fig. 1.4 True anomaly ϕ, eccentric anomaly u and their relation

$$\tan \frac{\phi}{2} = \sqrt{\frac{1 + e}{1 - e}} \tan \frac{u}{2}.$$

Introducing the coordinate system as shown in Fig. 1.4, we can express the cartesian coordinates (x, y) of a point on the orbit via the eccentric anomaly u. Namely,

$$x = a \cos u - ae = a(\cos u - e). \qquad (1.51)$$

As to y, we have

$$y = r \sin \phi = a(1 - e \cos u) \frac{2 \tan \frac{\phi}{2}}{1 + \tan^2 \frac{\phi}{2}} = a(1 - e \cos u) \frac{2 \sqrt{\frac{1+e}{1-e}} \tan \frac{u}{2}}{1 + \frac{1+e}{1-e} \tan \frac{u}{2}}.$$

Further simplification of the right hand side of the last formula gives

$$y = a \sqrt{1 - e^2} \sin u. \qquad (1.52)$$

Now we recall that $J = p_\phi = r^2 \dot{\phi}$ is an integral of motion. We can rewrite it in cartesian coordinates

$$J = r^2 \dot{\phi} = x \dot{y} - y \dot{x}.$$

Substituting here the formulae (1.51) and (1.52), we get the following differential equation

$$J = a^2\sqrt{1 - e^2}(1 - e\cos u)\frac{du}{dt}.$$ (1.53)

Separating the variables

$$\frac{Ja^2}{\sqrt{1 - e^2}}dt = du - ed\sin u$$

and integrating, one finds the famous Kepler's equation

$$u - e\sin u = n(t - t_0) = \zeta, \quad n \equiv \frac{Ja^2}{\sqrt{1 - e^2}}.$$ (1.54)

The linear function ζ is called *mean* anomaly.

Equation (1.54) can be solved in terms of power series and the solution is given by

$$u(e, \zeta) = \zeta + 2\sum_{m=1}^{\infty} \frac{J_m(me)}{m}\sin m\zeta.$$ (1.55)

Here $J_m(z)$ is the Bessel function and the series converges for $e < 1$.

The derivation of (1.55) is carried out as follows. It is clear that the solution is a function $u \equiv u(e, \zeta)$ and from the equation is clear that if we change $\zeta \to \zeta + 2\pi$ and simultaneously change $u \to u + 2\pi$ then the equation will remain invariant. Thus, $u(\zeta) - \zeta$ is periodic in ζ with period 2π, therefore it can be expanded in a Fourier series. Differentiating (1.54), we obtain

$$\frac{du}{d\zeta} = \frac{1}{1 - e\cos u} \equiv f(\zeta),$$

where $f(\zeta)$ must be periodic and, therefore, admits an expansion

$$f(\zeta) = \frac{1}{2}a_0 + \sum_{m=1}^{\infty} a_m\cos m\zeta + \sum_{m=1}^{\infty} b_m\sin m\zeta,$$

where the corresponding coefficients are

$$a_0 = \frac{1}{\pi}\int_0^{2\pi} f(\zeta)d\zeta, \quad a_m = \frac{1}{\pi}\int_0^{2\pi} f(\zeta)\cos m\zeta d\zeta, \quad b_m = \frac{1}{\pi}\int_0^{2\pi} f(\zeta)\sin m\zeta d\zeta.$$

From Kepler's equation one can see that $u(-\zeta) = -u(\zeta)$ and, therefore, the derivative $\frac{du}{d\zeta}$ is an even function of ζ. Consequently, the coefficients b_m vanish. Thus, we have

$$
\begin{aligned}
\frac{du}{d\zeta} &= \frac{1}{2\pi} \int_0^{2\pi} \frac{d\zeta}{1 - e \cos u} + \sum_{m=1}^{\infty} \frac{\cos m\zeta}{\pi} \int_0^{2\pi} \frac{\cos m\zeta \, d\zeta}{1 - e \cos u} \\
&= \frac{1}{2\pi} \int_0^{2\pi} du + \sum_{m=1}^{\infty} \frac{\cos m\zeta}{\pi} \int_0^{2\pi} \cos m(u - e \sin u) du .
\end{aligned}
$$

Now one needs to recall the definition of the Bessel function

$$
J_m(z) = \frac{1}{2\pi} \int_0^{2\pi} \cos(mx - z \sin x) dx = \sum_{k=0}^{\infty} \frac{(-1)^k (z/2)^{m+2k}}{k!(m+k)!} .
$$

This consideration of the Kepler problem shows that in general the most one could expect is to exhibit the ultimate solution of an integrable model only in terms of a transcendental equation with a solution given at best by a convergent power series.

1.3.2 Multi-body Systems

Here we present a few other physically interesting examples of integrable models. They all can be interpreted as systems of N interacting particles defined for arbitrary N. We expose only some general features of these models without attempting to discuss their solutions.

Neumann model. The Neumann model [9] describes an N-dimensional harmonic oscillator whose motion is constrained to an $N - 1$-dimensional sphere. The lagrangian is (the mass $m = 1$)

$$
L = \frac{1}{2} \sum_{i=1}^{N} (\dot{q}_i^2 - \omega_i^2 q_i^2) + \frac{1}{2} \Lambda \left(\sum_{i=1}^{N} q_i^2 - 1 \right),
$$

where we assume that all frequencies are different and $q \in \mathbb{R}^N$. Here Λ is the Lagrange multiplier and it ensures that the motion takes place on the sphere of unit radius. The equations of motion following from L are

$$
\ddot{q}_i + \omega_i^2 q_i - \Lambda q_i = 0 .
$$

The variable Λ enters the lagrangian without time derivative and therefore is not a dynamical variable. Varying the lagrangian with respect to Λ, one finds the constraint

$$C_1 \equiv q_i^2 - 1 = 0 \,.$$

Differentiating constraint C_1 with respect to time, we find another constraint

$$C_2 \equiv q_i \, p_i = 0 \,,$$

where $p_i = \dot{q}_i$ is the canonical momentum. From the equations of motion we can find Λ. To this end, we multiply the equations by q_i, sum over i and use the constraints

$$\Lambda = \ddot{q}_i q_i + \omega_i^2 q_i^2 = -\dot{q}_i^2 + \omega_i^2 q_i^2 \,.$$

Except C_1 and C_2 there are no other constraints. Indeed, differentiating C_2 with respect to time, using equations of motion, we find that on the constraint surface $C_1 = 0$

$$\dot{C}_2 = \dot{q}_i^2 + q_i \ddot{q}_i = \dot{q}_i^2 - \omega_i^2 q_i^2 + \Lambda = 0 \,.$$

To give the Hamiltonian description of the Neumann model, we have to resort to Dirac's method of dealing with constrained systems. We postpone the discussion of this method until Sect. 2.1.2 and here just state the necessary ingredients to formulate the model at the Hamiltonian level.

In the Dirac terminology the constraints we find are of the second class. Indeed, the Poisson bracket of the constraints restricted to the constraint surface $\mathcal{S} = \{C_1 = 0 = C_2\}$ is

$$\{C_i, C_j\} = \begin{pmatrix} 0 & -2 \\ 2 & 0 \end{pmatrix}_{ij} \,.$$

The next step is to define a new Poisson bracket called the Dirac bracket

$$\{F, G\}_{\mathrm{D}} = \{F, G\} - \{F, C_i\}\{C, C\}_{ij}^{-1}\{C_j, G\} \,.$$

Clearly, with respect to the Dirac bracket the constraints are in involution $\{C_i, C_j\}_D = 0$. By making use of the inverse matrix composed of the Poisson brackets of the constraints, one finds

$$\{F, G\}_{\mathrm{D}} = \{F, G\} - \frac{1}{2}\Big(\{F, C_1\}\{C_2, G\} - \{F, C_2\}\{C_1, G\}\Big) \,.$$

In particular, for the phase space coordinates we find on the constraint surface the following Dirac brackets

$$\begin{aligned}
\{q_i, q_j\}_{\mathrm{D}} &= 0 \,, \\
\{p_i, q_j\}_{\mathrm{D}} &= \delta_{ij} - q_i q_j \,, \\
\{p_i, p_j\}_{\mathrm{D}} &= q_i p_j - q_j p_i \,.
\end{aligned} \tag{1.56}$$

The hamiltonian of the Neumann model is

$$H = \frac{1}{2} \sum_{i=1}^{N} (p_i^2 + \omega_i^2 q_i^2) \, .$$

Computing the Dirac bracket of the dynamical variables with H, one finds on $\mathcal{S}$ the following dynamical equations

$$\dot{q}_i = \{H, q_i\}_{\mathrm{D}} = p_i \, ,$$
$$\dot{p}_i = \{H, p_i\}_{\mathrm{D}} = -\omega_i^2 q_i - q_i (\dot{q}_k^2 - \omega_k^2 q_k^2) \, ,$$

which are nothing else but the original equations of motion.

Finally, the Neumann model turns out to be integrable in the Liouville sense. The corresponding integrals of motion are the *Uhlenbeck integrals* [10]

$$F_i = q_i^2 + \sum_{\substack{j=1 \\ j \neq i}}^{N} \frac{(q_i p_j - q_j p_i)^2}{\omega_i^2 - \omega_j^2} \, .$$

They are in involution with respect to the Dirac bracket. On the constraint surface $\mathcal{S}$ the hamiltonian is expressed via the integrals as

$$H = \frac{1}{2} \sum_{i=1}^{N} \omega_i^2 F_i \, .$$

while the integrals satisfy the relation

$$\sum_{i=1}^{N} F_i = 1 \, .$$

One of the interesting aspects of the Neumann model revealed by Moser [11] is that it is related to the geodesic motion on an ellipsoid. Consequently, in the ellipsoidal coordinates the Hamilton-Jacobi equations separate and one can construct their solutions in terms of the hyperelliptic curve of genus $N - 1$. For the modern discussion of the Neumann model and its solutions we refer the reader to [12, 13].

Bose gas with delta-interaction. The so-called delta-interaction model is defined by the hamiltonian

$$H = \frac{1}{2m} \sum_{i=1}^{N} p_i^2 + \varkappa \sum_{i<j} \delta(q_i - q_j) \, , \tag{1.57}$$

where $\varkappa$ is a real coupling constant. For $\varkappa > 0$ the interaction is repulsive and for $\varkappa < 0$ it is attractive. A solution of the corresponding quantum-mechanical problem for the repulsive case was first obtained in the case of bosonic particles by Lieb and Liniger [14, 15], while the general case of distinguishable particles was solved by Yang [16, 17]. Expanded in many directions, this model serves as a prototype example of applications of the Bethe Ansatz techniques, the latter will be discussed in Chap. 5.

Toda chains. The dynamics of an integrable N-body system known as the (non-relativistic) Toda chain is specified by the hamiltonian[15]

$$H = \frac{1}{2m} \sum_{i=1}^{N} p_i^2 + V(q), \tag{1.58}$$

where the potential $V(q)$ distinguishes periodic and non-periodic cases. For the *periodic* Toda chain it is given by

$$V(q) = g^2 \sum_{i=1}^{N} e^{(q_i - q_{i+1})/\ell}, \quad q_{N+1} \equiv q_1,$$

while for the *non-periodic* one it is defined as

$$V(q) = g^2 \sum_{i=1}^{N-1} e^{(q_i - q_{i+1})/\ell},$$

where in both cases ℓ is a length parameter and g is a coupling constant. In the non-periodic case the first and the last particles do not interact with each other, while they do as any other particles in the periodic case. For q_i, $p_i \in \mathbb{R}$ the potential V is positive and the conservation of energy puts an upper bound on possible values of particle momenta and particles cannot escape to infinity in finite time. Also, in the periodic case the conservation of energy implies that distances between all the particles are bounded, i.e. particles undergo oscillations for any initial state. This accounted for the name *Toda molecule* for periodic chains. In the non-periodic case the potential energy decreases with $q_i - q_{i+1} \to -\infty$ which corresponds to a repulsive force acting between particles. Therefore, particles will scatter from any initial state, i.e. for the non-periodic Toda chains the scattering problem is well-defined. In this book we will not treat Toda chains[16] and restrict ourselves by noting that these dynamical systems can be obtained from the Ruijsenaars-Schneider models described below by means of a two-step limiting procedure, see the end of this subsection.

Calogero-Moser-Sutherland (CMS) models. Inverting the harmonic potential of the one-dimensional oscillator, one obtains a model with the hamiltonian

---

[15]In a general setting, the Toda chain can be associated with any root system.
[16]One can consult, e.g., [18, 19].

$$H = \frac{p^2}{2m} + \frac{\gamma^2}{mq^2}.$$

The corresponding dynamical system can be thought of as describing radial motion of a free particle on a two-dimensional plane with fixed angular momentum p_φ attributed to the coupling constant γ, the latter has the physical dimension of the Planck constant. The potential gives rise to centrifugal inverse-cube force. As any one-dimensional model with conserved energy, it can be elementary solved by quadratures. It is remarkable, however, that this model admits an integrable generalisation to many degrees of freedom

I.
$$H = \frac{1}{2m} \sum_{i=1}^{N} p_i^2 + \frac{\gamma^2}{2m} \sum_{i \neq j}^{N} \frac{1}{q_{ij}^2}. \tag{1.59}$$

The latter model describes N particles on a line interacting by the inverse-square potential. Here $q_{ij} = q_i - q_j$ is the difference between coordinates of i'th and j'th particle on a line. This mechanical system with n degrees of freedom is historically tied up with names of Calogero and Moser who solved it first in the quantum [20, 21] and classical [22] cases, respectively.

It has been shown by Sutherland [23, 24] that the model (1.59) can be further generalised to account for a periodic boundary conditions. The corresponding potential is a trigonometric generalisation of the one in (1.59) and the hamiltonian is

II.
$$H = \frac{1}{2m} \sum_{i=1}^{N} p_i^2 + \frac{\gamma^2}{2m} \sum_{i \neq j}^{N} \frac{1}{4\ell^2 \sin^2 \frac{1}{2\ell} q_{ij}}. \tag{1.60}$$

This is the Sutherland model. It can be viewed as an integrable deformation of (1.59) depending on an additional length parameter ℓ. Particles are confined here to a ring of circumference $2\pi\ell$, the decompactification limit $\ell \to \infty$ brings (1.60) back to the rational case (1.59). Sutherland used this model to study thermodynamical properties of quantum fluid based on (1.59). The Sutherland model has an interesting variant where the length ℓ is analytically continued to imaginary values $\ell \to i\ell$, giving rise to the hyperbolic model with the inverse-sinh-squared potential

III.
$$H = \frac{1}{2m} \sum_{i=1}^{N} p_i^2 + \frac{\gamma^2}{2m} \sum_{i \neq j}^{N} \frac{1}{4\ell^2 \sinh^2 \frac{1}{2\ell} q_{ij}}. \tag{1.61}$$

This time ℓ is naturally interpreted as an interaction length that sets the size of the region where interactions between particles are sizeable. In the limit $\ell \to \infty$ one again recovers the long-range model (1.59).

Evidently, the three models (1.59)–(1.61) are particular instances of the hamiltonian system with a pairwise potential $v(q) = v(-q)$

$$H = \frac{1}{2m} \sum_{i=1}^{N} p_i^2 + \sum_{i<j}^{N} v(q_{ij}) \, . \tag{1.62}$$

One can therefore ask a question on the most general function $v(q)$ for which the model defined by the hamiltonian is integrable in the Liouville sense. The answer turns out to be

IV. $$v(q) = \frac{\gamma^2}{m} \wp(q) \, , \tag{1.63}$$

where $\wp(q) \equiv \wp(q|\omega_1, \omega_2)$ is the Weierstrass elliptic function with half-periods ω_1 and ω_2, where we choose ω_1, $-i\omega_2$ to be any positive numbers, possibly infinite.[17] This potential defines an elliptic model from which the previous models follow as degenerate cases when one or both periods become infinite. Specifically, we have

Rational case: $\omega_1 = \infty$, $\omega_2 = i\infty$,

$$\wp(q) \to \frac{1}{q^2} \, .$$

Hyperbolic case: $\omega_1 = \infty$, $\omega_2 = i\pi\ell$,

$$\wp(q) \to \frac{1}{4\ell^2 \sinh^2 \frac{q}{2\ell}} + \frac{1}{12\ell^2} \, .$$

Trigonometric case: $\omega_1 = \pi\ell$, $\omega_2 = i\infty$,

$$\wp(q) \to \frac{1}{4\ell^2 \sin^2 \frac{q}{2\ell}} - \frac{1}{12\ell^2} \, .$$

According to [25], the rational, hyperbolic and trigonometric (potentials) models are marked as **I**, **II** and **III** respectively, while the most general elliptic case is refereed to as **IV**. In the following we abbreviate the systems **I** − **IV** as the CMS (Calogero-Moser-Sutherland) models. These CMS models are related to the root system of the Lie algebra A_{N-1} and can be generalised to other root systems [25].

Ruijsenaars-Schneider (RS) models. In [26] Ruijsenaars and Schneider (RS) introduced a new class of integrable N-particle models that can be viewed as relativistic generalisations of the CMS models. To describe the RS models, we introduce $\mu = mc$, where c is a deformation parameter playing the role of the speed of light; the rest energy of a particle of mass m is μc. The hamiltonian of the most general elliptic RS model then reads

---

[17]The $\wp$-function is homogeneous $\wp(\lambda q|\lambda\omega_1, \lambda\omega_2) = \lambda^{-2}\wp(q|\omega_1, \omega_2)$. With the assumption that ω_1, ω_2 has the physical dimension of length, this property allows one to use in $\wp(q)$ the dimensionful coordinate q.

$$H = \mu c \sum_{i=1}^{N} \cosh \frac{p_i}{\mu} \prod_{j \neq i}^{N} \sqrt{\sigma^2\left(\tfrac{i\gamma}{\mu}\right)\left(\wp\left(\tfrac{i\gamma}{\mu}\right) - \wp(q_{ij})\right)}. \tag{1.64}$$

As for the CMS models above, the variables (q_i, p_i) are the particle coordinates and momenta endowed with the canonical Poisson bracket (1.9) and $\gamma \in \mathbb{R}$ is the coupling constant with the physical dimension of angular momentum (the Planck constant): $[\gamma] = [\hbar]$. The function $\sigma(q) \equiv \sigma(q|\omega_1, \omega_2)$ is the Weierstrass σ-function. It is related to the elliptic ζ- and $\wp$-functions as

$$\zeta(q) = \sigma'(q)/\sigma(q), \quad \wp(q) = -\zeta'(q), \tag{1.65}$$

and when half-periods degenerate it reduces to

$$\begin{aligned}
\omega_1 = \infty, \omega_2 = i\infty, \quad & \sigma(q) \to q; \\
\omega_1 = \infty, \omega_2 = i\pi\ell, \quad & \sigma(q) \to 2\ell \sinh \tfrac{q}{2\ell} \exp\left(-\tfrac{q^2}{24\ell^2}\right); \\
\omega_1 = \pi\ell, \omega_2 = i\infty \quad & \sigma(q) \to 2\ell \sin \tfrac{q}{2\ell} \exp\left(\tfrac{q^2}{24\ell^2}\right).
\end{aligned} \tag{1.66}$$

By degenerating the elliptic functions in (1.64), we obtain the rational

$$H = \mu c \sum_{i=1}^{N} \cosh \frac{p_i}{\mu} \prod_{j \neq i}^{N} \sqrt{1 + \frac{\gamma^2}{\mu^2 q_{ij}^2}}, \tag{1.67}$$

hyperbolic

$$H = \mu c \sum_{i=1}^{N} \cosh \frac{p_i}{\mu} \prod_{j \neq i}^{N} \sqrt{1 + \frac{\sin^2 \frac{\gamma}{2\mu\ell}}{\sinh^2 \frac{q_{ij}}{2\ell}}} \tag{1.68}$$

and trigonometric

$$H = \mu c \sum_{i=1}^{N} \cosh \frac{p_i}{\mu} \prod_{j \neq i}^{N} \sqrt{1 + \frac{\sinh^2 \frac{\gamma}{2\mu\ell}}{\sin^2 \frac{q_{ij}}{2\ell}}} \tag{1.69}$$

RS models. Note that in deriving the hyperbolic and trigonometric hamiltonians an unessential overall constant has been discarded. In the hyperbolic case γ is naturally restricted to $-\pi\mu\ell \leqslant \gamma \leqslant \pi\mu\ell$, and for the rational and trigonometric cases it is an arbitrary real number.

The non-relativistic limit corresponds to sending the speed of light to infinity, $c \to \infty$, that also implies $\mu \to \infty$. Taking into account that for small z the elliptic functions behave as $\sigma(z) \approx z$ and $\wp(z) \approx 1/z^2$, for the elliptic hamiltonian $H = H_{RS}$ we find in this limit $H_{RS} - Nmc^2 \to H$, where H is the CMS hamiltonian **IV**. For degenerate cases, the non-relativistic limit to the CMS models follows either from this finding or, directly, from (1.67)–(1.68).

In what sense the RS models can be actually called *relativistic* requires an explanation [27]. It has been known since the early days of special relativity that a relativistic theory of particles cannot exist because it comes into a contradiction with the Einstein relativity principle. Any interacting theory depending on particle coordinates and momenta with a single time variable assumes an "action-at-a-distance" – the time development of coordinates and momenta of a particle at a given moment of time is determined by positions and momenta of all the other particles at the same moment. In other words, a particle must feel the state of other particles instantaneously that obviously contravenes finiteness of the speed of light. Thus, the RS model could not be relativistic invariant in the naive sense. What connects this model to the relativistic world is that it allows for a dynamical realisation of the generators of the Poincaré algebra, such that in the limit $c \to \infty$ the latter algebra turns into the (centrally extended) algebra of Galilean transformations. The construction of the Poincaré algebra goes as follows. In addition to the hamiltonian one defines two more dynamical quantities

$$
P = \mu \sum_{i=1}^{N} \sinh \frac{p_i}{\mu} \prod_{j \neq i}^{N} \sqrt{\sigma^2\left(\tfrac{i\gamma}{\mu}\right)\left(\wp\left(\tfrac{i\gamma}{\mu}\right) - \wp(q_{ij})\right)}, \tag{1.70}
$$

$$
B = -\frac{\mu}{c} \sum_{i=1}^{N} q_i . \tag{1.71}
$$

The algebra of Poisson brackets between H, P and B coincides with the Lie algebra of the Poincaré group in two dimensions

$$
\{H, P\} = 0 , \quad \{B, H\} = P , \quad \{B, P\} = H/c^2 , \tag{1.72}
$$

provided P and B are interpreted as the generators of the total momentum and the Lorentz boost respectively. In particular, according to the first bracket, P is an integral of motion. The vanishing of this bracket is due to the functional equation[18] satisfied by the $\wp$-function

$$
\begin{vmatrix} \wp(x) & \wp'(x) & 1 \\ \wp(y) & \wp'(y) & 1 \\ \wp(z) & \wp'(z) & 1 \end{vmatrix} = 0 , \tag{1.73}
$$

where the variables x, y, z fill the plane $x + y + z = 0$. In the non-relativistic limit (1.72) turns into the Galilean transformations with a central extension $\{B, P\} = mN$. However, one can show that neither q_i nor p_i transform under Poincaré generators as the usual coordinates and momenta associated to the Minkowski space, restricting

---

[18]This equation plays an essential role in proving classical and quantum integrability of the model [26, 28].

thereby the meaning of the relativistic terminology in this context. It is therefore more appropriate to regard the RS models as a one-parameter integrable deformation of the CMS systems.

Finally, we mention the relation of the RS models to the Toda theory arising in the special scaling limit. Consider, for instance, the hyperbolic RS model that is related to the non-periodic Toda chain, which both support scattering. First, we note that the hamiltonian in (1.68) remains positive if we make the substitution $\gamma/\mu\ell \to \pi - 2i \log(\gamma/\mu\ell\epsilon)$, the latter yields $\sin^2 \frac{\gamma}{2\mu\ell} \approx \frac{\gamma^2}{4\mu^2\ell^2\epsilon^2}$, as $\epsilon \to 0$. Second, replacing $q_j \to q_j - 2j\ell \ln \epsilon$, for small ϵ we get

$$\frac{1}{\sinh^2 \frac{q_{ij}}{2\ell}} \to \begin{cases} e^{1/\ell(q_j-q_i)}\epsilon^{2(i-j)}, \ j < i, \\ e^{1/\ell(q_i-q_j)}\epsilon^{2(j-i)}, \ j > i. \end{cases}$$

Thus, in the limit $\epsilon \to 0$ only nearest-neighbouring interaction terms survive and one recovers the hamiltonian of the *relativistic Toda chain* [29]

$$H/(\mu c) = \sum_{i=2}^{N-1} \cosh \frac{p_i}{\mu} \sqrt{1 + \frac{\gamma^2}{4\mu^2\ell^2} e^{(q_{i-1}-q_i)/\ell}} \sqrt{1 + \frac{\gamma^2}{4\mu^2\ell^2} e^{(q_i-q_{i+1})/\ell}}$$

$$+ \cosh \frac{p_1}{\mu} \sqrt{1 + \frac{\gamma^2}{4\mu^2\ell^2} e^{(q_1-q_2)/\ell}} + \cosh \frac{p_N}{\mu} \sqrt{1 + \frac{\gamma^2}{4\mu^2\ell^2} e^{(q_{N-1}-q_N)/\ell}}.$$

$$(1.74)$$

Finally, taking $\mu \to \infty$ one obtains the hamiltonian of the non-relativistic non-periodic Toda chain. As to the periodic chain, the latter is obtainable from the elliptic RS hamiltonian, see [29] for details.

1.3.3 Rigid Bodies

Historically, the description of the motion of rigid bodies was amongst the first problems of analytic mechanics. Of special interest, since 18th century, remains the dynamics of spinning tops, where the cases of Euler, Lagrange and Kowalevski provided prominent examples of completely integrable systems. In particular, the Kowalevski top brought into view the powerful methods of algebraic geometry to construct the solutions [30].

Following [2], below we outline the general approach to describe the motion of a rigid body and further consider the integrable case of Euler's top, which we solve completely. For the modern treatment of the most complicated Kowalevski case we refer the reader to [31].

Moving coordinate system. Let K and k be two oriented euclidean spaces isomorphic to $\mathbb{R}^3$. A *motion* of K relative to k is a mapping that smoothly depends on time t:

$$D_t : \ K \to k,$$

and preserves the metric and orientation. Every motion can be uniquely written as the composition of a rotation B_t (B_t maps the origin of K into the origin of k, implying that B_t is the linear mapping) and a translation $C_t: k \to k$. We will call K and k the moving and the stationary coordinate systems. Let $Q(t)$ and $q(t)$ be the radius-vectors of a point in moving and stationary coordinate systems, respectively. Then

$$q(t) = D_t Q(t) = B_t Q(t) + q_0(t),$$

where $q_0(t)$ describes the translation of the origin of k. Differentiating we get an addition formula for velocities ($B \equiv B_t$)

$$\dot{q} = \dot{B} Q + B \dot{Q} + \dot{q}_0.$$

Here the term $\dot{B} Q$ corresponds to the *transferred rotation*. Suppose a point does not move with respect to the moving frame, i.e. $\dot{Q} = 0$ and also that $q_0 = 0 = \dot{q}_0 = 0$. Then

$$\dot{q} = \dot{B} Q = \dot{B} B^{-1} q,$$

where $\dot{B} B^{-1} : k \to k$ is a linear operator on k. Since B is a rotation, it is an orthogonal transformation: $B B^t = \mathbb{1}$. Differentiating with respect to t, we get

$$\dot{B} B^t + B \dot{B}^t = 0 \quad \Longrightarrow \quad \dot{B} B^{-1} + (\dot{B} B^{-1})^t = 0,$$

so that $\dot{B} B^{-1}$ is skew-symmetric. On the other hand, every skew-symmetric operator from $\mathbb{R}^3$ to $\mathbb{R}^3$ is the operator of vector multiplication by a fixed vector ω:

$$\dot{q} = \omega \times q.$$

In general, ω depends on t. In the case of purely rotational motion but with $\dot{Q} \neq 0$ we will have

$$\dot{q} = \omega \times q + B \dot{Q} = \omega \times q + v',$$

where the first term is a transferred velocity and the second one is a relative velocity.

Rigid bodies. A rigid body is a system of point masses, constrained by holonomic relations expressed by the fact that the distance between points is constant.

If a rigid body moves freely then its center of mass moves uniformly and linearly. A rigid body rotates about its center of mass as if the center of mass were fixed at a stationary point O. Thus, rotation of a rigid body around a fixed point O constitutes a problem with three degrees of freedom.

The problem of rotation of a rigid body can be studied in more generality without assuming that the fixed point coincides with the center of mass of a body. The dynamical variables constitute the tangent bundle $T\mathrm{SO}(3) \simeq \mathrm{SO}(3) \times \mathbb{R}^3$ of the rotation group $\mathrm{SO}(3)$ which encodes the three rotation angles of the configuration manifold $\mathrm{SO}(3)$ and three angular velocities. Since the lagrangian function is chosen to be invariant under all rotations around O, by Noether's theorem the components of the angular momentum M are conserved: $\dot{M} = 0$. The total energy E is also conserved. Thus, for a rigid body rotating about a fixed point, in the absence of external forces, there are four integrals of motion: three components of M and the energy. Thus, motion happens on a two-dimensional surface M_f inside the six-dimensional manifold $T\mathrm{SO}(3)$:

$$M_f = \{M_1 = f_1, \quad M_2 = f_2, \quad M_3 = f_3, \quad E = f_4 > 0\}.$$

The two-dimensional manifold M_f is invariant: if the initial conditions of motion define a point on M_f, then this point corresponding to the position and velocity of the body remains in M_f for all times. The manifold M_f admits a globally defined vector field that is the field of velocities of the motion on $T\mathrm{SO}(3)$, it is orientable and compact (E is bounded). A two-dimensional compact orientable manifold admitting a globally defined vector field is isomorphic to a torus. This is the Liouville torus.[19] According to the Liouville theorem, motion on the torus will be characterised by two frequencies. If the ratio of these frequencies is not a rational number then in the course of evolution the body never returns to its initial state.

Consider a rigid body rotating around a fixed point O and denote by K a coordinate system rotating with the body around O: in K the body is at rest. Every vector in K is carried to k by an operator B. By the definition of the angular momentum we have

$$M = q \times m\dot{q} = m\, q \times (\omega \times q).$$

Denote by J and Ω the angular momentum and angular velocity in the moving frame K, respectively. By their definition, these quantities are related as

$$J = m\, Q \times (\Omega \times Q).$$

This formula motivates to consider a linear operator $A: K \to K$ given by

$$AX = m\, Q \times (X \times Q), \quad \forall X \in \mathbb{R}^3.$$

This operator is symmetric

$$\langle AX, Y \rangle = \langle m\, Q \times (X \times Q), Y \rangle = m\, \langle Q \times X, Q \times Y \rangle,$$

as the expression on the right hand side is a symmetric function of $X, Y \in \mathbb{R}^3$. Thus, we can write $A\Omega = J$. The operator A is called the *inertia tensor*. Being symmetric, the operator is diagonalisable and it defines three mutually orthogonal characteristic directions.

---

[19]We cannot however directly apply the Liouville theorem to derive this result, because the integrals M_i are not in involution. However, the theorem on non-commutative integrability mentioned in Sect. 1.1.3 applies.

The energy of a rotating body is its kinetic energy for which we have

$$E = \frac{m}{2} \langle \dot{q}, \dot{q} \rangle = \frac{m}{2} \langle \omega \times q, \omega \times q \rangle = \frac{1}{2} \langle m\, q \times (q \times \omega), \omega \rangle = \frac{1}{2} \langle M, \omega \rangle .$$

This can be also written in the form

$$E = \frac{1}{2} m \left(\omega^2 q^2 - \langle \omega, q \rangle^2 \right) = \frac{1}{2} I_{ij} \omega_i \omega_j . \tag{1.75}$$

Here $I_{ij} = m \left(q^2 \delta_{ij} - q_i q_j \right)$ is the inertia tensor in the frame k. The relation between the angular momenta and angular velocities in the stationary and moving coordinate systems

$$M = BJ , \quad \omega = B\Omega , \tag{1.76}$$

allows one to write the expression for the energy in terms of quantities associated to K, that is,

$$E = \frac{1}{2} \langle J, \Omega \rangle = \frac{1}{2} \langle A\Omega, \Omega \rangle , \tag{1.77}$$

where we have also used that the mapping B preserves the scalar product. In the basis of K where A is diagonal, $A = \{I_i\}$, the momentum J and the energy take a very simple form

$$J_i = I_i \Omega_i ,$$

$$E = \frac{1}{2} I_i \Omega_i^2 .$$

The axes of this particular coordinate system inside the body are called *the principal inertia axes*. We emphasise that we consider the situation where the body does not move with respect to the moving frame, so that $\dot{Q} = 0$. The momentum J inside the body changes with time, as well as the angular velocity Ω.

Euler's top. Euler's top is a rigid body without any particular symmetry rotating in the absence of any external force around a fixed point O that coincides with its center of mass. The conservation law of the angular momentum $M = BJ$ of the body in the stationary coordinate system yields

$$0 = \dot{M} = \dot{B}J + B\dot{J} = \dot{B}B^{-1}M + B\dot{J} = \omega \times M + B\dot{J} = B(\Omega \times J + \dot{J}) .$$

From here we find

$$\frac{dJ}{dt} = J \times \Omega = J \times A^{-1} J . \tag{1.78}$$

These are the *Euler equations*, they describe the evolution of the components of the angular momentum inside a rotating body. If one takes the coordinate system adjusted to the principal inertia axes, then Eq. (1.78) turns into the following system of equations

$$\frac{dJ_1}{dt} = a_1 J_2 J_3, \quad \frac{dJ_2}{dt} = a_2 J_3 J_1, \quad \frac{dJ_3}{dt} = a_3 J_1 J_2. \tag{1.79}$$

Here

$$a_1 = \frac{I_2 - I_3}{I_2 I_3}, \quad a_2 = \frac{I_3 - I_1}{I_1 I_3}, \quad a_3 = \frac{I_1 - I_2}{I_1 I_2}. \tag{1.80}$$

Let us take J_i as the dynamical variables. Consider the energy as the function of J_i

$$H = \frac{1}{2} \langle J, A^{-1} J \rangle = \frac{J_i^2}{2I_i}.$$

Evidently, the Euler equations are hamiltonian with respect to the following Poisson structure

$$\{J_i, J_j\} = \epsilon_{ijk} J_k. \tag{1.81}$$

It is straightforward to verify that H is conserved due to the Euler equations

$$\dot{H} = J_i \frac{\dot{J}_i}{I_i} = J_1 J_2 J_3 \left(\frac{a_1}{I_1} + \frac{a_2}{I_2} + \frac{a_3}{I_3} \right) = 0.$$

Similarly, one can verify the conservation of the length of the angular momentum vector

$$\dot{J}^2 = 2 J_i \dot{J}_i = 2 J_1 J_2 J_3 (a_1 + a_2 + a_3) = 0.$$

This clearly agrees with the fact that M is conserved and that $M^2 = J^2$. Thus, we have proven that the Euler equations have two quadratic integrals—the energy and $M^2 = J^2$. As a consequence, J lies on the intersection of an ellipsoid and a sphere:

$$2E = \frac{J_1^2}{I_1} + \frac{J_2^2}{I_2} + \frac{J_3^2}{I_3}, \quad J^2 = J_1^2 + J_2^2 + J_3^2. \tag{1.82}$$

One can further study the structure of the curves of intersection by fixing the ellipsoid $E > 0$ and changing the radius of the sphere.

Note that alternatively the Euler equations can be written as the evolution equations for the angular velocity Ω:

$$\frac{d\Omega_1}{dt} + \frac{I_3 - I_2}{I_1}\Omega_2\Omega_3 = 0,$$

$$\frac{d\Omega_2}{dt} + \frac{I_1 - I_3}{I_2}\Omega_3\Omega_1 = 0,$$

$$\frac{d\Omega_3}{dt} + \frac{I_2 - I_1}{I_3}\Omega_1\Omega_2 = 0.$$

From the conservation laws (1.82) we can express two angular velocities, for instance, Ω_1 and Ω_3,

$$\Omega_1^2 = \frac{1}{I_1(I_3 - I_1)}\Big((2EI_3 - J^2) - I_2(I_3 - I_2)\Omega_2^2\Big),$$

$$\Omega_3^2 = \frac{1}{I_3(I_3 - I_1)}\Big((J^2 - 2EI_1) - I_2(I_2 - I_1)\Omega_2^2\Big).$$

Then plugging these expressions into the Euler equation for Ω_2, we obtain

$$\frac{d\Omega_2}{dt} = \frac{1}{I_2\sqrt{I_1 I_3}}\sqrt{\Big((2EI_3 - J^2) - I_2(I_3 - I_2)\Omega_2^2\Big)\Big((J^2 - 2EI_1) - I_2(I_2 - I_1)\Omega_2^2\Big)}.$$

For definiteness we assume that $I_3 > I_2 > I_1$ and also that $J^2 > 2EI_2$. Then making the substitutions

$$\tau = t\sqrt{\frac{(I_3 - I_2)(J^2 - 2EI_1)}{I_1 I_2 I_3}}, \qquad s = \Omega_2\sqrt{\frac{I_2(I_3 - I_2)}{2EI_3 - J^2}}$$

and introducing the positive parameter $k^2 < 1$ by[20]

$$k^2 = \frac{(I_2 - I_1)(2EI_3 - J^2)}{(I_3 - I_2)(J^2 - 2EI_1)},$$

we obtain

$$\tau = \int_0^s \frac{ds}{\sqrt{(1 - s^2)(1 - k^2 s^2)}}.$$

The initial time $\tau = 0$ is chosen such that for $s = 0$ one has $\Omega_2 = 0$. Inverting the last integral, one gets the Jacobi elliptic function[21]

$$s = \operatorname{sn}\tau.$$

Using two other elliptic functions

---

[20]For a solution to exist the values of $J^2 = M^2$ must be bounded: $2EI_1 < J^2 < 2EI_3$.

[21]The treatment of Jacobi elliptic functions can be found, for instance, in [32, 33].

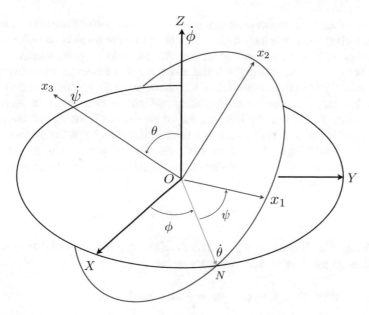

Fig. 1.5 The eulerian angles. Here XYZ and $x_1x_2x_3$ are the stationary and moving coordinate systems, respectively. The line ON is called the line of nodes

$$\mathrm{cn}^2\,\tau + \mathrm{sn}^2\,\tau = 1\,, \qquad \mathrm{dn}^2\,\tau + k^2\,\mathrm{sn}^2\,\tau = 1$$

we obtain the solution

$$\Omega_1 = \sqrt{\frac{2EI_3 - J^2}{I_1(I_3 - I_1)}}\,\mathrm{cn}\,\tau\,, \quad \Omega_2 = \sqrt{\frac{2EI_3 - J^2}{I_2(I_3 - I_1)}}\,\mathrm{sn}\,\tau\,, \quad \Omega_3 = \sqrt{\frac{J^2 - 2EI_1}{I_3(I_3 - I_1)}}\,\mathrm{dn}\,\tau\,.$$

The real period of all these three elliptic functions is given by 4K, where K is the complete elliptic integral of the first kind:

$$K = \int_0^1 \frac{\mathrm{d}s}{\sqrt{(1 - s^2)(1 - k^2 s^2)}}\,.$$

The period T in time t is, therefore,

$$T = 4K\sqrt{\frac{I_1 I_2 I_3}{(I_3 - I_2)(J^2 - 2EI_1)}}\,.$$

After this time both Ω and J will return to their original values. Thus, Ω and J perform a strictly periodic motion. What is remarkable, is that the top itself *does not return* to its original position in the stationary coordinate system k.

We have found that the angular momentum J moves periodically with the period T. On the other hand, we have argued that the Liouville torus is two-dimensional. This means that the actual motion of the body should be parameterised by two frequencies $\omega_{1,2}$. To understand how these two frequencies arise, let us express the angular velocity Ω in the moving frame via the eulerian angles and their derivatives.

Let XYZ and $x_1x_2x_3$ be the the stationary and moving coordinate frames, respectively, see Fig. 1.5. Denote by ON the line of nodes, that is the line of intersection of the planes XOY and x_1Ox_2. Clearly, the angular velocity of the system is compounded of angular velocities $\dot\theta$ about ON, $\dot\phi$ about OZ and $\dot\psi$ about Ox_3. First, we find the projections of these velocities on the coordinate axes of the moving frame. For the projections of $\dot\theta$ we have

$$\dot\theta_1 = \dot\theta \cos\psi \, , \quad \dot\theta_2 = -\dot\theta \sin\psi \, , \quad \dot\theta_3 = 0 \, .$$

The velocity $\dot\phi$ is directed along the axis Z of the stationary coordinate system. Its projections on the axes of the moving frame are

$$\dot\phi_1 = \dot\phi \sin\theta \sin\psi \, , \quad \dot\phi_2 = \dot\phi \sin\theta \cos\psi \, , \quad \dot\phi_3 = \dot\phi \cos\theta \, .$$

Finally, the velocity $\dot\psi$ is directed along x_3. Thus, collecting components along each axis, we obtain the angular velocity in the moving frame [34]

$$
\begin{aligned}
\Omega_1 &= \dot\phi \sin\theta \sin\psi + \dot\theta \cos\psi \, , \\
\Omega_2 &= \dot\phi \sin\theta \cos\psi - \dot\theta \sin\psi \, , \\
\Omega_3 &= \dot\phi \cos\theta + \dot\psi \, .
\end{aligned}
\tag{1.83}
$$

Recall that θ takes values from 0 to π, and ϕ and ψ from 0 to 2π. Substituting (1.83) into the expression for the kinetic energy $E = \frac{1}{2} I_i \Omega_i^2$, we obtain the latter in terms of the eulerian angles and their derivatives.

By using eulerian angles we can further relate the angular momenta in the moving and the stationary coordinate systems. Choosing the stationary coordinate system XYZ such that the angular momentum M is directed along the Z axis, we get

$$
\begin{aligned}
|M| \sin\theta \sin\psi &= I_1\Omega_1 \, , \\
|M| \sin\theta \cos\psi &= I_2\Omega_2 \, , \\
|M| \cos\theta &= I_3\Omega_3 \, ,
\end{aligned}
$$

where $|M|$ is the length of M. From here

$$\cos\theta = \frac{I_3\Omega_3}{|M|} \, , \qquad \tan\psi = \frac{I_1\Omega_1}{I_2\Omega_2} \, .$$

Substituting here the solutions for Ω_i allows one to find

$$\cos\theta = \sqrt{\frac{I_3(M^2 - 2EI_1)}{M^2(I_3 - I_1)}}\, \mathrm{dn}\,\tau,$$

$$\tan\psi = \sqrt{\frac{I_1(I_3 - I_2)}{I_2(I_3 - I_1)}}\frac{\mathrm{cn}\,\tau}{\mathrm{sn}\,\tau}.$$

Thus, both angles θ and ψ are periodic functions of time with the period T (the same period as for Ω). However, the angle ϕ *does not appear* in the formulas relating the angular momenta in the moving and the stationary coordinate systems. We can find it from the first two equations of (1.83), namely,

$$\dot{\phi} = \frac{\Omega_1 \sin\psi + \Omega_2 \cos\psi}{\sin\theta}.$$

This yields the following differential equation

$$\frac{d\phi}{d\tau} = |M|\frac{I_1\Omega_1^2 + I_2\Omega_2^2}{I_1^2\Omega_1^2 + I_2^2\Omega_2^2}.$$

The solution of this equation is found by integration but the integrand appears a rather involved combination of elliptic functions. Indeed, substituting the solution for Ω's in the last formula, we find

$$\frac{d\phi}{d\tau} = \frac{|M|}{I_1 \,\mathrm{cn}^2\,\tau + I_2 \,\mathrm{sn}^2\,\tau} = \frac{|M|}{I_1}\frac{1}{1 - \alpha^2 \,\mathrm{sn}^2\,\tau},$$

where we introduced $\alpha^2 \equiv 1 - I_2/I_1 < 0$, since by our assumptions $I_2 > I_1$. Thus,

$$\phi(\tau) = \phi_0 + \frac{|M|}{I_1}\int_0^\tau \frac{d\tau'}{1 - \alpha^2 \,\mathrm{sn}^2\,\tau'} = \phi_0 + \frac{|M|}{I_1}\Pi\left(\mathrm{am}(\tau, k), \alpha^2, k\right),$$

where $\Pi(\varphi, \alpha^2, k)$ is the *incomplete elliptic integral of the third kind*. Here $\varphi = \mathrm{am}(\tau, k)$ is the Jacobi amplitude. By using this explicit expression, one can show that the period of ϕ, which is denoted by T', is not commensurable with T. Consequently, this implies that the top never returns to its original state [35]. In fact, T and T' are the periods of motion on the Liouville torus we were looking for.

One comment is in order. The phase space of Euler's top has dimension six. We found four globally defined conserved quantities: the hamiltonian H and three components J_i of the angular momentum. One can verify that for $J \neq 0$ the 4×4 matrix of Poisson brackets of the set (J_i, H) has rank 2. Thus, according to the theorem on non-commutative integrability the motion should happen on the Liouville torus of dimension $6 - 4 = 2$. Since the Liouville torus is not one-dimensional, the motion of the top is partially but not completely degenerate.

1.4   Lax Pair and Classical r-Matrix

In this section we will introduce the cornerstone concepts of the modern theory of
integrable systems—the Lax representation and classical r-matrix. Generally, given
a dynamical system, to establish its Liouville integrability is a notoriously difficult
problem. However, if the corresponding equations of motion can be cast into a special
matrix form, known as the Lax representation, then the existence of an extended set
of conserved quantities is automatically guaranteed. These conserved quantities are
spectral invariants of the so-called Lax matrix and their involution property relies on
the special form of the Poisson bracket between components of the Lax matrix.

1.4.1   Lax Representation

Let L and M be two square matrices whose entries are functions on a phase space.
Consider the following matrix equation

$$\dot{L} = [M, L], \tag{1.84}$$

where as usual dot stands for the time derivative. If equation (1.84) is identically
satisfied as a consequence of hamiltonian equations for a given dynamical system,
then this dynamical system is said to admit a *Lax representation* (1.84) with L being
the corresponding *Lax matrix*. Such a pair of matrices L and M is often referred to
as *Lax pair*.

The importance of the Lax representation is that, once found, it allows for a simple
and universal construction of an extended set of conserved quantities as spectral
invariants of the corresponding Lax matrix. Indeed, consider

$$I_k = \operatorname{Tr} L^k .$$

for $k \in \mathbb{Z}$. We have

$$\dot{I}_k = k\operatorname{Tr}(L^{k-1}\dot{L}) = k\operatorname{Tr}(L^{k-1}[M, L]) = \operatorname{Tr}[M, L^k] = 0,$$

i.e. the I_k are time-independent as a consequence of the hamiltonian
equations implying (1.84). In fact, the matrix equation (1.84) can be readily solved
as

$$L(t) = g(t)L(0)g(t)^{-1},$$

where the invertible matrix $g(t)$ is determined from the equation

$$M(t) = \dot{g}g^{-1}.$$

By Newton's identities, integrals I_k are functions of the eigenvalues of the matrix L and vice versa. Since the eigenvalues of L are preserved in time, evolution of such a dynamical system is called *isospectral*.

We illustrate the concept of Lax representation on the simple example of a one-dimensional system with the following hamiltonian

$$H = \frac{1}{2}p^2 + \frac{1}{2}\omega^2 q^2 + \frac{\nu^2}{2q^2}. \tag{1.85}$$

This system can be called *Calogero oscillator*, as in the limiting cases $\nu \to 0$ and $\omega \to 0$ it reduces to the usual oscillator and the rational Calogero model, respectively. For this system we can take

$$L = \frac{1}{2}\begin{pmatrix} p & \omega q - \frac{\nu}{q} \\ \omega q - \frac{\nu}{q} & -p \end{pmatrix}, \qquad M = \frac{1}{2}\begin{pmatrix} 0 & -\omega - \frac{\nu}{q^2} \\ \omega + \frac{\nu}{q^2} & 0 \end{pmatrix}. \tag{1.86}$$

With this choice for L and M equation (1.84) is satisfied as a consequence of equations of motion $\dot{q} = p$ and $\dot{p} = -\omega^2 q + \frac{\nu^2}{q^3}$, and vice versa, satisfaction of (1.84) implies equations of motion for the Calogero oscillator. Notice that the conserved hamiltonian is expressed as $H = \mathrm{Tr}L^2 + \nu\omega$.

It should be emphasised that a Lax pair, if it exists, is not uniquely defined. First, the one and the same dynamical system might admit Lax pairs represented by $n \times n$ matrices of different size n. Second, there is a freedom related to transformations of the type

$$L' = gLg^{-1}, \qquad M' = gMg^{-1} + \dot{g}g^{-1}, \tag{1.87}$$

where g is an arbitrary invertible matrix possibly depending on dynamical variables. If here L, M is a Lax pair, then L', M' is another one for the same dynamical system. Indeed,

$$\dot{L}' = \dot{g}Lg^{-1} + g[M, L]g^{-1} - gLg^{-1}\dot{g}g^{-1} = [gMg^{-1} + \dot{g}g^{-1}, gLg^{-1}] \equiv [M', L'].$$

Note that M undergoes a gauge-type transformation. Lastly, for a fixed L shifting M by any polynomial of L will not influence the Lax equation (1.84).

Babelon-Viallet theorem and dynamical r-matrix. The Lax representation makes no reference to a Poisson structure. Spectral invariants of the Lax matrix are integrals of motion but without specifying this structure it is impossible to conclude anything about their involutive property.

A relation of integrals to the underlying Poisson structure gets established due to the *Babelon-Viallet theorem* [36]. According to this theorem, having the involutive property of the eigenvalues of $L \in \mathrm{Mat}_n(\mathbb{C})$ is equivalent to the existence of a function $\mathbf{r}$ on the phase space with values in $\mathrm{Mat}_n(\mathbb{C})^{\otimes 2}$ such that the Poisson bracket between the entries of L is

$$\{L_1, L_2\} = [\mathbf{r}_{12}, L_1] - [\mathbf{r}_{21}, L_2]. \tag{1.88}$$

Here and throughout the book L_1 and L_2 stand for two different embeddings of L in the tensor product $\mathrm{Mat}_n(\mathbb{C})^{\otimes 2}$, namely, $L_1 = L \otimes \mathbb{1}$ and $L_2 = \mathbb{1} \otimes L$, so that $\{L_1, L_2\} = \{L \otimes L\}$ represents a collection of all possible Poisson brackets between the entries of L. Correspondingly, the indices 1 and 2 of $\mathbf{r}_{12}$ refer to the first and second matrix components of $\mathrm{Mat}_n(\mathbb{C})^{\otimes 2}$, respectively. As an explicit matrix, $\mathbf{r}_{12} \equiv (r_{ij,kl})$, where $i, j = 1, \ldots, n$ correspond to the first matrix space and $k, l = 1, \ldots, n$ to the second one. The matrices on the right hand side of (1.88) are multiplied according to the standard rule of matrix multiplication. Thus, being written in components, formula (1.88) looks like

$$\{L_{ij}, L_{kl}\} = \mathbf{r}_{is,kl} L_{sj} - L_{is} \mathbf{r}_{sj,kl} - \mathbf{r}_{ks,ij} L_{sl} + L_{ks} \mathbf{r}_{sl,ij}, \quad \forall i, j, k, l \in 1, \ldots, n,$$

where we have separated the indices belonging two different matrix spaces of $\mathbf{r}_{12}$ by comma. Clearly, the use of the concise notation as in (1.88) saves a sufficient amount of work and space.

The matrix $\mathbf{r}$ is called *dynamical r-matrix*, which reflects the possibility for this matrix to depend on the phase space variables. Note that the bracket (1.88) is manifestly skew-symmetric. To obtain (1.88), following [36], we assume that L is diagonalisable,

$$L = S \Lambda S^{-1},$$

where Λ is a diagonal matrix whose entries Λ_i are prospective integrals of motion. Assuming that the phase space is equipped with a Poisson structure such that $\{\Lambda_i, \Lambda_j\} = 0$ for any i, j, we compute

$$
\begin{aligned}
\{L_1, L_2\} &= \{S_1 \Lambda_1 S_1^{-1}, S_2 \Lambda_2 S_2^{-1}\} = \\
&= \{S_1, S_2\} \Lambda_1 S_1^{-1} \Lambda_2 S_2^{-1} + S_1 \{\Lambda_1, S_2\} S_1^{-1} \Lambda_2 S_2^{-1} - S_1 \Lambda_1 S_1^{-1} \{S_1, S_2\} S_1^{-1} \Lambda_2 S_2^{-1} \\
&+ S_2 \{S_1, \Lambda_2\} \Lambda_1 S_1^{-1} S_2^{-1} - S_1 \Lambda_1 S_2 S_1^{-1} \{S_1, \Lambda_2\} S_1^{-1} S_2^{-1} - S_2 \Lambda_2 S_2^{-1} \{S_1, S_2\} S_1^{-1} \Lambda_1 S_1^{-1} \\
&- S_2 \Lambda_2 S_2^{-1} S_1 \{\Lambda_1, S_2\} S_2^{-1} S_1^{-1} + S_1 \Lambda_1 S_1^{-1} S_2 \Lambda_2 S_2^{-1} \{S_1, S_2\} S_1^{-1} S_2^{-1}.
\end{aligned}
$$

Introducing the notation

$$k_{12} = \{S_1, S_2\} S_1^{-1} S_2^{-1}, \qquad q_{12} = S_2 \{S_1, \Lambda_2\} S_1^{-1} S_2^{-1}, \qquad q_{21} = S_1 \{S_2, \Lambda_1\} S_1^{-1} S_2^{-1},$$

we have

$$
\begin{aligned}
\{L_1, L_2\} = {}&k_{12} L_1 L_2 + L_1 L_2 k_{12} - L_1 k_{12} L_2 - L_2 k_{12} L_1 \\
&- q_{21} L_2 + q_{12} L_1 - L_1 q_{12} + L_2 q_{21}.
\end{aligned}
$$

From the explicit form of k_{12} one sees that $k_{21} = -k_{12}$. This allows one to further rearrange the bracket as

$$\{L_1, L_2\} = [k_{12}L_2 - L_2 k_{12}, L_1] + [q_{12}, L_1] - [q_{21}, L_2]$$
$$= \tfrac{1}{2}[[k_{12}, L_2], L_1] - \tfrac{1}{2}[[k_{21}, L_1], L_2] + [q_{12}, L_1] - [q_{21}, L_2].$$

The last expression has precisely the form (1.88), where the corresponding r-matrix is

$$\mathbf{r}_{12} = q_{12} + \tfrac{1}{2}[k_{12}, L_2].$$

Note that $\mathbf{r}_{12}$ is not assumed to have any specific symmetry properties. Also, it is not uniquely defined: one can readily see that a shift $\mathbf{r}_{12} \to \mathbf{r}_{12} + [\sigma_{12}, L_2]$, where $\sigma_{12} = \sigma_{21}$, does not influence the right hand side of (1.88). Also, the bracket (1.88) does not change its form under symmetry transformations (1.87), although the r-matrix does.

Proceeding with our example of the Calogero oscillator, the r-matrix corresponding to L in (1.86) can be chosen as

$$\mathbf{r} = \frac{\omega q^2 + \nu}{2q(\omega q^2 - \nu)} \left(E_{12} \otimes E_{21} - E_{21} \otimes E_{12} \right), \tag{1.89}$$

which can be verified by straightforward calculation. The matrix is dynamical but depends on q only.

Concerning the Jacobi identity for (1.88), it yields the following constraint on the r-matrix

$$[L_1, [\mathbf{r}_{12}, \mathbf{r}_{13}] + [\mathbf{r}_{12}, \mathbf{r}_{23}] + [\mathbf{r}_{32}, \mathbf{r}_{13}] + \{L_2, \mathbf{r}_{13}\} - \{L_3, \mathbf{r}_{12}\}] + \text{cycl. perm} = 0.$$

In the case when $\mathbf{r}$ is independent of the dynamical variables, the last equation simplifies to

$$[L_1, [\mathbf{r}_{12}, \mathbf{r}_{13}] + [\mathbf{r}_{12}, \mathbf{r}_{23}] + [\mathbf{r}_{32}, \mathbf{r}_{13}]] + \text{cycl. perm} = 0.$$

In particular, the Jacobi identity will be satisfied if $\mathbf{r}$ obeys the following equation

$$[\mathbf{r}_{12}, \mathbf{r}_{13}] + [\mathbf{r}_{12}, \mathbf{r}_{23}] + [\mathbf{r}_{32}, \mathbf{r}_{13}] = 0. \tag{1.90}$$

Later on, in our analysis of various phase spaces and corresponding dynamical systems we will encounter further examples of r-matrices, both dynamical and non-dynamical, and equations similar to (1.90).

Another important point about the Poisson structure (1.88) is that it yields the Lax representation for evolution equations driven by any of the hamiltonians $H_k = \text{Tr} L^k$, $k \in \mathbb{Z}$. Indeed, from (1.88) one gets

$$\frac{dL}{dt_k} = \{H_k, L\} = [M_k, L],   \tag{1.91}$$

where $M_k = -k\mathrm{Tr}_1(\mathbf{r}_{21}L_1^{k-1})$ and t_k is the time evolution parameter along the hamiltonian flow triggered by H_k. One can verify that for the Calogero oscillator with $H = \mathrm{Tr}L^2$, the corresponding matrix M constructed in this way coincides with the one in (1.86).

1.4.2   Lax Representation with Spectral Parameter

In the canonical construction of integrals of motion as eigenvalues of the Lax matrix, the number of independent integrals cannot exceed the rank of this matrix. As a consequence, for a system with a sufficiently large number of degrees of freedom and with a Lax representation by low-rank matrices, the Liouville integrability is not guaranteed. Fortunately, there is a way to generalise the original approach based on equation (1.84) such that it will generate integrals in a number necessary for Liouville integrability and a priori unrelated to the rank of L. This far-reaching generalisation provides a basic mathematical structure for separation of variables and for construction of solutions of dynamical equations by powerful methods from algebraic geometry.

Introduce a variable $\lambda \in \mathbb{C}$, called in the following the *spectral parameter*. Let $M(\lambda)$, $L(\lambda)$ be two square matrices that are functions of the phase space variables and of the spectral parameter. A given dynamical system is said to admit a Lax representation with the spectral parameter, if there exist such $M(\lambda)$ and $L(\lambda)$ that the matrix equation

$$\dot{L}(\lambda) = [M(\lambda), L(\lambda)]   \tag{1.92}$$

is satisfied for any value of λ as a consequence of equations of motion for dynamical variables. And vice-versa, satisfaction of (1.92) for all λ implies the equations of motion for all dynamical variables. For a model which admits a Lax representation with the spectral parameter, the quantities

$$I_k(\lambda) = \mathrm{Tr}\, L^k(\lambda)   \tag{1.93}$$

are conserved in time for any λ. Assuming $I_k(\lambda)$ are rational functions of λ, the coefficients of their Laurent expansion around each pole are integrals of motion. This procedure yields an extended set of integrals, in many cases enough to conclude the Liouville integrability of the model, provided these integrals are in involution. Below we give some examples.

The Kepler problem. The Kepler problem, extensively discussed in Sect. 1.3, is integrable in the Liouville sense and, therefore, one can expect that it admits a spectral-parameter dependent Lax pair. One example of such a pair was constructed in [37]. The corresponding Lax matrix is characterised by having three simple poles at $\lambda_k, k = 1, 2, 3$, where λ_k are three arbitrary pairwise different constants. Explicitly, L and M matrices are

$$
L = \frac{1}{2} \begin{pmatrix} -\sum_{k=1}^{3} \frac{q_k \dot{q}_k}{\lambda - \lambda_k} & \sum_{k=1}^{3} \frac{q_k q_k}{\lambda - \lambda_k} \\ -\sum_{k=1}^{3} \frac{\dot{q}_k \dot{q}_k}{\lambda - \lambda_k} & \sum_{k=1}^{3} \frac{q_k \dot{q}_k}{\lambda - \lambda_k} \end{pmatrix}, \qquad M = \begin{pmatrix} 0 & 1 \\ \frac{k}{r^3} & 0 \end{pmatrix}.
$$

Here notations are the same as in Sect. 1.3. Newton's equation for q_k arises as the condition of vanishing of the residue of the pole at $\lambda = \lambda_k$. This Lax pair is, however, not fully satisfactory. Indeed, the quantities $\mathrm{Tr}(L^k)$ for k odd all vanish, while for k even they are expressed as powers of $\mathrm{Tr}(L^2)$, for which we find

$$
\mathrm{Tr}(L^2) = -\frac{J_1^2}{2(\lambda - \lambda_2)(\lambda - \lambda_3)} - \frac{J_2^2}{2(\lambda - \lambda_1)(\lambda - \lambda_3)} - \frac{J_3^2}{2(\lambda - \lambda_1)(\lambda - \lambda_2)}.
$$

Thus, the spectral invariants of L contain the conserved components of the angular momentum but not the hamiltonian of the Kepler problem. In fact, the information about the gravitational potential is contained in the matrix M, rather than in L.

The Neumann model. The Liouville integrability of the Neumann model can also be exhibited in terms of the spectral dependent Lax pair realised by 2×2 matrices. Namely, one can take for this model [38]

$$
L = \begin{pmatrix} v(\lambda) & u(\lambda) \\ -w(\lambda) & -v(\lambda) \end{pmatrix}, \qquad M = \begin{pmatrix} \sum_{k=1}^{N} q_k p_k & \sum_{k=1}^{N} q_k^2 \\ -\lambda - \sum_{k=1}^{N} q_k^2 & -\sum_{k=1}^{N} q_k p_k \end{pmatrix},
$$

where we introduced the notation

$$
u(\lambda) = \sum_{k=1}^{N} \frac{q_k^2}{\lambda - \omega_k^2}, \quad v(\lambda) = \sum_{k=1}^{N} \frac{q_k p_k}{\lambda - \omega_k^2}, \quad w(\lambda) = 1 + \sum_{k=1}^{N} \frac{p_k^2}{\lambda - \omega_k^2}.
$$

Note that there is yet another formulation of the Neumann model in terms of an $N \times N$ Lax pair [12].

Euler's top. Introduce two 3×3 anti-symmetric matrices

$$
J = \begin{pmatrix} 0 & -J_3 & -J_2 \\ J_3 & 0 & J_1 \\ J_2 & -J_1 & 0 \end{pmatrix}, \qquad \Omega = \begin{pmatrix} 0 & -\Omega_3 & -\Omega_2 \\ \Omega_3 & 0 & \Omega_1 \\ \Omega_2 & -\Omega_1 & 0 \end{pmatrix}.
$$

One can see that the Euler equations are equivalent to the following matrix equation

$$\dot{J} = [\Omega, J],$$

which is nothing else but the Lax representation (1.84), provided we identify $L = J$ and $M = \Omega$. We see, however, that this representation is not useful for generating integrals of motion. This is because the spectral invariants $\mathrm{Tr} L^k$ either vanish or are functions of J^2 and, therefore, do not contain the hamiltonian for any k. This situation can be cured by introducing a diagonal matrix D:

$$D = \begin{pmatrix} \frac{1}{2}(I_2 + I_3 - I_1) & 0 & 0 \\ 0 & \frac{1}{2}(I_1 + I_3 - I_2) & 0 \\ 0 & 0 & \frac{1}{2}(I_1 + I_2 - I_3) \end{pmatrix}.$$

One can see that

$$J = D\Omega + \Omega D.$$

Assuming that all entries of D are different, we introduce

$$L(\lambda) = D^2 + \frac{1}{\lambda} J, \quad M(\lambda) = \lambda D + \Omega. \tag{1.94}$$

Then we write the equation

$$\dot{L}(\lambda) = [M(\lambda), L(\lambda)] \tag{1.95}$$

which reduces to

$$\frac{1}{\lambda}\dot{J} = [\lambda D + \Omega, D^2 + \frac{1}{\lambda} J] = [\Omega, D^2] + [D, J] + \frac{1}{\lambda}[\Omega, J].$$

Further, we realise that

$$[\Omega, D^2] + [D, J] = \Omega D^2 - D^2 \Omega + D(D\Omega + \Omega D) - (D\Omega + \Omega D)D = 0.$$

Thus, the vanishing of the $1/\lambda$-term is equivalent to the Euler equations. The spectral-dependent Lax pair (1.94) produces the hamiltonian among the conserved quantities. Explicitly,

$$\mathrm{Tr} L(\lambda)^2 = \mathrm{Tr} D^4 - \frac{2}{\lambda^2} J^2,$$

$$\mathrm{Tr} L(\lambda)^3 = \mathrm{Tr} D^6 - \frac{3}{\lambda^2}\left(\frac{1}{4}(\mathrm{Tr} D)^2 J^2 - I_1 I_2 I_3 H\right).$$

Euler-Arnold equations. The three-dimensional Euler top admits a natural generalisation to the $\mathfrak{so}(n)$ Lie algebra. Let $\Omega \in \mathfrak{so}(n)$ and D is a diagonal matrix. Then

$$J = D\Omega + \Omega D$$

is also a $n \times n$ skew-symmetric matrix: $J^t = -J$. Equations

$$\dot{J} = [J, \Omega], \qquad J = D\Omega + \Omega D \qquad (1.96)$$

are known the *Euler-Arnold* equations. Assuming that all eigenvalues of D are different, we introduce the spectral-dependent Lax pair by the same Eq. (1.94). With this pair the Euler-Arnold equations (1.96) are equivalent to the spectral-dependent Lax equations (1.95), the latter are also known as *Manakov's* equations [39].

1.4.3 Building Up Integrable Systems

A general algorithm how to construct a Lax pair for a given integrable system is currently not known. However, there is a general procedure, due to Zakharov and Shabat, of how to construct consistent Lax pairs giving rise to integrable systems. This procedure yields spectral-dependent matrices $L(\lambda)$ and $M(\lambda)$ such that the matrix equation

$$\dot{L}(\lambda) = [M(\lambda), L(\lambda)] \qquad (1.97)$$

is, in fact, independent on λ and is equivalent to the equations of motion of an integrable system. Our exposition of the Zakharov-Shabat construction is close to that of [38]. We assume that $L(\lambda)$ and $M(\lambda)$ are $n \times n$ matrices.

The basic idea of the Zakharov-Shabat construction is to specify the analytic properties of the matrices $L(\lambda)$ and $M(\lambda)$ for $\lambda \in \mathbb{C}$. Let $f(\lambda)$ be a matrix-valued function which has poles at $\lambda = \lambda_k \neq \infty$ of order l_k. We can write

$$f(\lambda) = f_0 + \sum_k f_k(\lambda), \quad f_k(\lambda) = \sum_{s=-l_k}^{-1} f_{k,s}(\lambda - \lambda_k)^s,$$

where f_0 is a constant and $f_k(\lambda)$ is the polar part at the pole $\lambda = \lambda_k$. Around any λ_k the function $f(\lambda)$ admits the decomposition

$$f(\lambda) = f_+(\lambda) + f_-(\lambda),$$

where $f_+(\lambda)$ is regular at $\lambda = \lambda_k$ and $f_-(\lambda) = f_k(\lambda)$ is the polar part.

Assume that $L(\lambda)$ and $M(\lambda)$ are rational functions of λ. Let $\{\lambda_k\}$ be the set of poles of $L(\lambda)$ and $M(\lambda)$. Assuming that all λ_k are in the finite part of the complex plane, we can write

$$L(\lambda) = L_0 + \sum_k L_k(\lambda) , \qquad L_k(\lambda) = \sum_{s=-l_k}^{-1} L_{k,s}(\lambda - \lambda_k)^s$$

$$M(\lambda) = M_0 + \sum_k M_k(\lambda) , \qquad M_k(\lambda) = \sum_{s=-m_k}^{-1} M_{k,s}(\lambda - \lambda_k)^s .$$

Here $L_{k,s}$ and $M_{k,s}$ are matrices and we also assume that the λ_k do not depend on time.

Looking at the Lax equation (1.97), we see that at $\lambda = \lambda_k$ its left hand side exhibits a pole of order l_k, while the right hand side has a potential pole of the order $l_k + m_k$. Hence there are two type of equations. The first type does not contain the time derivatives and comes from setting to zero the coefficients of the poles of order greater than l_k on the right hand side of the Lax equation. This gives m_k constraints on the matrix M_k. The equations of the second type are obtained by matching the coefficients of the poles of order less or equal to l_k. These are equations for the dynamical variables because they involve time derivatives.

Consider the expansion of the matrix $L(\lambda)$ around $\lambda = \lambda_k$. Evidently, the matrix $Q(\lambda) = (\lambda - \lambda_k)^{l_k} L(\lambda)$ is regular around λ_k, i.e. we can write

$$Q(\lambda) = (\lambda - \lambda_k)^{l_k} L(\lambda) = Q_0 + (\lambda - \lambda_k)Q_1 + (\lambda - \lambda_k)^2 Q_2 + \cdots$$

Such a matrix can always be diagonalised by means of a regular similarity transformation

$$g(\lambda)Q(\lambda)g(\lambda)^{-1} = D(\lambda) = D_0 + (\lambda - \lambda_k)D_1 + \cdots .$$

Indeed, the condition of regularity means that around λ_k

$$g(\lambda) = g_0 + (\lambda - \lambda_k)g_1 + (\lambda - \lambda_k)^2 g_2 + \cdots ,$$
$$g(\lambda)^{-1} = h_0 + (\lambda - \lambda_k)h_1 + (\lambda - \lambda_k)^2 h_2 + \cdots$$

and, therefore,

$$\mathbb{1} = g(\lambda)g(\lambda)^{-1} =$$
$$= \Big(g_0 + (\lambda - \lambda_k)g_1 + (\lambda - \lambda_k)^2 g_2 + \cdots \Big)\Big(h_0 + (\lambda - \lambda_k)h_1 + (\lambda - \lambda_k)^2 h_2 + \cdots \Big)$$
$$= g_0 h_0 + (\lambda - \lambda_k)(g_0 h_1 + g_1 h_0) + \cdots$$

This allows to determine recurrently the coefficients h_i, $0 \leqslant i \leqslant \infty$, of the inverse matrix, for instance,

$$h_0 = g_0^{-1} , \qquad h_1 = -g_0^{-1} g_1 g_0^{-1} ,$$

and so on. We then have

$$g(\lambda)Q(\lambda)g(\lambda)^{-1} =$$
$$= \left(g_0 + (\lambda - \lambda_k)g_1 + \cdots\right)\left(Q_0 + (\lambda - \lambda_k)Q_1 + \cdots\right)\left(g_0^{-1} - (\lambda - \lambda_k)g_0^{-1}g_1g_0^{-1} + \cdots\right)$$
$$= g_0 Q_0 g_0^{-1} + (\lambda - \lambda_k)\left(g_0 Q_1 g_0^{-1} + g_1 Q_0 g_0^{-1} - g_0 Q_0 g_0^{-1}g_1g_0^{-1}\right) + \cdots$$

Thus, we see that g_0 should be chosen to diagonalize Q_0:

$$D_0 = g_0 Q_0 g_0^{-1}$$

and g_1 is found from the condition that

$$g_0 Q_1 g_0^{-1} + g_1 Q_0 g_0^{-1} - g_0 Q_0 g_0^{-1}g_1g_0^{-1} = g_0 Q_1 g_0^{-1} + [g_1 g_0^{-1}, D_0]$$

is diagonal. Since the commutator of a diagonal matrix with any matrix is off-diagonal, the matrix $[g_1 g_0^{-1}, D_0]$ is off-diagonal and g_1 is found by requiring that $[g_1 g_0^{-1}, D_0]$ cancels the off-diagonal entries of $g_0 Q_1 g_0^{-1}$. Thus, up to the first order in $\lambda - \lambda_k$ we will get

$$D(\lambda) = D_0 + (\lambda - \lambda_k)\left(g_0 Q_1 g_0^{-1}\right)_{ii} E_{ii} + \cdots$$

Continuing this procedure of perturbative diagonalisation, we conclude that by means of a regular similarity transformation around the pole $\lambda = \lambda_k$ the Lax matrix can be brought to the diagonal form

$$L(\lambda) \;\rightarrow\; A(\lambda) = \sum_{s=-l_k}^{-1} A_{k,s}(\lambda - \lambda_k)^s + \text{regular},$$

where $A_{k,s}$ are diagonal matrices. Clearly, the diagonalising matrix $g(\lambda)$ is defined up to right multiplication by an arbitrary analytic diagonal matrix.

Let $B(\lambda)$ be such that

$$M(\lambda) = g(\lambda)B(\lambda)g(\lambda)^{-1} + \dot{g}(\lambda)g(\lambda)^{-1},$$

where $g(\lambda)$ is a regular matrix which diagonalises $L(\lambda)$ around $\lambda = \lambda_k$. The Lax representation implies that

$$\dot{A}(\lambda) = [B(\lambda), A(\lambda)]. \tag{1.98}$$

Since $A(\lambda)$ is diagonal, it is made of the spectral invariants of L that are integrals of motion. As a result, $\dot{A}(\lambda) = 0$ and from (1.98) we deduce that $B(\lambda)$ is a diagonal matrix as well. Thus, with each pole λ_k we can associate two singular diagonal matrices

$$A_-^{(k)} \equiv \sum_{s=-l_k}^{-1} A_{k,s}(\lambda - \lambda_k)^s \, , \qquad B_-^{(k)} \equiv \sum_{s=-m_k}^{-1} B_{k,s}(\lambda - \lambda_k)^s \, .$$

These matrices allow one to reconstruct the corresponding pole parts of $L(\lambda)$ and $M(\lambda)$ as

$$L_k(\lambda) = (g^{(k)} A^{(k)} g^{(k)-1})_- \, , \qquad M_k(\lambda) = (g^{(k)} B^{(k)} g^{(k)-1})_- \, ,$$

where we set $g^{(k)} \equiv g(\lambda)$. Indeed, because $g^{(k)}$ is regular,

$$g^{(k)} = \sum_{s=0}^{l_k-1} g_{k,s}(\lambda - \lambda_k)^s + \text{higher powers} \, , \tag{1.99}$$

the matrices $L_k(\lambda)$ and $M_k(\lambda)$ depend only on the *singular* part of $A^{(k)}$ and $B^{(k)}$. In fact, in (1.99) only terms with $s = 0, \ldots, l_k - 1$ contribute to the singular parts $L_k(\lambda)$ and $M_k(\lambda)$.

The discussion above allows one to establish the independent degrees of freedom the Lax pair consists of. For every k these are two singular diagonal matrices $A^{(k)}$ and $B^{(k)}$, and a regular matrix $G^{(k)}$ of order $l_k - 1$,

$$G^{(k)} = \sum_{s=0}^{l_k-1} g_{k,s}(\lambda - \lambda_k)^s \, ,$$

defined up to right multiplication by a regular diagonal matrix. In addition there are two constant matrices L_0 and M_0. The L and M matrices are reconstructed from these data as

$$L(\lambda) = L_0 + \sum_k L_k(\lambda) \, , \qquad L_k(\lambda) = (G^{(k)} A_-^{(k)} G^{(k)-1})_-$$

$$M(\lambda) = M_0 + \sum_k M_k(\lambda) \, , \qquad M_k(\lambda) = (G^{(k)} B_-^{(k)} G^{(k)-1})_- \, .$$

Note that $g^{(k)}$ is fully determined by $G^{(k)}$. In other words, with $G^{(k)}$ one constructs $L(\lambda)$ and then diagonalizes it around the pole λ_k which produces the whole series $g^{(k)}$. This series is then used to build up M_k.

Taking into account that the matrices $L(\lambda)$ and $M(\lambda)$ are rational functions of λ, we can count the number of independent variables and the number of equations imposed on them by the Lax equation. The independent variables contained in L are L_0 and $L_{k,s}$, $s = 1, \cdots, l_k$ (i.e. for each k there are l_k matrices). The independent variables contained in M are M_0 and $M_{k,s}$, $s = 1, \cdots, m_k$ (i.e. for each k there are m_k matrices). Thus, a counting in units of n^2, which is the number of entries of a generic $n \times n$ matrix, gives

$$\text{number of variables} = \underbrace{2}_{L_0, M_0} + \sum_k l_k + \sum_k m_k = 2 + l + m$$

$$\text{number of equations} = \underbrace{1}_{\text{constant part}} + \underbrace{\sum_k (l_k + m_k)}_{\text{number of poles}} = 1 + l + m .$$

Clearly, there is one more variable than the number of equations which reflects the gauge invariance (1.87) of the Lax equation. On the Riemann surfaces of higher genus the situation changes and the number of equations is always bigger than the number of independent variables.

The general solution of the non-dynamic constraints on $M(\lambda)$ has the form

$$M = M_0 + \sum_k M_k , \qquad M_k = P^{(k)}(L, \lambda)_- , \qquad (1.100)$$

where $P^{(k)}(L, \lambda)$ is a polynomial in $L(\lambda)$ with coefficients rational in λ and $P^{(k)}(L, \lambda)_-$ is its singular part at $\lambda = \lambda_k$. Indeed, assuming that M_k is given by (1.100), we get

$$[M_k, L]_- = [P^{(k)}(L, \lambda)_-, L]_- = [P^{(k)}(L, \lambda) - P^{(k)}(L, \lambda)_+, L]_- = -[P^{(k)}(L, \lambda)_+, L]_- .$$

Since $[P^{(k)}(L, \lambda)_+, L]_-$ has poles of degree l_k and less, all non-dynamical constraints are solved. Let us show that (1.100) is the general solution. Recall that $A^{(k)}(\lambda)$ is a diagonal $n \times n$ matrix. Its powers

$$\left(A^{(k)}(\lambda) \right)^0, \quad \cdots , \quad \left(A^{(k)}(\lambda) \right)^{n-1}$$

span the space of all diagonal matrices. Thus, there exists a polynomial $P^{(k)}(A^{(k)}, \lambda)$ of degree $n - 1$ in $A^{(k)}$ with coefficients rationally dependent on λ, such that

$$B^{(k)}(\lambda) = P^{(k)}(A^{(k)}(\lambda), \lambda) .$$

Substituting this into the formula for M_k we get

$$M_k = (g^{(k)} B_-^{(k)} g^{(k)-1})_- = (g^{(k)} P^{(k)}(A^{(k)}(\lambda), \lambda) g^{(k)-1})_- = P^{(k)}(L, \lambda)_- .$$

The coefficients of $P^{(k)}$ are rational functions of the matrix elements of $A^{(k)}$ and $B^{(k)}$ and therefore they admit the well-defined Laurent expansion around $\lambda = \lambda_k$.

Evidently, from the point of view of dynamical systems, the Zakharov-Shabat constructions reveals the following:

(1) Dynamical variables are the entries of L. Choosing the number and the order of poles of the Lax matrix amounts to specifying a particular model.

(2) Choosing one of the polynomials $P^{(k)}(L, \lambda)$ is equivalent to specifying the dynamical flow (the hamiltonian).

Example: Euler's top. In the Sect. 1.4.2 we have introduced the following Lax pair for Euler's top

$$L(\lambda) = D^2 + \frac{1}{\lambda} J, \qquad M(\lambda) = \lambda D + \Omega.$$

The matrix $M(\lambda)$ has a pole at infinity. To bring this Lax pair in the framework of the Zakharov-Shabat construction, we note that if λ_k is a pole of $M(\lambda)$ but not of $L(\lambda)$, one can always make a polynomial redefintion $M(\lambda) \to M(\lambda) - P^{(k)}(L, \lambda)$ that eliminates from $M(\lambda)$ this pole without changing the Lax equation. In our present case we can take the following polynomial

$$P(L) = \lambda(\alpha L^2 + \beta L + \gamma),$$

that shifts the position of the pole from infinity to zero. Here the coefficients α, β, γ are

$$\alpha = -\frac{1}{I_1 I_2 I_3}, \qquad \beta = \frac{I_1^2 + I_2^2 + I_3^3}{2 I_1 I_2 I_3},$$
$$\gamma = \frac{(I_1 + I_2 + I_3)(I_2 + I_3 - I_1)(I_1 + I_2 - I_2)(I_1 + I_2 - I_3)}{16 I_1 I_2 I_3}.$$

With this choice we get

$$M(\lambda) \to \lambda D + \Omega - P(L) = \Omega - \alpha(D^2 J + J D^2) - \beta J - \frac{\alpha}{\lambda} J^2 = -\frac{\alpha}{\lambda} J^2,$$

because $\Omega - \alpha(D^2 J + J D^2) - \beta J = 0$. In this way we have obtained a new Lax pair

$$L(\lambda) = D^2 + \frac{1}{\lambda} J, \qquad M(\lambda) = -\frac{\alpha}{\lambda} J^2.$$

The Lax equation boils down to

$$\dot{L} = \frac{1}{\lambda} \dot{J} = [M, L] = -\frac{\alpha}{\lambda}[D^2, J^2].$$

The spectral parameter decouples from the last equation and we get

$$\dot{J} = -\frac{1}{I_1 I_2 I_3}[D^2, J^2].$$

Written in components this equation is equivalent to the Euler equations (1.79). The two non-zero eigenvalues of J are expressed via J^2 and they are integrals of motion for the trivial reason that J^2 belongs to the center of the Poisson structure (1.81).

References

1. Goldstein, H., Poole, C.P., Safko, J.L.: Classical Mechanics. Addison Wesley (2002)
2. Arnold, V.I.: Mathematical Methods of Classical Mechanics, vol. 60. Springer (1989)
3. Arnold, V.I., Kozlov, V.V., Neishtadt, A.I.: Mathematical Aspects of Classical and Celestial Mechanics. Springer, Berlin, Heidelberg (2006)
4. Mishchenko, A.S., Fomenko, A.T.: Generalized Liouville method of integration of Hamiltonian systems. Funct. Anal. Appl. **12**, 113–121 (1978)
5. Mishchenko, A.S., Fomenko, A.T.: Integration of Hamiltonian systems with non-commutative symmetries. Tr. Sem. Vekt. Tenzor Analiz Prilozh. Geom. Mekh. Fiz. **20**, 5–54 (1980)
6. Nekhoroshev, N.N.: Action-angle variables and their generalisations. Trans. Mosc. Math. Soc. **26**, 180–198 (1972)
7. Torres del Castillo, G.F., Miroón, C., Bravo Rojas, R.I.: Variational symmetries of Lagrangians. Revista Mexicana de Física E **59**, 140–147 (2013)
8. Torres del Castillo, G.F., Moreno-Ruiz, A.: Symmetries of the equations of motion that are not shared by the Lagrangian. arXiv:1705.08446 (2017)
9. Neumann, C.: De problemate quodam mechanico, quod ad primam integralium ultraellipticorum classem revocatur. Crelle J. **56**, 46 (1859)
10. Uhlenbeck, K.: Equivariant harmonic maps into spheres. In: Knill, K., Sealey (Eds.) Proceedings of the Tulane Conference on Harmonic Maps. Lecture Notes in Mathematics, vol. 949, pp. 146–158 (1982)
11. Moser, J.: Various aspects of hamiltonian systems. In: Proceedings of the C.I.M.E. Bressanone. Progress in Math., Birkhauser, vol. 8 (1978)
12. Avan, J., Talon, M.: Poisson structure and integrability of the Neumann-Moser-Uhlenbeck model. Int. J. Mod. Phys. A **5**, 4477–4488 (1990)
13. Babelon, O., Talon, M.: Separation of variables for the classical and quantum Neumann model. Nucl. Phys. B **379**, 321–339 (1992)
14. Lieb, E.H., Liniger, W.: Exact analysis of an interacting Bose gas. 1. The General solution and the ground state. Phys. Rev. **130**, 1605–1616 (1963)
15. Lieb, E.H., Liniger, W.: Exact analysis of an interacting bose gas II. The excitation spectrum. Phys. Rev. **130**, 1616 (1963)
16. Yang, C.N.: Some exact results for the many body problems in one dimension with repulsive delta function interaction. Phys. Rev. Lett. **19**, 1312–1314 (1967)
17. Yang, C.N.: S matrix for the one-dimensional N body problem with repulsive or attractive delta function interaction. Phys. Rev. **168**, 1920–1923 (1968)
18. Perelomov, A.: Integrable Systems of Classical Mechanics and Lie Algebras. Birkhauser Verlag Basel (1990)
19. Semenov-Tian-Shansky, M.A.: Quantization of open toda lattices. In: Arnold, V.I., Novikov, S.P. (Eds.) Dynamical Systems VII: Integrable Systems Nonholonomic Dynamical Systems (1993)
20. Calogero, F.: Solution of a three-body problem in one-dimension. J. Math. Phys. **10**, 2191–2196 (1969)
21. Calogero, F.: Solution of the one-dimensional N body problems with quadratic and/or inversely quadratic pair potentials. J. Math. Phys. **12**, 419–436 (1971)
22. Moser, J.: Three integrable Hamiltonian systems connected with isospectral deformations. Adv. Math. **16**, 197–220 (1975)
23. Sutherland, B.: Exact results for a quantum many body problem in one-dimension. Phys. Rev. A **4**, 2019–2021 (1971)
24. Sutherland, B.: Exact results for a quantum many body problem in one-dimension. 2. Phys. Rev., **A5**, 1372–1376 (1972)
25. Olshanetsky, M.A., Perelomov, A.M.: Classical integrable finite dimensional systems related to Lie algebras. Phys. Rept. **71**, 313 (1981)
26. Ruijsenaars, S.N.M., Schneider, H.: A New class of integrable systems and its relation to solitons. Annals Phys. **170**, 370–405 (1986)

27. Braden, H.W., Sasaki, R.: The Ruijsenaars-Schneider model. Prog. Theor. Phys. **97**, 1003–1018 (1997)
28. Ruijsenaars, S.N.M.: Complete integrability of relativistic Calogero-Moser systems and elliptic function identities. Commun. Math. Phys. **110**, 191 (1987)
29. Ruijsenaars, S.N.M.: Systems of calogero-moser type. In: Semenoff, G., Vinet, L. (Eds.) Particles and Fields, pp. 251–352. New York, NY, CRM Series in Mathematical Physics. Springer (1999)
30. Audin, M.: Spinning Tops: A Course on Integrable Systems (Cambridge Studies in Advanced Mathematics, Band 51). Cambridge University Press (1999)
31. Bobenko, A.I., Reyman, A.G., Semenov-Tian-Shansky, M.A.: The kowalewski top 99 years later: a lax pair, generalizations and explicit solutions. Comm. Math. Phys. **122**(2), 321–354 (1989)
32. Whittaker, E.T., Watson, G.N.: A Course of Modern Analysis. Cambridge University Press (1948)
33. Byrd, P.F., Friedman, M.D.: Handbook of Elliptic Integrals for Engineers and Physicists. Springer (1954)
34. Whittaker, E.T.: A Trease on the Analytic Dynamics of Particles & Rigid Bodies. Cambridge University Press (1988)
35. Landau, L.D., Lifshitz, E.M.: Mechanics, Third Edition: Volume 1. Butterworth-Heinemann, 3 edn (1976)
36. Babelon, O., Viallet, C.M.: Hamiltonian structures and lax equations. Phys. Lett. B **237**, 411–416 (1990)
37. Antonowicz, M., Rauch-Wojciechowski, S.: Lax representation for restricted flows of the KdV hierarchy and for the Kepler problem. Phys. Lett. A **171**, 303–310 (1992)
38. Babelon, O., Bernard, D., Talon, M.: Introduction to Classical Integrable Systems. Cambridge University Press (2003)
39. Manakov, S.V.: Note on the integration of Euler's equations of the dynamics of an n-dimensional rigid body. Funct. Anal. Appl. **10**, 93–94 (1976). (Funktsional. Anal. i Prilozhen. **10**(4), 328–329 (1976))

Chapter 2
Integrability from Symmetries

> *It is general wisdom in modern physics that problems are solvable because they have symmetries which are responsible for their privileged situation with respect to all other problems ... The traditional concept of symmetry is such that it leads to integrability wherever it applies.*
>
> Martin Gutzwiller
> Chaos in Classical and Quantum Mechanics

In this chapter we discuss a geometric approach to constructing integrable models. The starting point is a Poisson manifold with a rich group of symmetry transformations. Choosing an invariant hamiltonian and removing in a special way the degrees of freedom related to these symmetries, one obtains a new dynamical system on a smaller phase space which turns out to be integrable. The cotangent bundle of a Lie group and the Heisenberg double are representative examples of such Poisson manifolds with extended symmetries, and we explain their construction in great detail. Our discussion is paralleled by the treatment of the hamiltonian and Poisson reduction techniques, which we apply to these manifolds to obtain the integrable models of the CMS and RS type.

2.1 Phase Spaces with Symmetry

Here we introduce a few important phase spaces which will appear later in our treatment of concrete examples of integrable models.

We start with the euclidean phase space $\mathscr{P} = \mathbb{R}^{2N}$ introduced in Sect. 1.1. The Poisson bracket is defined by J of (1.7), the corresponding symplectic form is

© Springer Nature Switzerland AG 2019
G. Arutyunov, *Elements of Classical and Quantum Integrable Systems*,
UNITEXT for Physics, https://doi.org/10.1007/978-3-030-24198-8_2

$$\omega = dp_i \wedge dq^i = d\alpha , \tag{2.1}$$

where the 1-form α is

$$\alpha = p_i dq^i . \tag{2.2}$$

This shows that ω is not only closed but also exact. The symplectic manifold $\mathbb{R}^{2N}$ is an example of the cotangent bundle.

In general, a configuration space of a dynamical system is some manifold M. The role of the phase space $\mathcal{P}$ is played by the *cotangent bundle* T^*M over M, which is a vector bundle of 1-forms on M. On T^*M one can introduce a system of local coordinates (p_i, q^i), where q^i are coordinates on a local patch of M and p_i are coordinates in the fibre. Thus, α given by (2.2) is a special (canonical) 1-form on T^*M. The closed 2-form $\omega = d\alpha$ supplies T^*M with the structure of a symplectic manifold.

2.1.1 Coadjoint Orbits of a Lie Group

Let G be a Lie group and $\mathfrak{g}$ be its Lie algebra. Denote by $\mathfrak{g}^*$ the dual of $\mathfrak{g}$, i.e. the space of linear continuous functionals on $\mathfrak{g}$. Consider the algebra of smooth functions $\mathcal{F}(\mathfrak{g}^*)$ on $\mathfrak{g}^*$. If $f \in \mathcal{F}(\mathfrak{g}^*)$ then, according to the following definition,

$$\langle m, \nabla f(\ell) \rangle = \lim_{t \to 0} \frac{f(\ell + tm) - f(\ell)}{t} , \quad \ell, m \in \mathfrak{g}^* ,$$

the gradient ∇f takes values in the Lie algebra $\mathfrak{g}$. Here $\langle \cdot, \cdot \rangle$ denotes the natural pairing between $\mathfrak{g}$ and $\mathfrak{g}^*$. The Kirillov-Kostant Poisson bracket is defined as

$$\{f, h\}(\ell) = \langle \ell, [\nabla f(\ell), \nabla h(\ell)] \rangle \tag{2.3}$$

and it supplies the dual space to the Lie algebra with the structure of a Poisson manifold.[1]

Let us fix a basis $\{e_i\}$ in $\mathfrak{g}$, so that

$$[e_i, e_j] = f_{ij}^k e_k ,$$

where f_{ij}^k are the structure constants. Denote by e^i a basis in $\mathfrak{g}^*$ defined as $\langle e^i, e_j \rangle = \delta_j^i$. Then $\ell = \ell_i e^i$ and ℓ_i are coordinates[2] on $\mathfrak{g}^*$. In these coordinates

---

[1] In the literature this structure is often called *Lie-Poisson*.

[2] Normally coordinates on a manifold carry an upper index, here they occur with lower index as they are dual to coordinates on $\mathfrak{g}$, the latter naturally carry upper indices.

$$\nabla f = e_i \frac{\partial f}{\partial \ell_i}, \tag{2.4}$$

so that $\nabla \ell_i = e_i$. Therefore, the bracket between the coordinates is

$$\{\ell_i, \ell_j\} = \langle \ell, [e_i, e_j] \rangle = f_{ij}^k \ell_k. \tag{2.5}$$

It is clear that the Kirillov-Kostant bracket is degenerate as is seen, for instance, from the vanishing of the corresponding Poisson tensor at $\ell = 0$. There is a beautiful geometric description of symplectic leaves of the bracket (2.3) as orbits of the coadjoint representation of G [1]. Below we recall the corresponding construction.

Let $g \in G$ and $X \in \mathfrak{g}$. Denote by Ad_g and ad_X the adjoint representations of G and $\mathfrak{g}$, respectively. Assuming G is a matrix Lie group, one has

$$\mathrm{Ad}_g X = gXg^{-1}, \quad \mathrm{ad}_X Y = [X, Y], \quad Y \in \mathfrak{g}.$$

Then the coadjoint action (representation) of G in the dual space $\mathfrak{g}^*$ is defined as follows

$$\mathrm{Ad}_g^* \ell(X) = \ell(\mathrm{Ad}_{g^{-1}} X) = \ell(g^{-1} X g),$$

for any $X \in \mathfrak{g}$. The derivative map of this action at the group unity $g = e$ defines the coadjoint action (representation) of $\mathfrak{g}$ in $\mathfrak{g}^*$:

$$\mathrm{ad}_X^* \ell = \frac{d}{dt} \mathrm{Ad}_{e^{tX}}^* \ell \Big|_{t=0},$$

so that

$$\mathrm{ad}_X^* \ell(Y) = -\ell(\mathrm{ad}_X Y) = -\ell([X, Y]). \tag{2.6}$$

Under the coadjoint action the space $\mathfrak{g}^*$ splits into orbits. Consider an orbit

$$\mathcal{O}_n = \{\mathrm{Ad}_g^* n, \ g \in G\}$$

passing through a point $n \in \mathfrak{g}^*$ and denote by $G_n \subset G$ the stabiliser (stability group) of this point. Evidently, the orbit can be modelled as a homogenous space $G/G_n \approx \mathcal{O}_n$. Let us show that the tangent space to $\mathcal{O}_n$ at n is then naturally identified with the factor-space $\mathfrak{g}/\mathfrak{g}_n$, where $\mathfrak{g}_n$ is the Lie algebra of G_n

$$\mathfrak{g}_n = \{X \in \mathfrak{g} : \ \mathrm{ad}_X^* n = 0\}.$$

Any element $X \in \mathfrak{g}$ gives rise to a vector field on $\mathfrak{g}^*$ tangent to the orbits of the coadjoint action. For $f \in \mathcal{F}(\mathfrak{g}^*)$ this vector field is defined as

$$\xi_X f(\ell) = \frac{d}{dt} f\left(\mathrm{Ad}_{e^{-tX}}^* \ell\right)\Big|_{t=0} = -\langle \mathrm{ad}_X^* \ell, \nabla f(\ell) \rangle. \tag{2.7}$$

From here we find that

$$\xi_X = -\langle \mathrm{ad}^*_X \ell, e_i \rangle \frac{\partial}{\partial \ell_i} = \langle \ell, [X, e_i] \rangle \frac{\partial}{\partial \ell_i} = \langle e^j, [X, e_i] \rangle \ell_j \frac{\partial}{\partial \ell_i} .$$

These fields satisfy the relation

$$[\xi_X, \xi_Y] = \xi_{[X,Y]} , \tag{2.8}$$

implying that the map $X \to \xi_X$ is a homomorphism $\mathfrak{g} \to \mathfrak{X}(\mathfrak{g}^*)$. As follows from (2.7), for $X \in \mathfrak{g}_n$ the field ξ_X vanishes at the point n and vice versa. The tangent space to $\mathcal{O}_n$ at n is spanned by non-vanishing vector fields and is, therefore, isomorphic to $\mathfrak{g}/\mathfrak{g}_n$.

The fields ξ_X are hamiltonian, they are generated by the following linear function of ℓ

$$f_X(\ell) = \langle \ell, X \rangle .$$

Indeed, $\nabla f_X(\ell) = X$ and, therefore,

$$\{f_X, h\}(\ell) = \langle \ell, [X, \nabla h(\ell)] \rangle = -\langle \mathrm{ad}^*_X \ell, \nabla h(\ell) \rangle = \xi_X h(\ell) , \quad \forall h \in \mathcal{F}(\mathfrak{g}^*) .$$

According to (1.17), one can define on $\mathcal{O}_n$ a closed 2-form

$$\omega_\ell(\xi_X, \xi_Y) = \{f_X, f_Y\}(\ell) = \langle \ell, [X, Y] \rangle = f_{[X,Y]}(\ell) , \quad \ell \in \mathcal{O}_n . \tag{2.9}$$

In particular, at the point n

$$\omega_n(\xi_X, \xi_Y) = \langle n, [X, Y] \rangle . \tag{2.10}$$

The right hand side of this formula is a bilinear form on $\mathfrak{g}$ whose kernel coincides with $\mathfrak{g}_n$. Thus, it is non-degenerate on the factor space $\mathfrak{g}/\mathfrak{g}_n$, the latter being isomorphic to the tangent space to the orbit at n. Thus, on $\mathcal{O}_n$ the 2-form ω is non-degenerate and, therefore, any coadjoint orbit is a symplectic leaf of the Kirillov-Kostant bracket. As such, it is necessarily even-dimensional. The action of G on any orbit $\mathcal{O}$ is transitive and symplectic. Functions on $\mathfrak{g}^*$ invariant under the coadjoint action are obviously constant on any coadjoint orbit and, therefore, they are Casimir functions of the Kirillov-Kostant bracket.

2.1.2 Hamiltonian Reduction

One universal way to construct non-trivial dynamical systems is based on the idea of reduction. Let $\mathcal{P}$ be a symplectic manifold. The *hamiltonian or symplectic reduction* is a procedure of obtaining a new symplectic manifold $\mathcal{P}_r$ from $\mathcal{P}$ by means of reduction over the symplectic action of a Lie group G.

Given on $\mathcal{P}$ a dynamical system with a hamiltonian H invariant under a continuous symmetry group, Noether's theorem gives rise to integrals of motion corresponding to this symmetry. Physically, reduction consists in eliminating a number of degrees of freedom by setting these integrals to some constant values. Initial dynamics confined to the corresponding portion of the phase space is typically degenerate and to obtain a well-defined dynamical system, one has to factor out some further redundant degrees of freedom. Integrability of the reduced system, if present, is conventionally inherited from some simple and solvable dynamics on the initial phase space. Here we explain the basics of hamiltonian reduction traditionally founded on the geometric notion of the moment map. Further subtle details together with a number of important applications can be found in [2–4].

Hamiltonian action of a Lie group. Let $\mathcal{P}$ be a connected manifold. Suppose $\mathcal{P}$ is endowed with a smooth action of a Lie group G: $G \times \mathcal{P} \to \mathcal{P}$. Denote by $g \cdot x$ the image of $x \in \mathcal{P}$ under the action of g. Then for any g_1, g_2 from G we have $(g_1 g_2) \cdot x = g_1 \cdot (g_2 \cdot x)$ and $e \cdot x = x$, where e is the identity element. This action induces a representation of G in the space $\mathcal{F}(\mathcal{P})$:

$$T(g) f(x) = f(g^{-1} \cdot x), \quad f \in \mathcal{F}(\mathcal{P}). \tag{2.11}$$

Any element X of the Lie algebra $\mathfrak{g}$ of G gives rise to the corresponding vector field ξ_X according to

$$\left(\xi_X f\right)(x) = \frac{d}{dt} f\left(e^{-Xt} \cdot x\right)\Big|_{t=0}. \tag{2.12}$$

Since $T(g)$ is a representation, the map $X \to \xi_X$ is a Lie algebra homomorphism $\mathfrak{g} \to \mathfrak{X}(\mathcal{P})$ meaning that

$$[\xi_X, \xi_Y] = \xi_{[X,Y]}. \tag{2.13}$$

An example of the group action is provided by the coadjoint representation of G in $\mathcal{P} = \mathfrak{g}^*$. For this example the general formula (2.12) turns into (2.7).

Let $\mathcal{P}$ be a symplectic manifold with the 2-form ω. The action of G on $\mathcal{P}$ is called hamiltonian if for any $X \in \mathfrak{g}$ the corresponding vector field ξ_X is *hamiltonian*, i.e. there exists a single-valued function $f_X \in \mathcal{F}(\mathcal{P})$ such that

$$i_{\xi_X} \omega + d f_X = 0. \tag{2.14}$$

Since ξ_X is hamiltonian, it generates a symplectic transformation, i.e. the hamiltonian action of G on $\mathcal{P}$ is symplectic. Vice versa, at least locally, symplectic transformations by G are hamiltonian [5].

The hamiltonian function f_X is determined by (2.14) up to an arbitrary constant which can be chosen such that the dependence of f_X on X is linear.[3] Fixing these constants for all X, we obtain a well-defined linear map $X \to f_X$, where the corresponding functions f_X will satisfy the following relation

$$f_{[X,Y]} = \{f_X, f_Y\} + c(X, Y). \tag{2.15}$$

Here c is a bilinear skew-symmetric 2-form on $\mathfrak{g}$ that is a constant on $\mathscr{P}$, i.e. its value does not depend on a point $x \in \mathscr{P}$. The Jacobi identity for the Poisson bracket yields for c an equation

$$c([X, Y], Z) + c([Y, Z], X) + c([Z, X], Y) = 0,$$

meaning that c is a 2-cocycle: $\mathfrak{g} \wedge \mathfrak{g} \to \mathbb{R}$.

A trivial cocycle (coboundary) corresponds to $c(X, Y) = \ell([X, Y])$ for some $\ell \in \mathfrak{g}^*$ and it can always be elliminated from (2.15) by redefining the hamiltonian functions as $f_X \to f_X - \ell(X)$.

Moment map for hamiltonian action. For any $x \in \mathscr{P}$ the correspondence $X \to f_X(x)$ defines a linear functional on $\mathfrak{g}$ which is an element $\mu \in \mathfrak{g}^*$. Explicitly,

$$\langle \mu(x), X \rangle = f_X(x). \tag{2.16}$$

This relation defines a map μ from the symplectic manifold into the dual space to the Lie algebra

$$\mu: \ \mathscr{P} \to \mathfrak{g}^*, \tag{2.17}$$

known as the *moment map* [6]. This is a group-theoretic analogue of the angular momentum in classical mechanics from which it derives its name.

We further assume that the hamiltonian action of G on a connected manifold $\mathscr{P}$ is such that the hamiltonian functions can be chosen to satisfy

$$\{f_X, f_Y\} = f_{[X,Y]}. \tag{2.18}$$

In particular, this is always the case when the second cohomology class of $\mathfrak{g}$ is trivial. The important property (2.18) means that the linear map $X \to f_X$ is a homomorphism of the Lie algebra $\mathfrak{g}$ into the Lie algebra of hamiltonian functions.

If (2.18) holds then the moment map is unique. The quantity $\langle \mu, X \rangle$ is a Poisson algebra generator of the hamiltonian group action

$$\{\langle \mu, X \rangle, \cdot\} = \xi_X.$$

---

[3]Linearity of the map $X \to f_X$ is achieved by first picking particular hamiltonian functions for all elements in a basis of $\mathfrak{g}$ and extending to all $X \in \mathfrak{g}$ by linearity.

From linearity of the Poisson bracket, for any function f on the phase space one has

$$\langle \{\mu, f\}, X \rangle = \xi_X f . \tag{2.19}$$

Evidently, functions invariant under the group action Poisson-commute with the $\mathfrak{g}^*$-valued function $\mu(x)$. Below we point out the two most important and interrelated properties of the moment map.

First, the condition (2.18) implies that (2.17) is a Poisson map provided the algebra of functions on $\mathfrak{g}^*$ is equipped with the Kirillov-Kostant bracket.[4] The left hand side of (2.18) can be written in the form

$$\{f_X, f_Y\}(x) = \langle X \otimes Y, \{\mu \overset{\otimes}{,} \mu\}(x) \rangle . \tag{2.20}$$

Here the symbol of tensor product within the Poisson bracket indicates that its arguments are regarded as elements of two different vector spaces. For the right hand side of (2.18) one gets

$$f_{[X,Y]}(x) = \langle [X, Y], \mu(x) \rangle = \langle X \otimes Y, \langle \mu(x), [e_i, e_j] \rangle e^i \wedge e^j \rangle . \tag{2.21}$$

Thus, we conclude that

$$\{\mu \overset{\otimes}{,} \mu\}(x) = \langle \mu(x), [e_i, e_j] \rangle e^i \wedge e^j , \tag{2.22}$$

where the bracket in the left hand side is evaluated on $\mathscr{P}$. This shows that if we endow $\mathfrak{g}^*$ with the following Poisson bracket

$$\{\mu \overset{\otimes}{,} \mu\}_{\mathfrak{g}^*} = \langle \mu, [e_i, e_j] \rangle e^i \wedge e^j , \tag{2.23}$$

then (2.17) is a Poisson map. Obviously, (2.23) is the Kirillov-Kostant bracket (2.5) for the coordinates μ_i on $\mathfrak{g}^*$ amalgamated into

$$\mu = \mu_i e^i . \tag{2.24}$$

Second, (2.18) implies that the moment map is G-equivariant. This has the following meaning. Let $g(t)$ be a one-parametric subgroup corresponding to $X \in \mathfrak{g}$. According to the definition (2.12), a shift $x \to g(t)^{-1} \cdot x$ is generated by the vector field ξ_X. This vector field acts on the moment map as follows

$$\langle \xi_X \mu(x), Y \rangle = \xi_X f_Y(x) = \{f_X, f_Y\}(x) = f_{[X,Y]}(x) = \langle \mu(x), [X, Y] \rangle ,$$

from which we deduce that

---

[4]This fact provides the motivation to define the hamiltonian action of a Lie group G satisfying the condition (2.18) as *Poisson* [5]. In this book we will use, however, the notion of a Poisson action of G in a different context, see Sect. 2.1.6.

$$\xi_X \mu(x) = -\text{ad}_X^* \mu(x) \,. \tag{2.25}$$

The global version of this action is $\mu(g^{-1} \cdot x) = \text{Ad}_{g^{-1}}^* \mu(x)$ or, upon replacing $g \rightarrow g^{-1}$,

$$\mu(g \cdot x) = \text{Ad}_g^* \mu(x) \,. \tag{2.26}$$

Thus, the moment map intertwines the group action on $\mathcal{P}$ with the coadjoint action so that an orbit of G in $\mathcal{P}$ is mapped under μ into a coadjoint orbit in $\mathfrak{g}^*$.

To summarise, we have shown that the moment map (2.17) is an equivariant mapping of Poisson manifolds. In Sect. 2.1.6 we return to the discussion of the moment map but from a different angle.

Marsden and Weinstein theorem. For $m \in \mathfrak{g}^*$ denote by G_m its stabiliser (isotropy group) under the coadjoint action. Consider the inverse image $\mu^{-1}(m) \subset \mathcal{P}$. This subspace is invariant under the action of G_m. Indeed, for $g \in G_m$ and $x \in \mu^{-1}(m)$ one has

$$\mu(g \cdot x) = \text{Ad}_g^* \mu(x) = \text{Ad}_g^* m = m \,,$$

that is $g \cdot x \in \mu^{-1}(m)$. Thus, one can define the quotient

$$\mathcal{P}_r = \mu^{-1}(m)/G_m \,. \tag{2.27}$$

This quotient is usually referred to as the *reduced phase space*, see Fig. 2.1. If m is chosen such that the action of G_m on $\mu^{-1}(m)$ is free and proper,[5] then according to the well-known theorem $\mathcal{P}_r$ is a smooth manifold.

A theorem due to Marsden and Weinstein [2] asserts that $\mathcal{P}_r$ is a symplectic manifold with the symplectic structure inherited from ω on $\mathcal{P}$. To get an idea of the proof, let us evaluate ω on the vector field $\xi_i = \xi_{e_i}$ of a basis element e_i and an arbitrary vector field η. We have

$$\omega(\xi_i, \eta) = -d\mu_i(\eta) = -\eta \mu_i \,,$$

because ξ_i has the hamiltonian function $\mu_i(x)$, see (2.24). If we further assume that η is tangent to $\mu^{-1}(m)$, then

$$\omega_x(\xi_i, \eta) = -\eta \mu_i \Big|_{\mu(x)=m} = 0 \,.$$

Since ξ_i span at $x \in \mu^{-1}(m)$ the tangent space to the orbit of G through x, this space is a skew-orthogonal complement of the tangent space to the level set $\mu^{-1}(m)$

$$T_x(\mu^{-1}(m)) = T_x(G \cdot x)^\perp \,.$$

---

[5] An action $G \times \mathcal{P} \rightarrow \mathcal{P}$ is called free if there are no fixed points and proper if the map $(g, x) \rightarrow (x, g \cdot x)$ is proper (that is inverse images of compact sets are compact).

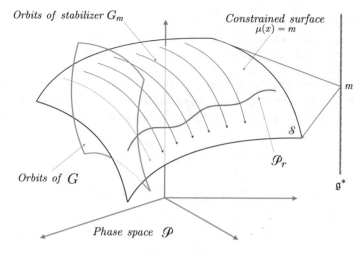

Fig. 2.1 Geometric picture of hamiltonian reduction

Since ω is non-degenerate,

$$T_x(\mu^{-1}(m))^\perp = \left(T_x(G \cdot x)^\perp\right)^\perp = T_x(G \cdot x),$$

i.e. these two tangent spaces are orthogonal complements of each other. Obviously,

$$T_x(\mu^{-1}(m)) \cap T_x(G \cdot x) = T_x(G_m \cdot x).$$

Thus, the kernel of ω restricted on the surface[6] $\mu^{-1}(m)$ is spanned at $x \in \mu^{-1}(m)$ by all the vectors tangent to the orbit of G_m at this point.[7] This allows one to define a non-degenerate closed 2-form ω as

$$\omega_x([\xi], [\eta]) = \omega_x(\xi, \eta),$$

where ξ, η are vectors tangent to $\mu^{-1}(m)$ at x, and $[\xi], [\eta]$ are their equivalence classes defined modulo vectors from $T(G_m \cdot x)$. This form does not depend on a point x along G_m, because ω is invariant under G_m: $\omega_{g \cdot x}(g_*\xi, g_*\eta) = \omega_x(\xi, \eta)$ for $g \in G_m$. This means that ω is well-defined on $\mathscr{P}_r$ endowing this space with the structure of a symplectic manifold. The dimension of $\mathscr{P}_r$ is

$$\dim \mathscr{P}_r = \dim \mathscr{P} - \dim \mathfrak{g} - \dim \mathfrak{g}_m, \qquad (2.28)$$

which is necessarily an even number.

---

[6]The restriction of a differential form on a surface means that one evaluates it only on vectors tangent to this surface.

[7]In particular, any orbit of G_m in $\mu^{-1}(m)$ is an isotropic submanifold, as the symplectic form evaluated on any two vector fields ξ, η tangent to an orbit vanishes: $\omega_x(\xi, \eta) = 0$, $x \in \mu^{-1}(m)$.

Invariant dynamics. So far our discussion was purely geometric and concerned only the construction of a symplectic manifold from another one equipped with a hamiltonian action of some Lie group G. Let us now make the extra assumption that on $\mathcal{P}$ a dynamical system is given with a hamiltonian H invariant under the action of G. More precisely, we will say that the triple $(\mathcal{P}, \{\ ,\ \}, H)$ defines a dynamical system invariant under G if $\mathcal{P}$ carries a hamiltonian action of G in the sense of (2.18) and $\xi_X H = 0$ for any $X \in \mathfrak{g}$, where $\mathfrak{g}$ is the Lie algebra of G. Since ξ_X is hamiltonian,

$$- \xi_X H = \{H, f_X\} = \frac{df_X}{dt} = 0 \,, \tag{2.29}$$

where t is the time parameter along the flow generated by H. Hence, the hamiltonian functions f_X are integrals of motion and, as a consequence, the moment map (2.16) remains constant along the trajectories of the dynamical system

$$\mu(x(t)) = \text{const.} \tag{2.30}$$

Note that Eq. (2.29) is a particular instance of Noether's theorem, see Sect. 1.2, corresponding to the hamiltonian action of the symmetry group G. From a geometric point of view, Eq. (2.30) means that if the starting point of a trajectory of the hamiltonian system is on the surface $\mu(x) = m$ in $\mathcal{P}$, then the whole trajectory belongs to this surface. To obtain a description of the emergent dynamics on the manifold $\mathcal{P}_r$ constructed by means of the Marsden-Weinstein theorem, we invoke the notion of the Dirac bracket.

Dirac bracket. The expansion of μ over a basis in $\mathfrak{g}^*$ yields a set of functions $\{\mu_i(x)\}$ on the phase space. In physics terminology these functions are called *constraints*, because fixing them to constant values m_i defines a surface $\mathcal{S}$ in the phase space

$$\mathcal{S} = \{x \in \mathcal{P} : \ _i(x) = m_i, \ \ i = 1, \ldots, \dim \mathfrak{g}\} \,, \tag{2.31}$$

called a *constraint surface*. This surface is the geometric location of points in $\mathcal{P}$ constituting the inverse image of the moment map evaluated at m: $\mathcal{S} = \mu^{-1}(m)$.

It should be emphasised that, in general, the concept of a constrained dynamical system is far more general than the one arising in the context of the hamiltonian group actions and it will be used later in this book in relation to the Poisson reduction technique. Thus, in anticipation of future developments it seems natural to connect our present discussion with the general classification of constraints due to Dirac [7].

According to (2.22), the Poisson brackets of $\mu_i(x)$'s reduced on the constraint surface take the form

$$\{\mu_i, \mu_j\}(x)\Big|_{x \in \mathcal{S}} = \langle \mu(x), [e_i, e_j] \rangle \Big|_{\mu(x)=m} = \langle \mathrm{ad}^*_{e_j} m, e_i \rangle \,. \tag{2.32}$$

We arrange basis vectors $\{e_i\}$ of $\mathfrak{g}$ such that the first $\dim \mathfrak{g}_m$ vectors with $i = 1, \ldots, \dim \mathfrak{g}_m$ constitute a basis of $\mathfrak{g}_m$, where $\mathfrak{g}_m \subset \mathfrak{g}$ is the Lie algebra of the stabiliser G_m of m. In the case $e_j \in \mathfrak{g}_m$ the bracket (2.32) vanishes for any $e_i \in \mathfrak{g}$. This observation motivates to split all the constraints into two sets

$$\{\mu_i(x)\} = \{\mu_\alpha(x), \ \alpha \in 1, \ldots, \dim \mathfrak{g}_m\} \cup \{\mu_{\bar{\alpha}}(x), \ \bar{\alpha} \in 1, \ldots, \dim(\mathfrak{g}/\mathfrak{g}_m)\}. \quad (2.33)$$

Constraints from the first set are called *first class*, they have the characteristic property that their Poisson bracket with any constraint vanishes on the constraint surface. Transformations of the phase space induced by the hamiltonian vector fields of μ_α are known in physics context as *gauge transformations*. These transformations form a *gauge* group which, by construction, coincides with the corresponding isotropy subgroup of the chosen value of the moment map.

The constraints from the second set are called *second class*, the matrix $\Psi_{\bar{\alpha}\bar{\beta}} = \{\mu_{\bar{\alpha}}, \mu_{\bar{\beta}}\}$ is invertible on the constraint surface. The hamiltonian vector fields $\xi_{\bar{\alpha}}$ corresponding to functions $\mu_{\bar{\alpha}}$ are in a sense "transversal" to $\mathcal{S}$.

Let us now imagine that the matrix $\Psi_{\bar{\alpha}\bar{\beta}}$ also remains non-degenerate in some neighbourhood of the constraint surface. Under this assumption, Dirac proposed a remarkable construction of a new bracket, *the Dirac bracket*, which for any two functions f and h on $\mathcal{P}$ is defined as

$$\{f, h\}_D = \{f, h\} - \{f, \mu_{\bar{\alpha}}\} \Psi^{-1}_{\bar{\alpha}\bar{\beta}} \{\mu_{\bar{\beta}}, h\}. \quad (2.34)$$

In the Appendix B we verify the fulfilment of the Jacobi identity for this bracket by straightforward calculations. With respect to the Dirac bracket, *any* function on the phase space Poisson commutes with all the second class constraints,

$$\{f, \mu_{\bar{\alpha}}\}_D = \{f, \mu_{\bar{\alpha}}\} - \{f, \mu_{\bar{\beta}}\} \Psi^{-1}_{\bar{\beta}\bar{\gamma}} \{\mu_{\bar{\gamma}}, \mu_{\bar{\alpha}}\} = \{f, \mu_{\bar{\alpha}}\} - \{f, \mu_{\bar{\alpha}}\} = 0.$$

It follows from this construction that the Dirac bracket $\{\mu_i, \mu_j\}_D$ reduced on the constraint surface vanishes there, so that *all* the constraints become of the first class with respect to the Dirac bracket.

Our immediate goal is to show that the inverse of the symplectic form on $\mathcal{P}_r$ defined in the Marsden-Weinstein theorem is naturally given by the Dirac bracket. We begin by noting that observables $\hat{f}$ on $\mathcal{P}_r$ can be defined as G_m-invariant functions on $\mathcal{S}$. To be able to compute the Poisson brackets of these observables, we need to extend them outside $\mathcal{S}$. For any such extension f of an observable $\hat{f}$, the condition of invariance under G_m on $\mathcal{S}$ is expressed as

$$\xi_\alpha f(x) = \{\mu_\alpha, f\}(x) = 0, \quad x \in \mathcal{S}, \quad (2.35)$$

where ξ_α is the hamiltonian vector field of μ_α. At the same time, this condition tells us that the hamiltonian vector field ξ_f of the extension f annihilates μ_α along $\mathcal{S}$:

$\xi_f \mu_\alpha = 0$. However, since in general $\xi_f \mu_{\bar{\alpha}} = \{f, \mu_{\bar{\alpha}}\} \neq 0$ on $\mathcal{S}$, this vector field does not preserve the constraint surface, which means geometrically that it is not tangent to $\mathcal{S}$. On the other hand, to determine the restriction of ω on $\mathcal{S}$, we need to evaluate ω on two vectors tangent to $\mathcal{S}$. Even though ξ_f is not in general tangent to the constraint surface, it is always possible to add to ξ_f a combination of the hamiltonian vector fields $\xi_{\bar{\alpha}}$ of the second class constraints $\mu_{\bar{\alpha}}$ to cancel its transversal component. Indeed, consider an $\mathcal{F}(\mathcal{P})$-linear combination $\xi_f + v_{\bar{\alpha}}\xi_{\bar{\alpha}}$, where the coefficients $v_{\bar{\alpha}}$ are some functions on $\mathcal{P}$. It follows from (2.35) and the definition of the first class constraints that this combination annihilates all μ_α along $\mathcal{S}$. Applying it to a second class constraint $\mu_{\bar{\beta}}$ and requiring the result to vanish on $\mathcal{S}$[8]

$$\xi_f \mu_{\bar{\beta}} + v_{\bar{\alpha}}\xi_{\bar{\alpha}}\mu_{\bar{\beta}} = \{f, \mu_{\bar{\beta}}\} + v_{\bar{\alpha}}\{\mu_{\bar{\alpha}}, \mu_{\bar{\beta}}\} = 0 \, ,$$

we find the coefficient functions

$$v_{\bar{\alpha}}(f) = -\{f, \mu_{\bar{\beta}}\}\Psi_{\bar{\beta}\bar{\alpha}}^{-1} . \tag{2.36}$$

The vector field $\xi_f + v_{\bar{\alpha}}(f)\xi_{\bar{\alpha}}$ is not in general hamiltonian, but we can always replace it by an equivalent hamiltonian field $\xi_{f+v_{\bar{\alpha}}(f)\mu_{\bar{\alpha}}}$ corresponding to a new extension

$$f \to \tilde{f} = f + v_{\bar{\alpha}}(f)\mu_{\bar{\alpha}} \, , \quad \tilde{f}|_{\mathcal{S}} = f|_{\mathcal{S}} = \hat{f} \, . \tag{2.37}$$

The field

$$\xi_{\tilde{f}} = \{f + v_{\bar{\alpha}}(f)\mu_{\bar{\alpha}}, \cdot \} \tag{2.38}$$

has the same property of being tangent to $\mathcal{S}$ along $\mathcal{S}$, and it is different from the previous linear combination by $\mu_{\bar{\alpha}}\{v_{\bar{\alpha}}(f), \cdot \}$ which vanishes at the points of $\mathcal{S}$. Further, one can see that the difference of two $\xi_{\tilde{f}}$ corresponding to two different extensions f is in the kernel of ω on $\mathcal{S}$. This follows from an observation that for any $\eta \in T\mathcal{S}$ and any f vanishing identically on $\mathcal{S}$, one has

$$\omega(\xi_f + v_{\bar{\alpha}}(f)\xi_{\bar{\alpha}}, \eta)|_{\mathcal{S}} = \eta(f + v_{\bar{\alpha}}(f)\mu_{\bar{\alpha}})|_{\mathcal{S}} = \eta(f)|_{\mathcal{S}} = 0 \, , \tag{2.39}$$

because η is tangent and f is constant (zero) along $\mathcal{S}$. This allows to unambiguously identify the hamiltonian vector field $\xi_{\hat{f}}$ of $\hat{f}$ on $\mathcal{P}_r$ with an equivalence class $[\xi_{\tilde{f}}]$ of $\xi_{\tilde{f}}$ modulo vectors from $TG_m \cap T\mathcal{S}$. The bracket of two functions on $\mathcal{P}_r$ is then

$$\{\hat{f}, \hat{h}\} = \omega([\xi_{\tilde{f}}], [\xi_{\tilde{h}}]) = \omega(\xi_{\tilde{f}}, \xi_{\tilde{h}})|_{\mathcal{S}}$$
$$= \{f + v_{\bar{\alpha}}(f)\mu_{\bar{\alpha}}, h + v_{\bar{\alpha}}(h)\mu_{\bar{\alpha}}\}|_{\mathcal{S}} = \{f, h\}_D|_{\mathcal{S}} \, .$$

---

[8]Or in the vicinity of $\mathcal{S}$ where $\Psi_{\alpha\beta}$ admits an inverse.

In words, to compute the Poisson bracket of two functions on the reduced space, one can pick up any G_m-invariant extensions, compute their Dirac bracket and then reduce the result on the constraint surface. The outcome of this procedure does not depend on the choice of extension, as any two extensions differ by an $\mathcal{F}(\mathcal{P})$-linear combination of constraints, the latter are always of the first class with respect to the Dirac bracket.

Finally, the evolution equation on $\mathcal{P}_r$ takes the form

$$\frac{df}{dt} = \{H, \hat{f}\}_D = \{H, \hat{f}\} \tag{2.40}$$

for any H invariant under G. Note that for functions invariant under the whole group G the Dirac bracket gives the same result as the original bracket.

More generally, it is also possible to compute the brackets on the reduced phase space by using non-invariant functions on $\mathcal{S}$. In this approach the phase space itself is singled out by imposing subsidiary conditions

$$\chi_\alpha(x) = 0, \quad \alpha = 1, \ldots, \dim \mathfrak{g}_m, \tag{2.41}$$

satisfying the requirement $\{\mu_\alpha, \chi_\beta\}|_\mathcal{S} \neq 0$. Imposition of these conditions is referred to as gauge fixing and (2.41) is called a gauge choice or simply a *gauge*. The full set of constraint functions (μ_i, χ_α) has a non-degenerate matrix of the Poisson brackets on the corresponding constraint surface, the latter coincides with $\mathcal{P}_r$. The inverse of this matrix is then used to define the corresponding Dirac bracket.

2.1.3 Cotangent Bundle of a Lie Group

The cotangent bundle of a Lie group is a classic example of a phase space with non-abelian symmetries. It provides the simplest illustration of some geometric concepts to enter our discussion later on, and it serves as a starting point for getting interesting classes of integrable models by means of the hamiltonian reduction technique. Here we give a comprehensive introduction into the structure of this manifold and its symmetries. Further details can be found in the literature including, for instance, [3, 8].

Symplectic form and Poisson brackets. Let G be a connected simply connected Lie group with Lie algebra $\mathfrak{g}$. To any element $g \in G$ one can associate two diffeomorphisms $G \to G$:

$$L_g(h) = gh, \quad R_g(h) = hg, \quad h \in G, \tag{2.42}$$

which are known as the left and the right shifts, respectively. The tangent bundle TG is trivial and its trivialization is achieved by one of the two push-forward maps $(L_g)_*$ or $(R_g)_*$. In both cases the fibre T_gG at g is identified with $T_eG \approx \mathfrak{g}$. The left- and right-invariant vector fields are defined by their values at the origin:

$$\xi_X^\ell = (L_g)_* X , \quad \xi_X^r = (R_g)_* X , \quad X \in \mathfrak{g} .$$

On $f \in \mathcal{F}(G)$ the left-invariant fields act according to the formula

$$\xi_X^\ell f(g) = \frac{d}{dt} f(ge^{tX}) \Big|_{t=0} = \mathcal{D}_X^r f(g) .$$

Here $\mathcal{D}_X^r$ is the differential of the right regular representation of G. Similarly, for the right-invariant fields we have

$$\xi_X^r f(g) = \frac{d}{dt} f(e^{tX} g) \Big|_{t=0} = \mathcal{D}_X^\ell f(g) ,$$

where up to the sign $\mathcal{D}_X^\ell$ is the differential of the left regular representation of G. The commutators of left- and right-invariant fields are

$$[\xi_X^\ell, \xi_Y^\ell] = \xi_{[X,Y]}^\ell , \quad [\xi_X^r, \xi_Y^r] = -\xi_{[X,Y]}^r , \quad [\xi_X^\ell, \xi_Y^r] = 0 . \quad (2.43)$$

In the following we will refer to $\mathcal{D}^\ell$ and $\mathcal{D}^r$ as the left and right differentials, respectively.

Analogously, pullbacks $(L_g)^*$ and $(R_g)^*$ are used to identify the cotangent space $T_g^* G$ with $T_e^* G \simeq \mathfrak{g}^*$, which defines an isomorphism of manifolds

$$T^* G \simeq G \times \mathfrak{g}^* .$$

Consider on $T^* G$ two differential 1-forms, θ and ϵ, such that

$$(L_g)^* \theta = \ell , \quad (R_g)^* \varepsilon = m , \quad \ell, m \in \mathfrak{g}^* .$$

The θ and ε are called the left- and the right-invariant forms, and they are cousins of (2.2) for the Lie group case. According to their definition,

$$(L_g)^* \theta(X) = \theta(\xi_X^\ell) = \ell(X) , \quad (R_g)^* \varepsilon(X) = \varepsilon(\xi_X^r) = m(X) , \quad (2.44)$$

for any $X \in \mathfrak{g}^*$. Assuming that G is a matrix Lie group, we introduce the Maurer-Cartan 1-forms $g^{-1} dg$ and $dg g^{-1}$ with values in $\mathfrak{g}$. One has

$$g^{-1} dg(\xi_X^\ell) = X , \quad dg g^{-1}(\xi_X^r) = X . \quad (2.45)$$

From (2.44) and (2.45) one finds

$$\theta(\xi_X^\ell) = \langle \ell, g^{-1} dg(\xi_X^\ell) \rangle , \quad \varepsilon(\xi_X^r) = \langle m, dg g^{-1}(\xi_X^r) \rangle ,$$

which yields the following explicit expressions for θ and ε

$$\theta = \langle \ell, g^{-1}dg \rangle, \qquad \varepsilon = \langle m, dgg^{-1} \rangle, \tag{2.46}$$

where for more transparency we adopted the notation $\langle \cdot, \cdot \rangle$ for the canonical pairing between $\mathfrak{g}$ and $\mathfrak{g}^*$. Hence, θ is a canonical 1-form at the point $(g, \ell) \in G \times \mathfrak{g}^*$ in the left parametrisation of the cotangent bundle, the form ε plays a similar role in the right parametrisation $(g, m) \in G \times \mathfrak{g}^*$. The left and right parametrisations are identified through

$$m = \mathrm{Ad}_g^* \ell. \tag{2.47}$$

With this identification, the value of ε at (g, m) coincides with that of θ at (g, ℓ).

The 2-form ω which endows T^*G with the structure of a symplectic manifold can be given either as $\omega = d\theta$ or $\omega = d\varepsilon$. Adopting $\omega = d\theta$, we then obtain an expression for ω in the left parametrisation

$$\omega = \langle d\ell \overset{\wedge}{,} g^{-1}dg \rangle - \langle \ell, g^{-1}dg \wedge g^{-1}dg \rangle, \tag{2.48}$$

where $d\ell = d\ell_i e^i$. Here $\wedge$ is the standard wedge product of differential forms which are the entries of the matrices $g^{-1}dg$ and $d\ell$.

Our next task is to invert ω and construct the corresponding Poisson bracket. Let $f \equiv f(g, \ell)$ be a function on $G \times \mathfrak{g}^*$. The differential df can be conveniently written with the help of the Maurer-Cartan form and the left-invariant vector fields

$$df = \langle e^i, g^{-1}dg \rangle \mathcal{D}_i^r f + d\ell_i \partial^i f, \quad \partial^i f = \frac{\partial f}{\partial \ell_i}. \tag{2.49}$$

We denote $\mathcal{D}_i^l$ and $\mathcal{D}_i^r$ the left and right differentials corresponding to the Lie algebra basis element e_i. In accordance with (2.43), these differential operators have the following commutators

$$[\mathcal{D}_i^l, \mathcal{D}_j^l] = -f_{ij}^k \mathcal{D}_k^l, \quad [\mathcal{D}_i^r, \mathcal{D}_j^r] = f_{ij}^k \mathcal{D}_k^r, \quad [\mathcal{D}_i^l, \mathcal{D}_j^r] = 0,$$

where f_{ij}^k are the structure constants of $\mathfrak{g}$ in a basis $\{e_i\}$. The relation between the left and right differential operators is

$$\mathcal{D}_X^l = \mathcal{D}_{g^{-1}Xg}^r, \quad X \in \mathfrak{g}. \tag{2.50}$$

Assuming the following ansatz for the hamiltonian vector field corresponding to f

$$\xi_f = \Psi^i \mathcal{D}_i^r + \chi_i \partial^i, \tag{2.51}$$

we compute

$$i_{\xi_f}\omega = \chi_i\langle e^i, g^{-1}dg\rangle - \langle d\ell, e_i\rangle\,\Psi^i - \langle\ell, e_i g^{-1}dg\rangle\,\Psi^i + \langle\ell, g^{-1}dge_i\rangle\,\Psi^i$$
$$= \langle\chi_i e^i + \Psi^i \mathrm{ad}^*_{e_i}\ell, g^{-1}dg\rangle - d\ell_i\,\Psi^i\,,$$

and from (1.15) we find

$$\Psi^i = \partial^i f\,, \qquad \chi_i = \langle\ell, [\nabla f(\ell), e_i]\rangle - \mathcal{D}^r_i f\,,$$

where we recall that $\nabla f(\ell) = \frac{\partial f}{\partial \ell_i}e_i$. Thus, the hamiltonian vector field is

$$\xi_f = \partial^i f\,\mathcal{D}^r_i - \mathcal{D}^r_i f\,\partial^i + \langle\ell, [\nabla f(\ell), e_i]\rangle\,\partial^i\,. \tag{2.52}$$

The knowledge of the hamiltonian vector field allows us to find an expression for the corresponding Poisson bracket

$$\{f, h\}(g, \ell) = \langle\ell, [\nabla f, \nabla h]\rangle - \langle\mathcal{D}^r f, \nabla h\rangle + \langle\mathcal{D}^r h, \nabla f\rangle\,, \tag{2.53}$$

where $\mathcal{D}^r f = e^i \mathcal{D}^r_i f \in \mathfrak{g}^*$.

For completeness, we also give an expression for ω in the right parametrisation

$$\omega = d\varepsilon = \langle dm, dgg^{-1}\rangle + \langle m, dgg^{-1} \wedge dgg^{-1}\rangle\,,$$

which leads to the following Poisson bracket

$$\{f, h\}(g, m) = -\langle m, [\nabla f, \nabla h]\rangle - \langle\mathcal{D}^l f, \nabla h\rangle + \langle\mathcal{D}^l h, \nabla f\rangle\,, \tag{2.54}$$

where $\mathcal{D}^l f = e^i \mathcal{D}^l_i f$ takes values in $\mathfrak{g}^*$ and $\nabla f = e_i \frac{\partial f}{\partial m_i}$.

Poisson brackets for generators of the coordinate ring. Our next goal is to adopt a coordinate approach to the description of Poisson algebras, in particular, to the algebra of functions on the cotangent bundle. This approach gives a simple and practical tool to deal with the reduction procedure, and it becomes compulsory when it comes to the issue of quantisation.

We thus aim to introduce in $\mathcal{F}(G \times \mathfrak{g}^*)$ a convenient coordinate system. To this end, we first identify $\mathcal{F}(G \times \mathfrak{g}^*) \simeq \mathcal{F}(G) \otimes \mathcal{F}(\mathfrak{g}^*)$, where in the right hand side a necessary completion of the tensor product is assumed. For our purposes G will always be a matrix Lie group, i.e. a closed subgroup in $\mathrm{GL}_N(\mathbb{C})$. Any group element g is then a matrix of $\mathrm{GL}_N(\mathbb{C})$. The entries of this matrix are functions on G which we regard as generators of the coordinate ring for $\mathcal{F}(G)$. For $\mathcal{F}(\mathfrak{g}^*)$, as before, we choose $\ell_i = \langle\ell, e_i\rangle$ as generators of the corresponding ring. In the following we will assign to (g, ℓ) a double meaning—when featuring in the Poisson brackets the entries of g and ℓ will be regarded as elements of the coordinate ring for $\mathcal{F}(T^*G)$, otherwise (g, ℓ) is a point in T^*G.

This being said, for the right differential of a matrix element $g_{\alpha\beta}$ (a tautological function on G), $1 \leq \alpha, \beta \leq N$, we will get $\mathcal{D}_i^r g_{\alpha\beta} = g_{\alpha\delta}(e_i)_{\delta\beta}$, or, in terms of matrix notation,

$$\mathcal{D}_i^r g = g e_i \, .$$

Then Poisson brackets (2.53) between the coordinate functions can be now evaluated and compactly presented in the following form

$$
\begin{aligned}
\{g \overset{\otimes}{,} g\} &= 0 \, , \\
\{\ell \overset{\otimes}{,} \ell\} &= \mathrm{ad}^*_{e_i} \ell \otimes e^i = \langle \ell, [e_i, e_j] \rangle \, e^i \wedge e^j \, , \\
\{\ell \overset{\otimes}{,} g\} &= e^i \otimes g e_i \, .
\end{aligned}
\tag{2.55}
$$

Here the symbol of tensor product on the left indicates that the arguments of the Poisson bracket are regarded as elements of the corresponding vector spaces. The order of the tensor product factors on the right is the same as on the left. The Poisson bracket between specific coordinate functions is obtained by projecting both sides of (2.55) on the basis elements of the respective tensor product factors.

The Poisson structure (2.55) is inherited from the left parametrisation of the cotangent bundle. To pass to the right parametrisation, one has to replace the variable ℓ by m. In this parametrisation the algebra of functions on $\mathfrak{g}^*$ is generated by $m_i = \langle m, e_i \rangle$. We can compute the corresponding Poisson brackets either directly from (2.54) or from (2.53) by considering m_i as the following functions of g and ℓ: $m_i(g, \ell) = \langle \mathrm{Ad}_g^* \ell, e_i \rangle = \langle \ell, \mathrm{Ad}_{g^{-1}} e_i \rangle$. We have

$$\nabla m_i = \mathrm{Ad}_{g^{-1}} e_i \, , \quad \mathcal{D}^r m_i = -\mathrm{ad}^*_{g^{-1}e_i g} \ell \, ,$$

which straightforwardly yields

$$
\begin{aligned}
\{g \overset{\otimes}{,} g\} &= 0 \, , \\
\{m \overset{\otimes}{,} m\} &= -\mathrm{ad}^*_{e_i} m \otimes e^i = -\langle m, [e_i, e_j] \rangle \, e^i \wedge e^j \, , \\
\{m \overset{\otimes}{,} g\} &= \mathrm{Ad}_g^* e^i \otimes g e_i = e^i \otimes e_i g \, .
\end{aligned}
\tag{2.56}
$$

Here we used the fact that an element $e^i \otimes e_i$ is $\mathrm{Ad}^* \otimes \mathrm{Ad}$-invariant, that is

$$\mathrm{Ad}_g^* e^i \otimes \mathrm{Ad}_g e_i = e^i \otimes e_i \, . \tag{2.57}$$

Indeed, a natural basis in the space $\mathfrak{g}^* \otimes \mathfrak{g}$ is spanned by $e^i \otimes e_j$, and in this basis the matrix elements of the left hand side of (2.57) are

$$\langle \mathrm{Ad}_g^* e^i, e_k \rangle \langle e^m, \mathrm{Ad}_g e_i \rangle = \langle e^i, \mathrm{Ad}_{g^{-1}} e_k \rangle \langle e^m, \mathrm{Ad}_g e_i \rangle = \rho(g^{-1})^i_k \rho_i^m(g) = \rho(g^{-1} g)^m_k = \delta^m_k \, ,$$

where $\rho(g)^j_i$ are matrix elements of the adjoint representation. Since $\langle e^i, e_k \rangle$ $\langle e^m, e_i \rangle = \delta^m_k$, the proclaimed invariance follows. Formulae (2.56) describe the

Poisson brackets between coordinate functions in the right parametrisation of the cotangent bundle. Finally, we note that the variables ℓ and m Poisson commute

$$\{m, \ell\} = 0. \tag{2.58}$$

Symmetries and moment maps. One more important fact about the space T^*G is that it is a group with the following multiplication law and inverse

$$(g, \ell)(h, m) = (gh, \mathrm{Ad}^*_{h^{-1}}\ell + m), \quad (g, \ell)^{-1} = (g^{-1}, -\mathrm{Ad}^*_g\ell).$$

Here G is embedded in T^*G as a subset of elements $(g, 0)$, while $\mathfrak{g}^*$ is understood as a subset $(e, \mathfrak{g}^*)$. Clearly, the above multiplication law endows T^*G with the structure of a semi-direct product $G \ltimes \mathfrak{g}^*$. It also induces the left and right actions of G on T^*G, namely,

$$\begin{aligned} h \cdot (g, \ell) &= (hg, \ell), \\ (g, \ell) \cdot h^{-1} &= (gh^{-1}, \mathrm{Ad}^*_h\ell). \end{aligned} \tag{2.59}$$

As is already obvious from (2.46) to (2.48), these transformations leave θ and ω invariant, i.e. they are symplectic. According to (2.12), any $X \in \mathfrak{g}$ gives rise to the corresponding vector fields

$$\begin{aligned} \xi^\ell_X f(g, \ell) &= \frac{d}{dt} f\big(e^{-tX} \cdot (g, \ell)\big)\Big|_{t=0} = -\langle e^i, g^{-1}Xg\rangle \mathcal{D}^r_i f, \\ \xi^r_X f(g, \ell) &= \frac{d}{dt} f\big((g, \ell) \cdot e^{tX}\big)\Big|_{t=0} = \langle e^i, X\rangle \mathcal{D}^r_i f - \langle \mathrm{ad}^*_X\ell, e_i\rangle \partial^i f. \end{aligned}$$

Comparing these expressions to the general form (2.52) of the hamiltonian vector field, we can find the hamiltonian functions on T^*G that trigger these actions, as well as the associated moment maps. A simple computation reveals then that the moment maps of the left and the right actions are

$$\mu_\ell(g, \ell) = -\mathrm{Ad}^*_g\ell, \quad \mu_r(g, \ell) = \ell.$$

The moment map μ_ℓ is nothing else but $-m$, where m is the corresponding variable for the right parametrisation of T^*G, while μ_r coincides with a projection π of an element (g, ℓ) on its second component, as shown to the left in Fig. 2.2. For the right action the moment map and the projection switch the roles, as illustrated to the right in same figure. In 2.1.6 we will rediscover the content of Fig. 2.2 from a different perspective based on the notion of dual pairs.

Further, in accordance with (2.55) and (2.56), the action of the moment maps on the coordinate functions g and ℓ is

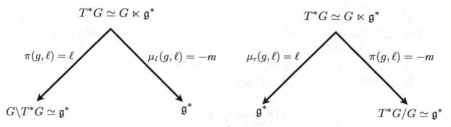

Fig. 2.2 Projections on the coset spaces and moment maps of the left and the right group actions on T^*G. All these maps are Poisson with respect to the Poisson structure on $\mathfrak{g}^*$ given by the Kirillov-Kostant bracket

$$\{\langle \mu_\ell, X\rangle, g\} = -Xg = \xi^\ell_X g\,, \quad \{\langle \mu_r, X\rangle, g\} = gX = \xi^r_X g\,,$$
$$\{\langle \mu_\ell, X\rangle, \ell\} = 0 = \xi^\ell_X \ell\,, \quad\quad \{\langle \mu_r, X\rangle, \ell\} = -\mathrm{ad}^*_X \ell = \xi^r_X \ell\,, \tag{2.60}$$

where $X \in \mathfrak{g}$.

The group G acts on T^*G by left and right translations which are two independent commuting actions, both symplectic. More generally, for any two subgroups H_ℓ, $H_r \subset G$ with Lie algebras $\mathfrak{b}_\ell$, $\mathfrak{b}_r \subset \mathfrak{g}$, we can define the action of $H_\ell \times H_r$ on T^*G

$$h_\ell \cdot (g, \ell) \cdot h_r^{-1} = (h_\ell g h_r^{-1}, \mathrm{Ad}^*_{h_r} \ell)\,, \tag{2.61}$$

where $h_\ell \in H_\ell$ and $h_r \in H_r$. This action is hamiltonian and the corresponding moment map $\mu : T^*G \to \mathfrak{h}^*_\ell \oplus \mathfrak{h}^*_r$ is

$$\mu(g, \ell) = (-P_{\mathfrak{h}^*_\ell}(\mathrm{Ad}_g \ell), 0) + (0, P_{\mathfrak{h}^*_r}(\ell))\,, \tag{2.62}$$

where $P_{\mathfrak{h}^*_\ell}$ and $P_{\mathfrak{h}^*_r}$ are projections of $\mathfrak{g}^*$ onto the duals of the subspaces $\mathfrak{b}_\ell$ and $\mathfrak{b}_r$, respectively.

In the following we will be mostly interested in the adjoint action

$$g \to hgh^{-1}\,, \quad \ell \to \mathrm{Ad}^*_h \ell\,, \tag{2.63}$$

which is obviously generated by the moment map

$$\mu(g, \ell) = \ell - \mathrm{Ad}^*_g \ell\,. \tag{2.64}$$

Coadjoint orbits from reduction. The hamiltonian reduction technique, i.e. a way of obtaining a symplectic manifold from another one by reducing over a symplectic action of a Lie group can be nicely illustrated in the present context by reconstructing the coadjoint orbits of G in $\mathfrak{g}^*$ with T^*G taken as an initial phase. To carry out a proper reduction, we first observe that $\{\mu_\ell, \mu_r\} = 0$, i.e. the vector fields generated by the left and right moment maps mutually commute implying that they are skew-orthogonal with respect to the symplectic form ω at any point of T^*G. This gives two foliations of T^*G, where for any $\ell, n \in \mathfrak{g}^*$, the corresponding leaves $\mu_l^{-1}(\ell)$ and $\mu_r^{-1}(n)$ are skew-orthogonal at any intersection point.

First, we consider the reduction under the left G-action. Fix $\ell = \ell_0$. Then $\mu_l(g, \ell) = -\mathrm{Ad}_g^* \ell = \ell_0$ and, therefore,

$$\mu_l^{-1}(\ell_0) = \{(g, -\mathrm{Ad}_{g^{-1}}^* \ell_0), \quad g \in G\} \simeq G.$$

If G_{ℓ_0} is an isotropy subgroup of ℓ_0, then the reduced phase space can be modelled as a left coset

$$\mathscr{P}_r = G_{\ell_0} \backslash G \simeq \mathscr{O}_{\ell_0},$$

where $\mathscr{O}_{\ell_0}$ is an orbit in $\mathfrak{g}^*$ through ℓ_0. Second, considering the reduction under the right G-action, one finds $\mu_r^{-1}(\ell_0) \simeq G \times \ell_0$. Here the reduced phase space can be described as a right coset

$$\mathscr{P}_r = G / G_{\ell_0} \simeq \mathscr{O}_{\ell_0}.$$

Realisation of T^*G for Lie algebras with inner product. In many important cases, which include simple Lie algebras and $\mathrm{Mat}_N(\mathbb{C})$, $\mathfrak{g}$ admits a bilinear symmetric non-degenerate Ad-invariant form $\langle\ ,\ \rangle_{\mathfrak{g}}$ (inner product) which allows one to identify $\mathfrak{g}^*$ with $\mathfrak{g}$. Assuming $\langle e_i, e_j \rangle_{\mathfrak{g}} = \delta_{ij}$, upon this identification $e^i \to e_i$ and the coadjoint representation becomes equivalent to the adjoint, while the Poisson structure in the left parametrisation turns into

$$\{g_1, g_2\} = 0\,,$$
$$\{\ell_1, \ell_2\} = [C_{12}, \ell_1] = -[C_{12}, \ell_2] = \frac{1}{2}[C_{12}, \ell_1 - \ell_2]\,, \tag{2.65}$$
$$\{\ell_1, g_2\} = g_2 C_{12}\,.$$

This time $\ell \in \mathfrak{g}$ and the subscripts 1 and 2 refer to different matrix spaces. Here the element

$$C = \sum_{i=1}^{\dim \mathfrak{g}} e_i \otimes e_i \tag{2.66}$$

is an Ad-invariant element in $\mathfrak{g} \otimes \mathfrak{g}$ conventionally called the *split* or *tensor Casimir*. Its invariant property can be expressed as

$$C(g \otimes g) = (g \otimes g)C\,, \tag{2.67}$$

which in terms of the new notation is the same as

$$C_{12} g_1 g_2 = g_2 g_1 C_{12}\,.$$

Since g_1 and g_2 are embeddings of g into different components of the tensor product of two matrix spaces, they commute, $g_1 g_2 = g_2 g_1$.

Similarly, for the Poisson structure in the right parametrisation one gets

$$\{g_1, g_2\} = 0,$$
$$\{m_1, m_2\} = -\frac{1}{2}[C_{12}, m_1 - m_2], \qquad (2.68)$$
$$\{m_1, g_2\} = C_{12} g_2.$$

To summarise, the cotangent bundle T^*G is a Poisson manifold with Poisson brackets (2.65) or (2.68), where $m = g\ell g^{-1}$. This manifold carries the hamiltonian action (2.63) of G generated by the moment map (2.64),

$$\mu = \ell - g\ell g^{-1} \in \mathfrak{g}^*.$$

The moment map obeys the Poisson algebra relations (2.23).

Matrix unities and split Casimir. The space $\mathrm{Mat}(\mathbb{R})$ has a bilinear form

$$\langle A, B \rangle = \mathrm{Tr}(AB) \qquad (2.69)$$

and a natural basis of matrix unities. Denote by $E_{ij} \in \mathrm{Mat}(\mathbb{R})$ a matrix unit, i.e. a matrix which has only one non-trivial matrix element equal to 1 standing on the intersection of i's row with j's column

$$(E_{ij})_{kl} = \delta_{ik}\delta_{jl}. \qquad (2.70)$$

For the commutator and product one has

$$[E_{ij}, E_{kl}] = \delta_{jk}E_{il} - \delta_{il}E_{kj}, \quad E_{ij}E_{kl} = \delta_{jk}E_{il}.$$

Note that this basis is not orthogonal with respect to (2.69), rather $\mathrm{Tr}(E_{ij}E_{kl}) = \delta_{il}\delta_{jk}$. Further, the matrix realisation of C is

$$C = \sum_{i,j=1}^{N} E_{ij} \otimes E_{ji} \qquad (2.71)$$

and the matrix elements of C are

$$C_{ij,kl} = \sum_{a,b=1}^{N} (E_{ab})_{ij} \otimes (E_{ba})_{kl} = \delta_{il}\delta_{jk} = \mathrm{Tr}(E_{ij}E_{kl}). \qquad (2.72)$$

For any matrix A one has

$$A = E_{ij}\text{Tr}(E_{ji}A) \quad \rightarrow \quad A_1 = \text{Tr}_2(C_{12}A_2)\,.$$

The formula above can be, for instance, applied to determine the action of the moment map on the coordinate functions from the matrix bracket (2.65). In particular,

$$\{\text{Tr}_1(\mu_1 X_1), \ell_2\} = -[\text{Tr}_1(C_{12}X_1), \ell_2] = -[X_2, \ell_2]\,, \quad X \in \mathfrak{g}\,,$$

which is equivalent to the Poisson realisation of the action of the fundamental vector field ξ_X on the coordinate (matrix) function ℓ

$$\{\text{Tr}(\mu X), \ell\} = -[X, \ell] = \xi_X \ell\,, \tag{2.73}$$

conform the second line of (2.60).

We further note that when pulling through any element $B_{12} \in \text{Mat}_N(\mathbb{C})^{\otimes 2}$, the split Casimir acts as a *permutation* of matrix spaces

$$C_{12}B_{12} = B_{21}C_{12}\,.$$

Finally, we mention one technical point useful for practical calculations. Given an element

$$h = \sum_{ijkl} h_{ij,kl} E_{ij} \otimes E_{kl}\,, \tag{2.74}$$

the individual coefficients $h_{ij,kl}$ are extracted as

$$h_{ij,kl} = \text{Tr}\big(h\,(E_{ji} \otimes E_{lk})\big)\,, \tag{2.75}$$

where the trace is taken over the space $\text{Mat}_N(\mathbb{C})^{\otimes 2}$, which is equivalent to the application of Tr_{12} which corresponds to tracing separatelly over the first and the second matrix spaces.

2.1.4 Poisson-Lie Groups

Let G be a Lie group and a Poisson manifold, then G is called a Poisson-Lie group if multiplication $G \times G \rightarrow G$ is a Poisson mapping, where the space $G \times G$ is equipped with the product Poisson structure. If $\{\,,\,\}$ is a Poisson bracket on G, then the condition of a Poisson-Lie group implies the following relation

$$\{f, h\}(g_1 g_2) = \{R_{g_2} f, R_{g_2} h\}(g_1) + \{L_{g_1} f, L_{g_1} h\}(g_2)\,, \tag{2.76}$$

where L_g and R_g are the left and right shifts introduced in (2.42). Any Poisson bracket on a Lie group can be written in the basis of left-invariant vector fields or, equivalently, the differential operators of right shifts[9] as

$$\{f, h\}(g) = J^{ij}(g)\mathcal{D}^r_i f \mathcal{D}^r_j h\,,\tag{2.77}$$

where $1 \leq i \leq \dim \mathfrak{g}$ and J^{ij} is the Poisson tensor in the basis of left-invariant vector fields. The condition (2.76) results into the following relation for the Poisson tensor

$$J^{ij}(g_1 g_2) = J^{ij}(g_2) + J^{kl}(g_1)\langle \mathrm{Ad}^*_{g_2} e^i, e_k \rangle \langle \mathrm{Ad}^*_{g_2} e^j, e_l \rangle\,.\tag{2.78}$$

It is convenient to introduce the following element

$$J(g) = J^{ij}(g)\, e_i \wedge e_j \in \mathfrak{g} \wedge \mathfrak{g}\,,\tag{2.79}$$

in terms of which Eq. (2.78) takes the form

$$J(g_1 g_2) = J(g_2) + \mathrm{Ad}_{g_2^{-1}} \otimes \mathrm{Ad}_{g_2^{-1}} J(g_1)\,.\tag{2.80}$$

Setting here $g_2 = e$ results into $J(e) = 0$, i.e. the Poisson tensor must necessarily vanish at the identity element and, as a result, the Poisson bracket degenerates at the group origin. Further, setting $g_2 = g_1^{-1}$, we deduce that

$$J(g^{-1}) = -\mathrm{Ad}_g \otimes \mathrm{Ad}_g\, J(g)\,.$$

From the point of view of group cohomology, J is a 1-cochain: $G \to \mathfrak{g} \wedge \mathfrak{g}$. The standard coboundary operator $\mathcal{D}$ for the right action of G will act on 1-cochains as [9]

$$\mathcal{D}J(g_1, g_2) = J(g_2) - J(g_1 g_2) + \mathrm{Ad}_{g_2^{-1}} \otimes \mathrm{Ad}_{g_2^{-1}} J(g_1)\,.\tag{2.81}$$

For a Poisson tensor satisfying the Poisson-Lie condition (2.80), the right hand side of (2.81) vanishes, i.e. J is a 1-cocycle on G with values in $\mathfrak{g} \wedge \mathfrak{g}$. This cocycle on G gives rise to the corresponding Lie algebra cocycle $\delta : \mathfrak{g} \to \mathfrak{g} \wedge \mathfrak{g}$. Explicitly,

$$\delta(X) = \frac{d}{dt}J^{ij}(e^{tX})\Big|_{t=0} e_i \wedge e_j = \partial_k J^{ij}(e)\,\langle e^k, X \rangle\, e_i \wedge e_j\,, \quad X \in \mathfrak{g}\,.\tag{2.82}$$

The action of the corresponding coboundary operator in the Lie algebra cohomology, which we also denote as $\mathcal{D}$, is

$$\mathcal{D}\delta(X, Y) = (\mathrm{ad}_X \otimes \mathbb{1} + \mathbb{1} \otimes \mathrm{ad}_X)\delta(Y) - (\mathrm{ad}_Y \otimes \mathbb{1} + \mathbb{1} \otimes \mathrm{ad}_Y)\delta(X) - \delta([X, Y])\,,$$

---

[9] Alternatively, the same bracket can be expressed via the operators of left shifts.

and $\mathfrak{D}\delta = 0$, as can be derived from (2.81). Thus, δ is a map which satisfies the property

$$\delta\big([X, Y]\big) = (\mathrm{ad}_X \otimes \mathbb{1} + \mathbb{1} \otimes \mathrm{ad}_X)\delta(Y) - (\mathrm{ad}_Y \otimes \mathbb{1} + \mathbb{1} \otimes \mathrm{ad}_Y)\delta(X) \,. \quad (2.83)$$

The map δ obeying (2.83) is called *co-commutator*.

We point out for completeness that the same bracket (2.77) can be equivalently written in the basis of differential operators of left shifts

$$\{f, h\}(g) = J_\ell^{ij}(g)\mathcal{D}_i^\ell f \mathcal{D}_j^\ell h \,, \quad (2.84)$$

The Poisson-Lie condition yields

$$J_\ell(g_1 g_2) = J_\ell(g_1) + \mathrm{Ad}_{g_1} \otimes \mathrm{Ad}_{g_1} J_\ell(g_2) \,. \quad (2.85)$$

The relation of $J_\ell(g)$ to $J(g) \equiv J_r(g)$ in (2.77) is

$$J_\ell(g) = \mathrm{Ad}_g \otimes \mathrm{Ad}_g J(g) \quad (2.86)$$

and it transforms (2.80) into (2.85).

One should also keep in mind that the inversion map $i(g) = g^{-1}$ is anti-Poisson, that is,

$$\{f \circ i, h \circ i\} = -\{f, h\} \circ i \,. \quad (2.87)$$

The co-commutator allows one to equip the dual space $\mathfrak{g}^*$ with the structure of Lie algebra. Namely, for any $\ell, n \in \mathfrak{g}^*$ we define the commutator $[\ ,\ ]_*$ in $\mathfrak{g}^*$ by means of

$$\langle [\ell, n]_*, X \rangle = \langle \ell \otimes n, \delta(X) \rangle \,, \quad (2.88)$$

or, more explicitly,

$$[\ell, n]_* = e^k \, \partial_k J^{ij}(e)\ell_i n_j \,, \quad (2.89)$$

In the dual bases $\{e^i\}$ and $\{e_i\}$ the structure constants λ_k^{ij} define the co-commutator and cocycle

$$[e^i, e^j]_* = \lambda_k^{ij} e^k \,, \quad \delta(e_k) = \lambda_k^{ij} \, e_i \wedge e_j \,, \quad (2.90)$$

and they are $\lambda_k^{ij} = \partial_k J^{ij}(e)$. If $[e_i, e_j] = f_{ij}^k e_k$, then Eq. (2.83) reduces to the following relation between the structure constants of $\mathfrak{g}$ and $\mathfrak{g}^*$

$$\lambda_m^{kl} f_{ij}^m = f_{im}^k \lambda_j^{ml} - f_{jm}^k \lambda_i^{ml} - f_{im}^l \lambda_j^{mk} + f_{jm}^l \lambda_i^{mk} \,. \quad (2.91)$$

The fulfilment of the Jacobi identity for the bracket $[\ ,\ ]_*$ is not immediately obvious. The Jacobi identity for (2.77) is equivalent to

$$\sum_{(k,n,m)} \left(J^{kj} \mathcal{D}_j^r J^{nm} + f_{ij}^k J^{in} J^{jm} \right) = 0 . \tag{2.92}$$

Using this relation together with $J(e) = 0$, one can show that the Jacobi identity for (2.89) is satisfied.

A Lie algebra $\mathfrak{g}$ with a 1-cocycle $\delta : \mathfrak{g} \to \mathfrak{g} \otimes \mathfrak{g}$ such that the dual map $\delta^* : \mathfrak{g}^* \otimes \mathfrak{g}^* \to \mathfrak{g}^*$ is a Lie bracket on $\mathfrak{g}^*$, is called a *Lie bialgebra*. A Lie bialgebra $(\mathfrak{g}, \mathfrak{g}^*)$ is an infinitesimal structure underlying the notion of a Poisson-Lie group. As it turns out, if G is connected and simply connected, then every Lie bialgebra structure on $\mathfrak{g}$ gives rise to a unique Poisson-Lie structure on G [10, 11]. The Lie bialgebra arising from a Poisson-Lie group is called *tangent*.

Coadjoint actions. With the notion of a Lie bialgebra we can associate two coadjoint actions. The first one is already defined in (2.6) the coadjoint action $\mathrm{ad}_{\mathfrak{g}}^*$ of $\mathfrak{g}$ in $\mathfrak{g}^*$:

$$\langle \mathrm{ad}_X^* \ell, Y \rangle = -\langle \ell, [X, Y] \rangle . \tag{2.93}$$

The second one is an action $\mathrm{ad}_{\mathfrak{g}^*}^*$ of $\mathfrak{g}^*$ in $\mathfrak{g}$ defined through

$$\langle n, \mathrm{ad}_\ell^* X \rangle = -\langle [\ell, n]_*, X \rangle = -\langle \ell \otimes n, \delta(X) \rangle = -\langle \ell, \mathrm{ad}_n^* X \rangle , \tag{2.94}$$

for $X \in \mathfrak{g}$ and $\ell, n \in \mathfrak{g}^*$. Which of these actions is meant in any concrete case should be clear from the corresponding subscript of ad^*.

To appreciate some properties of the coadjoint action, let us consider

$$\langle n, \mathrm{ad}_\ell^* [X, Y] \rangle = -\langle [\ell, n]_*, [X, Y] \rangle = -\langle \ell \otimes n, \delta([X, Y]) \rangle .$$

Using (2.83), we obtain

$$\langle n, \mathrm{ad}_\ell^* [X, Y] \rangle = \langle \mathrm{ad}_X^* \ell \otimes n + \ell \otimes \mathrm{ad}_X^* n, \delta(Y) \rangle - \langle \mathrm{ad}_Y^* \ell \otimes n + \ell \otimes \mathrm{ad}_Y^* n, \delta(X) \rangle .$$

This can be further transformed with the help of (2.94)

$$\begin{aligned}
\langle n, \mathrm{ad}_\ell^* [X, Y] \rangle &= -\langle n, \mathrm{ad}_{\mathrm{ad}_X^* \ell}^* Y \rangle - \langle \mathrm{ad}_X^* n, \mathrm{ad}_\ell^* Y \rangle + \langle n, \mathrm{ad}_{\mathrm{ad}_Y^* \ell}^* X \rangle + \langle \mathrm{ad}_Y^* n, \mathrm{ad}_\ell^* X \rangle \\
&= -\langle n, \mathrm{ad}_{\mathrm{ad}_X^* \ell}^* Y \rangle + \langle n, [X, \mathrm{ad}_\ell^* Y] \rangle + \langle n, \mathrm{ad}_{\mathrm{ad}_Y^* \ell}^* X \rangle - \langle n, [Y, \mathrm{ad}_\ell^* X] \rangle ,
\end{aligned}$$

where in the second line (2.93) was also used. This shows that ad_ℓ^* is not in general a differentiation with respect to the Lie bracket on $\mathfrak{g}$, rather,

$$\mathrm{ad}_\ell^* [X, Y] = [\mathrm{ad}_\ell^* X, Y] + [X, \mathrm{ad}_\ell^* Y] + \mathrm{ad}_{\mathrm{ad}_Y^* \ell}^* X - \mathrm{ad}_{\mathrm{ad}_X^* \ell}^* Y , \tag{2.95}$$

for any $X, Y \in \mathfrak{g}$ and $\ell \in \mathfrak{g}^*$. In the case of abelian $\mathfrak{g}^*$ the last two terms in the above formula are absent and ad_ℓ^* acts as differentiation.

Analogously, for the $\text{ad}_\mathfrak{g}^*$-action we will get from (2.93), (2.83) and (2.94)

$$\langle \text{ad}_X^*[\ell_1, \ell_2]_*, Y \rangle = -\langle \ell_1 \otimes \ell_2, \delta([X, Y]) \rangle = \qquad (2.96)$$
$$\langle [\text{ad}_X^* \ell_1, \ell_2]_* + [\ell_1, \text{ad}_X^* \ell_2], Y \rangle - \langle [\text{ad}_Y^* \ell_1, \ell_2]_* + [\ell_1, \text{ad}_Y^* \ell_2], X \rangle \,,$$

which implies the following relation

$$\text{ad}_X^*[\ell_1, \ell_2]_* = [\text{ad}_X^* \ell_1, \ell_2]_* + [\ell_1, \text{ad}_X^* \ell_2]_* + \text{ad}_{\text{ad}_{\ell_2}^* X}^* \ell_1 - \text{ad}_{\text{ad}_{\ell_1}^* X}^* \ell_2 \,, \quad (2.97)$$

that is ad_X^* is not a derivation with respect to the Lie bracket on $\mathfrak{g}^*$.

Coboundary Lie bialgebra and the Sklyanin bracket. The simplest example of a Poisson Lie group is a Lie group G with a trivial Poisson bracket on $\mathscr{F}(G)$. Since $J(g)$ is identically zero, $[\ell, n]_* = 0$, that is $\mathfrak{g}^*$ is abelian. A less trivial example arises when taking the following cocycle

$$J(g) = \text{Ad}_{g^{-1}} \otimes \text{Ad}_{g^{-1}} r - r \equiv \text{Ad}_{g^{-1}}^{\otimes 2} r - r \,, \qquad (2.98)$$

where[10]

$$r \equiv r^{ij} e_i \wedge e_j \equiv \tfrac{1}{2} r^{ij} (e_i \otimes e_j - e_j \otimes e_i) \qquad (2.99)$$

is a constant element of $\mathfrak{g} \wedge \mathfrak{g}$. The associated Lie algebra cocycle is

$$\delta(X) = [r, X \otimes \mathbb{1} + \mathbb{1} \otimes X] \,, \qquad (2.100)$$

and the commutator in $\mathfrak{g}^*$ reads as

$$[\ell, n]_* = e^k \langle \ell \otimes n, [r, e_k \otimes \mathbb{1} + \mathbb{1} \otimes e_k] \rangle \,. \qquad (2.101)$$

The matrix r in (2.99) defines a linear operator $\mathbf{r} : \mathfrak{g}^* \to \mathfrak{g}$, which acts on $\ell \in \mathfrak{g}^*$ as

$$\mathbf{r}\, \ell = r^{ij} e_i \langle \ell, e_j \rangle \,. \qquad (2.102)$$

This allows one to write

$$\langle [\ell, n]_*, X \rangle = \langle \ell \otimes n, [r, X \otimes \mathbb{1} + \mathbb{1} \otimes X] \rangle = \langle \ell, [\mathbf{r}(n), X] \rangle - \langle n, [\mathbf{r}(\ell), X] \rangle \,, \quad (2.103)$$

---

[10]To render $J(g)$ skew-symmetric, it is enough to require that the symmetric part of $r = r^{ij} e_i \otimes e_j$ is Ad-invariant, so that it decouples from $J(g)$.

from where, by using the notion of the coadjoint action, we get the following form of the commutator in $\mathfrak{g}^*$

$$[\ell, n]_* = \mathrm{ad}^*_{\mathbf{r}(\ell)} n - \mathrm{ad}^*_{\mathbf{r}(n)} \ell \,. \tag{2.104}$$

The structure constants in the basis $\{e^i\}$ are

$$\lambda^{ij}_k = r^{is} f^j_{sk} - r^{js} f^i_{sk} \,. \tag{2.105}$$

For (2.98) the Poisson bracket, also known as the *Sklyanin bracket*, is

$$\{f, h\} = r^{ij} \mathcal{D}^\ell_i f \mathcal{D}^\ell_j h - r^{ij} \mathcal{D}^r_i f \mathcal{D}^r_j h \,. \tag{2.106}$$

For the generators of the coordinate ring this bracket yields

$$\{g_1, g_2\} = [r_{12}, g_1 g_2] \,. \tag{2.107}$$

It remains to find the conditions imposed on r by the Jacobi identity.

Classical and modified classical Yang-Baxter equations. To make our treatment a bit more general, we consider the following Poisson bracket

$$\{f, h\} = r^{ij} \mathcal{D}^\ell_i f \mathcal{D}^\ell_j h - r'^{ij} \mathcal{D}^r_i f \mathcal{D}^r_j h \,, \tag{2.108}$$

where r and r' are two skew-symmetric matrices. This bracket corresponds to the following Poisson tensor

$$J(g) = \mathrm{Ad}^{\otimes 2}_{g^{-1}} r - r' \,, \tag{2.109}$$

so that $\mathcal{D}^r_j J(g) = [\mathrm{Ad}^{\otimes 2}_{g^{-1}} r, e_j \otimes \mathbb{1} + \mathbb{1} \otimes e_j]$. Pairing the left hand side of (2.92) with basis elements in $\mathfrak{g}^{\otimes 3}$, we can rewrite the Jacobi identity in the form

$$[\mathrm{Ad}^{\otimes 2}_{g^{-1}} r_{23}, J_{12} + J_{13}] + [\mathrm{Ad}^{\otimes 2}_{g^{-1}} r_{13}, J_{12} - J_{23}] - [\mathrm{Ad}^{\otimes 2}_{g^{-1}} r_{12}, J_{13} + J_{23}]$$
$$+ [J_{12}, J_{13}] + [J_{12}, J_{23}] + [J_{13}, J_{23}] = 0 \,.$$

Substituting here the expression (2.109) results into the following equation

$$\mathrm{Ad}^{\otimes 3}_g [\![r', r']\!] - [\![r, r]\!] = 0 \,, \tag{2.110}$$

where

$$[\![r, r]\!] \equiv [r_{12}, r_{13}] + [r_{13}, r_{23}] + [r_{12}, r_{23}] \,. \tag{2.111}$$

If $r' = 0$, then the only possible solution is

$$[\![r, r]\!] = 0. \tag{2.112}$$

This is the so-called *classical Yang-Baxter equation* (CYBE) for r. The same conclusion applies for r' if $r = 0$. Finally, if $r' = r$, i.e. the Poisson-Lie condition for (2.108) is met, Eq. (2.110) implies that $[\![r, r]\!]$ is an Ad-invariant element in $\mathfrak{g}^{\otimes 3}$. An r-matrix satisfying (2.112) is called *quasi-triangular* and if, in addition, it is skew-symmetric, then it is *triangular*.

Assume that $\mathfrak{g}$ admits a non-degenerate symmetric bilinear Ad-invariant form $\langle \cdot , \cdot \rangle$ allowing to identify $\mathfrak{g}$ with $\mathfrak{g}^*$. In this case there exists a canonical Ad-invariant element in $\wedge^3 \mathfrak{g}$. This element is proportional to $[C_{12}, C_{13}]$, where C is the split Casimir (2.66) and $\{e_i\}$ is a basis of $\mathfrak{g}$. An equation

$$[\![r, r]\!] = -c^2 [C_{12}, C_{13}] \tag{2.113}$$

is known as *modified classical Yang-Baxter equation* (mCYBE). Here $c^2 \in \mathbb{R}$ is a constant. If $c \neq 0$, by rescaling of r one can always reach $c^2 = \pm 1$. Accordingly, there are three different classes of skew-symmetric solutions of (2.116): $c = i, c = 1$ and $c = 0$, where the latter case corresponds to the CYBE. Solutions of the mCYBE (r-matrices) corresponding to $c = i$ are called *non-split* and to $c = 1$ *split*, respectively. This definition originates from the classification of real forms of semi-simple complex Lie algebras. A real semi-simple Lie algebra $\mathfrak{g}$ is said to be *split* if any of its maximal $\mathbb{R}$-diagonalisable subalgebras is a Cartan subalgebra. This means that the rank of $\mathfrak{g}$ coincides with the rank of its complexification $\mathfrak{g}^{\mathbb{C}}$. Any complex semi-simple Lie algebra has a unique (up to isomorphism) split real form [12].

An r-matrix $r \in \mathfrak{g} \wedge \mathfrak{g}$ gives rise to an operator $\mathbf{r} : \mathfrak{g} \to \mathfrak{g}$ that for $X \in \mathfrak{g}$ acts on X as[11]

$$\mathbf{r} X = r^{ij} e_i \langle X, e_j \rangle , \tag{2.114}$$

This operator is skew-symmetric

$$\langle \mathbf{r} X, Y \rangle = -\langle X, \mathbf{r} Y \rangle .$$

By using (2.114), Eq. (2.113) can be rewritten in an operator form. Pairing the expression

$$[r_{12}, r_{13}] + [r_{13}, r_{23}] + [r_{12}, r_{23}] = -c^2 [e_i, e_j] \otimes e_i \otimes e_j \tag{2.115}$$

with arbitrary $X, Y \in \mathfrak{g}$ in the second and third spaces of the tensor product, respectively, yields the following relation

---

[11] In (2.102) we defined $\mathbf{r}$ as an operator $\mathbf{r} : \mathfrak{g}^* \to \mathfrak{g}$. We hope this will not lead to confusion. Our present definition requires to fix on $\mathfrak{g}$ an invariant symmetric form.

$$[\mathbf{r} X, \mathbf{r} Y] - \mathbf{r}([\mathbf{r} X, Y] + [X, \mathbf{r} Y]) = -c^2[X, Y]. \tag{2.116}$$

This equation can be put in the form

$$(\mathbf{r} \pm c\mathbb{1})([\mathbf{r} X, Y] + [X, \mathbf{r} Y]) = [(\mathbf{r} \pm c\mathbb{1})X, (\mathbf{r} \pm c\mathbb{1})Y]. \tag{2.117}$$

Now we observe that under the isomorphism $\mathfrak{g} \simeq \mathfrak{g}^*$, the commutator (2.104) in $\mathfrak{g}^*$ turns into

$$[X, Y]_{\mathbf{r}} = [\mathbf{r} X, Y] + [X, \mathbf{r} Y]. \tag{2.118}$$

As a consequence of Eq. (2.116) for $\mathbf{r}$, this bracket satisfies the Jacobi identity and, therefore, defines a new Lie algebra structure on the vector space $\mathfrak{g}$, which we call the $\mathbf{r}$-bracket [13]. We refer to the Lie algebra structure on $\mathfrak{g}$ defined by the $\mathbf{r}$-bracket as $\mathfrak{g}_{\mathbf{r}}$. For Eq. (2.117) we then have

$$(\mathbf{r} \pm c\mathbb{1})([X, Y]_{\mathbf{r}}) = [(\mathbf{r} \pm c\mathbb{1})X, (\mathbf{r} \pm c\mathbb{1})Y]. \tag{2.119}$$

This shows that the maps $\mathbf{r} \pm c\mathbb{1}$ are Lie algebra homomorhisms $\mathfrak{g}_{\mathbf{r}} \to \mathfrak{g}$ for $c = 1$ and $\mathfrak{g}_{\mathbf{r}} \to \mathfrak{g}^{\mathbb{C}}$ for $c = i$, where $\mathfrak{g}^{\mathbb{C}}$ is the complexification of $\mathfrak{g}$.

Quasi-Frobenius Lie algebras and solutions of CYBE. Concerning the CYBE equation, we point out the following important statement [10]. Let $r = r^{ij} e_i \wedge e_j$, where e_i is a basis in $\mathfrak{g}$ and the matrix r^{ij} is non-degenerate. Denote by ω_{ij} the inverse of r, that is $\omega_{ik} r^{kj} = \delta_i^j$. Let $\omega = \omega_{ij} e^i \wedge e^j$ be the bilinear form on $\mathfrak{g}$, where e^i is the dual basis in $\mathfrak{g}^*$. Then r satisfies the CYBE if and only if $\omega : \mathfrak{g} \wedge \mathfrak{g} \to \mathbb{C}$ is a 2-cocycle, i.e. if and only if

$$\omega(X, [Y, Z]) + \omega(Y, [Z, X]) + \omega(Z, [X, Y]) = 0, \quad \forall X, Y, Z \in \mathfrak{g}. \tag{2.120}$$

The proof of this statement is straightforward. The CYBE is equivalent to

$$f^i_{mk} r^{jm} r^{lk} + f^j_{km} r^{im} r^{lk} + f^l_{km} r^{ik} r^{jm} = 0,$$

where f^k_{ij} are the corresponding structure constants. Pairing this equation with $\omega_{pi} \omega_{rj} \omega_{sl}$, we obtain

$$\omega_{pi} f^i_{rs} + \omega_{ri} f^i_{sp} + \omega_{si} f^i_{pr} = 0,$$

which in turn is equivalent to (2.120). A Lie algebra $\mathfrak{g}$ equipped with a non-degenerate 2-cocycle ω is called *quasi-Frobenius*. If ω is a coboundary, i.e. there exists a linear form $\ell : \mathfrak{g} \to \mathbb{C}$ such that $\omega(X, Y) = \ell([X, Y])$, then $\mathfrak{g}$ is called *Frobenius* Lie algebra. Thus, with any quasi-Frobenius Lie algebra one can associate a non-degenerate solution of the classical Yang-Baxter equation [10]. Further discussion of the quasi-Frobenius Lie algebras and solutions of the CYBE can be found in [14, 15].

2.1.5 Double

Given a Lie bialgebra $\mathfrak{g}$, one can introduce a new Lie algebra, namely, the double of $\mathfrak{g}$. Below we discuss the corresponding construction and give two important examples.

Let $(\mathfrak{g}, \mathfrak{g}^*)$ be a Lie bialgebra and let $\mathscr{D} = \mathfrak{g} \oplus \mathfrak{g}^*$. We equip $\mathscr{D}$ with a non-degenerate symmetric bilinear form

$$\langle (X_1, \ell_1), (X_2, \ell_2) \rangle_{\mathscr{D}} = \ell_1(X_2) + \ell_2(X_1) . \tag{2.121}$$

With respect to this form the subalgebras $\mathfrak{g}$ and $\mathfrak{g}^*$ are isotropic.

There is a unique Lie algebra structure on $\mathscr{D}$ such that $\mathfrak{g}$ and $\mathfrak{g}^*$ are subalgebras and the form (2.121) is ad-invariant. The corresponding Lie bracket is given by

$$[(X_1, \ell_1), (X_2, \ell_2)] = \left([X_1, X_2] + \mathrm{ad}^*_{\ell_1} X_2 - \mathrm{ad}^*_{\ell_2} X_1, [\ell_1, \ell_2]_* + \mathrm{ad}^*_{X_1} \ell_2 - \mathrm{ad}^*_{X_2} \ell_1 \right). \tag{2.122}$$

In Appendix C.1.1 we demonstrate the fulfilment of the Jacobi identity and the ad-invariance of the scalar product (2.121). The Lie algebra $\mathscr{D}$ is called the *double* of $(\mathfrak{g}, \mathfrak{g}^*)$. A connected Lie group D with Lie algebra $\mathscr{D}$ is called the double Lie group or simply the double.

Let G and G^* be the subgroups in D corresponding to the subalgebras $\mathfrak{g}$ and $\mathfrak{g}^*$. In the vicinity of the unit, an element $d \in D$ admits factorisations

$$d = gg^* = g'^* g' ,$$

where $g, g' \in G$ and $g^*, g'^* \in G^*$.

It turns out that the double itself can be supplied with the structure of a Lie bialgebra in a canonical way. To describe this structure, we introduce projectors P and P^* which project onto the spaces $\mathfrak{g}$ and $\mathfrak{g}^*$, respectively, that is,

$$P(X, \ell) = (X, 0) , \quad P^*(X, \ell) = (0, \ell) .$$

In fact, the projector P^* coincides with the conjugate of P with respect to (2.121). To these projectors we associate two elements $\mathscr{R}_\pm \in \mathscr{D} \otimes \mathscr{D}$ as

$$\langle a \otimes b, \mathscr{R}_+ \rangle_{\mathscr{D} \otimes \mathscr{D}} = \langle a, P^*(b) \rangle_{\mathscr{D}} ,$$
$$\langle a \otimes b, \mathscr{R}_- \rangle_{\mathscr{D} \otimes \mathscr{D}} = -\langle a, P(b) \rangle_{\mathscr{D}} . \tag{2.123}$$

In terms of dual bases $\{e_i\}$ and $\{e^i\}$ these elements are

$$\mathscr{R}_+ = (0, e^i) \otimes (e_i, 0) , \quad \mathscr{R}_- = -(e_i, 0) \otimes (0, e^i) . \tag{2.124}$$

As is shown in Appendix C.1.2, $\mathscr{R}_\pm$ are solutions of the CYBE (2.112). The difference of $\mathscr{R}_+$ and $\mathscr{R}_-$ corresponds to the identity operator $P + P^* = \mathbb{1}$, and is equal to the following $\mathrm{Ad}_{\mathscr{D}}$-invariant element in $\mathscr{D} \otimes \mathscr{D}$

$$\mathscr{C} = (0, e^i) \otimes (e_i, 0) + (e_i, 0) \otimes (0, e^i).$$

The sum

$$\mathscr{R} = \mathscr{R}_+ + \mathscr{R}_- = (0, e^i) \otimes (e_i, 0) - (e_i, 0) \otimes (0, e^i) \in \mathscr{D} \wedge \mathscr{D}, \quad (2.125)$$

is a skew-symmetric solution of the mCYBE and we have $\mathscr{R}_\pm = \frac{1}{2}(\mathscr{R} \pm \mathscr{C})$.

Thus, owing to these properties of $\mathscr{R}$, formula (2.100) with $r = \mathscr{R}$ defines on $\mathscr{D}$ the structure of a Lie bialgebra. The structure of a Poisson-Lie group on D which corresponds to this Lie bialgebra structure is defined as follows. Consider a basis of $\mathscr{D}$ formed by the elements $E_i = (0, e_i)$ and $E^i = (e^i, 0)$, which are paired as $\langle E^i, E_j \rangle_\mathscr{D} = \delta^i_j$. The desired structure of a Poisson-Lie group on D is then given by

$$\{f, h\} = \mathscr{D}^r_{E^i} f \mathscr{D}^r_{E_i} h - \mathscr{D}^l_{E_i} f \mathscr{D}^l_{E^i} h, \quad (2.126)$$

for $f, h \in \mathscr{F}(D)$.

Now we consider some examples of the double construction.

Complexified Lie algebra as a double. Consider a real Lie algebra $\mathfrak{g}$ equipped with a bilinear symmetric non-degenerate ad-invariant form $\langle \cdot, \cdot \rangle$. Let $\mathfrak{g}^\mathbb{C} = \mathfrak{g} \otimes_\mathbb{R} \mathbb{C}$ be the complexification of $\mathfrak{g}$, so that elements of $\mathfrak{g}^\mathbb{C}$ are $X + iY$, where $X, Y \in \mathfrak{g}$. Inside $\mathfrak{g}^\mathbb{C}$ the subalgebra $\mathfrak{g}$ is identified as a set of fixed points under an anti-linear involution τ.[12] Regarding $\mathfrak{g}^\mathbb{C}$ as a vector space over $\mathbb{R}$ (the realification), and supply it with the following real non-degenerate form $\langle \cdot, \cdot \rangle_{\mathfrak{g}^\mathbb{C}}$:

$$\langle X + iY, X' + iY' \rangle_{\mathfrak{g}^\mathbb{C}} = \text{Im} \langle X + iY, X' + iY' \rangle = \langle Y, X' \rangle + \langle Y', X \rangle. \quad (2.127)$$

With respect to this form $\mathfrak{g}$ is an isotropic subalgebra of the realification. Let $\mathfrak{p} \subset \mathfrak{g}^\mathbb{C}$ be an isotropic subalgebra complementary to $\mathfrak{g}$, so that $\mathfrak{g}^\mathbb{C}$ admits the direct sum decomposition

$$\mathfrak{g}^\mathbb{C} = \mathfrak{g} \oplus \mathfrak{p} \quad (2.128)$$

over $\mathbb{R}$. Evidently, (2.127) allows one to consider $\mathfrak{p}$ as a model of $\mathfrak{g}^*$ in the double construction.

Consider the surjective map $\varrho : \mathfrak{g}^\mathbb{C} \to \mathfrak{g}$ defined as $\varrho(X) = \frac{1}{2i}(X - \tau(X))$, $X \in \mathfrak{g}^\mathbb{C}$. The kernel of ϱ coincides with $\mathfrak{g} \subset \mathfrak{g}^\mathbb{C}$ and, therefore, ϱ restricted to $\mathfrak{p}$ induces a linear isomorphism $\varrho|_\mathfrak{p} \equiv \varrho_\mathfrak{p} : \mathfrak{p} \to \mathfrak{g}$. Thus, for any $X \in \mathfrak{g}$ one can find $A \in \mathfrak{p}$, such that $X = \varrho(A) = \frac{1}{2i}(A - \tau(A))$. Let us determine the inverse map $\varrho_\mathfrak{p}^{-1}$, see Fig. 2.3. We write

$$A = i\left(\frac{1}{2i}(A + \tau(A)) + \frac{1}{2i}(A - \tau(A))\right) = (\mathbf{r} + i)X = \varrho_\mathfrak{p}^{-1}(X),$$

---

[12] An anti-linear involution τ is a Lie algebra automorphism of $\mathfrak{g}^\mathbb{C}$ such that $\tau(iX) = -i\tau(X)$ and $\tau^2 = 1$.

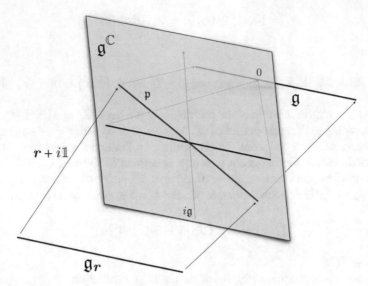

Fig. 2.3 The complex Lie algebra $\mathfrak{g}^{\mathbb{C}}$ as a double of a real form $\mathfrak{g}$. Both $\mathfrak{g}$ and its complementary subalgebra $\mathfrak{p}$ are maximal isotropic with respect to the bilinear form (2.127)

that is

$$\varrho_{\mathfrak{p}}^{-1} = \mathbf{r} + i\mathbb{1}. \tag{2.129}$$

Here we have introduced an operator $\mathbf{r} : \mathfrak{g} \to \mathfrak{g}$, which acts as

$$\mathbf{r}\left(\tfrac{1}{2i}(A - \tau(A))\right) = \tfrac{1}{2}(A + \tau(A)). \tag{2.130}$$

The requirement that the subalgebra $\mathfrak{p}$ is isotropic is equivalent for $\mathbf{r}$ to be skew-symmetric with respect to the bilinear form on $\mathfrak{g}$

$$\langle A, A' \rangle_{\mathfrak{g}^{\mathbb{C}}} = \langle \mathbf{r}X + iX, \mathbf{r}Y + iY \rangle_{\mathfrak{g}^{\mathbb{C}}} = \langle X, \mathbf{r}Y \rangle + \langle Y, \mathbf{r}X \rangle = 0, \quad A, A' \in \mathfrak{p}. \tag{2.131}$$

Let us take two arbitrary $X, Y \in \mathfrak{g}$

$$X = \tfrac{1}{2i}(A - \tau(A)), \quad Y = \tfrac{1}{2i}(B - \tau(B)),$$

where A, B are representatives of X and Y in $\mathfrak{p}$. Using (2.130), we compute the left hand side of (2.116)

$$[\mathbf{r}X, \mathbf{r}Y] - \mathbf{r}([\mathbf{r}X, Y] + [X, \mathbf{r}Y]) =$$
$$= \tfrac{1}{4}[A + \tau(A), B + \tau(B)] - \tfrac{1}{4i}\mathbf{r}([A + \tau(A), B - \tau(B)] + [A - \tau(A), B + \tau(B)])$$
$$= \tfrac{1}{4}[A + \tau(A), B + \tau(B)] - \tfrac{1}{2i}\mathbf{r}([A, B] - \tau[A, B]) = [X, Y].$$

Here we used the fact that τ is the Lie algebra homomorphism and that an element $\frac{1}{2i}([A, B] - \tau[A, B]) \in \mathfrak{g}$ has $[A, B]$ as its representative on $\mathfrak{p}$. It is this last point where we rely on the assumption that $\mathfrak{p}$ is the Lie algebra. Thus, $\mathbf{r}$ satisfies the mCYBE (2.116) corresponding to the non-split case $c^2 = -1$.

In this way we have shown that an isotropic subalgebra $\mathfrak{p}$ complementary to $\mathfrak{g}$ yields a skew-symmetric non-split solution of the mCYBE (2.116). The inverse is also true—a skew-symmetric non-split $\mathbf{r} : \mathfrak{g} \to \mathfrak{g}$ defines an isotropic subalgebra $\mathfrak{p}$ complementary to $\mathfrak{g}$ in $\mathfrak{g}^\mathbb{C}$. Indeed, define $\mathfrak{p}$ as an image of $\mathfrak{g}$ in $\mathfrak{g}^\mathbb{C}$ under the linear map $\mathbf{r} + i\mathbb{1}$. Any $Z = X + iY \in \mathfrak{g}^\mathbb{C}$, $X, Y \in \mathfrak{g}$ can be uniquely represented as

$$Z = X - \mathbf{r}\,Y + (\mathbf{r} + i\mathbb{1})(Y),$$

where $X - \mathbf{r}\,Y \in \mathfrak{g}$ and $(\mathbf{r} + i\mathbb{1})(Y) \in \mathfrak{p}$. Due to (2.131) the vector space $\mathfrak{p}$ is isotropic in $\mathfrak{g}^\mathbb{C}$. Finally, if we endow $\mathfrak{g}$ with the $\mathbf{r}$-bracket (2.118), then under the map $\mathbf{r} + i\mathbb{1}$ the $\mathbf{r}$-bracket supplies $\mathfrak{p}$ with the Lie algebra structure.

Thus, the operator $\mathbf{r} : \mathfrak{g} \to \mathfrak{g}$ corresponding to the Lie algebra structure of $\mathfrak{p}$ induces, by means of the bilinear form on $\mathfrak{g}$, the r-matrix cocycle (2.100) and vice versa, an r-matrix cocycle on $\mathfrak{g}$ sets up the Lie algebra structure on $\mathfrak{p} = \mathrm{Im}(\mathbf{r} + i\mathbb{1})$. Therefore, $(\mathfrak{g}, \mathfrak{p})$ is the Lie bialgebra and $\mathscr{D} = \mathfrak{g} \oplus \mathfrak{p}$ is its double.

As an example, take $\mathfrak{g}$ to be the *compact real form* of a complex simple Lie algebra $\mathfrak{g}^\mathbb{C}$. Let $(h_i, e_{\pm\alpha})$ be the Cartan-Weyl basis of $\mathfrak{g}^\mathbb{C}$. The generators h_i span the Cartan subalgebra of $\mathfrak{g}^\mathbb{C}$, while $e_{\pm\alpha}$ are root vectors for roots $\alpha \in \Delta_+$, where Δ_+ is the sets of positive roots. An involution τ that singles out the compact form is defined by its action on the elements of the Cartan-Weyl basis

$$\tau(h_i) = -h_i, \quad \tau(e_\alpha) = -e_{-\alpha}, \quad \tau(e_{-\alpha}) = -e_\alpha. \tag{2.132}$$

The compact $\mathfrak{g}$ is spanned over $\mathbb{R}$ by the elements

$$-ih_i, \quad \tfrac{1}{2i}(e_\alpha + e_{-\alpha}), \quad \tfrac{1}{2}(e_\alpha - e_{-\alpha}). \tag{2.133}$$

Consider the Iwasawa decomposition of the realification a complex simple Lie algebra

$$\mathfrak{g}^\mathbb{C} = \mathfrak{g} \oplus \mathfrak{a} \oplus \mathfrak{n}_+, \tag{2.134}$$

where $\mathfrak{a}$ is the real Cartan subalgebra and $\mathfrak{n}_+$ is the nilpotent sublagebra corresponding to Δ_+. Define the form (2.127) with the help of the Cartan-Killing form $\langle \cdot, \cdot \rangle$ and take

$$\mathfrak{p} = \mathfrak{a} \oplus \mathfrak{n}_+. \tag{2.135}$$

Assuming that the Cartan-Killing form is normalised such that $\langle e_\alpha, e_{-\alpha} \rangle = 1$, the operator **r** in (2.130) gives rise to the following r-matrix[13] $r \in \mathfrak{g} \wedge \mathfrak{g}$

$$r = i \sum_{\alpha \in \Delta_+} (e_\alpha - e_{-\alpha}) \wedge (e_\alpha + e_{-\alpha}). \tag{2.136}$$

This r-matrix is a solution of (2.113) with $c = i$. Under the map (2.129) the basis elements (2.133) are mapped to

$$h_i, \quad e_\alpha, \quad i e_\alpha, \quad \alpha \in \Delta_+. \tag{2.137}$$

The span of h_i over $\mathbb{R}$ coincides with the real Cartan subalgebra $\mathfrak{a}$, while the span of $(e_\alpha, i e_\alpha)$ gives the nilpotent subalgebra $\mathfrak{n}_+$. According to the classification of real forms of semi-simple Lie algebras, $\mathfrak{g}$ is non-split, so that $r \in \mathfrak{g} \wedge \mathfrak{g}$ in (2.136) is a non-split solution of the mCYBE.

Double of a factorisable Lie bialgebra. Let $(\mathfrak{g}, \mathfrak{g}^*)$ be a Lie bialgebra such that the Lie algebra structure on $\mathfrak{g}^*$ is given by the co-commutator (2.100) with a split r-matrix $r \in \mathfrak{g} \wedge \mathfrak{g}$, while $\mathfrak{g}$ admits a bilinear symmetric non-degenerate ad-invariant form $\langle \cdot, \cdot \rangle$. A Lie bialgebra $(\mathfrak{g}, \mathfrak{g}^*)$ with these properties is called *factorisable* [16, 17].[14] For instance, any simple complex Lie algebra (or its unique real split form) is a factorisable Lie bialgebra with the canonical r-matrix

$$r = 2 \sum_{\alpha \in \Delta_+} e_\alpha \wedge e_{-\alpha}, \tag{2.138}$$

being a solution of (2.113) with $c^2 = 1$.

It turns out that the double of a factorisable Lie bialgebra has another, more transparent realisation than (2.122), which we now describe. Let $(\mathfrak{g}, \mathfrak{g}^*)$ be a Lie bialgebra, we use the corresponding invariant form to identify $\mathfrak{g}^* \simeq \mathfrak{g}$. By definition, $\mathscr{D} = \mathfrak{g} \oplus \mathfrak{g}^* \simeq \mathfrak{g} \oplus \mathfrak{g}_{\mathbf{r}}$, where the last isomorphism emphasises that the Lie algebra structure of $\mathscr{D}$ restricted on the second factor $\mathfrak{g}^* \simeq \mathfrak{g}$ coincides with that given by the **r**-bracket.

Consider now a vector space $\mathscr{D}^{\mathbb{R}} = \mathfrak{g} \oplus \mathfrak{g}$ supplied with the Lie algebra structure of the direct sum of two copies of the Lie algebra $\mathfrak{g}$. This means that the Lie bracket in $\mathscr{D}^{\mathbb{R}}$ is defined as

$$[(X, Y), (X', Y')] = ([X, X'], [Y, Y']),$$

where (X, Y) and (X', Y') are two arbitrary elements from $\mathfrak{g} \oplus \mathfrak{g}$. Thus, as a vector space $\mathscr{D}^{\mathbb{R}}$ is isomorphic to $\mathscr{D}$ but it has a different Lie algebra structure. Below we will show that there exists a Lie algebra homomorphism

---

[13] We recall that the wedge product is defined with the coefficient $\frac{1}{2}$, as in (2.99).

[14] In terms of an r-matrix that satisfies the CYBE, the definition of factorisable Lie bialgebra means that $r_{12} + r_{21} \in S^2 \mathfrak{g}$ defines a non-degenerate invariant scalar product on $\mathfrak{g}^*$.

$$\rho : \mathscr{D} \to \mathscr{D}^{\mathbb{R}}. \tag{2.139}$$

The Lie algebra $\mathscr{D}^{\mathbb{R}}$ is called the *real double* $\mathscr{D}^{\mathbb{R}}$ of $\mathfrak{g}$.

According to our assumptions and (2.114), a split r-matrix $r \in \mathfrak{g} \wedge \mathfrak{g}$ with $c^2 = 1$ gives rise to two linear operators

$$\mathbf{r}_{\pm} = \tfrac{1}{2}(\mathbf{r} \pm \mathbb{1}), \tag{2.140}$$

which are Lie algebra homomorphisms of $\mathfrak{g}_r$ into $\mathfrak{g}$,

$$\mathbf{r}_{\pm}[X, Y]_{\mathbf{r}} = [\mathbf{r}_{\pm}X, \mathbf{r}_{\pm}Y].$$

We denote the images of $\mathfrak{g}_r$ under $\mathbf{r}_{\pm}$ as $\mathfrak{g}_{\pm} \subset \mathfrak{g}$. Obviously, $\mathfrak{g}_{\pm}$ are subalgebras in $\mathfrak{g}$. Note that in contrast to (2.118), we have found it convenient to change the definition of the $\mathbf{r}$-bracket by a factor of $1/2$

$$[X, Y]_{\mathbf{r}} = \tfrac{1}{2}([\mathbf{r}\, X, Y] + [X, \mathbf{r}\, Y]) \tag{2.141}$$

that in turn leads to the appearance of the same factor in (2.140). With this definition we have

$$\mathbf{r}_+ - \mathbf{r}_- = \mathbb{1}. \tag{2.142}$$

The map (2.139) is then constructed as

$$\rho(X, Y) = (X + \mathbf{r}_+ Y, X + \mathbf{r}_- Y) \in \mathscr{D}^{\mathbb{R}}, \tag{2.143}$$

where $(X, Y) \in \mathscr{D}$. Under (2.143) the Lie subalgebra $\mathfrak{g}$ is mapped to the diagonal subalgebra of elements $(X, X) \in \mathscr{D}^{\mathbb{R}}$, while $\mathfrak{g}^*$ is embedded in $\mathscr{D}^{\mathbb{R}}$ as the subspace (X_+, X_-), where we adopt the notation $X_{\pm} = \mathbf{r}_{\pm} X \in \mathfrak{g}$. Due to (2.142), each $X \in \mathfrak{g}$ has a unique decomposition

$$X = X_+ - X_-, \tag{2.144}$$

where $(X_+, X_-) \in \text{Im } \mathfrak{g}^*$. In Appendix C.1.3 we prove that ρ is a Lie algebra homomorphism. In a sense, ρ untangles the non-trivial Lie structure of $\mathscr{D}$.

The space $\mathscr{D}^{\mathbb{R}}$ admits a bilinear symmetric form $\langle\!\langle \cdot , \cdot \rangle\!\rangle$

$$\langle\!\langle (X_1, X_2), (Y_1, Y_2) \rangle\!\rangle = \langle X_1, Y_1 \rangle - \langle X_2, Y_2 \rangle, \tag{2.145}$$

where on the right hand side the pairing is done with respect to the bilinear form on $\mathfrak{g}$. It is straightforward to verify that

$$\langle\!\langle \rho(d_1), \rho(d_2) \rangle\!\rangle = \langle d_1, d_2 \rangle_{\mathscr{D}}, \tag{2.146}$$

where $d_{1,2} \in \mathscr{D}$.

Any element $(X, Y) \in \mathscr{D}^{\mathbb{R}}$ can be uniquely factorised as

$$(X, Y) = (A, A) + (\mathbf{r}_+ B, \mathbf{r}_- B),\qquad(2.147)$$

where

$$A = \mathbf{r}_+ Y - \mathbf{r}_- X,\quad B = X - Y.$$

We introduce the projector P on the diagonal subalgebra

$$P(X, Y) = (\mathbf{r}_+ Y - \mathbf{r}_- X,\ \mathbf{r}_+ Y - \mathbf{r}_- X).$$

The operator P^* adjoint to P with respect to the form (2.145) is a projector on the complementary subalgebra

$$P^*(X, Y) = (\mathbf{r}_+ X - \mathbf{r}_+ Y,\ \mathbf{r}_- X - \mathbf{r}_- Y),$$

so that $P + P^* = \mathbb{1}$. Define the operator $\widehat{\mathscr{R}}_{\mathbb{R}}$ as the difference of these projectors

$$\widehat{\mathscr{R}}_{\mathbb{R}} = P - P^*.\qquad(2.148)$$

Obviously, $\widehat{\mathscr{R}}_{\mathbb{R}}$ is skew-symmetric and its action is given by

$$\widehat{\mathscr{R}}_{\mathbb{R}}(X, Y) = (2\mathbf{r}_+ Y - \mathbf{r} X,\ \mathbf{r} Y - 2\mathbf{r}_- X).\qquad(2.149)$$

One can check that this operator satisfies the mCYBE and, therefore, it endows $\mathscr{D}^{\mathbb{R}}$ with the structure of a factorisable Lie bialgebra. Through the correspondence

$$\langle\!\langle a \otimes b, \mathscr{R}_{\mathbb{R}} \rangle\!\rangle_{\mathscr{D}^{\mathbb{R}} \otimes \mathscr{D}^{\mathbb{R}}} = \langle\!\langle a, \widehat{\mathscr{R}}_{\mathbb{R}}(b) \rangle\!\rangle,\quad a, b \in \mathscr{D}^{\mathbb{R}},$$

the operator $\widehat{\mathscr{R}}_{\mathbb{R}}$ defines an element $\mathscr{R}_{\mathbb{R}} \in \mathscr{D}^{\mathbb{R}} \wedge \mathscr{D}^{\mathbb{R}}$. We find for $\mathscr{R}_{\mathbb{R}}$ the following explicit expression

$$-\tfrac{1}{2}\mathscr{R}_{\mathbb{R}} = r^{ij}(e_i, 0) \wedge (e_j, 0) + r_+^{ij}(e_i, 0) \wedge (0, e_j)\qquad(2.150)$$
$$+ r_-^{ij}(0, e_i) \wedge (e_j, 0) + r^{ij}(0, e_i) \wedge (0, e_j).$$

One can further recognise that $\mathscr{R}_{\mathbb{R}}$ is nothing else but the image of $\mathscr{R}$ in (2.125) under the map ρ. Explicitly,

$$\mathscr{R}_{\mathbb{R}} = -(\rho \otimes \rho)\mathscr{R},$$

where the minus sign on the right hand side is due to our choice of the bilinear form (2.145). Here we have also introduced the notation

$$r_{\pm} = r_{\pm}^{ij} e_i \otimes e_j = \mathbf{r}_{\pm}(e_i) \otimes e_i \,. \tag{2.151}$$

In particular,

$$r_+ - r_- = e_i \otimes e_i = C \,, \tag{2.152}$$

where C is the split Casimir.

Let G and G^* be connected and simply connected Lie groups corresponding to the (tangent) Lie algebras $\mathfrak{g}$ and $\mathfrak{g}^*$. The Lie algebra homomorphisms $\mathbf{r}_{\pm}$ and ρ are lifted to Lie group homomorphisms which we denote by the same letters. A connected and simply connected Lie group D that has the double of a factorisable Lie algebra $(\mathfrak{g}, \mathfrak{g}^*)$ as its tangent Lie bialgebra, can be thus modelled as the direct product of two copies of G: $D(G) = G \times G$. We denote the embedding of G^* into $G \times G$ under the Lie group homomorphism ρ as $G_+ \times G_-$, where $G_{\pm} \subset G$ are subgroups corresponding to the Lie algebras $\mathfrak{g}_{\pm}$. The corresponding representative will be denoted as (h_+, h_-), $h_{\pm} \in G_{\pm}$, and we allow ourselves to write with some abuse of notation $(h_+, h_-) \in G^*$, as a reminder of the origin of this element. The group G is embedded into $G \times G$ as the diagonal subgroup: (g, g), $g \in G$.

In the following we will need an explicit formula for the coadjoint action Ad^* of G^* on $\mathfrak{g}$ which is an integrated version of the $\mathrm{ad}^*_{\mathfrak{g}^*}$ representation, see (2.93). This can be obtained through the real double construction by exploiting the Ad-invariance of the bilinear form (2.145). Let $(h_+, h_-) \in G^*$ and $(Y_+, Y_-) \in \mathfrak{g}^*$. For $X \in \mathfrak{g}$ we have

$$\langle\!\langle \mathrm{Ad}^*_{(h_+, h_-)}(X, X), (Y_+, Y_-) \rangle\!\rangle = \langle\!\langle (X, X), (h_+^{-1} Y_+ h_+, h_-^{-1} Y_- h_-) \rangle\!\rangle$$
$$= \langle\!\langle (h_+ X h_+^{-1}, h_- X h_-^{-1}), (Y_+, Y_-) \rangle\!\rangle \,.$$

The corresponding action is then defined by applying to $(h_+ X h_+^{-1}, h_- X h_-^{-1})$ the projection P on $\mathfrak{g}$, which gives [18]

$$\mathrm{Ad}^*_{(h_+, h_-)} X = (h_- X h_-^{-1})_+ - (h_+ X h_+^{-1})_- \,. \tag{2.153}$$

The infinitesimal version of this relation is

$$\mathrm{ad}^*_{(Y_+, Y_-)} X = [Y_-, X]_+ - [Y_+, X]_- = \tfrac{1}{2}[\mathbf{r}(Y), X] - \tfrac{1}{2}\mathbf{r}([Y, X]) \,, \tag{2.154}$$

where $Y_{\pm} \in \mathfrak{g}_{\pm}$ and $Y = Y_+ - Y_-$. This is the same formula as (C.8).

In a similar fashion we can obtain the coadjoint action of G on $\mathfrak{g}^*$. Taking $(X_+, X_-) \in \mathfrak{g}^*$ and $g \in G$, we get

$$\mathrm{Ad}^*_g (X_+, X_-) = \big((g X g^{-1})_+, (g X g^{-1})_- \big) \,, \tag{2.155}$$

where $X = X_+ - X_-$. The corresponding infinitesimal version of this action is

$$\text{ad}^*_Y(X_+, X_-) = \big(\mathbf{r}_+([Y, X]), \mathbf{r}_-([Y, X])\big), \quad Y \in \mathfrak{g}. \tag{2.156}$$

On $X = X_+ - X_-$ this action simplifies to

$$\text{ad}^*_Y X = (\mathbf{r}_+ - \mathbf{r}_-)[Y, X] = [Y, X] \tag{2.157}$$

and coincides with (C.7).

2.1.6 Poisson Reduction

Here we discuss the far-reaching generalisation of the hamiltonian reduction procedure that is applicable to more general group actions than the symplectic ones [19–21].

Poisson reduction and dual pairs. Let $\mathscr{P}$ be a symplectic manifold with a form ω and let Φ be some foliation of $\mathscr{P}$ such that the quotient space $\mathscr{N}_\Phi$ of $\mathscr{P}$ over an equivalence relation set up by Φ is a manifold. Denote by $\mathscr{F}_\Phi$ the space of functions on $\mathscr{P}$ that are constant along the leaves of Φ. We require this set to be closed under the Poisson bracket on $\mathscr{P}$,

$$\{\mathscr{F}_\Phi, \mathscr{F}_\Phi\} \subset \mathscr{F}_\Phi.$$

Let $T\Phi$ be a bundle of vectors tangent to the leaves of Φ, and let $T\Phi^\perp$ be its orthogonal complement with respect to ω. Denote by χ_f the hamiltonian vector field of $f \in \mathscr{F}_\Phi$. For any $v \in T\Phi$ it follows from (1.15) that $\omega(v, \chi_f) = vf = 0$, as f is constant along Φ. This means that χ_f lies in $T\Phi^\perp$. In fact, by counting dimensions one sees that $T\Phi^\perp$ is spanned at each point by the hamiltonian vector fields of functions in $\mathscr{F}_\Phi$. Further, for any $f, h \in \mathscr{F}_\Phi$ one has

$$[\chi_f, \chi_h] = \chi_{\{f,h\}} \tag{2.158}$$

and since $\mathscr{F}_\Phi$ is a Lie subalgebra under the Poisson bracket, by the Frobenius theorem, the distribution of planes $T\Phi^\perp$ is integrable and, therefore, is tangent to another foliation $\Phi^\perp$ called *polar* to Φ. Thus, the requirement that $\mathscr{F}_\Phi$ is a Poisson subalgebra results in the existence of a polar foliation.

Analogously, denote by $\mathscr{F}_{\Phi^\perp}$ a set of functions on $\mathscr{P}$ which are constant along $\Phi^\perp$. The hamiltonian vector fields of $\mathscr{F}_{\Phi^\perp}$ are tangent to Φ and, since the latter is a foliation, $\mathscr{F}_{\Phi^\perp}$ is also a Poisson subalgebra. A foliation polar to $\Phi^\perp$ coincides with Φ and a pair $(\Phi, \Phi^\perp)$ is called a *bi-foliation* of $\mathscr{P}$, see Fig. 2.4.

We can view $\mathscr{F}_\Phi$ as functions on $\mathscr{N}_\Phi$. Choosing in $\mathscr{F}_\Phi$ a functionally independent basis $\{\phi_i\}$, $1 \le i \le m$, we get

$$\{\phi_i, \phi_j\} = J_{ij}(\phi_1, \dots \phi_m) \equiv \{\phi_i, \phi_j\}_{\mathscr{N}_\Phi}, \tag{2.159}$$

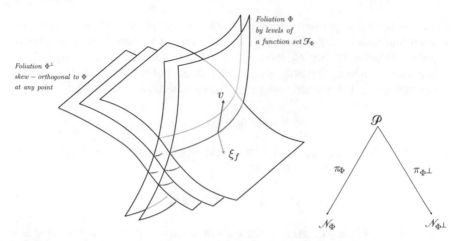

Fig. 2.4 A bi-foliation of a symplectic manifold $\mathscr{P}$ and the corresponding dual pair. The first foliation Φ is by levels of a function set $\mathscr{F}_\Phi$ closed under the Poisson bracket on $\mathscr{P}$. The hamiltonian vector fields of functions from $\mathscr{F}_\Phi$ span an integrable distribution $T\Phi^\perp$. The leaves tangent to this distribution form the second foliation $\Phi^\perp$ of $\mathscr{P}$

since $\mathscr{F}_\Phi$ is a Poisson subalgebra. Although the bracket (2.159) was computed on $\mathscr{P}$, the result of this computation can serve as the definition of a Poisson bracket on $\mathscr{N}_\Phi$ with J_{ij} being the corresponding Poisson tensor. Thus, $\mathscr{N}_\Phi$ becomes a Poisson manifold and the projection $\pi_\Phi : \mathscr{P} \to \mathscr{N}_\Phi$ is a Poisson map. Similarly, we define a quotient space $\mathscr{N}_{\Phi^\perp}$ of $\mathscr{P}$ by the equivalence relation set up by the foliation $\Phi^\perp$. For a basis $\{\varphi_{i'}\}$, $1 \le i' \le m'$, we have

$$\{\varphi_{i'}, \varphi_{j'}\} = J'_{i'j'}(\varphi_1, \ldots \varphi_{m'}) \equiv \{\varphi_{i'}, \varphi_{j'}\}_{\mathscr{N}_{\Phi^\perp}} .$$

This defines the Poisson structure on $\mathscr{N}_{\Phi^\perp}$ such that a projection $\pi_{\Phi^\perp} : \mathscr{P} \to \mathscr{N}_{\Phi^\perp}$ is a Poisson map. The Poisson structures arising on the reduced manifolds $\mathscr{N}_\Phi$ and $\mathscr{N}_{\Phi^\perp}$ are typically degenerate and the ultimate goal of the reduction procedure consists in determining the corresponding symplectic leaves. This can be done with the help of dual pairs.

In general, for a symplectic manifold $\mathscr{P}$ a pair of Poisson maps $\mathscr{N}_1 \xleftarrow{\pi_1} \mathscr{P} \xrightarrow{2} \mathscr{N}_2$ is called a *dual pair* if the pulled back function sets $\mathscr{F}_1 = \pi_1^*(C^\infty(\mathscr{N}_1))$ and $\mathscr{F}_2 = \pi_2^*(C^\infty(\mathscr{N}_2))$ are in involution with each other [19]. Dual pairs provide an efficient technique for describing symplectic leaves of a Poisson manifold realised as one of the components of a dual pair.

By definition of $\mathscr{F}_{\Phi^\perp}$ we have

$$\{\mathscr{F}_\Phi, \mathscr{F}_{\Phi^\perp}\} = 0 .$$

Hence, the Poisson maps π_Φ and $\pi_{\Phi^\perp}$ form a dual pair for $\mathscr{P}$. The intersection $\mathscr{F}_\Phi \cap \mathscr{F}_{\Phi^\perp}$ can be viewed as a set of functions on $\mathscr{N}_\Phi$ which Poisson commute with all functions on $\mathcal{N}_\Phi$ or, alternatively, as functions in $\mathcal{N}_{\Phi^\perp}$ which Poisson commute with all functions on $\mathcal{N}_{\Phi^\perp}$. Thus, $\mathcal{F}_\Phi \cap \mathcal{F}_{\Phi^\perp}$ is a set of Casimir functions common to both Poisson manifolds $\mathcal{N}_\Phi$ and $\mathcal{N}_{\Phi^\perp}$. For this reason there is a correspondence between the symplectic leaves of $\mathcal{N}_\Phi$ and $\mathcal{N}_{\Phi^\perp}$. Namely, for $x \in \mathcal{N}_\Phi$, each connected component of $\pi_\Phi^{-1}(x)$ projects under $\pi_{\Phi^\perp}$ to a symplectic leaf $\mathcal{O}$ in $\mathcal{N}_{\Phi^\perp}$

$$\mathcal{O} = \pi_{\Phi^\perp}(\pi_\Phi^{-1}(x)) \subset \mathcal{N}_{\Phi^\perp},$$

see Fig. 2.5. Vice versa, for $x \in \mathcal{N}_{\Phi^\perp}$, each connected component of $\pi_{\Phi^\perp}^{-1}(x)$ projects under π_Φ to a symplectic leaf $\mathcal{O}$ in $\mathcal{N}_\Phi$

$$\mathcal{O} = \pi_\Phi(\pi_{\Phi^\perp}^{-1}(x)) \subset \mathcal{N}_\Phi.$$

Now we explain in which sense the Poisson reduction procedure described here is a generalisation of the hamiltonian reduction discussed in Sect. 2.1.2. Suppose $\mathcal{P}$ carries the action $G \times \mathcal{P} \to \mathcal{P}$ of some Lie group G with Lie algebra $\mathfrak{g}$. The vector fields ξ_X for $X \in \mathfrak{g}$ form an integrable distribution corresponding to a foliation of $\mathcal{P}$ by orbits of the G-action. Choosing a basis e_i in $\mathfrak{g}$ for the vector fields $\xi_i \equiv \xi_{e_i}$ we have

$$[\xi_i, \xi_j] = f_{ij}^k \xi_k, \tag{2.160}$$

where f_{ij}^k are the corresponding structure constants of $\mathfrak{g}$. In the context of the Poisson reduction discussed above we will identify the foliation of $\mathcal{P}$ by G-orbits with $\Phi^\perp$, so that $\mathcal{N}_{\Phi^\perp}$ has a natural interpretation as a coset $\mathcal{N}_{\Phi^\perp} = \mathcal{P}/G$. A set $\mathcal{F}_{\Phi^\perp}$ is spanned by all functions which are invariant under the G-action. Importantly, regarding G as a symmetry over which we want to reduce, its action on $\mathcal{P}$ *is not assumed* to preserve the symplectic structure, i.e. the vector fields ξ_i are not necessarily hamiltonian.

From the point of view of a foliation Φ polar to $\Phi^\perp$, hamiltonian vector fields of $\mathcal{F}_\Phi$ are tangent to the leaves of $\Phi^\perp$. By the Frobenius criterion, there exists a function basis $\{\phi_i\}$ in $\mathcal{F}_\Phi$ such that

$$[\chi_{\phi_i}, \chi_{\phi_j}] = c_{ij}^k(x)\chi_{\phi_k}. \tag{2.161}$$

Thus, we have two integrable distributions: one is (2.160) and the other is (2.161), both give rise to the same foliation $\Phi^\perp$. Thus, any hamiltonian vector field χ_f, $f \in \mathcal{F}_\Phi$, should admit an expansion over a basis ξ_i

$$\chi_f = A^i(f)\xi_i, \tag{2.162}$$

and vice versa. Here $A^i(f)$ are coordinate-dependent coefficients that are used to define a Lie algebra valued element $A(f) = A^i(f)e_i \in \mathfrak{g}$. The requirement (2.158) then implies the following condition for A

$$A(\{f, h\}) = \xi_{A(f)} A(h) - \xi_{A(h)} A(f) + [A(f), A(h)], \qquad (2.163)$$

for any $f, h \in \mathcal{F}_\Phi$. In the basis ϕ_i Eq. (2.162) reads

$$\{\phi_i, \} = A^j(\phi_i)\xi_j \equiv A_i^j(x)\xi_j. \qquad (2.164)$$

Evidently, in the case when the map A is x-independent, the G-action is equivalent to the hamiltonian action, and $\mu = (A^{-1})_i^j \phi_j(x)e^i \in \mathfrak{g}^*$ is the standard moment map of the hamiltonian action. The dual pair associated to this hamiltonian action is

$$G\backslash \mathcal{P} \xleftarrow{\Phi^\perp} \mathcal{P} \xrightarrow{\Phi} \mathfrak{g}^*, \qquad (2.165)$$

where π_Φ coincides with μ. When A is not a constant and satisfies (2.163), we have to resort to a more general Poisson reduction procedure. Below we discuss a concrete situation when $\mathcal{P}$ enjoys the Poisson action of a Poisson-Lie group G.

Moment map for a Poisson-Lie group action. The action of a Poisson-Lie group G on a Poisson manifold $\mathcal{P}$ is called *Poisson* if the corresponding map $G \times \mathcal{P} \to \mathcal{P}$ is a mapping of Poisson manifolds, the space $G \times \mathcal{P}$ being equipped with the product Poisson structure.[15] For any $f, h \in \mathcal{F}(\mathcal{P})$, $x \in \mathcal{P}$ and $g \in G$ this definition means that

$$\{f_x, h_x\}_G(g) + \{f_g, h_g\}_{\mathcal{P}}(x) = \{f, h\}_{\mathcal{P}}(g \cdot x), \qquad (2.166)$$

where $f_x(g) = f(g \cdot x)$, $f_g(x) = f(g \cdot x)$ and the subscript denotes the manifold on which the Poisson bracket is computed. To find an infinitesimal criterion for the Poisson action, we consider a one-parametric subgroup $g = e^{-tX}$, $X \in \mathfrak{g}$. Then under the action of a fundamental vector field ξ_X, see (2.12), the Poisson bracket on G undergoes a change

$$\frac{d}{dt}\{f_x, h_x\}_G(e^{-tX})\Big|_{t=0} = \frac{d}{dt}J^{ij}(e^{-tX})\Big|_{t=0}\mathcal{D}_i^r f_x(e)\mathcal{D}_j^r h_x(e), \qquad (2.167)$$

In (2.167) we also used that $J(e) = 0$ where e is the group identity. Furthermore, we have

$$\mathcal{D}_i^r f_x(e) = \frac{d}{dt} f(e^{te_i} \cdot x)\Big|_{t=0} \equiv -\xi_i f(x), \qquad (2.168)$$

where $\xi_i \equiv \xi_{e_i}$ is a vector field on $\mathcal{P}$ corresponding to the basis element e_i of $\mathfrak{g}$. Let $\delta : \mathfrak{g} \to \mathfrak{g} \wedge \mathfrak{g}$ be the standard cocycle associated with the Poisson-Lie tensor J, see (2.82). Obviously,

$$\frac{d}{dt}J^{ij}(e^{-tX})\Big|_{t=0} e_i \wedge e_j = -\delta(X). \qquad (2.169)$$

---

[15] See (1.10) for the definition of a Poisson map.

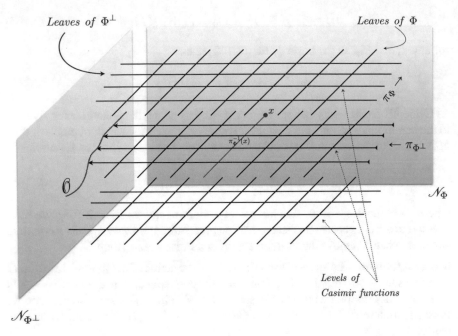

Fig. 2.5 Construction of a symplectic leaf of $\mathcal{O} \in \mathcal{N}_{\Phi^\perp}$. Surfaces strained on intersecting orbits of polar foliations correspond to different levels of Casimir functions

Introduce an opposite cocycle $\delta^{\mathrm{op}}(X) = -\delta(X)$ and supply the space $\mathfrak{g}^*$ with the structure of the Lie algebra corresponding to δ^{op}, that is, define the Lie bracket $[\ ,\ ]_*$ on $\mathfrak{g}^*$ as

$$\langle [\ell, n]_*, X \rangle = \langle \ell \otimes n, \delta^{\mathrm{op}}(X) \rangle = -\langle \ell \otimes n, \delta(X) \rangle. \qquad (2.170)$$

Let G^* be a connected and simply connected Lie group with the tangent Lie algebra $\mathfrak{g}^*$. Defining G^* with the help of the cocycle opposite to the standard one will be motivated in a moment.

In a local coordinate system $\{x^\alpha\}$ on $\mathscr{P}$ this vector field has the form $\xi_i = \xi_i^\alpha \partial_\alpha$. As follows from (2.166), (2.167) and (2.170), the infinitesimal form of the Poisson action can be represented in the form

$$\xi_X\{f, h\} - \{\xi_X f, h\} - \{f, \xi_X h\} = \langle X, [\xi^\alpha, \xi^\beta]_* \rangle \, \partial_\alpha f \partial_\beta h, \qquad (2.171)$$

where $\xi^\alpha = e^k \xi_k^\alpha \in \mathfrak{g}^*$. It is now clear that the hamiltonian action is a special case of a more general Poisson action. For the hamiltonian action the Poisson bracket on G is chosen to vanish, so that $\mathfrak{g}^*$ is abelian. In this case Eq. (2.171) reduces to the invariance condition for the Poisson bracket under the action of a hamiltonian vector field

$$\xi_X\{f,h\} = \{\xi_X f, h\} + \{f, \xi_X h\}. \tag{2.172}$$

Indeed, in the hamiltonian case $\xi_X = \{f_X, \cdot\}$ and Eq. (2.172) holds due to the Jacobi identity.

Now we aim at rewriting Eq. (2.171) in a more suggestive manner that will allow us to define a natural Poisson analogue of the hamiltonian moment map. Let $\Psi^{\alpha\beta}$ be a Poisson tensor on $\mathscr{P}$, that is

$$\{f,g\}_{\mathscr{P}} = \Psi^{\alpha\beta}\partial_\alpha f \partial_\beta h. \tag{2.173}$$

Taking into account that $\xi_X = \langle X, e^k\rangle\,\xi_k$, for (2.171) we get

$$\xi_X \Psi^{\alpha\beta} - \Psi^{\gamma\beta}\partial_\gamma\xi_X^\alpha - \Psi^{\alpha\gamma}\partial_\gamma\xi_X^\beta = \langle X, [\xi^\alpha,\xi^\beta]_*\rangle.$$

We are primarily interested in the situation when $\mathscr{P}$ is symplectic. Let $\omega_{\alpha\beta}$ be the inverse of $\Psi^{\alpha\beta}$, that is $\Psi^{\alpha\gamma}\omega_{\gamma\beta} = \delta^\alpha_\beta$. Multiplying the previous relation with $\omega_{\mu\alpha}\omega_{\beta\nu}$ and summing over α and β, we arrive at

$$\langle X, e^k\rangle\left(\xi_k^\alpha\partial_\alpha\omega_{\mu\nu} + \omega_{\mu\alpha}\partial_\nu\xi_k^\alpha + \omega_{\alpha\nu}\partial_\mu\xi_k^\alpha\right) = \langle X, [\omega_{\alpha\mu}\xi^\alpha, \omega_{\beta\nu}\xi^\beta]_*\rangle. \tag{2.174}$$

Taking into account that ω is closed, that is

$$\partial_\alpha\omega_{\mu\nu} + \partial_\nu\omega_{\alpha\mu} + \partial_\mu\omega_{\nu\alpha} = 0, \tag{2.175}$$

we exclude $\partial_\alpha\omega_{\mu\nu}$ from (2.174) and get

$$\langle X, e^k\rangle\left(\partial_\mu(\omega_{\alpha\nu}\xi_k^\alpha) - \partial_\nu(\omega_{\alpha\mu}\xi_k^\alpha)\right) = \langle X, [\omega_{\alpha\mu}\xi^\alpha, \omega_{\beta\nu}\xi^\beta]_*\rangle. \tag{2.176}$$

Introduce a connection one-form $\Omega = \Omega_\mu dx^\mu$ on $\mathscr{P}$ with coefficients

$$\Omega_\mu(x) = \omega_{\nu\mu}(x)\xi^\nu(x) = -\omega_{\mu\nu}e^k\xi_k^\nu(x) \in \mathfrak{g}^*. \tag{2.177}$$

From (2.176) one gets

$$\partial_\mu\Omega_\nu - \partial_\nu\Omega_\mu - [\Omega_\mu, \Omega_\nu]_* = 0. \tag{2.178}$$

In the specific case of an abelian $\mathfrak{g}^*$, the commutator term in (2.178) vanishes. Assuming that $\mathscr{P}$ is simply connected, it follows from the Poincaré lemma that there exists a function f such that $\xi^\beta\omega_{\beta\alpha} = \partial_\alpha f$. This expression is nothing else but the definition (1.13) of the hamiltonian vector field of f.

In the general case, Eq. (2.178) means that the connection Ω with values in $\mathfrak{g}^*$ has vanishing curvature and for this reason can locally be represented in the form

$$\Omega_\mu = \partial_\mu m\, m^{-1}. \tag{2.179}$$

Since $\mathcal{M}(x) \in G^*$, this construction defines a local map

$$\mathcal{M} : \mathcal{P} \to G^*, \tag{2.180}$$

which is called *the moment map of the Poisson action* of G on $\mathcal{P}$, or, a *non-abelian moment map* [21]. From (2.179) one gets

$$\Psi^{\alpha\beta}\partial_\alpha\mathcal{M}\,\mathcal{M}^{-1} = -e^k\Psi^{\alpha\beta}\omega_{\alpha\gamma}\xi_k^\gamma = e^k\xi_k^\beta. \tag{2.181}$$

Multiplying both sides of this relation with $\partial_\alpha f$ and recalling the definition of the Poisson bracket on $\mathcal{P}$, one obtains

$$\{\mathcal{M}, f\}_{\mathcal{P}}\mathcal{M}^{-1} = e^k\xi_k^\alpha\partial_\alpha f. \tag{2.182}$$

Thus, $\mathcal{M}$ is a generator of the group action in the Poisson algebra

$$\xi_X f = \langle X, \{\mathcal{M}, f\}_{\mathcal{P}}\mathcal{M}^{-1}\rangle. \tag{2.183}$$

It is worth emphasising that the product of $\{\mathcal{M}, f\}_{\mathcal{P}}$ and $\mathcal{M}^{-1}$ on the right hand side of the last formula is understood in the sense of the composition law in G^*, while $\langle\,,\,\rangle$ is a pairing between the dual spaces $\mathfrak{g}$ and $\mathfrak{g}^*$.

Denote by $\mathcal{F}^G \subset \mathcal{F}$ functions on $\mathcal{P}$ invariant under the G-action. Then, it follows from (2.183) that for $f \in \mathcal{F}^G$, the Poisson bracket of f with $\mathcal{M}$ vanishes, i.e.,

$$\{\mathcal{F}^G, \mathcal{M}\} = 0. \tag{2.184}$$

It follows from (2.171) that $\mathcal{F}^G$ is a Poisson subalgebra. Formula (2.184) then suggests an existence of the dual pair

$$G\backslash\mathcal{P} \xleftarrow{\;\pi\;} \mathcal{P} \xrightarrow{\;\mathcal{M}\;} G^*, \tag{2.185}$$

generalising the dual pair (2.165) corresponding to the case of hamiltonian action. The projection π on the coset space $G\backslash\mathcal{P}$ is a Poisson map provided we naturally identify the space of functions on this coset with $\mathcal{F}^G$. To declare that (2.185) is a dual pair indeed, we need to equip G^* with the Poisson bracket $\{\,,\,\}_{G^*}$ such that the momentum map (2.180) is Poisson,[16]

$$\{\mathcal{M}^* f, \mathcal{M}^* h\}_{\mathcal{P}}(x) = \{f, h\}_{G^*}(\mathcal{M}(x)). \tag{2.186}$$

From (2.181) and (2.182) for the Poisson bracket between the entries of the moment map we get

---

[16] Relaxing this condition leads to a more general Poisson structure on G^* that includes an extension by central terms [22].

$$\{m_1, m_2\}_{\mathscr{P}}(x) = -\xi_i^{\alpha}\xi_j^{\beta}\omega_{\alpha\beta}\, e^i m(x) \otimes e^j m(x). \qquad (2.187)$$

On the other hand, considering the entries of m as generators for the coordinate ring of $\mathscr{F}(G^*)$, without loss of generality, we can assume that the Poisson bracket on G^* is

$$\{m_1, m_2\}_{G^*} = \Phi_{ij}(m)\, e^i m \otimes e^j m, \qquad (2.188)$$

which corresponds to representing its right hand side in the basis of left differentials $\mathscr{D}^{\ell}$, conform (2.84). The requirement for m to be a Poisson map then reduces to

$$\Phi_{ij}\big(m(x)\big) = -\xi_i^{\alpha}\xi_j^{\beta}\omega_{\alpha\beta}(x),$$

or, in other words,[17]

$$\{m_1, m_2\}_{\mathscr{P}}(x) = \Phi_{ij}\big(m(x)\big)\, e^i m(x) \otimes e^j m(x). \qquad (2.189)$$

Obviously, tensor Φ cannot be arbitrary as, for instance, it should intertwine the action of G on $\mathscr{P}$ with that on G^*

$$\xi_i m(x) = \Phi_{ij}(m)e^j m. \qquad (2.190)$$

The simplest way [23] to reveal the conditions on Φ is to consider

$$[\xi_X, \xi_Y]f = \langle X \otimes Y, \big(\{\{m_1, m_2\}, f\}m_1^{-1}m_2^{-1}$$
$$- \{m_1, f\}m_1^{-1}\{m_1, m_2\}m_1^{-1}m_2^{-1} - \{m_2, f\}m_2^{-1}\{m_1, m_2\}m_1^{-1}m_2^{-1}\big)\rangle,$$

where the Jacobi identity for the bracket on $\mathscr{P}$ was used. With the help of (2.189) one finds

$$\{\{m_1, m_2\}, f\}m_1^{-1}m_2^{-1} = \frac{\partial \Phi_{ij}}{\partial m_{kl}}\{m_{kl}, f\}e^i \otimes e^j$$
$$+ \Phi_{ij}\, e^i\{m, f\}m^{-1} \otimes e^j + \Phi_{ij}e^i \otimes e^j\{m, f\}m^{-1}.$$

Introducing the concise notation $\ell = \{m, f\}m^{-1} \in \mathfrak{g}^*$, we therefore get

$$[\xi_X, \xi_Y]f = \langle X \otimes Y, \frac{\partial \Phi_{ij}}{\partial m_{kl}}(\ell m)_{kl}\, e^i \otimes e^j - \Phi_{ij}\big([\ell, e^i]_* \otimes e^j + e^i \otimes [\ell, e^j]_*\big)\rangle. \qquad (2.191)$$

Let us now require the map $X \to \xi_X$ given by

$$\langle X, \{m, \cdot\}_{\mathscr{P}}m^{-1}\rangle = \xi_X$$

---

[17]In the case of a hamiltonian group action the analogue of (2.189) is (2.22) or (2.18).

to be a Lie algebra homomorphism $\mathfrak{g} \to \mathfrak{X}(\mathcal{P})$. This means the fulfilment of the relation

$$[\xi_X, \xi_Y]f = \xi_{[X,Y]}f = \langle [X, Y], \ell \rangle = \langle X \otimes Y, \delta_*(\ell) \rangle, \qquad (2.192)$$

where we have introduced the notation

$$\delta_*(\ell) = \ell([e_i, e_j]) e^i \wedge e^j \qquad (2.193)$$

and regard δ_* as a map $\delta_* : \mathfrak{g}^* \to \mathfrak{g}^* \wedge \mathfrak{g}^*$. Introducing $\Phi = \Phi_{ij}e^i \wedge e^j$ and comparing (2.191) with (2.192), we conclude that

$$\frac{d}{dt}\Phi(e^{t\ell}m)|_{t=0} = \delta_*(\ell) + [\ell \otimes \mathbb{1} + \mathbb{1} \otimes \ell, \Phi(m)]_* \qquad (2.194)$$

for any $\ell \in \mathfrak{g}^*$. This is a condition for the Poisson tensor Φ we are looking for.

Assume that Φ is such that it satisfies the Poisson-Lie property (2.85)

$$\Phi(m_1 m_2) = \Phi(m_1) + \mathrm{Ad}_{m_1} \otimes \mathrm{Ad}_{m_1} \Phi(m_2) \qquad (2.195)$$

and gives rise to a Lie algebra cocycle δ_* according to the standard definition

$$\delta_*(\ell) = \frac{d}{dt}\Phi(e^{t\ell})|_{t=0} = \ell([e_i, e_j]) e^i \wedge e^j, \qquad (2.196)$$

Specifying in (2.195) the group elements as $m_1 = e^{t\ell}$, $m_2 = m$, differentiating over t and further putting $t = 0$, we recover (2.194). In other words, (2.195) can be viewed as an integrated version of (2.194).

Summarising, Eq. (2.195) defines on G^* a Poisson-Lie structure with the cocycle δ_* that corresponds to the standard Lie bracket on $\mathfrak{g}$. Thus, G^* is the Poisson-Lie dual of G. Note that the tangent Lie algebra of G^* was defined with the help of δ^{op} in order for the map (2.183) to be the Lie algebra homomorphism. Of course, one can also use the standard cocycle δ to define G^*. In the latter case, the map (2.183) will be the anti-homomorphism and the Poisson-Lie structure on G^* will be determined by the opposite cocycle $-\delta_*^{\mathrm{op}}$. Finally, in the case of abelian G^*, it is convenient to parametrise $m = \exp \mu$. Then formula (2.183) reduces to the standard hamiltonian action (2.19) generated by the moment map $\mu \in \mathfrak{g}^*$.

Poisson-Lie group action on a product manifold. Let $\mathcal{P}_1$ and $\mathcal{P}_2$ be two Poisson manifolds with brackets $\{\cdot, \cdot\}_{\mathcal{P}_1}$ and $\{\cdot, \cdot\}_{\mathcal{P}_2}$ which carry the Poisson action of a Poisson-Lie group G. Let $m_i : \mathcal{P}_i \to G^*$ be the corresponding non-abelian moment maps which we assume to be Poisson. In this situation one can define the Poisson action of G on the product manifold $\mathcal{P} = \mathcal{P}_1 \times \mathcal{P}_2$ [24]. This is done by taking the product[18]

---

[18]The product is naturally taken in G^*.

$$m = m_1 m_2,$$

and allowing it to act on $\mathcal{F}(\mathcal{P})$ by means of the formula

$$\xi_X f = \langle X, \{m, f\}_{\mathcal{P}} m^{-1} \rangle, \quad f \in \mathcal{F}(\mathcal{P}), \quad (2.197)$$

where ξ_X is a vector field corresponding to $X \in \mathfrak{g}$ and $\langle \cdot, \cdot \rangle$ is the canonical pairing between $\mathfrak{g}$ and $\mathfrak{g}^*$. We have

$$\xi_X f = \langle X, \{m_1, f\}_{\mathcal{P}_1} m_1^{-1} + m_1 \{m_2, f\}_{\mathcal{P}_2} m_2^{-1} m_1^{-1} \rangle. \quad (2.198)$$

Let $\xi_X^{(1)}$ and $\xi_X^{(2)}$ be the fundamental vector fields induced by the group action on $\mathcal{P}_1$ and P_2, respectively. Formula (2.198) is equivalent to the statement that at a point $x = (x_1, x_2) \in \mathcal{P}$, where $x_1 \in \mathcal{P}_1$ and $x_1 \in \mathcal{P}_2$, the vector field ξ_X is defined as

$$\xi_X(x) = \xi_X^{(1)}(x_1) + \xi_{\mathrm{Ad}^*_{m_1^{-1}(x_1)} X}^{(2)}(x_2), \quad (2.199)$$

where Ad^*_h, $h \in G^*$ is the coadjoint action of G^* on G.

Our goal is to show that the map $X \to \xi_X$, where ξ_X is defined by (2.199), is the Lie algebra homomorphism, so that ξ_X is the fundamental vector field of the group action on G. To this end, consider

$$[\xi_X, \xi_Y] = [\xi_X^{(1)} + \xi_{\mathrm{Ad}^*_{m_1^{-1}} X}^{(2)}, \xi_Y^{(1)} + \xi_{\mathrm{Ad}^*_{m_1^{-1}} Y}^{(2)}] = \xi_{[X,Y]}^{(1)} + \xi_Z^{(2)},$$

where

$$Z = \xi_X^{(1)} \mathrm{Ad}^*_{m_1^{-1}} Y - \xi_Y^{(1)} \mathrm{Ad}^*_{m_1^{-1}} X + [\mathrm{Ad}^*_{m_1^{-1}} X, \mathrm{Ad}^*_{m_1^{-1}} Y],$$

and we have used that the maps $X \to \xi_X^{(1,2)}$ are the Lie algebra homomorphisms. For any $n \in \mathfrak{g}^*$ one has[19]

$$\langle \xi_X^{(1)} \mathrm{Ad}^*_{m_1^{-1}} Y, n \rangle = \langle Y, [\xi_X^{(1)} m_1 m_1^{-1}, m_1 n m_1^{-1}]_* \rangle = -\langle \mathrm{Ad}^*_{m_1^{-1}} (\mathrm{ad}^*_{\xi_X^{(1)} m_1 m_1^{-1}} Y), n \rangle,$$

so that

$$\xi_X^{(1)} \mathrm{Ad}^*_{m_1^{-1}} Y = -\mathrm{Ad}^*_{m_1^{-1}} (\mathrm{ad}^*_{\xi_X^{(1)} m_1 m_1^{-1}} Y).$$

Hence, for Z we find

$$Z = [\mathrm{Ad}^*_{m_1^{-1}} X, \mathrm{Ad}^*_{m_1^{-1}} Y] + \mathrm{Ad}^*_{m_1^{-1}} (\mathrm{ad}^*_{\xi_Y^{(1)} m_1 m_1^{-1}} X) - \mathrm{Ad}^*_{m_1^{-1}} (\mathrm{ad}^*_{\xi_X^{(1)} m_1 m_1^{-1}} Y). \quad (2.200)$$

---

[19]Sometimes we write the action of $\mathrm{Ad}_m : \mathfrak{g}^* \to \mathfrak{g}^*$ in the matrix form as $n \to m n m^{-1}$, $n \in \mathfrak{g}^*$.

Below we will show that the last expression is nothing else but

$$Z = \text{Ad}^*_{m_1^{-1}}[X, Y],$$

so that

$$[\xi_X, \xi_Y] = = \xi^{(1)}_{[X,Y]} + \xi^{(2)}_{\text{Ad}^*_{m_1^{-1}}[X,Y]} = \xi_{[X,Y]}$$

and our statement is proved.

Expression (2.200) is of special interest. Since all the quantities involved belong to one and the same manifold, we can define Z for any Poisson action of G with the moment map m being the mapping of Poisson manifolds

$$Z = [\text{Ad}^*_{m^{-1}} X, \text{Ad}^*_{m^{-1}} Y] + \text{Ad}^*_{m^{-1}}(\text{ad}^*_{\xi_Y} mm^{-1} X) - \text{Ad}^*_{m^{-1}}(\text{ad}^*_{\xi_X} mm^{-1} Y). \quad (2.201)$$

Consider a one-parametric subgroup passing through the group identity $m = e^{-t\ell}$, where ℓ is an arbitrary element of $\mathfrak{g}^*$. This defines a curve $Z(t) \in \mathfrak{g}$. We first construct a differential equation satisfied by $Z(t)$ and then find its unique solution corresponding to the initial condition at $t = 0$.

We start with

$$\frac{d}{dt}\langle \text{Ad}^*_{m^{-1}} X, n \rangle = -\langle \text{Ad}^*_{m^{-1}} X, [\ell, n]_* \rangle = \langle \text{ad}^*_\ell(\text{Ad}^*_{m^{-1}} X), n \rangle$$

for arbitrary $n \in \mathfrak{g}^*$, so that

$$\frac{d}{dt}\text{Ad}^*_{m^{-1}} X = \text{ad}^*_\ell(\text{Ad}^*_{m^{-1}} X). \quad (2.202)$$

Next, from (2.190) one gets

$$\xi_X mm^{-1} = \langle X, \Phi(m) \rangle \in \mathfrak{g}^*, \quad (2.203)$$

where $\Phi \in \mathfrak{g}^* \wedge \mathfrak{g}^*$ is the Poisson tensor of the Poisson-Lie group G^*. This expression vanishes at $t = 0$ due to vanishing of Φ at the group identity. To proceed, we need to know

$$\frac{d}{dt}\Phi(e^{-it\ell}) = \lim_{\Delta t \to 0} \frac{\Phi(e^{-t\ell}e^{-\Delta t\ell}) - \Phi(e^{-t\ell})}{\Delta t} = (\text{Ad}_m \otimes \text{Ad}_m)\frac{d}{dt}\Phi^{-i\ell}\Big|_{t=0}$$
$$= -\langle \ell, [e_i, e_j] \rangle \, me^i m^{-1} \wedge me^j m^{-1},$$

where we have used the Poisson-Lie property (2.195) and (2.196). Hence,

$$\frac{d}{dt}(\xi_X m m^{-1}) = -\langle \ell, [\mathrm{Ad}^*_{m^{-1}} X, e_j] \rangle \, m e^j m^{-1} =$$

$$\langle \mathrm{ad}^*_{\mathrm{Ad}^*_{m^{-1}} X} \ell, e_j \rangle \, m e^j m^{-1} = \mathrm{Ad}_m(\mathrm{ad}^*_{\mathrm{Ad}^*_{m^{-1}} X} \ell) \,. \tag{2.204}$$

Introducing the concise notation

$$X' = \mathrm{Ad}^*_{m^{-1}} X \,, \quad Y' = \mathrm{Ad}^*_{m^{-1}} Y \,,$$

we compute

$$\begin{aligned}
\frac{dZ(t)}{dt} &= [\mathrm{ad}^*_\ell X', Y'] + [X', \mathrm{ad}^*_\ell Y'] + \\
&\quad + \mathrm{Ad}^*_{m^{-1}}(\mathrm{ad}^*_{\mathrm{Ad}_m(\mathrm{ad}^*_{Y'}\ell)} X) - \mathrm{Ad}^*_{m^{-1}}(\mathrm{ad}^*_{\mathrm{Ad}_m(\mathrm{ad}^*_{X'}\ell)} Y) \\
&\quad + \mathrm{ad}_\ell\big(\mathrm{Ad}^*_{m^{-1}}(\mathrm{ad}^*_{\xi_Y m m^{-1}} X) - \mathrm{Ad}^*_{m^{-1}}(\mathrm{ad}^*_{\xi_X m m^{-1}} Y)\big) \,,
\end{aligned}$$

where in the first and the last line we used (2.202) and the middle line was obtained with the help of (2.204). Taking into account that

$$\mathrm{Ad}^*_{m^{-1}}(\mathrm{ad}^*_{\mathrm{Ad}_m(\mathrm{ad}^*_{X'}\ell)} Y) = \mathrm{ad}^*_{\mathrm{ad}^*_{X'}\ell} Y' \,,$$

we obtain

$$\begin{aligned}
\frac{dZ(t)}{dt} &= [\mathrm{ad}^*_\ell X', Y'] + [X', \mathrm{ad}^*_\ell Y'] + \mathrm{ad}^*_{\mathrm{ad}^*_{Y'}\ell} X' - \mathrm{ad}^*_{\mathrm{ad}^*_{X'}\ell} Y' \\
&\quad + \mathrm{ad}_\ell\big(\mathrm{Ad}^*_{m^{-1}}(\mathrm{ad}^*_{\xi_Y m m^{-1}} X) - \mathrm{Ad}^*_{m^{-1}}(\mathrm{ad}^*_{\xi_X m m^{-1}} Y)\big) \,.
\end{aligned}$$

It remains to note that, according to (2.95), one has

$$\mathrm{ad}^*_\ell [X', Y'] = [\mathrm{ad}^*_\ell X', Y'] + [X', \mathrm{ad}^*_\ell Y'] + \mathrm{ad}^*_{\mathrm{ad}^*_{Y'}\ell} X' - \mathrm{ad}^*_{\mathrm{ad}^*_{X'}\ell} Y' \,, \quad \forall X', Y' \in \mathfrak{g} \,.$$

Thus, the differential equation for $Z(t)$ takes the form

$$\frac{dZ(t)}{dt} = \mathrm{ad}^*_\ell Z(t) \,. \tag{2.205}$$

This equation should be supplied with the initial condition at $t = 0$

$$Z(0) = [X, Y] \,. \tag{2.206}$$

Due to (2.202), a unique solution of (2.205), (2.206) for any t is

$$Z(t) = \mathrm{Ad}^*_{m^{-1}}[X, Y] \,. \tag{2.207}$$

This result implies the fulfilment of the following important relation

$$\xi_X \mathrm{Ad}^*_{m^{-1}} Y - \xi_Y \mathrm{Ad}^*_{m^{-1}} X + [\mathrm{Ad}^*_{m^{-1}} X, \mathrm{Ad}^*_{m^{-1}} Y] = \mathrm{Ad}^*_{m^{-1}}[X, Y], \quad (2.208)$$

where we recall that ξ_X is a fundamental vector field of the Poisson group action related to the corresponding moment map m by means of (2.183).

2.1.7 Heisenberg Double

Let G be a Lie group corresponding to a factorisable Lie algebra $\mathfrak{g}$. The double $D = D(G)$ of G is identified with $G \times G$. In the factorisable setting the Poisson-Lie structure (2.126) is realised with the help of the r-matrix $\mathscr{R}_{\mathbb{R}}$, the corresponding Poisson bracket is

$$\{f, h\}_- = \tfrac{1}{2} \langle\!\langle \mathcal{D}^l f, \widehat{\mathscr{R}_{\mathbb{R}}}(\mathcal{D}^l h) \rangle\!\rangle - \tfrac{1}{2} \langle\!\langle \mathcal{D}^r f, \widehat{\mathscr{R}_{\mathbb{R}}}(\mathcal{D}^r h) \rangle\!\rangle, \quad (2.209)$$

where $f, h \in \mathcal{F}(D)$. As was observed in [25], one can define on D another Poisson structure, namely,

$$\{f, h\}_+ = \tfrac{1}{2} \langle\!\langle \mathcal{D}^l f, \widehat{\mathscr{R}_{\mathbb{R}}}(\mathcal{D}^l h) \rangle\!\rangle + \tfrac{1}{2} \langle\!\langle \mathcal{D}^r f, \widehat{\mathscr{R}_{\mathbb{R}}}(\mathcal{D}^r h) \rangle\!\rangle. \quad (2.210)$$

The Lie group D equipped with the bracket $\{\cdot, \cdot\}_+$ is called the *Heisenberg double* of G. The Heisenberg double is a symplectic manifold[20] that can be regarded as a deformation of the cotangent bundle T^*G. As we will show, the Heisenberg double admits a variety of symmetry transformations which makes it an interesting phase space for obtaining non-trivial integrable models by means of a suitable reduction procedure. In the following we denote the double of G supplied with the Poisson-Lie structure (2.209) as D_- and the Heisenberg double as D_+.

Our immediate goal is to understand the structure of $D_\pm$ in more detail. To this end, we parametrise an element of D by a pair (x, y), where $x, y \in G$. From the general expressions (2.209) and (2.210) we can deduce the Poisson brackets between the coordinate functions corresponding to x and y. A simple calculation reveals the following structure

$$\begin{aligned}
\{x_1, x_2\}_\pm &= -\left(\tfrac{1}{2} r x_1 x_2 \pm x_1 x_2 \tfrac{1}{2} r\right), \\
\{y_1, y_2\}_\pm &= -\left(\tfrac{1}{2} r y_1 y_2 \pm y_1 y_2 \tfrac{1}{2} r\right), \\
\{x_1, y_2\}_\pm &= -\left(r_+ x_1 y_2 \pm x_1 y_2 r_+\right), \\
\{y_1, x_2\}_\pm &= -\left(r_- y_1 x_2 \pm y_1 x_2 r_-\right),
\end{aligned} \quad (2.211)$$

---

[20]One can consider the more general situation where the double is defined as connected but not simply connected Lie group. In this case it admits a stratification, each stratification cell being a symplectic manifold [26].

where the subscript indices label the corresponding matrix spaces and the r-matrices are given by (2.151). Due to the permutation property of the split Casimir C_{12}, the first two lines for the "+" bracket can be equivalently written as

$$\{x_1, x_2\}_+ = -\left(r_+x_1x_2 + x_1x_2r_-\right) = -\left(r_-x_1x_2 + x_1x_2r_+\right),$$
$$\{y_1, y_2\}_+ = -\left(r_+y_1y_2 + y_1y_2r_-\right) = -\left(r_-y_1y_2 + y_1y_2r_+\right).$$

Almost every element of D can be factorised[21] as

$$(x, y) = (\mathcal{L}_+, \mathcal{L}_-)(g^{-1}, g^{-1}) = (\mathcal{L}_+g^{-1}, \mathcal{L}_-g^{-1}). \tag{2.212}$$

Here $(\mathcal{L}_+, \mathcal{L}_-)$ is a representative of an element from G^* corresponding to the embedding $G^* \hookrightarrow G \times G$, which at the infinitesimal level is given by the map (2.139). Similarly, (g, g) is the image of $g \in G$ under the diagonal embedding $G \hookrightarrow G \times G$. The matrix elements of $\mathcal{L}_\pm$ and g give a new system of generators in $\mathcal{F}(D)$. They are rational functions of x and y with singularities at those points where factorisation (2.212) fails. In Appendix C we explain how to derive the Poisson brackets between these generators and quote below the corresponding result. First, we have

$$\{\mathcal{L}_{+1}, \mathcal{L}_{+2}\} = -\tfrac{1}{2}[r, \mathcal{L}_{+1}\mathcal{L}_{+2}], \qquad \{\mathcal{L}_{+1}, \mathcal{L}_{-2}\} = -[r_+, \mathcal{L}_{+1}\mathcal{L}_{-2}],$$
$$\{\mathcal{L}_{-1}, \mathcal{L}_{-2}\} = -\tfrac{1}{2}[r, \mathcal{L}_{-1}\mathcal{L}_{-2}], \qquad \{\mathcal{L}_{-1}, \mathcal{L}_{+2}\} = -[r_-, \mathcal{L}_{-1}\mathcal{L}_{+2}]. \tag{2.213}$$

These brackets are the same for both Poisson structures (2.209) and (2.210). Next,

$$\{\mathcal{L}_{+1}, g_2\}_+ = \mathcal{L}_{+1}g_2r_+, \qquad \{\mathcal{L}_{+1}, g_2\}_- = 0,$$
$$\{\mathcal{L}_{-1}, g_2\}_+ = \mathcal{L}_{-1}g_2r_-, \qquad \{\mathcal{L}_{-1}, g_2\}_- = 0, \tag{2.214}$$

and, finally,

$$\{g_1, g_2\}_\pm = \mp\tfrac{1}{2}[r, g_1g_2].$$

We immediately observe that since $\{\mathcal{L}_\pm, g\}_- = 0$, the Poisson-Lie structure $\{\ ,\ \}_-$ renders D_- the direct product of Poisson manifolds.

One can also factorise an element $(x, y) \in D$ in the opposite order, that is,

$$(x, y) = (g', g')(\mathcal{L}_+'^{-1}, \mathcal{L}_-'^{-1}) = (g'\mathcal{L}_+'^{-1}, g'\mathcal{L}_-'^{-1}). \tag{2.215}$$

Evaluation of the Poisson brackets for the system of generators $(g', \mathcal{L}_\pm')$ yields

---

[21]For our treatment D is assumed to be connected and simply connected. We multiply $(\mathcal{L}_+, \mathcal{L}_-)$ by the inverse $(g, g)^{-1}$ so that to have the standard definition of the right action of G on a manifold which in the present case is D, compare, for instance, with formulae (2.59).

$$\{\mathcal{L}'_{+1}, \mathcal{L}'_{+2}\}_{\pm} = \mp\tfrac{1}{2}[r, \mathcal{L}'_{+1}\mathcal{L}'_{+2}], \qquad \{\mathcal{L}'_{+1}, \mathcal{L}'_{-2}\}_{\pm} = \mp[r_+, \mathcal{L}'_{+1}\mathcal{L}'_{-2}],$$
$$\{\mathcal{L}'_{-1}, \mathcal{L}'_{-2}\}_{\pm} = \mp\tfrac{1}{2}[r, \mathcal{L}'_{-1}\mathcal{L}'_{-2}], \qquad \{\mathcal{L}'_{-1}, \mathcal{L}'_{+2}\}_{\pm} = \mp[r_-, \mathcal{L}'_{-1}\mathcal{L}'_{+2}], \tag{2.216}$$

together with

$$\{\mathcal{L}'_{+1}, g'_2\}_+ = \quad \mathcal{L}'_{+1}g'_2 r_+, \quad \{\mathcal{L}'_{+1}, g'_2\}_- = 0,$$
$$\{\mathcal{L}'_{-1}, g'_2\}_+ = \quad \mathcal{L}'_{-1}g_2 r_-, \quad \{\mathcal{L}'_{-1}, g'_2\}_- = 0, \tag{2.217}$$

and

$$\{g'_1, g'_2\}_{\pm} = -\tfrac{1}{2}[r, g'_1 g'_2]. \tag{2.218}$$

Further analysis shows that for the Heisenberg double the following trivial cross relations are trivial

$$\{\mathcal{L}_{\pm1}, \mathcal{L}'_{\pm2}\}_+ = 0, \quad \{g_1, g'_2\}_+ = 0. \tag{2.219}$$

The remaining cross relations for D_+ are found as follows. First, using the decompositions $x = g'\mathcal{L}'^{-1}_+$, $y = g'\mathcal{L}'^{-1}_-$, we compute

$$\{\mathcal{L}'_{\pm1}, x_2\}_+ = x_2 r_{\pm} \mathcal{L}'_{\pm1}, \quad \{\mathcal{L}'_{\pm1}, y_2\}_+ = y_2 r_{\pm} \mathcal{L}'_{\pm1}.$$

Substituting here the opposite decomposition $x = \mathcal{L}_+ g^{-1}$, $y = \mathcal{L}_- g^{-1}$ and using (2.219), we obtain

$$\{\mathcal{L}'_{\pm1}, g_2\}_+ = -r_{\pm}\mathcal{L}'_{\pm1}g_2. \tag{2.220}$$

Analogously, the usage of

$$\{\mathcal{L}_{\pm1}, x_2\}_+ = -r_{\pm}\mathcal{L}_{\pm1}x_2, \quad \{\mathcal{L}_{\pm1}, y_2\}_+ = -r_{\pm}\mathcal{L}_{\pm1}y_2,$$

yields

$$\{\mathcal{L}_{\pm1}, g'_2\}_+ = -r_{\pm}\mathcal{L}_{\pm1}g'_2. \tag{2.221}$$

Poisson-Lie subgroups. From now on we focus on the Poisson structure $\{\cdot, \cdot\}_+$ corresponding to the Heisenberg double and omit the subscript "+" in the notation of the Poisson brackets. As it should be clear from the previous discussion, the subgroup G featuring in both decompositions (2.212) and (2.215) is a Poisson-Lie group with the Sklyanin bracket

$$\{h_1, h_2\} = -\tfrac{1}{2}[r, h_1 h_2] = -[r_{\pm}, h_1 h_2], \tag{2.222}$$

the latter is inherited from the Poisson structure of the double. In the present context we adopt the unifying notation h to denote either g or g'. Clearly, the projections $\pi(\pi') : D \to G$ defined as

$$\pi(x, y) = g, \quad \pi'(x, y) = g'$$

are Poisson mappings.

The group $G^* \simeq G_+ \times G_- \subset D$ that provides the complementary factor to G in the decompositions (2.212) and (2.215) is also a Poisson-Lie subgroup with the Poisson structure given by

$$\{u_{+1}, u_{+2}\} = -\tfrac{1}{2}[r, u_{+1}u_{+2}], \qquad \{u_{+1}, u_{-2}\} = -[r_+, u_{+1}u_{-2}], \tag{2.223}$$
$$\{u_{-1}, u_{-2}\} = -\tfrac{1}{2}[r, u_{-1}u_{-2}], \qquad \{u_{-1}, u_{+2}\} = -[r_-, u_{-1}u_{+2}].$$

Again, we adopt here the notation (u_+, u_-) which stands either for $(\mathcal{L}_+, \mathcal{L}_-)$ or $(\mathcal{L}'_+, \mathcal{L}'_-)$. Naturally, the projections $\pi_*(\pi'_*) : D \to G_+ \times G_-$ defined as

$$\pi_*(x, y) = (\mathcal{L}_+, \mathcal{L}_-), \quad \pi'_*(x, y) = (\mathcal{L}'_+, \mathcal{L}'_-)$$

are Poisson mappings.

Now we present an alternative and, perhaps, more practical way to think about the Poisson-Lie group G^*. The group G^* can be embedded into G by a map σ

$$\sigma(u_+, u_-) = u_+ u_-^{-1} = u. \tag{2.224}$$

Conversely, assuming σ is a global diffeomorphism $G^* \simeq G$, for a given $u \in G$ the element (u_+, u_-) is defined as the unique solution of the factorisation problem $u = u_+ u_-^{-1}$. In a similar fashion one can define $\sigma'(u_+, u_-) = u_+^{-1} u_-$.

Under the map (2.224) the product in G^* induces a new product in G which will be denoted as $\star$. This product is constructed as follows. First, we take two elements $v, u \in G$, factorise them as $v = v_+ v_-^{-1}$, $u = u_+ u_-^{-1}$, and then embed into D as (v_+, v_-) and (u_+, u_-), respectively. Second, we multiply these elements in D to get $(v_+, v_-)(u_+, u_-) = (v_+ u_+, v_- u_-)$. A further application of the map (2.224) gives a new element in G which we associate with the new product of v and u

$$v \star u = v_+ u_+ u_-^{-1} v_-^{-1} = v_+ u v_-^{-1}. \tag{2.225}$$

Note that the inverse element $I(v)$ with respect to this product is also defined through the lift in the double

$$I(v) \star v = v_+^{-1} v v_- = e. \tag{2.226}$$

Further, under the map (2.224) the Poisson structure (2.223) induces the following Poisson structure on G

$$\{u_1, u_2\} = -\tfrac{1}{2} r u_1 u_2 - \tfrac{1}{2} u_1 u_2 r + u_1 r_- u_2 + u_2 r_+ u_1 \,. \tag{2.227}$$

This structure is Poisson-Lie with respect to the product (2.225), that is

$$\{v_1 \star u_1, v_2 \star u_2\}_{G^* \times G} = \{u_1, u_2\}_G (v \star u)\,,$$

where on the right hand side the subscript G refers to the bracket (2.227), while $\{\cdot, \cdot\}_{G^* \times G}$ on the left hand side refers to the product Poisson structure, where the bracket on G^* is taken as in (2.223). In this way we model the Poisson-Lie group G^* as a group G with the Poisson brackets (2.227) and the composition law (2.225). In the context of Poisson-Lie theory (2.223) and (2.227) are known as the *Semenov-Tian-Shansky bracket* [25].

Dual pair. The Poisson-Lie structures induced on G and G^* by the factorisation problem in the double can naturally be understood in terms of Poisson reductions and dual pairs associated with the corresponding group actions on D.

We start with the group G and define its actions on D by means of left and right shifts realised on generators (x, y) as

$$\begin{aligned} G \times D \to D: & \quad (x, y) \to (hx, hy)\,, \\ D \times G \to D: & \quad (x, y) \to (xh^{-1}, yh^{-1})\,. \end{aligned} \tag{2.228}$$

It is readily verified that these actions are Poisson, provided (2.222) is taken as a Poisson bracket on G. Functions on D invariant under these actions are constructed by taking projections on the corresponding coset spaces, the latter are modelled by G. Namely,

$$\mathcal{L} \equiv \sigma \circ \pi_*(x, y) = x y^{-1}\,, \qquad \mathcal{R} \equiv \sigma' \circ \pi'_*(x, y) = x^{-1} y\,.$$

The function $\mathcal{L}$ is invariant under right shifts and, as we will show, is the moment map of the left Poisson G-action. Correspondingly, $\mathcal{R}$ is invariant under left shifts and coincides with the moment map of the right G-action. In the basis $(\mathcal{L}_\pm, g)$ the generators $\mathcal{L}$ and $\mathcal{R}$ read as

$$\mathcal{L} = \mathcal{L}_+ \mathcal{L}_-^{-1}\,, \qquad \mathcal{R} = g \mathcal{L}_+^{-1} \mathcal{L}_- g^{-1}\,, \tag{2.229}$$

while in the basis $(\mathcal{L}'_\pm, g')$ the same generators take the form

$$\mathcal{L} = g' \mathcal{L}_+'^{-1} \mathcal{L}'_- g'^{-1}\,, \qquad \mathcal{R} = \mathcal{L}'_+ \mathcal{L}_-'^{-1}\,. \tag{2.230}$$

To clarify the meaning of $\mathcal{L}$ and $\mathcal{R}$ as moment maps for the transformations (2.228), we start with the following formulae

$$\begin{aligned} \{\mathcal{L}_{\pm 1}, x_2\} = -r_\pm \mathcal{L}_{\pm 1} x_2\,, & \qquad \{\mathcal{L}'_{\pm 1}, x_2\} = x_2 r_\pm \mathcal{L}'_{\pm 1}\,, \\ \{\mathcal{L}_{\pm 1}, y_2\} = -r_\pm \mathcal{L}_{\pm 1} y_2\,, & \qquad \{\mathcal{L}'_{\pm 1}, y_2\} = y_2 r_\pm \mathcal{L}'_{\pm 1}\,. \end{aligned} \tag{2.231}$$

Rewriting these formulae as Poisson relations between the coordinate functions of the double, we get

$$\{(\mathcal{L}_{+1}, \mathcal{L}_{-1}), (x_2, y_2)\}(\mathcal{L}_{+1}, \mathcal{L}_{-1})^{-1} = (-r_+, -r_-)(x_2, y_2),$$
$$\{(\mathcal{L}'_{+1}, \mathcal{L}'_{-1}), (x_2, y_2)\}(\mathcal{L}'_{+1}, \mathcal{L}'_{-1})^{-1} = (x_2, y_2)(r_+, r_-).$$

Let us now take $X \in \mathfrak{g}$ and form the following scalar products

$$\langle\!\langle (X_1, X_1), \{(\mathcal{L}_{+1}, \mathcal{L}_{-1}), (x_2, y_2)\}(\mathcal{L}_{+1}, \mathcal{L}_{-1})^{-1}\rangle\!\rangle = \langle\!\langle (X_1, X_1), (-r_+, -r_-)\rangle\!\rangle (x_2, y_2),$$
$$\langle\!\langle (X_1, X_1), \{(\mathcal{L}'_{+1}, \mathcal{L}'_{-1}), (x_2, y_2)\}(\mathcal{L}'_{+1}, \mathcal{L}'_{-1})^{-1}\rangle\!\rangle = (x_2, y_2)\langle\!\langle (X_1, X_2), (r_+, r_-)\rangle\!\rangle.$$

Taking into account that

$$\langle r_{+12}, X_1\rangle = -\mathbf{r}_-(X), \quad \langle r_{-12}, X_2\rangle = -\mathbf{r}_+(X), \quad \mathbf{r}_+(X) - \mathbf{r}_-(X) = X, \quad (2.232)$$

we rewrite the formulae above as

$$\langle X, \{m, (x, y)\}m^{-1}\rangle = (-Xx, -Xy), \qquad m \equiv (\mathcal{L}_+, \mathcal{L}_-) \in G^*,$$
$$\langle X, \{m, (x, y)\}m^{-1}\rangle = (xX, yX), \qquad m \equiv (\mathcal{L}'_+, \mathcal{L}'_-) \in G^*. \qquad (2.233)$$

Here on the right hand side we recognise the action of the fundamental vector field[22] ξ_X, $X \in \mathfrak{g}$, corresponding to the left and right Poisson actions (2.228) of the Poisson-Lie group G on the manifold $\mathcal{P} = D$. Comparing (2.233) to the general definition (2.183) of the non-abelian moment map, we conclude that $(\mathcal{L}_+, \mathcal{L}_-)$ is the moment map for the left action (2.228), while $(\mathcal{L}'_+, \mathcal{L}'_-)$ has a similar interpretation for the right action. The variables $\mathcal{L}$ and $\mathcal{R}$ are obtained from these by applying the map σ and, therefore, they can also be interpreted as the corresponding moment maps for (2.228). The validity of such an interpretation can be also checked in a direct manner. Indeed, we have

$$\{\mathcal{L}_1, x_2\} = -(r_+\mathcal{L}_1 - \mathcal{L}_1 r_-)x_2, \qquad \{\mathcal{R}_1, x_2\} = x_2(r_+\mathcal{R}_1 - \mathcal{R}_1 r_-),$$
$$\{\mathcal{L}_1, y_2\} = -(r_+\mathcal{L}_1 - \mathcal{L}_1 r_-)y_2, \qquad \{\mathcal{R}_1, y_2\} = y_2(r_+\mathcal{R}_1 - \mathcal{R}_1 r_-), \qquad (2.234)$$

which gives, for example,

$$I(\mathcal{L}_1) \star \{\mathcal{L}_1, x_2\} = \mathcal{L}_+^{-1}\{\mathcal{L}_1, x_2\}\mathcal{L}_- = (\mathcal{L}_{-1}^{-1} r_- \mathcal{L}_{-1} - \mathcal{L}_{+1}^{-1} r_+ \mathcal{L}_{+1})x_2.$$

Here on the left hand side we applied the $\star$-product (2.225) and used formula (2.226) for the inverse of $\mathcal{L}$. Pairing the last expression with an arbitrary $Y \in \mathfrak{g}$ yields

$$\langle Y, I(\mathcal{L}) \star \{\mathcal{L}, x\}\rangle = \left(\mathbf{r}_-(\mathcal{L}_+ Y \mathcal{L}_+^{-1}) - \mathbf{r}_+(\mathcal{L}_- Y \mathcal{L}_-^{-1})\right)x = -(\mathrm{Ad}_{\mathcal{L}}^* Y)x,$$

---

[22]Pay attention to the minus sign in the definition (2.12) of the fundamental vector field.

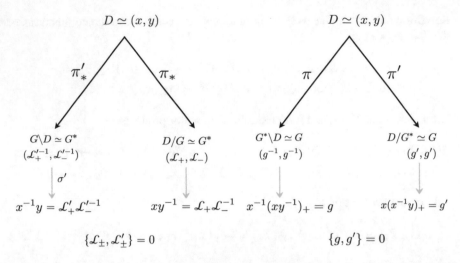

Fig. 2.6 Polar decompositions of the Heisenberg double. The dual pairs on the left figure corresponds to left and right actions of the Poisson-Lie group G. The dual pairs on the right figure arises due to left and right actions of the Poisson-Lie group G^*. In each case the respective invariants are in involution

where we recognised the coadjoint action (2.153) of an element $\mathcal{L} \equiv (\mathcal{L}_+, \mathcal{L}_-) \in G^*$ on $Y \in \mathfrak{g}$. Finally, choosing $Y = \mathrm{Ad}^*_{\mathcal{L}^{-1}} X$, we obtain

$$\langle X, \{\mathcal{L}, x\} \star I(\mathcal{L}) \rangle = \langle \mathrm{Ad}^*_{\mathcal{L}^{-1}} X, I(\mathcal{L}) \star \{\mathcal{L}, x\} \rangle = -Xx. \qquad (2.235)$$

Equation (2.235) is precisely the formula (2.183) which renders the relationship between the non-abelian moment map $\mathcal{m} = \mathcal{L}$ and the vector field ξ_X corresponding to the infinitesimal action (2.228) of G on D by the left shift. Analogously,

$$\langle X, \{\mathcal{R}, x\} \star I(\mathcal{R}) \rangle = \langle \mathrm{Ad}^*_{\mathcal{R}^{-1}} X, I(\mathcal{R}) \star \{\mathcal{R}, x\} \rangle = xX \qquad (2.236)$$

corresponds to the second line in (2.228).

Since $\mathcal{L}$ and $\mathcal{R}$ are invariant functions on D, they can be naturally viewed as generators of the function algebra on respective coset spaces. The Poisson structures on these coset spaces induced from D_+ appear identical and read

$$\{\mathcal{L}_1, \mathcal{L}_2\} = -\tfrac{1}{2} r \, \mathcal{L}_1 \mathcal{L}_2 - \tfrac{1}{2} \mathcal{L}_1 \mathcal{L}_2 \, r + \mathcal{L}_1 \, r_- \mathcal{L}_2 + \mathcal{L}_2 \, r_+ \mathcal{L}_1, \qquad (2.237)$$

$$\{\mathcal{R}_1, \mathcal{R}_2\} = -\tfrac{1}{2} r \, \mathcal{R}_1 \mathcal{R}_2 - \tfrac{1}{2} \mathcal{R}_1 \mathcal{R}_2 \, r + \mathcal{R}_1 \, r_- \mathcal{R}_2 + \mathcal{R}_2 \, r_+ \mathcal{R}_1. \qquad (2.238)$$

This unique structure is obviously the same as (2.227), that is the Poisson-Lie structure on G^* modelled by G. Importantly, $\mathcal{L}$ and $\mathcal{R}$ are in involution, $\{\mathcal{L}, \mathcal{R}\} = 0$, and they give rise to the first dual pair in Fig. 2.6.

According to the general theory of dual pairs, a symplectic leave of the Poisson bracket (2.237) passing through a point $\mathcal{L}$ is described as the connected component of the following set

$$\mathcal{O}_{\mathcal{L}} = \sigma \circ \pi_*(\pi_*'^{-1} \circ \sigma'^{-1}(x^{-1}y)) = \sigma \circ \pi_*(hx, hy) = hxy^{-1}h^{-1} = h\mathcal{L}h^{-1}, \quad h \in G,$$

and, therefore, coincides with the adjoint orbit of G through $\mathcal{L}$. Moreover, the adjoint action of G equipped with the Poisson-Lie structure (2.222) on G with the bracket (2.237) is Poisson. Casimir functions of the Semenov-Tian-Shansky bracket (2.237) coincide with the set of central functions on G generated by

$$C_k = \langle e^i, \mathcal{L}^k e_i \rangle = \operatorname{Tr} \mathcal{L}^k, \quad k \in \mathbb{N}. \tag{2.239}$$

The adjoint action $\mathcal{L} \to h\mathcal{L}h^{-1}$ is, in fact, the coadjoint action of G on G^*.

Analogously, the following actions of the Poisson-Lie group G^* with the bracket (2.223) on the Heisenberg double

$$G^* \times D \to D : \quad (x, y) \to (u_+x, u_-y), \tag{2.240}$$
$$D \times G^* \to D : \quad (x, y) \to (xu_+^{-1}, yu_-^{-1}), \tag{2.241}$$

are Poisson and the corresponding moment maps are generated by the right and left invariants, respectively,

$$\mathcal{L}_* = x(x^{-1}y)_+ = g', \qquad \mathcal{R}_* = x^{-1}(xy^{-1})_+ = g. \tag{2.242}$$

The moment maps $\mathcal{L}_*$ and $\mathcal{R}_*$ are in involution and give rise to the the second dual pair on Figure 2.6. The symplectic leaf of the Sklyanin bracket (2.222) is a set

$$\pi(\pi'^{-1}(g')) = \pi(xu_+^{-1}, yu_-^{-1}) = u_+x^{-1}(xu_+^{-1}u_-y^{-1})_+, \tag{2.243}$$

which coincides with the orbit of the *dressing transformation*[23] [25] through g

$$\mathcal{O}_g = u_+g\,(g^{-1}u_+^{-1}u_-g)_+, \tag{2.244}$$

In fact, the dressing transformation defines a homomorphism of G^* into a transformation group of G

$$(u_+, u_-) \cdot g \equiv g_u = u_+g\,(g^{-1}u_+^{-1}u_-g)_+ = u_-g\,(g^{-1}u_+^{-1}u_-g)_-. \tag{2.245}$$

Indeed, one has

---

[23] Dressing transformations arose in the context of soliton theory, see [27–29].

$$(gu)_v = v_+ g_u (g_u^{-1} v_+^{-1} v_- g_u)_+$$
$$= v_+ u_+ g \left(g^{-1} u_+^{-1} u_- g \right)_+ \left((g^{-1} u_+^{-1} u_- g)_+^{-1} g^{-1} u_+^{-1} v_+^{-1} v_- u_- g \left(g^{-1} u_+^{-1} u_- g \right)_- \right)_+$$
$$= (v_+ u_+) g \left(g^{-1} (v_+ u_+)^{-1} (v_- u_-) g \right)_+ .$$

Hence,

$$(v_+, v_-) \cdot (u_+, u_-) \cdot g = (v_+ u_+, v_- u_-) \cdot g ,$$

which shows that the dressing transformations obey the composition law of G^*. Moreover, (2.245) induces the coadjoint action (2.153) of G^* on $\mathfrak{g}$. Indeed, for $X \in \mathfrak{g}$ we obviously get

$$(u_+, u_-) \cdot X = u_+ X u_+^{-1} + u_+ (-X u_+^{-1} u_- + u_+^{-1} u_- X)_+$$
$$= u_+ X u_+^{-1} + (-u_+ X u_+^{-1} + u_- X u_-^{-1})_+ = (u_- X u_-^{-1})_+ - (u_+ X u_+^{-1})_- = \mathrm{Ad}^*_{(u_+, u_-)} X .$$

Importantly, the action of G^* on G by dressing transformations is Poisson. This can be, for instance, shown by exhibiting the corresponding moment map. Under the infinitesimal dressing transformation triggered by the one-parametric subgroup $(u_+, u_-) = (e^{tX_+}, e^{tX_-})$ with $X_\pm = \mathbf{r}_\pm(X)$ one has

$$\delta_{(X_+, X_-)} g = \mathbf{r}_+(X) g - g \mathbf{r}_+(g^{-1} X g) , \quad X = X_+ - X_- . \tag{2.246}$$

The moment map takes value in the dual of G^*, i.e. in G, and in fact coincides with g. Indeed, we have

$$\langle X, \{m, g\} m^{-1} \rangle = \langle X_1, \{g_1, g_2\} g_1^{-1} \rangle = \langle X_1, -r_{-12} g_2 + g_2 g_1 r_{-12} g_1^{-1} \rangle.$$

Taking into account that

$$\langle X_1, g_1 r_{-12} g_1^{-1} \rangle = r_-^{ij} e_j \langle e_i, g^{-1} X g \rangle = -\mathbf{r}_+(g^{-1} X g) , \quad \langle X_1, r_{-12} \rangle = -\mathbf{r}_+(X) ,$$

we get

$$\langle X, \{m, g\} m^{-1} \rangle = \delta_{(X_+, X_-)} g .$$

More on the theory of dressing transformations and Poisson group actions can be found in the work [23–25, 30–32].

Adjoint action. In addition to left and right actions (2.228) there is one more interesting way to let G act on the Heisenberg double. We introduce new generators

$$m_+ = \mathcal{L}_+ \mathcal{L}'_+ , \quad m_- = \mathcal{L}_- \mathcal{L}'_- ,$$

which inherit from the Poisson structure of the double the following relations

$$\{m_{+1}, m_{+2}\} = -\tfrac{1}{2}[r, m_{+1}m_{+2}], \qquad \{m_{+1}, m_{-2}\} = -[r_+, m_{+1}m_{-2}],$$
$$\{m_{-1}, m_{-2}\} = -\tfrac{1}{2}[r, m_{-1}m_{-2}], \qquad \{m_{-1}, m_{+2}\} = -[r_-, m_{-1}m_{+2}]. \tag{2.247}$$

These are obviously the same as the Poisson relations of G^*. Further, let us define $m \in G$

$$m = m_+(m_-)^{-1} = \mathcal{L}_+ \mathcal{R} \mathcal{L}_-^{-1}. \tag{2.248}$$

Obviously, we get for m the Poisson-Lie relations of G^*, when the latter group is modelled by G,

$$\{m_1, m_2\} = -r_+ m_1 m_2 - m_1 m_2 r_- + m_1 r_- m_2 + m_2 r_+ m_1, \tag{2.249}$$

confer (2.227). In close analogy to the moment map interpretation of the generators $\mathcal{L}$ and $\mathcal{R}$, we can think of m as a non-abelian moment map that gives rise to a certain Poisson action of the Poisson-Lie group G on D. Since we are mostly interested in the realisation of D in terms of generators $(\mathcal{L}_\pm, g)$, we evaluate the following brackets

$$\{m_1, \mathcal{L}_2\} = -r_+ m_1 \mathcal{L}_2 - m_1 \mathcal{L}_2 r_- + m_1 r_- \mathcal{L}_2 + \mathcal{L}_2 r_+ m_1,$$
$$\{m_1, g_2\} = \mathcal{L}_{+1}(g_2 r_+ \mathcal{R}_1 - r_+ \mathcal{R}_1 g_2 + \mathcal{R}_1 r_- g_2 - g_2 \mathcal{R}_1 r_-)\mathcal{L}_{-1}^{-1}. \tag{2.250}$$

Taking $Y \in \mathfrak{g}$, from here we find

$$\langle Y, \boldsymbol{I}(m) \star \{m, \mathcal{L}\}\rangle = -[\text{Ad}_m^* Y, \mathcal{L}],$$
$$\langle Y, \boldsymbol{I}(m) \star \{m, g\}\rangle = -[\text{Ad}_{\mathcal{L}'}^* Y, g].$$

Choosing now $Y = \text{Ad}_{m^{-1}}^* X$, we will get for $\mathcal{L}$

$$\xi_X \mathcal{L} = \langle X, \{m, \mathcal{L}\} \star \boldsymbol{I}(m)\rangle = -[X, \mathcal{L}], \tag{2.251}$$

where we recognise in ξ_X the fundamental vector field corresponding to the standard adjoint action $\mathcal{L} \to h\mathcal{L}h^{-1}$, $h = e^X \in G$. For g we will get

$$\xi_X g = \langle X, \{m, g\} \star \boldsymbol{I}(m)\rangle = -[\text{Ad}_{\mathcal{L}^{-1}}^* X, g]. \tag{2.252}$$

The map $X \to \xi_X$, where ξ_X acts by means of (2.251) and (2.252), is the homomorphism of Lie algebras. This is immediately seen for (2.251). Indeed,

$$[\xi_X, \xi_Y]\mathcal{L} = -[Y, \xi_X \mathcal{L}] + [X, \xi_Y \mathcal{L}]$$
$$= [Y, [X, \mathcal{L}]] - [X, [Y, \mathcal{L}]] = -[[X, Y], \mathcal{L}] = \xi_{[X,Y]}\mathcal{L},$$

as follows from the Jacobi identity. As to (2.252), by using the Jacobi identity we have

$$[\xi_X, \xi_Y]g = -[\xi_X(\mathrm{Ad}^*_{\mathcal{L}^{-1}}Y) - \xi_Y(\mathrm{Ad}^*_{\mathcal{L}^{-1}}X) + [\mathrm{Ad}^*_{\mathcal{L}^{-1}}X, \mathrm{Ad}^*_{\mathcal{L}^{-1}}Y], g].$$

On the other hand, according to (2.208),

$$\xi_X(\mathrm{Ad}^*_{\mathcal{L}^{-1}}Y) - \xi_Y(\mathrm{Ad}^*_{\mathcal{L}^{-1}}X) + [\mathrm{Ad}^*_{\mathcal{L}^{-1}}X, \mathrm{Ad}^*_{\mathcal{L}^{-1}}Y] = \mathrm{Ad}^*_{\mathcal{L}^{-1}}[X, Y],$$

because $\mathcal{L}$ is the moment map for the Poisson action (2.251). Thus, $[\xi_X, \xi_Y]g = \xi_{[X,Y]}g$.

We consider (2.251) and (2.252) as the Poisson-Lie analogue of the adjoint action of G on T^*G. Further justification for such an interpretation follows from considering the relation of the Heisenberg double to the cotangent bundle T^*G.

Relation to the cotangent bundle. Having advanced into the structure of the Heisenberg double, we can now explain in which sense this phase space can be viewed as a deformation of the cotangent bundle T^*G.

The formula $\mathcal{L} = \mathcal{L}_+\mathcal{L}_-^{-1}$ gives an embedding $G^* \hookrightarrow G$ and, alternatively, having $\mathcal{L} \in G$ the components $\mathcal{L}_\pm$ are found by solving the factorisation problem in G for which we assume a unique solution. The Poisson structure of D_+ in terms of generators $(\mathcal{L}, g)$ is then

$$\frac{1}{\varkappa}\{\mathcal{L}_1, \mathcal{L}_2\} = -r_+\mathcal{L}_1\mathcal{L}_2 - \mathcal{L}_1\mathcal{L}_2 r_- + \mathcal{L}_1 r_-\mathcal{L}_2 + \mathcal{L}_2 r_+\mathcal{L}_1,$$

$$\frac{1}{\varkappa}\{\mathcal{L}_1, g_2\} = g_2\,\mathcal{L}_{+1}C_{12}\mathcal{L}_{-1}^{-1}, \qquad\qquad (2.253)$$

$$\frac{1}{\varkappa}\{g_1, g_2\} = -[r, g_1g_2].$$

Since rescaling of the Poisson bracket is always possible, we introduce here the corresponding parameter $\varkappa \in \mathbb{C}$ whose actual meaning will be clarified soon. Notice that the set $(\mathcal{L}, g)$ does not form a closed Poisson algebra, because the right hand side of the middle bracket in (2.253) cannot be written in terms of generators $(\mathcal{L}, g)$ only. The situation appears to be different for the generators $(\mathcal{R}, g)$, for which we find

$$\frac{1}{\varkappa}\{\mathcal{R}_1, \mathcal{R}_2\} = -r_+\,\mathcal{R}_1\mathcal{R}_2 - \mathcal{R}_1\mathcal{R}_2\,r_- + \mathcal{R}_1\,r_-\mathcal{R}_2 + \mathcal{R}_2\,r_+\mathcal{R}_1,$$

$$\frac{1}{\varkappa}\{\mathcal{R}_1, g_2\} = -(r_+\mathcal{R}_1 - \mathcal{R}_1 r_-)g_2, \qquad\qquad (2.254)$$

$$\frac{1}{\varkappa}\{g_1, g_2\} = -[r, g_1g_2].$$

Fig. 2.7 Hierarchy of phase spaces

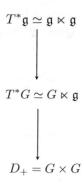

$$T^*\mathfrak{g} \simeq \mathfrak{g} \ltimes \mathfrak{g}$$

$$T^*G \simeq G \ltimes \mathfrak{g}$$

$$D_+ = G \times G$$

This difference in the relations of $\mathcal{L}$ and $\mathcal{R}$ with g has its explanation in the fact that the generator $g = x^{-1}(xy^{-1})_+$ exhibits simple behaviour under the right shifts of (x, y) by $h^{-1} \in G$

$$g \to hg, \tag{2.255}$$

conform (2.234), while for the left shifts this is not the case. Under the right shifts we also have

$$\mathcal{R} \to h\mathcal{R}h^{-1}. \tag{2.256}$$

Transformations (2.255) and (2.256), with h obeying the Poisson relations (2.222), are of course Poisson mappings of the structure (2.254) inherited from the second line in (2.228). However, under left shifts g does not transform in a simple way.

The parameter $\varkappa$ entering the Poisson brackets as an arbitrary constant allows for the interpretation of a non-trivial deformation parameter, if we assume that the coordinate functions also exhibit some non-trivial scaling with $\varkappa$. Most importantly, a connection to the Poisson structure of the cotangent bundle arises in the limit $\varkappa \to 0$, provided we assume the following behaviour of $\mathcal{L}_\pm$ in this limit

$$\mathcal{L}_\pm = \mathbb{1} + \varkappa\ell_\pm + \ldots, \quad \ell_\pm = r_\pm\ell,$$

while g remains unchanged. In this scaling limit

$$\mathcal{L} = \mathbb{1} + \varkappa\ell + \ldots, \quad \mathcal{R} = \mathbb{1} - \varkappa g\ell g^{-1} + \ldots = \mathbb{1} - \varkappa m + \ldots \tag{2.257}$$

and the Poisson structures (2.253) and (2.254) turn into (2.65) and (2.68), respectively. From this point of view, formulae (2.253) and (2.254) are deformed counterparts of (2.65) and (2.68) describing the Poisson structure of the cotangent bundle, correspondingly in the left and right parametrisations. Furthermore, the space T^*G

itself can be considered as the deformation of the cotangent bundle $T^*\mathfrak{g}$ of the Lie algebra $\mathfrak{g}$. The Poisson structure of $T^*\mathfrak{g}$ is described by

$$\begin{aligned}
\{q_1, q_2\} &= 0\,, \\
\{\ell_1, \ell_2\} &= \tfrac{1}{2}[C_{12}, \ell_1 - \ell_2]\,, \\
\{\ell_1, q_2\} &= C_{12}\,.
\end{aligned} \tag{2.258}$$

To obtain (2.258), one first introduces an affine scaling parameter $1/\kappa$ on the left hand side of (2.65). Then expanding $g = \mathbb{1} + \kappa q + \dots$ one finds in the limit $\kappa \to 0$ the desired result. The Poisson structure (2.258) is the closest Lie-algebraic analogue of the fundamental Poisson bracket (1.9) on $\mathbb{R}^{2N}$, where the momentum space parametrised by ℓ carries a non-trivial (linear) Poisson bracket. Thus, we have an interesting hierarchy of phase spaces depicted in Fig. 2.7, where passing from one space to another in order of increasing complexity means the replacement (deformation) of a linear space $\mathfrak{g} \simeq \mathbb{R}^N$ with a group manifold G. As we will discuss in the next section, this hierarchy of phase spaces leads to a natural hierarchy of integrable models which are obtained from the corresponding spaces by a properly settled Poisson reduction.

Relation to the moduli space of flat connections. Finally, we introduce yet another set of generators of the Heisenberg double which makes a remarkable connection to the Poisson structure on the moduli space of flat connections on a punctured torus. Since the corresponding moduli space and related geometric structures are only tangential to our treatment, we restrict our discussion to an essential algebraic aspect of it and refer the interested reader to the original literature [33, 34]. Let us define

$$A = \mathcal{L}\,, \quad B = \mathcal{L}_+ g \mathcal{L}_-^{-1} = \mathcal{L} \star g\,. \tag{2.259}$$

The generators $A, B \in G$ exhibit the following Poisson relations

$$\begin{aligned}
\frac{1}{\varkappa}\{A_1, A_2\} &= -r_- A_1 A_2 - A_1 A_2 r_+ + A_1 r_- A_2 + A_2 r_+ A_1\,, \\
\frac{1}{\varkappa}\{A_1, B_2\} &= -r_- A_1 B_2 - A_1 B_2 r_- + A_1 r_- B_2 + B_2 r_+ A_1\,, \\
\frac{1}{\varkappa}\{B_1, A_2\} &= -r_+ B_1 A_2 - B_1 A_2 r_+ + B_1 r_- A_2 + A_2 r_+ B_1\,, \\
\frac{1}{\varkappa}\{B_1, B_2\} &= -r_- B_1 B_2 - B_1 B_2 r_+ + B_1 r_- B_2 + B_2 r_+ B_1\,.
\end{aligned} \tag{2.260}$$

Here (A, B) can be interpreted as a pair of monodromies of a flat connection on a punctured torus around its two fundamental cycles [33]. The monodromies are not gauge invariants as they undergo an adjoint action of the group of residual gauge transformations which coincides with G

$$A \to h A h^{-1}\,, \quad B \to h B h^{-1}\,. \tag{2.261}$$

If G is Poisson-Lie group with the bracket (2.222), then transformations (2.261) are Poisson maps for the structure (2.260). The moment map of this action is nothing else but the generator m introduced in (2.248), which in terms of monodromies takes the form

$$m = \mathcal{L}_+ g \mathcal{L}_+^{-1} \mathcal{L}_- g^{-1} \mathcal{L}_-^{-1} = BA^{-1}B^{-1}A \tag{2.262}$$

and generates the following infinitesimal transformations of (A, B)

$$\frac{1}{\varkappa}\{m_1, A_2\} = -(r_+ m_1 - m_1 r_-)A_2 + A_2(r_+ m_1 - m_1 r_-),$$
$$\frac{1}{\varkappa}\{m_1, B_2\} = -(r_+ m_1 - m_1 r_-)B_2 + B_2(r_+ m_1 - m_1 r_-), \tag{2.263}$$

to be compared with (2.234). The Poisson brackets (2.260) can be viewed as giving the Poisson structure to the manifold $G \times G$ albeit the G-factors here (monodromies) are not the same[24] as in the realisation of the double as the direct product $D = G \times G$. This new description of the Heisenberg double in terms of generators (A, B) is however advantageous because it unravels the nature of (2.251) and (2.252) as infinitesimal transformations corresponding to the conventional adjoint action of G. Finally, we note that expanding m in the limit $\varkappa \to 0$

$$m = \mathbb{1} + \varkappa \mu + \dots, \tag{2.264}$$

from (2.262) we find for μ the expression $\mu = \ell - g\ell g^{-1}$, which coincides with the moment map of the adjoint action of G on T^*G.

2.2 Integrable Models from Reduction

In this section we illustrate using a number of explicit examples how to obtain an integrable model starting from trivial dynamics on the initial phase space and reducing it by the symplectic action of a Lie group. The dynamical system arising on the reduced phase comes naturally in the form of a Lax pair and its equations of motion are solved in terms of an appropriately formulated factorisation problem.

We already introduced three rather generic phase spaces related to a Lie group G with Lie algebra $\mathfrak{g}$, namely, $T^*\mathfrak{g}$, T^*G and $D_+(G)$, all being the members of the deformation hierarchy in Fig. 2.7. Application to these spaces of a properly formulated reduction procedure leads to interesting and wide classes of integrable models, among them are those mentioned in Sect. 1.3. For instance, considering T^*G for a real split simple Lie group G and the group action (2.61), where H_l and H_r are some suitably chosen subgroups, one finds on the reduced space integrable systems

---

[24]Note, in particular, the relation $(x, y) = (Ay, y) \in D$.

identical to open Toda chains. This way of constructing Toda chains together with the proof of their Liouville integrability is well described in the literature, for instance, in [35].

In this book to exemplify the reduction techniques in the context of integrable systems, we will focus on the case of CMS and RS models. As we will see, these models exhibit a natural hierarchical structure which, in fact, derives its origin from the hierarchy of phase spaces in Fig. 2.7. Historically, the idea of constructing the CMS systems via hamiltonian reduction goes back to [4], where the models with rational and trigonometric potentials, see (1.59) and (1.60), have been obtained from $T^*\mathfrak{g}$ and T^*G, respectively. We prefer to start our treatment directly with the space T^*G and to come back to $T^*\mathfrak{g}$ afterwards, as the application of the corresponding reduction procedure to the latter space represents a relatively simple limiting case of the former. Subsequently, we move on to the more complicated space $D_+(G)$ which will be handled in the framework of Poisson reduction.

Our further introductory remark concerns reality conditions. In this book we will mainly deal with algebraic (holomorphic) integrable systems. Such a system is a complex algebraic manifold $\mathscr{P}$ with an associated non-degenerate closed holomorphic $(2, 0)$-form ω and an abelian subvariety of $\mathscr{P}$, lagrangian with respect to ω [36–38]. In this context the complex canonical variables p_i and q_i are treated as holomorphic coordinates on the corresponding phase space. Although this approach makes the content of the corresponding integrable model, as describing the time evolution of a real physical system less transparent, it is helpful in many other aspects. This concerns, in particular, the reduction techniques, where working in a holomorphic setting considerably simplifies the analysis. Once a reduction has been performed and an algebraic integrable system was constructed, one can impose suitable reality conditions, compatible with natural physical requirements, such as positivity of the hamiltonian, etc. For the derivation of the CMS and RS models by means of the suitably formulated reduction procedure in the category of real phase spaces, the reader is invited to consult the work [39–42].

2.2.1   RS and CMS Models from T^*G

Let $G = \mathrm{GL}_N(\mathbb{C})$ and $\mathfrak{g} = \mathrm{Mat}_N(\mathbb{C})$. Consider $T^*G \simeq G \times \mathfrak{g}$ as an initial phase space with the Poisson structure (2.65) and let G act on this space by the adjoint action (2.63). Before performing the reduction, we introduce on our Poisson manifold a new coordinate system.

Factorisations and Frobenius group. Any matrix $\ell \in \mathfrak{g}$ belonging to an orbit of G of maximal dimension can be represented as

$$\ell = TQT^{-1},\tag{2.265}$$

where Q is a diagonal matrix with entries q_i, where $q_i \neq q_j$, $1 \leq i, j \leq N$. By acting with elements from the Weyl group we can order q_i lexicographically. The matrix T is not uniquely defined by this decomposition, as it can be multiplied from the right by any invertible diagonal matrix. It is convenient to eliminate this ambiguity by imposing on T a condition

$$Te = e, \tag{2.266}$$

where e is a N-dimension vector with all entries equal to 1. Matrices satisfying (2.266) form a subgroup $F \subset G$. The corresponding Lie algebra $\mathfrak{f}$ has complex dimension $N(N-1)$ and a natural basis

$$f_{ij} = E_{ii} - E_{ij}, \tag{2.267}$$

where E_{ij} are the matrix unities introduced in (2.70). The structure constants of $\mathfrak{f}$ in this basis can be read off from the commutation relations

$$[f_{ij}, f_{kl}] = \delta_{ik}(f_{il} - f_{ij}) + \delta_{il}(f_{kj} - f_{kl}) + \delta_{jk}(f_{ij} - f_{il}).$$

We can regard (T, Q) as new coordinates on $\mathfrak{g}^* \simeq \mathfrak{g}$ and the formula (2.265) as a change of variables. In particular, T is a natural parameter on the orbit through Q.

Consider now the Kirilliov-Kostant bracket for ℓ and find out which Poisson structure it induces for (T, Q). First, recall that functions on $\mathfrak{g}$ invariant under adjoint action are Casimir functions for this bracket. Thus, q_i Poisson-commute with ℓ, T and between themselves. Second, we have

$$\{T_{ij}, T_{kl}\} = \frac{\delta T_{ij}}{\delta \ell_{mn}} \frac{\delta T_{kl}}{\delta \ell_{rs}} \{\ell_{mn}, \ell_{rs}\}. \tag{2.268}$$

The variations needed to compute this bracket are found from varying (2.265)

$$[T^{-1}\delta T, Q] + \delta Q = T^{-1}\delta \ell \, T. \tag{2.269}$$

Projecting this matrix equation on the diagonal, one immediately gets

$$\frac{\delta q_i}{\delta \ell_{mn}} = T_{im}^{-1} T_{ni}. \tag{2.270}$$

On the other hand, $T^{-1}\delta T$ can be found from (2.269) only up to an arbitrary diagonal matrix, this diagonal matrix is then fixed by requiring $\delta Te = 0$, which is an infinitesimal version of the condition (2.266). An easy calculation yields

$$\frac{\delta T_{ij}}{\delta \ell_{mn}} = \sum_{a \neq j} \frac{1}{q_{ja}} (T_{ia} T_{nj} T_{am}^{-1} + T_{ij} T_{na} T_{jm}^{-1}), \tag{2.271}$$

where $q_{ij} = q_i - q_j$. With these formula at hand, we compute (2.268) and get the following elegant result

$$\{T_1, T_2\} = T_1 T_2 \, r_{12}(q) \,, \tag{2.272}$$

where there appears a skew-symmetric matrix $r \in \mathfrak{f} \wedge \mathfrak{f}$ given by

$$r(q) = \sum_{i \neq j}^{N} \frac{1}{q_{ij}} f_{ij} \otimes f_{ji} \,. \tag{2.273}$$

This r-matrix is a solution of the classical Yang-Baxter equation [43], in accordance with the general theory. In this context, the variables q_i are merely parameters labelling different solutions of this equation. When two q's coincide the Poisson bracket (2.272) becomes singular which reflects the breakdown of the factorisation (2.265).

The origin of the classical r-matrix becomes clear if we note that $\mathfrak{f}$ is a Frobenius Lie algebra, see the discussion around the formula (2.120). Here the non-degenerate 2-cocycle $\omega: \wedge^2 \mathfrak{f} \to \mathbb{C}$ is a coboundary given by

$$\omega(X, Y) = \mathrm{Tr}(Q[X, Y]) \,, \quad X, Y \in \mathfrak{f} \,. \tag{2.274}$$

According to [10], to any Frobenius Lie algebra one can associate a skew-symmetric solution of the CYBE by inverting the corresponding 2-cocycle. Consequently, one can check that the inversion of (2.274) yields the r-matrix (2.273). In fact, ω defines the Kirillov symplectic form on the coadjoint orbit of maximal dimension parametrized by Q, and (2.272) is the corresponding Poisson structure.

Now we turn our attention to the right parametrisation of T^*G and note that

$$m = g \ell g^{-1} = g T Q T^{-1} g^{-1} \equiv U Q U^{-1} \,, \tag{2.275}$$

where to render this factorisation unique, we impose on U the condition $U e = e$. Thus, U is also an element of the Frobenius group. The Poisson structure for U is inherited from that for m and, repeating the same manipulations as above, we obtain

$$\{U_1, U_2\} = -U_1 U_2 \, r_{12}(q) \,, \tag{2.276}$$

with the same r-matrix (2.273). The variables q_i Poisson commute with U. From (2.275) we cannot conclude that U coincides with gT because we imposed on U the constraint $U e = e$. It is clear, however, that these two matrices must be related as $gTP = U$ for some diagonal P. From here we obtain the following parametrisation for the group element

$$g = U P^{-1} T^{-1} \,. \tag{2.277}$$

Further, it follows from (2.58) that the entries of T and U are in involution. Taking into account that

$$\{q_i, g_{mn}\} = \frac{\delta q_i}{\delta \ell_{kl}}\{\ell_{kl}, g_{mn}\} = T_{in}^{-1}(UP^{-1})_{mi},$$

we derive $\{P_i, q_j\} = P_i \delta_{ij}$. Thus, introducing the variables p_i as $P_i = e^{p_i}$, we conclude that the Poisson bracket between q and p is canonical

$$\{p_i, q_j\} = \delta_{ij}. \tag{2.278}$$

The remaining brackets between P and U, T are computed in a similar way and we find

$$\begin{aligned}\{U_1, P_2\} &= -U_1 \bar{r}_{12}(q) P_2, \\ \{T_1, P_2\} &= -T_1 \bar{r}_{12}(q) P_2,\end{aligned} \tag{2.279}$$

where we have introduced a new matrix

$$\bar{r}(q) = \sum_{i \neq j}^{N} \frac{1}{q_{ij}} f_{ij} \otimes E_{jj}. \tag{2.280}$$

The Jacobi identity leads to a set relations between the matrices r and $\bar{r}$ which will be discussed later. To summarise, we have obtained a parametrisation of a locus of the cotangent bundle by means of coordinate functions

$$(Q, P, T, U) \tag{2.281}$$

with non-trivial Poisson relations (2.272), (2.276), (2.278) and (2.279). Here T and U are coordinate functions on two independent copies of the Frobenius group F. Under (2.63) matrices T, U and the diagonal P transform as follows

$$T \to hTd_{(T,h)}, \quad U \to hUd_{(U,h)}, \quad P \to P\, d_{(U,h)}d_{(T,h)}^{-1},$$

where $d_{(T,h)}$ and $d_{(U,h)}$ are compensating diagonal matrices needed to maintain the Frobenius condition (2.266) for the transformed elements. In the particular case when $h \in F$, these transformations simplify to

$$T \to hT, \quad U \to hU, \tag{2.282}$$

while P stays invariant.

Construction of the reduced phase space. Until now we involved ourselves in the discussion of T^*G from the point of view of various coordinate systems adopted for the future reduction procedure. Our next goal will be to fix a suitable level for the moment map and construct the corresponding reduced phase space following the procedure of Sect. 2.1.2. Let us choose in $\mathfrak{g}$ the following element

$$n = e \otimes e^t - \mathbb{1}. \tag{2.283}$$

This is an $N \times N$-matrix with zeros on the diagonal and ones on all other positions. We then consider a level of the moment map corresponding to this element

$$\ell - g\ell g^{-1} = \gamma(e \otimes e^t - \mathbb{1}), \tag{2.284}$$

where γ is an arbitrary (complex) constant. This choice is motivated by the fact that n has the largest possible isotropy group under coadjoint action, such that the corresponding coadjoint orbit is among the ones of minimal dimension. In turn, this gives the possibility to have a sufficiently large involutive family on the reduced phase space and thereby assure Liouville integrability of an emergent dynamical system on the latter. We will come to the discussion of the isotropy group of (2.283) later on and now we explain how to solve (2.284).

In terms of coordinates (2.281) Eq. (2.284) turns into

$$TQT^{-1} - UQU^{-1} = \gamma(e \otimes e^t - \mathbb{1}). \tag{2.285}$$

In particular, the matrix P completely decouples. Since $Te = e$, the last equation can be written as

$$\gamma W + QW - WQ = \gamma e \otimes (e^t U), \tag{2.286}$$

where we introduced $W = T^{-1}U$. This equation can be now elementary solved and the corresponding solution renders the matrix W as the following function of q_i and U

$$W(q, U) = \sum_{i,j=1}^{N} \frac{\gamma}{\gamma + q_{ij}} b_j E_{ij}, \tag{2.287}$$

where b is a column with elements $b_j = (e^t U)_j$. Up to now we did not use the condition that U is an element of the Frobenius group. Imposing this condition translates into the requirement that $W = T^{-1}U$ must also belong to the Frobenius group, i.e. it must satisfy $We = e$. This yields a system of equations for the coefficients b_j

$$\sum_{j=1}^{N} \frac{\gamma}{\gamma + q_{ij}} b_j = 1, \quad 1 \le i \le N.$$

Introducing a matrix V with entries

$$V_{ij} = \frac{\gamma}{\gamma + q_{ij}},$$

we observe that V is a Cauchy matrix. Recall that in general a Cauchy matrix $\mathscr{C}$ is an $N \times N$-matrix with entries

$$\mathscr{C}_{ij} = \frac{1}{x_i - y_j}.$$

Such a matrix has an inverse[25]

$$\mathscr{C}_{ij}^{-1} = (x_j - y_i) \frac{\displaystyle\prod_{k \neq j}^{N} (y_i - x_k) \prod_{k \neq i}^{N} (x_j - y_k)}{\displaystyle\prod_{k \neq j}^{N} (x_j - x_k) \prod_{k \neq i}^{N} (y_i - y_k)}. \tag{2.288}$$

The determinant of the Cauchy matrix is

$$\det \mathscr{C} = \frac{\displaystyle\prod_{i=2}^{N} \prod_{j=1}^{i-1} (x_i - x_j)(y_j - y_i)}{\displaystyle\prod_{i=1}^{N} \prod_{j=1}^{N} (x_i - y_j)}.$$

According to (2.288), the inverse of V is

$$V_{ij}^{-1} = \frac{1}{\gamma(q_{ji} + \gamma)} \frac{\displaystyle\prod_{a=1}^{N} (q_{ai} + \gamma) \prod_{a=1}^{N} (q_{ja} + \gamma)}{\displaystyle\prod_{a \neq i}^{N} q_{ai} \prod_{a \neq j}^{N} q_{ja}}. \tag{2.289}$$

We then find

$$b_j(q) = \sum_{k=1}^{N} V_{jk}^{-1} = \frac{1}{\gamma} \frac{\displaystyle\prod_{a=1}^{N} (q_{aj} + \gamma)}{\displaystyle\prod_{a \neq j}^{N} q_{aj}} = \frac{\displaystyle\prod_{a \neq j}^{N} (q_{aj} + \gamma)}{\displaystyle\prod_{a \neq j}^{N} q_{aj}}. \tag{2.290}$$

Further, we note that $\sum_{j=1}^{N} b_j = N$, in accordance with the definition $b_j = \sum_{i=1}^{N} U_{ij}$ and the fact that $U \in F$. In this way we find the entries of the matrix W as the following functions of q

---

[25] For the entries of the inverse matrix we adopt the notation $(\mathscr{C}^{-1})_{ij} \equiv \mathscr{C}_{ij}^{-1}$.

$$W_{ij}(q) = V_{ij}b_j = \frac{\displaystyle\prod_{a\neq i}^{N}(q_{aj} + \gamma)}{\displaystyle\prod_{a\neq j}^{N} q_{aj}}. \tag{2.291}$$

For future convenience we can immediately obtain the inverse of W. Indeed, from the previous formula we see that

$$W_{ij}^{-1}(q) = \frac{1}{b_i}V_{ij}^{-1} = \frac{\displaystyle\prod_{a\neq i}^{N}(q_{ja} + \gamma)}{\displaystyle\prod_{a\neq j}^{N} q_{ja}} = W_{ij}(-q), \tag{2.292}$$

where in the last step we substituted (2.289). Also, we replace $e^t U = b^t$ by an alternative condition

$$e^t T = e^t U U^{-1} T = b^t(q) W^{-1}(q) = b^t(-q). \tag{2.293}$$

Summarising, Eq. (2.284) is equivalent to the following constraints on the phase space variables

$$T^{-1}U = W(q), \quad e^t T = b^t(-q). \tag{2.294}$$

Here $T, U \in F$, the quantities $W(q), b(q)$ are given by (2.291) and (2.290), respectively.

Let us now determine the isotropy group G_n corresponding to (2.283). Under conjugation by $h \in G$, the element n remains invariant provided h obeys the following two conditions

$$he = \lambda e, \quad e^t h = \lambda e^t, \tag{2.295}$$

where $\lambda \in \mathbb{C}^*$. It is convenient to separate here a GL_1-subgroup that acts on e by dilatations and define the following subgroup $F_e \subset F$ as

$$F_e \equiv \{h \in G : he = e, \quad e^t h = e^t\}. \tag{2.296}$$

As a manifold, the group F_e can be parametrised as

$$h(z) = \left(\begin{array}{c|c} \displaystyle\sum_{i,j}^{N-1} z_{ij} - (N-2) & 1 - \displaystyle\sum_{k=1}^{N-1} z_{kj} \\ \hline 1 - \displaystyle\sum_{k=1}^{N-1} z_{ik} & z_{ij} \end{array}\right), \tag{2.297}$$

where the coordinates $z = (z_{ij})$ comprise a $(N-1) \times (N-1)$-matrix. In fact, since n has $N-1$ coincident eigenvalues, $G_n \simeq \mathrm{GL}_1 \times \mathrm{GL}_{N-1}$ and $F_e \simeq \mathrm{GL}_{N-1}$. Therefore, the orbit in $\mathfrak{g}^*$ we are interested in, is isomorphic to the coset

$$\mathcal{O}_n \simeq \mathrm{GL}_N / (\mathrm{GL}_1 \times \mathrm{GL}_{N-1}). \tag{2.298}$$

This orbit has a complex dimension $2(N-1)$ and is among the orbits of minimal positive dimension.

Having determined the isotropy group of the moment map, we can now come back to solving the constraints (2.294). The first of these constraints yields a solution for U in terms of T

$$U = T W(q).$$

As to the second constraint, the general solution for T is given by the product $T = h T_0$ of an arbitrary $h \in F_e$ and a particular solution $T_0 \in F$ of the second equation, which we can choose, for instance, as

$$T_0(q) = \begin{pmatrix} b_1(-q) & b_2(-q) - 1 & b_3(-q) - 1 & \dots & b_N(-q) - 1 \\ 0 & 1 & 0 & & 0 \\ 0 & 0 & 1 & & 0 \\ \vdots & & & \ddots & \vdots \\ 0 & 0 & 0 & \dots & 1 \end{pmatrix}. \tag{2.299}$$

Thus, $\mathcal{S} = \mu^{-1}(n)$ can be identified with a set of points $(\ell, g) \subset T^*G$, where

$$\ell = h T_0(q) Q T_0(q)^{-1} h^{-1}, \qquad g = h T_0(q) W(q) P^{-1} T_0(q)^{-1} h^{-1}. \tag{2.300}$$

Here Q, P are unrestricted (modulo Weyl ordering) and h is an arbitrary element of F_e. The reduced phase space $\mathcal{P}_r$ is then obtained in the standard fashion by taking a quotient of the level set of the moment map over the action of its stability subgroup coinciding in the present case with F_e

$$\mathcal{P}_r = \mathcal{S}/F_e. \tag{2.301}$$

Naturally, (Q, P) can be regarded as coordinates on $\mathcal{P}_r$. Since these coordinates inherit from T^*G the canonical Poisson bracket (2.278), the reduced phase space is symplectic. The formal counting of its dimension goes as follows. The adjoint action (2.63) of G on T^*G is not effective, as GL_1 acts trivially. Formula (2.28) then gives

$$\dim \mathcal{P}_r = \dim(T^*G) - \dim(\mathrm{GL}_N/\mathrm{GL}_1) - \dim(F_e) = 2N^2 - (N^2 - 1) - (N-1)^2 = 2N.$$

This completes the construction of the reduced phase space.

Dirac bracket and Frobenius invariants. Now we discuss the procedure of computing the Poisson structure on the reduced space by using the concept of the Dirac bracket and show that this naturally leads to the construction of the Lax pair from our geometric context.

Let $\mathfrak{f}_e$ be the Lie algebra of F_e and let $\mathfrak{d}$ denote the one-dimensional Lie subalgebra corresponding to GL_1. The Lie algebra $\mathfrak{g}$ can be decomposed into a direct sum of vector spaces

$$\mathfrak{g} = \mathfrak{f}_e \oplus \mathfrak{c} \oplus \mathfrak{a} \oplus \mathfrak{d} = \mathfrak{f} \oplus \mathfrak{a} \oplus \mathfrak{d}. \tag{2.302}$$

Here $\mathfrak{a}$ and $\mathfrak{c}$ are two abelian $N-1$-dimensional Lie subalgebras. In this decomposition the summand $\mathfrak{f}_e \oplus \mathfrak{c}$ coincides with the Frobenius Lie algebra $\mathfrak{f}$. We note for completeness that $\mathfrak{f} \oplus \mathfrak{d}$ is isomorphic to the maximal parabolic subalgebra $\mathfrak{p}$ of $\mathfrak{g}$

$$\mathfrak{p} = \left\{ \left(\begin{array}{c|c} a & \mathbf{x}^t \\ \hline 0 & A \end{array} \right) \in \mathrm{GL}_N \,\Big|\, \mathbf{x} \in \mathbb{C}^{N-1}, \, a \in \mathbb{C}^\times, \, A \in \mathrm{GL}_{N-1} \right\}. \tag{2.303}$$

From the point of view of the Dirac classification in Sect. 2.1.2, constraints $\mu_X(x)$ are of the first class for $X \in \mathfrak{f}_e$ and of the second class for $X \in \mathfrak{c} \oplus \mathfrak{a}$. To compute the Dirac bracket (2.34), we first need to invert the matrix Ψ of the second class constraints

$$\Psi_{ij}(x) = \langle \mu(x), [e_i, e_j] \rangle, \tag{2.304}$$

where e_i are basis elements of $\mathfrak{g}$ which reside in the subspace $\mathfrak{c} \oplus \mathfrak{a}$. Schematically, the matrix Ψ has the following structure

$$\Psi = \begin{pmatrix} \Psi_{\mathfrak{cc}} & \Psi_{\mathfrak{ca}} \\ \Psi_{\mathfrak{ac}} & \Psi_{\mathfrak{aa}} \end{pmatrix} = \begin{pmatrix} 0 & \Psi_{\mathfrak{ca}} \\ \Psi_{\mathfrak{ac}} & 0 \end{pmatrix}, \tag{2.305}$$

where the subscript denotes the subspace to which the corresponding Lie algebra elements belong. Here the diagonal blocks $\Psi_{\mathfrak{aa}}$ and $\Psi_{\mathfrak{cc}}$ vanish because $\mathfrak{a}$ and $\mathfrak{c}$ are abelian Lie subalgebras. Such an off-diagonal structure of Ψ persists even outside the points of $\mathscr{P}_r$. As a consequence, the Dirac bracket (2.34) specifies to

$$\{f, h\}_{\mathrm{D}} = \{f, h\} - \{f, \mu_{\mathfrak{c}}\} \Psi_{\mathfrak{ca}}^{-1} \{\mu_{\mathfrak{a}}, h\} - \{f, \mu_{\mathfrak{a}}\} \Psi_{\mathfrak{ac}}^{-1} \{\mu_{\mathfrak{c}}, h\}. \tag{2.306}$$

Functions on $\mathscr{P}_r$ can be regarded as F_e-invariant functions on the level set $\mu^{-1}(n)$ and we can extend them to the whole T^*G with maintaining this invariance. Remarkably, formula (2.306) suggests that if we additionally require these extension to be also invariant under $\mathfrak{c}$, i.e. $\{f, \mu_X\} = 0$ for any $X \in \mathfrak{c}$, then for such extensions the Dirac bracket will simply give the same result as the original Poisson bracket. This combined invariance is nothing else but the invariance under the action of the Frobenius group and function on T^*G invariant under F will be naturally called *Frobenius invariants*. For Frobenius invariants the projection on $\mathscr{P}_r$ can be taken either before or after evaluating their Poisson brackets with the same result for the latter.

It is rather easy to construct a basis of Frobenius invariants. Introduce a matrix

$$L = T^{-1}gT = T^{-1}UP^{-1},$$ (2.307)

Obviously, L is invariant under (2.282) and therefore its entries are Frobenius invariants on T^*G. Since L has a structure $L = WP^{-1}$, where $W = T^{-1}U \in F$, there are precisely dim F such invariants. With the help of (2.272), (2.276) and (2.279) one finds the following Poisson relations

$$\{Q_1, L_2\} = L_2\overline{C}_{12},$$
$$\{T_1, L_2\} = T_1L_2\bar{r}_{12}(q) - T_1r_{12}(q)L_2,$$

and also

$$\{L_1, L_2\} = r_{12}(q)L_1L_2 - L_1L_2(r_{12}(q) - \bar{r}_{12}(q) + \bar{r}_{21}(q))$$ (2.308)
$$+ L_1\bar{r}_{21}(q)L_2 - L_2\bar{r}_{12}(q)L_1,$$

where we introduced the concise notation

$$\overline{C} = \sum_{i=1}^{N} E_{ii} \otimes E_{ii}.$$ (2.309)

We see that the bracket of L's is closed, i.e. its right hand side is solely given in terms of L, without any involvement of T. This happens due to the hamiltonian nature of the G-action. Indeed, computing the bracket of the moment map with L, we get

$$\{\mu_{ij}, L_{kl}\} = (T_{lj}^{-1} - T_{kj}^{-1})L_{kl}.$$ (2.310)

As shown in Appendix D, the action of the Frobenius subalgebra is generated by $\mu_{ii} - \mu_{ji}, i \neq j$, and we get

$$\{\mu_{ii} - \mu_{ji}, L_{kl}\} = 0.$$

This is in accordance with the fact that L is the Frobenius invariant, i.e. $\xi_X L = 0$, where ξ_X is the hamiltonian vector field corresponding to $X \in \mathfrak{f}$. Then, acting with ξ_X on the bracket, we get

$$\xi_X\{L_1, L_2\} = \{\mu_X, \{L_1, L_2\}\} = \{\{\mu_X, L_1\}, L_2\} + \{L_1, \{\mu_X, L_2\}\} = 0, \quad X \in \mathfrak{f},$$

so that the bracket of L's should also be a Frobenius invariant and cannot contain a non-invariant variable T.

Further, we note that the involutive family $I_k = \mathrm{Tr}\, g^k$ on T^*G can be written via the single variable L

$$I_k = \mathrm{Tr}\, L^k.$$ (2.311)

As a consequence, spectral invariants of L must be in involution with respect to (2.308). According to the discussion in Sect. 1.4.1, to comply with this fact, the bracket (2.308) should admit the representation (1.88) with some, possibly dynamical, r-matrix $\mathbf{r}$. And indeed, (2.308) can be cast in this form, if one defines

$$\mathbf{r}_{12} = \tfrac{1}{2}(L_2 r_{12}(q) + r_{12}(q)L_2) - L_2 \bar{r}_{12}(q). \tag{2.312}$$

Another interesting observation is the following. Calculating the Poisson bracket for W, we find that it coincides with the Sklyanin bracket defining on F the structure of a Poisson-Lie group:

$$\{W_1, W_2\} = [r_{12}(q), W_1 W_2]. \tag{2.313}$$

We also add here the Poisson relations between W and P

$$\{W_1, P_2\} = [\bar{r}_{12}(q), W_1]P_2, \tag{2.314}$$

from where we observe that the following functions

$$J_k = \operatorname{Tr} W^k \tag{2.315}$$

Poisson commute with P. Moreover, these functions Poisson commute with Q and also between themselves, the latter property follows from (2.313).

It remains to find equations for the dynamical r-matrices $r(q)$ and $\bar{r}(q)$ implied by the Jacobi identities. For this we can use any of the Poisson brackets established above, the simplest choice is to use the formulae (2.272) and (2.279). As was already mentioned, the Jacobi identity for (2.272) reduces to the standard Yang-Baxter equation for $r \equiv r(q)$

$$[r_{12}, r_{13}] + [r_{12}, r_{23}] + [r_{13}, r_{23}] = 0. \tag{2.316}$$

As to (2.279), considering the Jacobi identities involving various combinations of T, U and P, we find two more equations both containing $\bar{r} \equiv \bar{r}(q)$

$$[\bar{r}_{12}, \bar{r}_{13}] + \{\bar{r}_{12}, p_3\} - \{\bar{r}_{13}, p_2\} = 0,$$
$$[r_{12}, \bar{r}_{13}] + [r_{12}, \bar{r}_{23}] + [\bar{r}_{13}, \bar{r}_{23}] + \{r_{12}, p_3\} = 0. \tag{2.317}$$

Both equations are quadratic in $\bar{r}$ and involve the Poisson bracket of the r-matrices with p_i, which is a manifestation of the genuine dynamical nature of these r-matrices. Equations (2.316) and (2.317) are satisfied by the explicit expressions (2.273) and (2.280). We also point out that r has a special property, namely, it is nilpotent, i.e. $r^2 = 0$. This property will become important when it will come to the issue of quantisation. Finally, the form of the bracket (2.308) suggests to introduce the following r-matrix combination

$$\underline{r}_{12} = r_{12} + \bar{r}_{21} - \bar{r}_{12}. \tag{2.318}$$

This r-matrix satisfies a closed equation of the dynamical type

$$[\underline{r}_{12}, \underline{r}_{13}] + [\underline{r}_{12}, \underline{r}_{23}] + [\underline{r}_{13}, \underline{r}_{23}] + \{\underline{r}_{12}, p_3\} - \{\underline{r}_{13}, p_2\} + \{\underline{r}_{23}, p_1\} = 0. \tag{2.319}$$

We will say more about this equation in the section devoted to quantisation of the RS model. Meanwhile, we point out that (2.273) and (2.280) yield for $\underline{r}$ the following explicit expression

$$\underline{r} = \sum_{i \neq j}^{N} \frac{1}{q_{ij}} (E_{ij} \otimes E_{ji} - E_{ii} \otimes E_{jj}). \tag{2.320}$$

Invariant dynamics and Lax representation. Let us assume that on T^*G a dynamical system is defined with a hamiltonian H being an element of the ring generated by $I_k = \mathrm{Tr}\, g^k$. Since H is an invariant function under the adjoint action (2.63) of G on T^*G, we deal with an invariant dynamical system and $\dot{\mu} = 0$. Further, the hamiltonian Poisson commutes with g, in other words $\dot{g} = 0$. Therefore, under the corresponding hamiltonian flow the matrix function L will evolve as

$$\dot{L} = \frac{d}{dt}(T^{-1}gT) = [-T^{-1}\dot{T}, L]. \tag{2.321}$$

This equation has the form of the Lax representation (1.84) with the matrix $M = -T^{-1}\dot{T}$, which gives a natural reason to call L in (2.307) a Lax matrix. From this consideration it is also clear that M will take values in the Frobenius Lie algebra $\mathfrak{f}$ for *any choice* of the invariant hamiltonian.

The Lax representation (2.321) of an invariant hamiltonian flow holds on the initial phase space T^*G. Let us now see how this representation descends on the reduced phase space. On $\mathscr{P}_r$ the variable $W = T^{-1}U$ turns into the expression (2.291) and, therefore, L becomes the following matrix function of the canonical variables

$$L = W(q)P^{-1} = \sum_{i,j=1}^{N} \frac{\gamma}{q_{ij} + \gamma} b_j(q)e^{-p_j} E_{ij}. \tag{2.322}$$

It is straightforward to verify that the Poisson brackets between the entries of L computed with the help of (2.278) coincide with (2.308). One can also check that W given by (2.291) reproduce the Poisson brackets (2.313) and (2.314). We do not give an explicit proof of this statement here since in Sect. 3.3.1 we will show that the same function W realises the representation for the corresponding quantum algebra.

As a concrete example of an invariant dynamical system, we can take, for instance, the one defined by the following G-invariant function

$$H = \tfrac{1}{2}\mathrm{Tr}(g + g^{-1}) = \tfrac{1}{2}\mathrm{Tr}(L + L^{-1}). \tag{2.323}$$

On the reduced space this H turns into

$$H = \tfrac{1}{2} \sum_{i=1}^{N} \left(e^{p_i} W_{ii}(-q) + e^{-p_i} W_{ii}(q) \right), \tag{2.324}$$

where

$$W_{ii}(q) = \prod_{a \neq i}^{N} \frac{q_{ai} + \gamma}{q_{ai}}, \quad W_{ii}(-q) = \prod_{a \neq i}^{N} \frac{q_{ai} - \gamma}{q_{ai}}.$$

Upon performing the canonical transformation

$$q_i \to q_i, \quad p_i \to p_i + \frac{1}{2} \log \frac{W_{ii}(q)}{W_{ii}(-q)}, \tag{2.325}$$

this H becomes

$$H = \sum_{i=1}^{N} \cosh p_i \left(W_{ii}(q) W_{ii}(-q) \right)^{1/2} = \sum_{i=1}^{N} \cosh p_i \prod_{j \neq i}^{N} \sqrt{1 - \frac{\gamma^2}{q_{ij}^2}}. \tag{2.326}$$

For p and q real, the positive-definite hamiltonian corresponds to the choice γ to be purely imaginary. Thus, upon $\gamma \to i\gamma$, the hamiltonian turns into

$$H = \sum_{i=1}^{N} \cosh p_i \prod_{j \neq i}^{N} \sqrt{1 + \frac{\gamma^2}{q_{ij}^2}}, \quad \gamma \in \mathbb{R}, \tag{2.327}$$

which is nothing else but the hamiltonian of the rational RS model [44]. The involutive family (2.311) renders this model Liouville integrable.

In general, we can take for H any function from the ring generated by (2.311), obtaining on the reduced space a concrete integrable system form the rational RS hierarchy. For instance, choosing $H = I_k$ for some k and by using (2.308) one finds for the corresponding equations of motion the Lax representation (1.84) with the matrix M

$$M = k \sum_{i \neq j}^{N} \frac{1}{q_{ij}} (L^k)_{ij} f_{ij} \in \mathfrak{f}. \tag{2.328}$$

Now we are ready to show that solving the hamiltonian equations amounts to the factorisation (diagonalisation) problem (2.265). On the initial phase space equations of motion for ℓ are

$$\dot{\ell} = \{I_k, \ell\} = -kg^k, \tag{2.329}$$

and these can be immediately integrated because g itself is an integral of motion

$$\ell(t) = -kg^k t + \ell_0 .$$

Without loss of generality we can assume that at $t = 0$ an element $\ell_0 \in \mathfrak{g}$ is a point in the reduced phase space, i.e. we identify $\ell_0 = \mathrm{diag}(q_1(0), \ldots, q_N(0)) = Q(0)$. We then set up a t-dependent factorisation (diagonalisation) problem

$$\ell(t) = -kg^k t + \ell_0 = T(t)Q(t)T(t)^{-1} ,$$

where $T(0) = \mathbb{1}$. With this initial condition for T, we get $L(0) = T^{-1}(0)gT(0) = g$ and the factorisation problem specifies to

$$Q(0) - kL^k(0)t = T(t)Q(t)T(t)^{-1} . \tag{2.330}$$

The matrix $L(0)$ depends on the initial values of coordinates and momenta and, therefore, the left hand side of (2.330) contains the full dependence on the initial data and it has a linear dependence on time. A solution of the equations of motion for $q(t)$ is then obtained by finding eigenvalues of $Q(0) - kL^k(0)t$, which is a purely algebraic problem. Thus, solving differential equations of time development has been reduced to algebraic operations, hence to quadratures. For generic particle number the solution cannot be given in an analytic form, but it can be constructed and verified numerically, for instance, for the simplest two- and three-particle cases.

Hyperbolic CMS model from T^*G. In our previous treatment we have solved the moment map Eq. (2.283) and defined invariant dynamics on the initial phase space in such a way that on the reduced space we obtained the rational RS model. Here we show that one can give an alternative description of the reduced space and choose another involutive family so that the reduced dynamics is that of the hyperbolic Calogero-Moser system. This interrelation between RS and CMS models is an example of the so-called *Ruijsenaars duality* [45, 46] and our discussion here will reveal its geometric origin [47].

We start with noting that instead of diagonalising ℓ, we can diagonalise g and write under the same assumptions as before that

$$g = TQT^{-1} , \tag{2.331}$$

where $T \in F$, i.e. $Te = e$, and Q is a diagonal matrix with pairwise different non-vanishing eigenvalues. Computation of the Poisson brackets of T, Q and ℓ yields

$$\begin{aligned}
\{T_1, \ell_2\} &= T_1 T_2 s_{12} T_2^{-1} , \\
\{Q_1, \ell_2\} &= -Q_1 T_2 \overline{C}_{12} T_2^{-1} ,
\end{aligned} \tag{2.332}$$

where

$$s_{12} = -\sum_{i \neq j}^{N} \frac{Q_i}{Q_{ij}} (E_{ii} - E_{ij}) \otimes E_{ji} , \qquad (2.333)$$

and $Q_{ij} = Q_i - Q_j$. The second formula in (2.332) implies the relation

$$s_{12} - Q_1^{-1} s_{12} Q_1 + \overline{C} = C , \qquad (2.334)$$

that is satisfied by (2.333).

A Frobenius invariant, which is a candidate Lax matrix for the hyperbolic CMS model $L \equiv \mathscr{L}^{(h)}$, is now introduced as

$$\mathscr{L}^{(h)} = -T^{-1} \ell T , \qquad (2.335)$$

where the minus sign on the right hand side is chosen for convenience. In terms of T, Q, $\mathscr{L}^{(h)}$ the Poisson algebra relations of T^*G read as

$$\{T_1, \mathscr{L}_2^{(h)}\} = -T_1 s_{12} , \qquad \{Q_1, \mathscr{L}_2^{(h)}\} = Q_1 \overline{C}_{12} , \qquad (2.336)$$

and

$$\{\mathscr{L}_1^{(h)}, \mathscr{L}_2^{(h)}\} = [r_{12}^{(h)}, \mathscr{L}_1^{(h)}] - [r_{21}^{(h)}, \mathscr{L}_2^{(h)}] , \qquad (2.337)$$

where

$$r_{12}^{(h)} = s_{12} - \tfrac{1}{2} C_{12} . \qquad (2.338)$$

In the present case a relevant involutive family on T^*G is generated by the Casimir functions of the Kirillov-Konstant bracket

$$I_k = (-1)^k \operatorname{Tr} \ell^k = \operatorname{Tr}(\mathscr{L}^{(h)})^k . \qquad (2.339)$$

Again, commutativity of the eigenvalues of $\mathscr{L}^{(h)}$ is consistent with the r-matrix form of the bracket (2.337).

We choose the same n as in (2.283) for a fixed value of the moment map. In terms of the new variables equation (2.284) will then read as

$$Q \mathscr{L}^{(h)} Q^{-1} - \mathscr{L}^{(h)} + \gamma \mathbb{1} = \gamma e \otimes e^t T . \qquad (2.340)$$

Separating here the diagonal part, we see that this equation has a solution if and only if

$$e^t T = e^t , \qquad (2.341)$$

implying that $T \in F_e$. The diagonal part of $\mathscr{L}^{(h)}$, which we parametrise as $-p_i E_{ii}$, is not fixed by this equation, while its non-diagonal part is fixed uniquely. Thus, we arrive at the following solution

$$\mathscr{L}^{(h)} = -\sum_{i=1}^{N} p_i E_{ii} + \gamma \sum_{i \neq j}^{N} \frac{Q_j}{Q_{ij}} E_{ij} . \tag{2.342}$$

Introducing the coordinates $q_i = \log Q_i$, the last formula can be cast in the form

$$\mathscr{L}^{(h)} = -\sum_{i=1}^{N} p_i E_{ii} + \gamma \sum_{i \neq j}^{N} \frac{e^{-\frac{1}{2}q_{ij}}}{2 \sinh \frac{1}{2} q_{ij}} E_{ij} . \tag{2.343}$$

The variables p and q have the canonical Poisson bracket (2.278), as follows, for instance, from the second relation in (2.336). One can now recognise in (2.343) the Lax matrix of the Calogero-Moser model with the hyperbolic-type potential.

For any invariant hamiltonian[26] H the dynamical equation for the Lax matrix (2.335) takes the form of the Lax equation

$$\frac{d\mathscr{L}^{(h)}}{dt} = \{H, \mathscr{L}^{(h)}\} = [M, \mathscr{L}^{(h)}], \quad M = -T^{-1}\dot{T} . \tag{2.344}$$

Here the element $T \equiv T(t)$ takes values in the space of solutions of the moment map equation (2.340) and, therefore, in addition to the standard Frobenius condition $Te = e$, it satisfies (2.341) for all t. As a result, M can always be chosen to satisfy two conditions

$$Me = 0, \quad e^t M = 0, \tag{2.345}$$

which in terms of matrix indices read as

$$\sum_{j=1}^{N} M_{ij} = 0, \quad \sum_{j=1}^{N} M_{ji} = 0, \quad \forall i , \tag{2.346}$$

that is M belongs to the Lie algebra of the stabiliser group of the moment map.

As an example, consider the hamiltonian $H = \frac{1}{2}\text{Tr}(\mathscr{L}^{(h)})^2$. The Poisson bracket of H with $\mathscr{L}^{(h)}$ is easily computed from (2.332) or from (2.337) and one finds

$$M = -\text{Tr}_1(\mathscr{L}_1^{(h)} s_{21}) = \gamma \sum_{i \neq j}^{N} \frac{Q_i Q_j}{Q_{ij}^2} (E_{ii} - E_{ij}) . \tag{2.347}$$

---

[26]Recall that an invariant hamiltonian is a Casimir function for the middle bracket in (2.65) so that $\dot{\ell} = 0$ with respect to such a hamiltonian.

Evidently, this matrix satisfies the conditions (2.345). As a consequence, M commutes with the element $e \otimes e^t = \sum_{ij} E_{ij}$. In this respect we point out that there arises a freedom to redefine the Lax matrix without changing the form of the Lax equation or the matrix M. Indeed, introducing

$$\mathcal{L}_\alpha^{(h)} = -\sum_{i=1}^{N} p_i E_{ii} + \gamma \sum_{i \neq j}^{N} \frac{Q_j}{Q_{ij}} E_{ij} + \alpha \sum_{i \neq j}^{N} E_{ij}, \tag{2.348}$$

where α is an arbitrary constant, we see that $\mathcal{L}_\alpha^{(h)}$ obeys the same Lax equation, because the terms added to the original Lax matrix do not depend on the dynamical variables and commute with M. In particular, taking $\alpha = \frac{\gamma}{2}$, we obtain the Lax matrix often used in the literature

$$\mathcal{L}_{\gamma/2}^{(h)} = -\sum_{i=1}^{N} p_i E_{ii} + \gamma \sum_{i \neq j}^{N} \frac{Q_i + Q_j}{2 Q_{ij}} E_{ij} = -\sum_{i=1}^{N} p_i E_{ii} + \frac{\gamma}{2} \sum_{i \neq j}^{N} \coth \tfrac{1}{2} q_{ij} E_{ij}.$$
$$\tag{2.349}$$

Concerning such a redefinition freedom, the RS model is less degenerate: its matrix M, say, for $k = 1$ in (2.328), is an element of the Frobenius Lie algebra

$$M = \sum_{i \neq j}^{N} \frac{\gamma}{q_{ij}(q_{ij} + \gamma)} b_j e^{-p_j} (E_{ii} - E_{ij}), \tag{2.350}$$

but it does not belong to the stabiliser of the corresponding moment map, since this time solving the moment map equation imposes on T in $M = -T^{-1}\dot{T}$ a different (dynamical) condition (2.293).

To conclude, resolving the same reduced phase space in terms of variables adapted for different involutive families on T^*G, we obtained two different integrable hierarchies—the rational RS and the hyperbolic CMS. This is precisely a manifestation of the Ruijsenaars duality. For investigation of various aspects of this duality including its geometric origin, see [40, 48–51].

2.2.2 RS Model from Heisenberg Double

Now we come to the top phase space in the hierarchy on Fig. 2.7, the Heisenberg double $D_+(G)$. The adjoint action of G on $D_+(G)$ is Poisson rather than hamiltonian, if we endow G with the structure of a Poisson-Lie group. Therefore, this time we have to use the Poisson reduction technique. As we will show, with a proper choice for the value of the corresponding moment map, an integrable system arising on the reduced phase space coincides with the hyperbolic RS model. This model can be viewed as a deformation of the rational RS model discussed earlier, which in our present context comes as a simple acknowledgement of the fact that $D_+(G)$ is a deformation of T^*G.

Reduced phase space and the Lax matrix. We choose to work with the description of the Heisenberg double in terms of variables (A, B) obeying the Poisson relations (2.260). For the case of interest $G = \mathrm{GL}_N(\mathbb{C})$, the r-matrices featuring (2.260) are

$$r_+ = +\frac{1}{2} \sum_{i=1}^{N} E_{ii} \otimes E_{ii} + \sum_{i<j}^{N} E_{ij} \otimes E_{ji},$$

$$r_- = -\frac{1}{2} \sum_{i=1}^{N} E_{ii} \otimes E_{ii} - \sum_{i>j}^{N} E_{ij} \otimes E_{ji}. \tag{2.351}$$

These r-matrices satisfy the CYBE and have the following properties $r_+ - r_- = C$ and $r_{\pm 21} = -r_{\mp 12}$. The construction of the non-abelian moment map $\mathcal{m}$ for the Poisson action of G has been extensively discussed in Sect. 2.1.7; in terms of A, B it reads as

$$\mathcal{m} = BA^{-1}B^{-1}A. \tag{2.352}$$

Since this time the moment map is group-valued and we are seeking for the generalisation of the reduction scheme that led us to the rational RS model, it is natural to fix

$$\mathcal{m} = \exp(\gamma n), \tag{2.353}$$

where n is the Lie algebra element (2.283). We are thus led to find all A, B that solve the following matrix equation

$$BA^{-1}B^{-1}A = e^{-\gamma}\mathbb{1} - e^{-\gamma}\frac{1 - e^{N\gamma}}{N}e \otimes e^t, \tag{2.354}$$

where on the right hand side we worked out an explicit form of the exponential $\exp(n)$. In the following we adopt the concise notation

$$t = e^{-\gamma}, \quad \beta = -e^{-\gamma}\frac{1 - e^{N\gamma}}{N} = -\frac{t}{N}(1 - t^{-N}). \tag{2.355}$$

To solve (2.354), very similar to the previous case, we introduce a convenient representation

$$A = TQT^{-1}, \tag{2.356}$$

$$B = UP^{-1}T^{-1}, \tag{2.357}$$

where this time Q is a group- rather than Lie algebra-valued element. As before, $T, U \in F$. In terms of new variables (2.354) takes the form

$$UQ^{-1}U^{-1}TQT^{-1} = t\mathbb{1} + \beta e \otimes e^t .$$

With the same notation $W = T^{-1}U$, the last equation becomes

$$Q^{-1}W^{-1}QW = t\mathbb{1} + \beta e \otimes e^t U , \tag{2.358}$$

where we used the fact that $U \in F$. As a next step, we write

$$Q^{-1}W^{-1}Q - tW^{-1} = \beta e \otimes e^t UW^{-1} = \beta e \otimes e^t T .$$

This equation can be elementary solved for W^{-1} and we get

$$W^{-1} = \sum_{i,j=1}^{N} \frac{\beta}{Q_i^{-1} - tQ_j^{-1}} \frac{c_j}{Q_j} E_{ij} , \tag{2.359}$$

where we introduced $c_j = (e^t T)_j$. Again, the condition $W^{-1} \in F$ gives a set of equations to determine the coefficients c_j:

$$\sum_{j=1}^{N} V_{ij} \frac{c_j}{Q_j} = 1 , \quad \forall i .$$

Here V is a Cauchy matrix with entries

$$V_{ij} = \frac{\beta}{Q_i^{-1} - tQ_j^{-1}} .$$

We apply the inverse of V

$$V_{ij}^{-1} = \frac{1}{\beta(Q_i^{-1} - t^{-1}Q_j^{-1})} \frac{\prod\limits_{a=1}^{N} (tQ_i^{-1} - Q_a^{-1}) \prod\limits_{a=1}^{N} (t^{-1}Q_j^{-1} - Q_a^{-1})}{\prod\limits_{a \neq i}^{N} (Q_i^{-1} - Q_a^{-1}) \prod\limits_{a \neq j}^{N} (Q_j^{-1} - Q_a^{-1})} ,$$

to obtain the following formula for the coefficients c_j

$$c_j = Q_j \sum_{j=1}^{N} V_{ij}^{-1} = \frac{(1-t)}{\beta} \frac{\prod\limits_{a \neq j}^{N} (Q_j^{-1} - t^{-1}Q_a^{-1})}{\prod\limits_{a \neq j}^{N} (Q_j^{-1} - Q_a^{-1})} = N \frac{1-t}{1-t^N} \prod_{a \neq j}^{N} \frac{Q_j - tQ_a}{Q_j - Q_a} ,$$

$$\tag{2.360}$$

where we substituted β from (2.355). Finally, inverting W^{-1} we find W itself

$$W_{ij}(Q) = \frac{Q_i}{c_i}(V^{-1})_{ij} = \frac{\displaystyle\prod_{a\neq i}^{N}(Q_j^{-1} - tQ_a^{-1})}{\displaystyle\prod_{a\neq j}^{N}(Q_j^{-1} - Q_a^{-1})}. \tag{2.361}$$

It is obvious that Eq. (2.354) is equivalent to the following two constraints, cf. (2.294),

$$U = TW(Q), \quad e^t T = c^t, \tag{2.362}$$

where $T, U \in F$, and the quantities $W(Q), c(Q)$ are given by (2.361) and (2.360), respectively. A particular solution $T_0 \in F$ of the equation $e^t T = c^t$ can be conveniently chosen as

$$T_0(Q) = \begin{pmatrix} c_1 & c_2 - 1 & c_3 - 1 & \dots & c_N - 1 \\ 0 & 1 & 0 & & 0 \\ 0 & 0 & 1 & & 0 \\ \vdots & & & \ddots & \vdots \\ 0 & 0 & 0 & \dots & 1 \end{pmatrix}.$$

It is worthwhile to notice that

$$\sum_{j=1}^{N} c_j = N.$$

On the constrained surface $T = hT_0$, where h is an arbitrary element from F_e.

In analogy with the rational case, we can take the combination $L = W(Q)P^{-1}$ as the Lax matrix associated with a family of invariant dynamical systems. Explicitly,

$$L = \sum_{i,j=1}^{N} \frac{(1-t)Q_i}{Q_i - tQ_j} \prod_{a\neq j}^{N} \frac{tQ_j - Q_a}{Q_j - Q_a} P_j^{-1} E_{ij}. \tag{2.363}$$

As we will see later, under proper reality conditions this L is nothing else but the Lax matrix of the RS family with the hyperbolic (trigonometric) potential. Note that the (A, B)-variables take on the constrained surface the following form

$$A(P, Q, h) = hT_0QT_0^{-1}h^{-1}, \quad B(P, Q, h) = hT_0LT_0^{-1}h^{-1}, \quad h \in F_e.$$

The reduced phase space can be singled out by, for instance, fixing the gauge $h = 1$.

Poisson structure on the reduced phase space. Now we turn to the analysis of the Poisson structure of the reduced phase space. Similar to the case of T^*G and by using the obvious analogue of (2.270), we find from (2.260) the following formula

$$\{Q_j, B\} = \varkappa B \sum_{kl} T_{lj} Q_j T_{jk}^{-1} E_{lk} \,. \tag{2.364}$$

Next, we need to determine the bracket between Q_j and P_i. We have

$$\{Q_j, P_i\} = \frac{\delta P_i}{\delta A_{mn}} \{Q_j, A_{mn}\} + \frac{\delta P_i}{\delta B_{mn}} \{Q_j, B_{mn}\} \,.$$

Here the first bracket on the right hand side vanishes because all Q_j commute with A.[27] To compute the second bracket, we consider the variation of $B = U P^{-1} T^{-1}$

$$U^{-1} \delta B \, T P = U^{-1} \delta U - P^{-1} \delta P \,.$$

Note that this formula does not include the variation δT. This is because T is solely determined by A and the same is true for its variation. The condition $\delta U e = 0$ allows one to find

$$\frac{\delta P_i}{\delta B_{mn}} = - \sum_r P_i U_{im}^{-1} (T P)_{nr} \,.$$

We thus have

$$\{Q_j, P_i\} = -\varkappa \sum_r P_i U_{im}^{-1} (T P)_{nr} (BT)_{mj} Q_j T_{jn}^{-1} = -\varkappa Q_i P_i \delta_{ij} \,. \tag{2.365}$$

This formula suggests to employ the exponential parametrisation for both P and Q, that is, to set

$$P_i = \exp p_i \,, \qquad Q_i = \exp \varkappa q_i \,. \tag{2.366}$$

Here we made a particular choice by putting the entire dependence on $\varkappa$ into Q. From (2.365) we therefore obtain $\{p_i, q_j\} = \delta_{ij}$. At this point it is tempting to conclude that (p_i, q_i) are canonical variables. This is, however, not so. Although $\{q_i, q_j\} = 0$, an analysis in Appendix D.1.2 shows that $\{p_i, p_j\} \neq 0$, see (D.28). Only upon taking the contribution of the second class constraints, do the variables (p_i, q_i) become canonical coordinates on the reduced phase space with respect to the Dirac bracket. As we will discuss below, from the reductionistic point of view the origin of this non-canonicity lies in the fact that the action of G on the double is not hamiltonian.

An invariant extension of the Lax matrix away from the reduced space is naturally given by the following Frobenius invariant

$$L = T^{-1} B T \,, \tag{2.367}$$

---

[27] The spectral invariants of A are central in the Poisson subalgebra of A, the latter is described by the Semenov-Tian-Schansky bracket $\{A_1, A_2\}$ given by the first line in (2.260).

where T is an element of the Frobenius group entering the factorisation (2.356). The Poisson bracket of Q_j with components of L is computed in a straightforward manner

$$\{Q_j, L_{mn}\} = \{Q_j, (T^{-1}BT)_{mn}\} = (T^{-1}B)_{mp} \sum_{kl} T_{lj} Q_j T_{jk}^{-1} (E_{lk})_{ps} T_{sn} = \varkappa L_{mn} Q_n \delta_{jn} \,,$$

which is perfectly compatible with the form (2.363) of the Lax matrix on the reduced space. In matrix form the previous formula reads as

$$\frac{1}{\varkappa} \{Q_1, L_2\} = Q_1 L_2 \overline{C}_{12} \,, \tag{2.368}$$

where $\overline{C}$ is defined in (2.309).

As to the brackets between the entries of L, this time they cannot be represented in terms of L alone but also involve T. Ultimately, such a structure is a consequence of the fact that the action of the Poisson-Lie group G on the phase space is Poisson rather than hamiltonian, so that there is an obstruction for the Poisson bracket of two Frobenius invariants to also be such an invariant, see formula (2.171). In addition, computing the Dirac brackets of L one cannot neglect a non-trivial contribution from the second class constraints and, therefore, the analysis of the Poisson structure for L requires as an intermediate step to understand the nature of constraints (2.354) imposed in the process of reduction. We save the details of the corresponding analysis for Appendix D and present here the final result for the Poisson bracket between the entries of the Lax matrix on the reduced phase space[28]

$$\frac{1}{\varkappa} \{L_1, L_2\} = r_{12}(Q)L_1 L_2 - L_1 L_2 \underline{r}_{12}(Q) + L_1 \bar{r}_{21}(Q)L_2 - L_2 \bar{r}_{12}(Q)L_1 \,. \tag{2.369}$$

Clearly, the bracket (2.369) has the same form as the corresponding bracket (2.308) for the rational model but with new dynamical r-matrices for which we get the following explicit expressions

$$r(Q) = \sum_{i \neq j}^{N} \left(\frac{Q_j}{Q_{ij}} E_{ii} - \frac{Q_i}{Q_{ij}} E_{ij} \right) \otimes (E_{jj} - E_{ji}) \,,$$

$$\bar{r}(Q) = \sum_{i \neq j}^{N} \frac{Q_i}{Q_{ij}} (E_{ii} - E_{ij}) \otimes E_{jj} \,, \tag{2.370}$$

$$\underline{r}(Q) = \sum_{i \neq j}^{N} \frac{Q_i}{Q_{ij}} (E_{ij} \otimes E_{ji} - E_{ii} \otimes E_{jj}) \,,$$

where similarly to the rational case we introduced the notation $Q_{ij} = Q_i - Q_j$.

---

[28]The quadratic and linear forms of the r-matrix structure for the RS model have been investigated in [52–54].

Concerning the properties of these matrices and the Lax matrix, we note the following. First, $\underline{r}$ is given via r and $\bar{r}$ by the same formula (2.318). Second, matrix $r = r(Q)$ is degenerate, $\det r = 0$, and it obeys a characteristic equation $r^2 = -r$. Moreover, in contrary to the rational case, r is not symmetric, rather it has the property

$$r_{12} + r_{21} = C_{12} - 1 \otimes 1 . \tag{2.371}$$

Third, it is a matter of straightforward calculation to verify that the Lax matrix (2.363) obeys the Poisson algebra relations (2.369), provided the bracket between the components of Q and P is given by (2.365).[29] Finally, since the form of the Poisson algebra of the Lax matrix is the same for the rational and hyperbolic cases, the matrices (2.370) satisfy the same system of Eqs. (2.316), (2.317) and (2.319). For this reason, $I_k = \mathrm{Tr} L^k$ are in involution with respect to (2.370). This property of I_k is, of course, inherited from the same property for $\mathrm{Tr} B^k$ on the original phase space (2.260).

Finally, in terms of variables (p_i, q_i) and upon rescaling[30] $\gamma \to \varkappa \gamma$ the Lax matrix (2.363) takes the form

$$L = e^{-\frac{\varkappa \gamma N}{2}} \sinh \frac{\varkappa \gamma}{2} \sum_{i,j=1}^{N} \frac{e^{\frac{\varkappa}{2}(q_{ij}+\gamma)}}{\sinh \frac{\varkappa}{2}(q_{ij}+\gamma)} \prod_{a \neq j}^{N} \frac{\sinh \frac{\varkappa}{2}(q_{aj}+\gamma)}{\sinh \frac{\varkappa}{2} q_{aj}} e^{-p_j} E_{ij} . \tag{2.372}$$

Since the Poisson algebra of L is quadratic, the overall coefficient in front of the sum in the last formula is unessential and can be neglected.

Invariant dynamics. To define an invariant dynamical system, we can choose the hamiltonian to coincide with one of the I_k's or a linear combination thereof. Choosing $H = I_k$, one finds from (2.369), (2.370) that the evolution equation for the Lax matrix (2.363) takes the form of the Lax representation (1.84) with the following matrix M

$$M = k \sum_{i \neq j}^{N} \frac{Q_j}{Q_{ij}} (L^k)_{ij} f_{ij} . \tag{2.373}$$

As in the rational case, M takes values in the Lie algebra of the Frobenius group.

Let us take a closer look at the member of the RS integrable hierarchy defined by the following hamiltonian

$$H = \tfrac{1}{2} \mathrm{Tr}(B + B^{-1})\Big|_{\mathscr{P}_r} = \tfrac{1}{2} \mathrm{Tr}(L + L^{-1}) . \tag{2.374}$$

---

[29] In Appendix D we give an alternative form of the Poisson structure (2.369) useful for constructing the spin generalisations of the RS models [55–58].

[30] This rescaling can be explained by noting that the moment map (2.352) does not involve any rescaling by $\varkappa$ and, therefore, it generates the fundamental vector field corresponding to the action of a group element g of the form $g = e^{\varkappa X}$, where $X \in \mathfrak{g}$. Hence, the fixed value of the moment map should be taken as $\mathfrak{m} = e^{\varkappa \gamma n}$ which is obtained by rescaling γ in (2.353) as $\gamma \to \varkappa \gamma$.

Explicitly,

$$H = \tfrac{1}{2} \sum_{i=1}^{N} \left(e^{p_i} W_{ii}^{-1} + e^{-p_i} W_{ii} \right), \tag{2.375}$$

where

$$W_{ii} = \prod_{a \neq i}^{N} \frac{Q_a - t Q_i}{Q_a - Q_i}, \qquad W_{ii}^{-1} = \prod_{a \neq i}^{N} \frac{Q_a - t^{-1} Q_i}{Q_a - Q_i}.$$

Upon performing a canonical transformation

$$q_i \rightarrow q_i, \qquad p_i \rightarrow p_i + \frac{1}{2} \log \frac{W_{ii}}{W_{ii}^{-1}}, \tag{2.376}$$

together with $\gamma \rightarrow i\gamma$, formula (2.375) turns into

$$H = \sum_{i=1}^{N} \left(W_{ii} W_{ii}^{-1} \right)^{1/2} \cosh p_i = \sum_{i=1}^{N} b_i \cosh p_i, \tag{2.377}$$

where b_i are functions of coordinates alone and they have the following explicit form

$$b_i = \prod_{j \neq i}^{N} \sqrt{1 + \frac{\sin^2 \frac{\varkappa \gamma}{2}}{\sinh^2 \frac{\varkappa}{2} q_{ij}}}. \tag{2.378}$$

Formula (2.377) is the hamiltonian of the hyperbolic RS model with real phase space variables. It is the same as (1.68), where one needs to identify the length parameter $\ell = 1/\varkappa$ and set $\mu = c = 1$. The equations of motion driven by this hamiltonian are [59]

$$\dot{q}_i = \{H, q_i\},$$

$$\dot{p}_i = \{H, p_i\} = -\sum_{j=1}^{N} \cosh p_j \frac{\partial b_j}{\partial q_i}.$$

To find their explicit form, we need the relations

$$\frac{\partial b_j}{\partial q_i} = -b_j \frac{\frac{\varkappa}{2} \sin^2 \frac{\varkappa \gamma}{2} \coth \frac{\varkappa}{2} q_{ij}}{\sinh^2 \frac{\varkappa}{2} q_{ij} + \sin^2 \frac{\varkappa \gamma}{2}}, \qquad i \neq j,$$

$$\frac{\partial b_i}{\partial q_i} = -b_i \sum_{j \neq i}^{N} \frac{\frac{\varkappa}{2} \sin^2 \frac{\varkappa \gamma}{2} \coth \frac{\varkappa}{2} q_{ij}}{\sinh^2 \frac{\varkappa}{2} q_{ij} + \sin^2 \frac{\varkappa \gamma}{2}}, \tag{2.379}$$

where the first relation also implies an identity

$$b_i \frac{\partial b_j}{\partial q_i} + b_j \frac{\partial b_i}{\partial q_j} = 0, \quad i \neq j.$$

Taking into account (2.379), we then get Hamilton's equations

$$\dot{q}_i = b_i \sinh p_i,$$

$$\dot{p}_i = \frac{\varkappa}{2} \sin^2 \frac{\varkappa\gamma}{2} \sum_{j \neq i}^{N} \coth \frac{\varkappa}{2} q_{ij} \frac{b_i \cosh p_i + b_j \cosh p_j}{\sinh^2 \frac{\varkappa}{2} q_{ij} + \sin^2 \frac{\varkappa\gamma}{2}}. \tag{2.380}$$

In particular, the second equations imply that $\sum_i \dot{p}_i = 0$.

Equation (2.380) can be solved in the same way as in the rational case. Equations of motion induced by the hamiltonian (2.374) on the initial phase space are

$$\dot{A} = -\tfrac{1}{2}(B - B^{-1})A, \quad \dot{B} = 0, \tag{2.381}$$

and they are elementary solved as

$$A(t) = e^{-\frac{1}{2}(B-B^{-1})t} A(0).$$

Choosing $A(0) = \mathrm{diag}(e^{q_1}, \ldots, e^{q_N}) \equiv Q(0)$ as an initial condition, the solution $q_i(t) = \log Q_i(t)$ is found by diagonalising the matrix $A(t)$ with the Frobenius matrix $T(t)$:

$$A(t) = T(t)Q(t)T(t)^{-1} = e^{-\frac{1}{2}(B-B^{-1})t} Q(0), \quad T(0) = 1.$$

As is clear from (2.367), $B = L(0)$, which allows one to express a solution in terms of (q_i, p_i) taken at $t = 0$.

Limiting cases. According to our findings, the hyperbolic RS model admits a description in terms of the Lax matrix (2.372) obeying the Poisson algebra (2.369). There are two limiting cases when the model reduces to the rational one or to the hyperbolic CMS system. To get the rational model, one simply takes $\varkappa \to 0$ with all the other variables kept finite. In this limit the Lax matrix (2.372) and relations (2.369) straightforwardly reduce to (2.322) and (2.308), while (2.380) yield

$$\dot{q}_i = b_i \sinh p_i,$$

$$\dot{p}_i = \gamma^2 \sum_{j \neq i}^{N} \frac{b_i \cosh p_i + b_j \cosh p_j}{q_{ij}(q_{ij}^2 + \gamma^2)}, \tag{2.382}$$

which are equations of motion for the rational model with the Hamiltonian (2.327) and b_i given by

$$b_i = \prod_{j \neq i}^{N} \sqrt{1 + \frac{\gamma^2}{q_{ij}^2}}, \quad \gamma \in \mathbb{R}.$$

As to the hyperbolic CMS model, it is obtained by the following limiting procedure. Namely, we perform a canonical transformation $q_i \to q_i/\kappa$, $p_i \to \kappa p_i$ and simultaneously make a rescaling $\varkappa \to \kappa\varkappa$. Note that under these manipulations the variables $Q_i = \exp(\varkappa q_i)$ remain invariant, as do all the r-matrices of the hyperbolic model. Then we take $\kappa \to 0$. In this limit the Lax matrix (2.363) expands as $L = \mathbb{1} + \kappa \mathcal{L}^{(h)} + o(\kappa)$, where[31]

$$\mathcal{L}^{(h)} = -\sum_{i=1}^{N}(p_i + c_i)E_{ii} + \gamma \sum_{i \neq j}^{N} \frac{e^{\frac{1}{2}q_{ij}}}{2\sinh\frac{1}{2}q_{ij}}E_{ij}, \quad c_i = \sum_{j \neq i} \frac{e^{\frac{1}{2}q_{ij}}}{2\sinh\frac{1}{2}q_{ij}}. \quad (2.383)$$

Further, the quadratic structure (2.369) turns into a linear one

$$\{\mathcal{L}_1^{(h)}, \mathcal{L}_2^{(h)}\} = [r_{12}(Q) - \bar{r}_{12}(Q), \mathcal{L}_1^{(h)}] + [r_{12}(Q) + \bar{r}_{21}(Q), \mathcal{L}_2^{(h)}]. \quad (2.384)$$

Using (2.371), the last expression can be put in the form (2.337) with the r-matrix

$$r_{12}^{(h)} = r_{12}(Q) - \bar{r}_{12}(Q) - \tfrac{1}{2}(C_{12} - \mathbb{1} \otimes \mathbb{1}). \quad (2.385)$$

Note that the canonical transformation $p_i \to p_i - c_i$ with q_i unchanged removes c_i from $\mathcal{L}^{(h)}$ without changing the Poisson bracket between the entries of the latter. Then, performing a similarity transformation $\mathcal{L}^{(h)} \to Q^{-1}\mathcal{L}^{(h)}Q$, we obtain for the transformed Lax matrix formula (2.343) and recover (2.337) with $r^{(h)}$ given by (2.338). Thus, our results on the hyperbolic CMS model obtained from the reduction procedure are fully reproduced as the limiting case of the corresponding RS model.

Introduction of spectral parameter. Here we introduce a Lax matrix depending on a spectral parameter and discuss the associated algebraic structures and an alternative way to exhibit commuting integrals.

To start with, we point out one important identity satisfied by the Lax matrix (2.363). According to the moment map equation (2.358), we have

$$tQ^{-1}WQ = W\left[\mathbb{1} + \frac{\beta}{t} e \otimes e^t U\right]^{-1}, \quad (2.386)$$

where β is related to the coupling constant parameter $t = e^{-\gamma}$ by means of (2.355). The inverse on the right hand side of (2.386) can be computed with the help of the Sherman-Morrison formula[32] and we get

---

[31] Without loss of generality, for further comparison we put $\varkappa = 1$.

[32] The Sherman-Morrison formula is $(A + u \otimes v^t)^{-1} = A^{-1} - \frac{A^{-1}(u \otimes v^t)A^{-1}}{1 + v^t A^{-1} u}$, valid for any two vectors u, v and an invertible matrix A. This formula then implies that $\det(A + u \otimes v^t) = (1 + v^t A^{-1} u)\det A$.

$$tQ^{-1}WQ = W\left[\mathbb{1} - \frac{1-t^N}{N}e \otimes e^t U\right] = W - \frac{1-t^N}{N}e \otimes c^t W, \qquad (2.387)$$

where we used the fact that W is a Frobenius matrix, so that $We = e$. Here the vector c has components (2.360) and satisfies the relation $e^t T = c^t$. Multiplying both sides of (2.362) with P^{-1} we obtain the following identity

$$tQ^{-1}LQ = L - \frac{1-t^N}{N}e \otimes c^t L, \qquad (2.388)$$

for the Lax matrix (2.363).

Evidently, we can consider

$$L' = tQ^{-1}LQ \qquad (2.389)$$

as another Lax matrix since the evolution equation of the latter is of the Lax form

$$\dot{L}' = [M', L'], \quad M' = Q^{-1}MQ - Q^{-1}\dot{Q}, \qquad (2.390)$$

where M is defined by the hamiltonian flow of L. Note that one can add to M' any function of L' without changing the evolution equation for L', which defines a class of equivalent M''s. Now, it turns out that due to the special dependence of L on the momentum, M and M' fall in the same equivalence class. To understand this point, let us take the simplest hamiltonian $H = \text{Tr}L$ for which the matrix M is given by (2.373) for $k = 1$,

$$M = \sum_{i \neq j}^{N} \frac{Q_j}{Q_{ij}}L_{ij}(E_{ii} - E_{ij}), \qquad (2.391)$$

It follows from (2.368) that for the flow generated by this hamiltonian

$$Q^{-1}\dot{Q} = Q^{-1}\{H, Q\} = -\sum_{i=1}^{N} L_{ii}E_{ii}.$$

Therefore,

$$M' = Q^{-1}MQ - Q^{-1}\dot{Q} = \sum_{i \neq j}^{N} \frac{Q_j}{Q_{ij}}L_{ij}\left(E_{ii} - \frac{Q_j}{Q_i}E_{ij}\right) + \sum_{i=1}^{N} L_{ii}E_{ii}.$$

Taking into account that $Q_j/(Q_{ij}Q_i) = 1/Q_{ij} - 1/Q_i$, we then find

$$M' = \sum_{i \neq j}^{N} \frac{Q_j}{Q_{ij}}L_{ij}(E_{ii} - E_{ij}) + \sum_{i \neq j}^{N} Q_i^{-1}L_{ij}Q_j E_{ij} + \sum_{i=1}^{N} L_{ii}E_{ii} = M + L'.$$

Hence, M' is in the same equivalence class as M and, therefore, we can take the dynamical matrix M to be the same for both L and L'.

The above observation motivates to introduce a Lax matrix depending on a spectral parameter just as a linear combination of L and L'. Namely, we define

$$L(\lambda) = L - \frac{1}{\lambda}L',\tag{2.392}$$

where $\lambda \in \mathbb{C}$ is the spectral parameter. The matrix $L(\lambda)$ has a pole at zero and the original matrix L is obtained from $L(\lambda)$ in the limit $\lambda \to \infty$, in particular,

$$H = \lim_{\lambda \to \infty} \mathrm{Tr} L(\lambda) = \mathrm{Tr} L.\tag{2.393}$$

The evolution equation for $L(\lambda)$ must, therefore, be of the form

$$\dot{L}(\lambda) = \{H, L(\lambda)\} = [M, L(\lambda)],\tag{2.394}$$

where M is the expression (2.391).

The next task is to compute the Poisson brackets between the components of (2.392). We aim at finding the structure similar to (2.369), namely,

$$\{L_1(\lambda), L_2(\mu)\} = r_{12}(\lambda, \mu)L_1(\lambda)L_2(\mu) - L_1(\lambda)L_2(\mu)\underline{r}_{12}(\lambda, \mu)$$
$$+ L_1(\lambda)\bar{r}_{21}(\mu)L_2(\mu) - L_2(\mu)\bar{r}_{12}(\lambda)L_1(\lambda),\tag{2.395}$$

where $r(\lambda, \mu)$, $\underline{r}(\lambda, \mu)$ and $\bar{r}(\lambda)$ are some spectral-parameter-dependent r-matrices. We show how to derive these r-matrices in Appendix D. Our considerations are essentially based on the identity (2.388). To state the corresponding result, we need the matrix

$$\sigma_{12} = \sum_{i \neq j}^{N}(E_{ii} - E_{ij}) \otimes E_{jj}.\tag{2.396}$$

The *minimal solution* for the spectral-dependent r-matrices realising the Poisson algebra (2.395) is then found to be

$$r_{12}(\lambda, \mu) = \frac{\lambda r_{12} + \mu r_{21}}{\lambda - \mu} + \frac{\sigma_{12}}{\lambda - 1} - \frac{\sigma_{21}}{\mu - 1},$$
$$\bar{r}_{12}(\lambda) = \bar{r}_{12} + \frac{\sigma_{12}}{\lambda - 1},\tag{2.397}$$
$$\underline{r}_{12}(\lambda, \mu) = r_{12}(\lambda, \mu) + \bar{r}_{21}(\mu) - \bar{r}_{12}(\lambda) = \frac{\lambda\underline{r}_{12} + \mu\underline{r}_{21}}{\lambda - \mu}.$$

The matrices r and $\underline{r}$ are skew-symmetric in the sense that

$$r_{12}(\lambda, \mu) = -r_{21}(\mu, \lambda), \quad \underline{r}_{12}(\lambda, \mu) = -\underline{r}_{21}(\mu, \lambda). \tag{2.398}$$

Further, one can establish implications of the Jacobi identity satisfied by (2.395) for these r-matrices. Introducing the dilatation operator acting on the spectral parameter

$$D_\lambda = \lambda \frac{\partial}{\partial \lambda},$$

we find that the r-matrix $r(\lambda, \mu)$ does not satisfy the standard CYBE but rather the following modification thereof

$$[r_{12}(\lambda, \mu), r_{13}(\lambda, \tau)] + [r_{12}(\lambda, \mu), r_{23}(\mu, \tau)] + [r_{13}(\lambda, \tau), r_{23}(\mu, \tau)] = \tag{2.399}$$
$$= -(D_\lambda + D_\mu) r_{12}(\lambda, \mu) + (D_\lambda + D_\tau) r_{13}(\lambda, \tau) - (D_\tau + D_\mu) r_{23}(\mu, \tau).$$

Following [60], we refer to (2.399) as *shifted classical Yang-Baxter equation*. This equation can be rewritten it in the form of the standard Yang-Baxter equation

$$[\hat{r}_{12}(\lambda, \mu), \hat{r}_{13}(\lambda, \tau)] + [\hat{r}_{12}(\lambda, \mu), \hat{r}_{23}(\mu, \tau)] + [\hat{r}_{13}(\lambda, \tau), \hat{r}_{23}(\mu, \tau)] = 0.$$

for the matrix differential operator

$$\hat{r}(\lambda, \mu) = r(\lambda, \mu) - D_\lambda + D_\mu. \tag{2.400}$$

There are also two more equations involving the matrix $\bar{r}$

$$[r_{12}(\lambda, \mu), \bar{r}_{13}(\lambda) + \bar{r}_{23}(\mu)] + [\bar{r}_{13}(\lambda), \bar{r}_{23}(\mu)] + P_3^{-1}\{r_{12}(\lambda, \mu), P_3\} =$$
$$= -(D_\lambda + D_\mu) r_{12}(\lambda, \mu) + (D_\lambda \bar{r}_{13}(\lambda) - D_\mu \bar{r}_{23}(\mu)) \tag{2.401}$$

and

$$[\bar{r}_{12}(\lambda), \bar{r}_{13}(\lambda)] + P_3^{-1}\{\bar{r}_{12}(\lambda), P_3\} - P_2^{-1}\{\bar{r}_{13}(\lambda), P_2\} = -D_\lambda(\bar{r}_{12}(\lambda) - \bar{r}_{13}(\lambda)). \tag{2.402}$$

One can check that relations (2.399), (2.401) and (2.402) guarantee the fulfilment of the Jacobi identity for the brackets (2.368) and (2.369). Note that $\underline{r}$ is scale-invariant: $(D_\lambda + D_\mu)\underline{r}(\lambda, \mu) = 0$, implying that it depends on the ratio λ/μ. This property does not hold, however, for r and $\bar{r}$.

The solution we found for the spectral-dependent dynamical r-matrices is minimal in the sense that there is a freedom to modify these r-matrices without changing the Poisson bracket (D.51). First of all, there is a trivial freedom of shifting r and $\underline{r}$ as

$$r_{12} \to r_{12} + f(\lambda/\mu) \mathbb{1} \otimes \mathbb{1}, \quad \underline{r}_{12} \to \underline{r}_{12} + f(\lambda/\mu) \mathbb{1} \otimes \mathbb{1}, \tag{2.403}$$

where f is an arbitrary function of the ratio of the spectral parameters. This redefinition affects neither the bracket (2.369) nor Eqs. (2.399), (2.401), (2.402).

Second, one can redefine $\bar{r}$ and r as

$$\begin{aligned} r(\lambda, \mu) &\to r(\lambda, \mu) - s(\lambda) \otimes \mathbb{1} + \mathbb{1} \otimes s(\mu) \\ \bar{r}(\lambda) &\to \bar{r}(\lambda) - s(\lambda) \otimes \mathbb{1} , \end{aligned} \tag{2.404}$$

where $s(\lambda)$ is an arbitrary matrix function of the spectral parameter. Owing to the structure of the bracket (2.395) the redefinition (2.404) produces no effect on this bracket, as $\underline{r}$ remains unchanged, while the matrix s decouples from (D.51). For generic $s(\lambda)$, redefintion (2.404) affects,[33] however, equations (2.399), (2.401) and (2.402). In particular, there exists a choice of $s(\lambda)$ which turns the shifted Yang-Baxter equations for $\bar{r}$ and r into the conversional ones, where the derivative terms on the right hand side of (2.399), (2.401) and (2.402) are absent. One can take, for instance,

$$s(\lambda) = \frac{1}{N} \sum_{i \neq j}^{N} \frac{Q_i}{Q_{ij}} (E_{ii} - E_{ij}) + \frac{1}{\lambda - 1} \frac{1}{N} \sum_{i \neq j}^{N} (E_{ii} - E_{ij}) . \tag{2.405}$$

With the last choice the matrix $\bar{r}(\lambda)$ becomes

$$\bar{r}(\lambda) = \frac{1}{\lambda - 1} \sum_{i \neq j} \frac{\lambda Q_i - Q_j}{Q_{ij}} (E_{ii} - E_{ij}) \otimes \left(E_{jj} - \frac{1}{N} \mathbb{1} \right) ,$$

while for $r(\lambda, \mu)$ one finds

$$r_{12}(\lambda, \mu) = \frac{\lambda r_{12}^{m} + \mu r_{21}^{m}}{\lambda - \mu} + \frac{\rho_{12}}{\lambda - 1} - \frac{\rho_{21}}{\mu - 1} , \tag{2.406}$$

where

$$\rho_{12} = \sum_{i \neq j} (E_{ii} - E_{ij}) \otimes \left(E_{jj} - \frac{1}{N} \mathbb{1} \right)$$

and the modified r-matrix is

$$\begin{aligned} r_{12}^{m} = &\sum_{i \neq j}^{N} \left(\frac{Q_j}{Q_{ij}} E_{ii} - \frac{Q_i}{Q_{ij}} E_{ij} \right) \otimes (E_{jj} - E_{ji}) \\ &- \frac{1}{N} \sum_{i \neq j} \frac{Q_i}{Q_{ij}} (E_{ii} - E_{ij}) \otimes \mathbb{1} + \frac{1}{N} \sum_{i \neq j} \frac{Q_i}{Q_{ij}} \mathbb{1} \otimes (E_{ii} - E_{ij}) . \end{aligned} \tag{2.407}$$

The modified r-matrix still solves the CYBE and obeys the same relation (2.371).

---

[33] An example of such a redefinition that does not affect the shifted Yang-Baxter equation corresponds to the choice $s(\lambda) = f(\lambda) \mathbb{1}$, where f is an arbitrary function of λ.

There is no symmetry operating on r-matrices that would allow one to remove from these matrices the scale-non-invariant terms. Clearly, the r-matrices satisfying the shifted version of the Yang-Baxter equations have a simpler structure than their cousins subjected to the standard Yang-Baxter equations. This fact plays an important role when it comes to quantisation of the corresponding model and the associated algebraic structures. We also point out that the r-matrices we found here through considerations in Appendix D also follow from the elliptic r-matrices of [60] upon their hyperbolic degeneration, albeit modulo the shift symmetries (2.403) and (2.404).

From (2.395) one then finds

$$\{\mathrm{Tr}_1 L_1(\lambda), L_2(\mu)\} = [\mathrm{Tr}_1 L_1(\lambda)(r_{12}(\lambda, \mu) + \bar{r}_{21}(\mu)), L_2(\mu)],$$

which, upon taking the limit $\lambda \to \infty$, yields the Lax equation (2.394) with M given by (2.391). The conserved quantities are, therefore, $I_k(\lambda) = \mathrm{Tr} L(\lambda)^k$, $k \in \mathbb{Z}$. The determinant $\det(L(\lambda) - \zeta \mathbb{1})$, which generates $I_k(\lambda)$ in the power series expansion over the parameter ζ, defines the *classical spectral curve*

$$\det(L(\lambda) - \zeta \mathbb{1}) = 0, \quad \zeta, \lambda \in \mathbb{C}. \tag{2.408}$$

2.3   Infinite-Dimensional Phase Spaces

So far we have considered finite-dimensional phase spaces with symmetries. Reduction of degrees of freedom related to these symmetries led us to the construction of new phase spaces and a non-trivial class of dynamical systems defined over them. These systems come along equipped with a Lax pair that is sufficient to prove their kinematic integrability. On the other hand, there is no room within this construction for the emergence of a spectral parameter. The situation is thus not fully satisfactory, because only the knowledge of a spectral-dependent Lax representation allows for the application of powerful analytical tools to solve the model, including, for instance, the linearisation of the hamiltonian flows on the Jacobian of the corresponding spectral curve. Yet another problem is that the most general CMS and RS models with elliptic potentials escaped our treatment.

Fortunately, it is possible to extend the present construction by replacing the initial finite-dimensional phase spaces with their properly chosen affine versions. This allows one to include in our reduction scheme the models with elliptic potential, which automatically come equipped with a spectral parameter. Upon degeneration these models produce the spectral-dependent Lax representations for the corresponding rational and hyperbolic (trigonometric) cousins. Interesting examples of infinite-dimensional phase spaces constitute cotangent bundles over centrally extended loop algebras or loop group, as well as the so-called affine Heisenberg double. As before, a suitably formulated reduction procedure applied to these spaces yields a finite-dimensional phase space and, in some cases, the Lax representation comes with a spectral parameter, the latter being inherited from the loop parameter of an initial phase space.

To keep the book within a reasonable size, we restrict ourselves to a brief description of the corresponding phase spaces and types of integrable models that follow from them upon reduction. We will not describe the reduction itself, referring the reader to the existing literature.

Let $\mathfrak{g}$ be a finite-dimensional Lie algebra with a non-degenerate invariant symmetric bilinear form $\langle \cdot, \cdot \rangle$ and $\mathcal{L}\mathfrak{g}$ be the corresponding loop algebra $\mathcal{L}\mathfrak{g}$. This bilinear form is used to identify $\mathfrak{g}^* \simeq \mathfrak{g}$. The affine Kac-Moody algebra $\widehat{\mathcal{L}\mathfrak{g}}$ is defined as a central extension of $\mathcal{L}\mathfrak{g}$ by a one-dimensional center [61]. It has a realisation as a space of pairs $(X(x), c)$, where c is a complex number and X is a mapping of the circle S^1 into $\mathfrak{g}$, that is a 2π-periodic $\mathfrak{g}$-valued function of x. The Lie bracket of $\widehat{\mathcal{L}\mathfrak{g}}$ is given by [62]

$$[(X(x), c), (Y(x), c')] = \Big([X(x), Y(x)], \omega(X, Y) \Big),$$

where

$$\omega(X, Y) = \int_{S^1} \langle X, \partial Y \rangle, \quad \partial = \frac{\partial}{\partial x}. \tag{2.409}$$

is a 2-cocycle $\omega : \mathcal{L}\mathfrak{g} \wedge \mathcal{L}\mathfrak{g} \to \mathbb{C}$, so that ω satisfies (2.120). The dual space $\widehat{\mathcal{L}\mathfrak{g}}^*$ to $\widehat{\mathcal{L}\mathfrak{g}}$ is spanned by pairs (ℓ, k), where $\ell \equiv \ell(x)\mathrm{d}x$ is a one-form with values in $\mathfrak{g}^* \simeq \mathfrak{g}$ and $k \in \mathbb{C}$. The pairing between $\widehat{\mathcal{L}\mathfrak{g}}^*$ and $\widehat{\mathcal{L}\mathfrak{g}}$ is introduced as

$$\langle (\ell, k), (X, c) \rangle = \int_{S^1} \langle \ell, X \rangle + kc. \tag{2.410}$$

Let $\mathcal{L}G$ be the loop group corresponding to $\mathcal{L}\mathfrak{g}$. The adjoint action of $\mathcal{L}G$ on $\widehat{\mathcal{L}\mathfrak{g}}$ is defined as

$$(X(x), c) \to \Big(h(x)X(x)h(x)^{-1}, c - \int_{S_1} \langle X, h^{-1}\partial h \rangle \Big). \tag{2.411}$$

The pairing (2.410) leads then to the following form of the coadjoint action of the loop group on $\widehat{\mathcal{L}\mathfrak{g}}^*$

$$(\ell(x), k) \to \Big(h(x)\ell(x)h(x)^{-1} + k\partial h(x)h(x)^{-1}, k \Big), \tag{2.412}$$

that is k remains unchanged.

Given these notions, we can now list a few phase spaces and the corresponding integrable models derived from them. For the rest of this section $\mathfrak{g} = \mathrm{Mat}_N(\mathbb{C})$ and $G = \mathrm{GL}_N(\mathbb{C})$.

Cotangent bundle over a loop group. The cotangent bundle $T^*\widehat{\mathcal{L}G}$ over a centrally extended loop group $\widehat{\mathcal{L}G}$ is a symplectic manifold.[34] In terms of coordinate functions $g(x) \in G$ and $\ell(x) \in \mathfrak{g}^* \simeq \mathfrak{g}$ the corresponding Poisson structure reads as follows

$$\{g_1(x), g_2(y)\} = 0,$$

$$\{\ell_1(x), \ell_2(y)\} = \frac{1}{2}[C_{12}, \ell_1(x) - \ell_2(y)]\delta^{(1)}(x - y) + kC_{12}\partial\delta^{(1)}(x - y), \quad (2.413)$$

$$\{\ell_1(x), g_2(y)\} = g_2(y)C_{12}\delta^{(1)}(x - y).$$

This structure is an affine analogue of T^*G, see (2.65). In particular, the middle bracket is the current algebra realisation of the Kirillov-Kostant bracket on $\widehat{\mathcal{L}\mathfrak{g}}$. In the above formulae

$$\delta^{(1)}(x) = \lim_{\epsilon \to +0} \frac{1}{2\pi} \sum_{k \in \mathbb{Z}} e^{ikx - |k|\epsilon} = \sum_{k \in \mathbb{Z}} \delta(x + 2\pi k) \quad (2.414)$$

is the periodic delta-function, $\delta^{(1)}(x + 2\pi) = \delta^{(1)}(x)$.

The group $\mathcal{L}G$ acts on $T^*\widehat{\mathcal{L}G}$ by means of the following transformations

$$g(x) \to h(x)g(x)h(x)^{-1},$$
$$\ell(x) \to h(x)\ell(x)h(x)^{-1} + k\partial h(x)h(x)^{-1}, \quad (2.415)$$

where $h(x) \in \mathcal{L}G$. It is not hard to verify that this action is hamiltonian and that the corresponding moment map $\mu : T^*\widehat{\mathcal{L}G} \to \widehat{\mathcal{L}\mathfrak{g}}^*$ is

$$\mu(x) = \ell(x) - g(x)\ell(x)g(x)^{-1} - k\partial g(x)g(x)^{-1}. \quad (2.416)$$

If we fix the moment map to the following singular value

$$\mu = \gamma(\boldsymbol{e} \otimes \boldsymbol{e}^t - \mathbb{1})\delta(x), \quad (2.417)$$

then the integrable model arising on the corresponding reduced phase space is the trigonometric RS system with the coupling constant γ [63]. The group element $g(x)$ on the reduced space essentially coincides with the Lax matrix; the dependence of g on x appears to be trivial and can be removed by passing to a gauge-equivalent Lax matrix. The fact that the reduced phase space is finite-dimensional is explained by noting that the stability group of a singular moment map is infinite-dimensional and large enough to remove all but a finite number degrees of freedom.

The space $T^*\widehat{\mathcal{L}G}$ is a deformation of $T^*\widehat{\mathcal{L}\mathfrak{g}}$. Let us introduce in (2.413) a parameter κ by multiplying the left hand side of the brackets in (2.413) with $1/\kappa$. Now, expanding $g(x) = \mathbb{1} + \kappa q(x) + \ldots$ and rescaling $k \to \kappa k$, we find that in the limit

---

[34] We assume that the level k is fixed.

$\kappa \to 0$ the formulae (2.413) turn into the Poisson relations of $T^*\widehat{\mathcal{L}\mathfrak{g}}$. The corresponding reduced model coincides with the trigonometric CMS system, see Fig. 2.8.

Cotangent bundle over double loop group. Let T_τ be a torus endowed with the standard complex structure and periods 1 and τ with $\mathrm{Im}\tau > 0$. Denote by $\mathcal{L}_\tau G$ a group of smooth mappings from T_τ into G. The elements of $\mathcal{L}_\tau G$ are G-valued matrices on the torus

$$g(z+1, \bar{z}+1) = g(z, \bar{z}), \quad g(z+\tau, \bar{z}+\bar{\tau}) = g(z, \bar{z}).$$

The group $\mathcal{L}_\tau G$ admits a central extension $\widehat{\mathcal{L}_\tau G}$ [64]. The dual space to the double-loop Lie algebra $\mathcal{L}_\tau \mathfrak{g}$ is spanned by the $\mathfrak{g}^* \simeq \mathfrak{g}$-valued functions $\ell(z, \bar{z})$ on the torus:

$$\ell(z+1, \bar{z}+1) = \ell(z, \bar{z}), \quad \ell(z+\tau, \bar{z}+\bar{\tau}) = \ell(z, \bar{z}).$$

We regard $g(z) \equiv g(z, \bar{z})$ and $\ell(z) \equiv \ell(z, \bar{z})$ as the coordinate functions on the cotangent bundle $T^*\widehat{\mathcal{L}_\tau G}$. In terms of these functions the Poisson structure of $T^*\widehat{\mathcal{L}_\tau G}$ is given by

$$\{g_1(z), g_2(w)\} = 0,$$
$$\{\ell_1(z), \ell_2(w)\} = \frac{1}{2}[C_{12}, \ell_1(z) - \ell_2(w)]\delta^{(2)}(z-w) - k\,C_{12}\bar{\partial}\delta^{(2)}(z-w), \quad (2.418)$$
$$\{\ell_1(z), g_2(w)\} = g_2(w)C_{12}\delta^{(2)}(z-w),$$

where $\bar{\partial} \equiv \frac{\partial}{\partial \bar{z}}$ and $\delta^{(2)}(z)$ is the two-dimensional doubly-periodic delta-function with periods 1 and τ. The hamiltonian action of the group $\mathcal{L}_\tau G$ is

$$g(z) \to h(z)g(z)h(z)^{-1},$$
$$\ell(z) \to h(z)\ell(z)h(z)^{-1} + k\bar{\partial}h(z)h(z)^{-1}$$

where $h(z) \equiv h(z, \bar{z}) \in \mathcal{L}_\tau G$. The corresponding momentum map is

$$\mu(z) = \ell(z) - g(z)\ell(z)g(z)^{-1} - k\bar{\partial}g(z)g(z)^{-1}.$$

The hamiltonian reduction that leads to the RS model with the elliptic potential has been described in [65]. On the reduced phase space the Lax matrix becomes a quasi-periodic meromorphic function of the spectral parameter z with a single pole in the fundamental domain. For the quantum version of this reduction the reader may consult [60]. The classical r-matrix structure of the elliptic CMS and RS models has been investigated earlier in [54, 66, 67].

Affine Heisenberg double. Here we present the Poisson structure of the Heisenberg double associated to the centrally extended loop algebra $\widehat{\mathcal{L}G}$. Let $A(x)$ and $B(x)$ be 2π-periodic functions of $x \in \mathbb{R}$ with values in G:

$$A(x + 2\pi) = A(x), \quad B(x + 2\pi) = B(x).$$

The matrix elements of $A(x)$ and $B(x)$ can be regarded as generators of the algebra of functions on the affine Heisenberg double. The Poisson algebra relations between these generators are

$$\frac{1}{\varkappa}\{A_1(x), A_2(y)\} = -r_{\mp}(x - y)A_1(x)A_2(y) - A_1(x)A_2(y)r_{\pm}(x - y)$$
$$+ A_1(x)r_-(x - y - 2\Delta)A_2(y) + A_2(y)r_+(x - y + 2\Delta)A_1(x),$$

$$\frac{1}{\varkappa}\{A_1(x), B_2(y)\} = -r_-(x - y)A_1(x)B_2(y) - A_1(x)B_2(y)r_-(x - y - 2\Delta)$$
$$+ A_1(x)r_-(x - y - 2\Delta)B_2(y) + B_2(y)r_+(x - y)A_1(x),$$

$$\frac{1}{\varkappa}\{B_1(x), A_2(y)\} = -r_+(x - y)B_1(x)A_2(y) - B_1(x)A_2(y)r_+(x - y + 2\Delta)$$
$$+ B_1(x)r_-(x - y)A_2(y) + A_2(y)r_+(x - y + 2\Delta)B_1(x),$$

$$\frac{1}{\varkappa}\{B_1(x), B_2(y)\} = -r_{\mp}(x - y)B_1(x)B_2(y) - B_1(x)B_2(y)r_{\pm}(x - y)$$
$$+ B_1(x)r_-(x - y)B_2(y) + B_2(y)r_+(x - y)B_1(x),$$

where $\varkappa$ and Δ are complex numbers with Im $\Delta > 0$. The matrices $r_{\pm}(x)$ are defined by their Fourier series expansions as

$$r_+(x) = r_+ + C \lim_{\epsilon \to +0} \sum_{k>0} e^{-ikx - k\epsilon}, \quad r_-(x) = r_- - C \lim_{\epsilon \to +0} \sum_{k>0} e^{ikx - k\epsilon},$$

where $x \in \mathbb{R}$ and $\epsilon \to +0$ is the regularisation parameter. The zero modes $r_{\pm}$ are given by (2.351). It can be easily checked that

$$r_+(x) - r_-(x) = 2\pi C \delta^{(1)}(x), \quad Cr_+(x)C = -r_-(-x),$$

where $\delta^{(1)}(x)$ is the periodic delta-function (2.414). In the domain of convergence $r_{\pm}(x)$ coincide with the standard trigonometric r-matrices associated to the affine Lie algebra[35] corresponding to the A_{N-1} series of simple Lie algebras

$$r_{\pm}(x) = \lim_{\epsilon \to 0} \frac{r_+ e^{ix} - r_-}{e^{ix} - 1 \pm \epsilon}. \tag{2.419}$$

Note that the Poisson relations $\{A_1, A_2\}$ are slightly different from $\{B_1, B_2\}$, the latter bracket does not involve the shift by Δ.

---

[35]In the homogeneous gradation.

If we set $B(x) = g(x)$, $2\Delta = \varkappa k$ and assume the expansions

$$A(x) = \mathbb{1} + \varkappa \ell(x) + \dots,$$

then in the limit $\varkappa \to 0$ we recover the Poisson structure (2.413) of $T^*\widehat{\mathcal{L}G}$. The Heisenberg double carries a Poisson action of the loop group

$$A(x) \to h(x + \Delta)A(x)h^{-1}(x - \Delta),$$
$$B(x) \to h(x + \Delta)B(x)h^{-1}(x + \Delta),$$

where $h \in \mathcal{L}G$ and the Poisson-Lie structure on $\mathcal{L}G$ is given by

$$\{h_1(x), h_2(y)\} = -[r_\pm(x - y), h_1(x)h_2(y)]. \tag{2.420}$$

The corresponding non-abelian moment map is

$$\mathfrak{m}(x) = B(x - \Delta)A^{-1}(x + \Delta)B^{-1}(x + \Delta)A(x + \Delta) \tag{2.421}$$

and it satisfies the following Poisson algebra

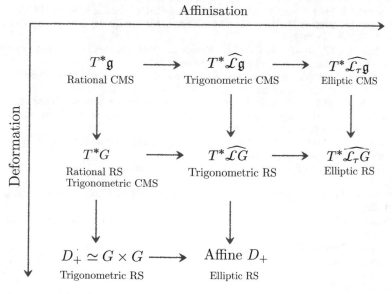

Fig. 2.8 Hierarchy of phase spaces and corresponding integrable models. Passing from one initial phase space to another in order of increasing complexity corresponds to either deformation or affinisation. Elliptic models come equipped with a spectral parameter

$$\{m_1(x), m_2(y)\} = -r_+(x - y)m_1(x)m_2(y) - m_1(x)m_2(y)r_-(x - y)$$
$$+ m_1(x)r_-(x - y)m_2(y) + m_2(y)r_+(x - y)m_1(x). \quad (2.422)$$

Fixing the moment map as described in [68] and performing the Poisson reduction of the double by the action of $\mathcal{L}G$, we recover on the corresponding reduced phase space the same elliptic RS model [68].

The full hierarchy of phase spaces that are obtained from the cotangent bundle $T^*\mathfrak{g}$ either by deformations or by passing to an affine cousin is summarised on Fig. 2.8. The lower right corner of the figure is empty and it would correspond to the double affine Heisenberg double, whose existence remains currently an open question.

References

1. Kirillov, A.A.: Lectures on the Orbit Method. AMS (2004)
2. Marsden, J., Weinstein, A.: Reduction of symplectic manifolds with symmetry. Rep. Math. Phys. **5**, 121–130 (1974)
3. Abraham, R., Marsden, J.E.: Foundations of Mechanics. AMS (1978)
4. Kazhdan, D., Kostant, B., Sternberg, S.: Hamiltonian group actions and dynamical systems of calogero type. Commun. Pure Appl. Math. **31**(4), 481–507 (1978)
5. Arnold, V.I.: Mathematical Methods of Classical Mechanics, vol. 60. Springer (1989)
6. Souriau, J.-M.: Structure des systmes dynamiques. Dunod (1970)
7. Dirac, P.A.M.: Generalized hamiltonian dynamics. Can. J. Math. **2**, 129–148 (1950)
8. Babelon, O., Bernard, D., Talon, M.: Introduction to Classical Integrable Systems. Cambridge University Press (2003)
9. de Azcrraga, J.A., Izquierdo, J.M.: Lie Groups, Lie Algebras, Cohomology and some Applications in Physics. Cambridge University Press (2011)
10. Drinfeld, V.G.: Hamiltonian structures of lie groups, lie bialgebras and the geometric meaning of the classical Yang-Baxter equations. Sov. Math. Dokl. **27**, 68–71 (1983)
11. Chari, V., Pressley, A.: A Guide to Quantum Groups. University Press, Cambridge, UK (1994)
12. Onishchik, A.L., Vinberg, E.B. (eds.): Lie Groups and Lie Algebras III, vol. 41. Springer, Berlin, Heidelberg, New York (1994)
13. Semenov-Tian-Shansky, M.A.: What is a classical r-matrix? Funct. Anal. Appl. **17**, 259–272 (1983). [Funkt. Anal. Pril. 17N4, 17 (1983)]
14. Stolin, A.: On rational solutions of Yang-Baxter equation for $\mathfrak{sl}(n)$. Math. Scand. **69**, 57–80 (1991)
15. Stolin, A.: Constant solutions of Yang-Baxter equation for $\mathfrak{sl}(2)$ and $\mathfrak{sl}(3)$. Math. Scand. **69**, 81–88 (1991)
16. Reshetikhin, N.Y., Semenov-Tian-Shansky, M.A.: Quantum R-matrices and factorization problems. J. Geom. Phys. **5**, 533–550 (1988)
17. Semenov-Tian-Shansky, M.A.: Poisson Lie groups, quantum duality principle, and the quantum double. Theor. Math. Phys. **93**, 1292–1307 (1992). [Teor. Mat. Fiz. 93N2, 302 (1992)]
18. Semenov-Tian-Shansky, M.A.: Poisson groups and dressing transformations. J. Math. Sci. **46**, 1641–1657 (1989)
19. Weinstein, A.: The local structure of poisson manifolds. J. Differ. Geom. **18**(3), 523–557 (1983)
20. Marsden, J.E., Ratiu, T.: Reduction of poisson manifolds. Lett. Math. Phys. **11**(2), 161–169 (1986)
21. Lu, J.-H.: Momentum mappings and reduction of Poisson actions. In: Symplectic Geometry, Groupoids, and Integrable Systems (Berkeley, CA, 1989). Springer, New York (1991)
22. Delduc, F., Lacroix, S., Magro, M., Vicedo, B.: On q-deformed symmetries as Poisson-Lie symmetries and application to Yang-Baxter type models. J. Phys. A **49**(41), 415402 (2016)

23. Babelon, O., Bernard, D.: Dressing symmetries. Commun. Math. Phys. **149**, 279–306 (1992)
24. Flaschka, H., Ratiu, T.: A convexity theorem for poisson actions of compact lie groups. Annales scientifiques de l'École Normale Supérieure, Ser. 4, **29**(6), 787–809 (1996)
25. Semenov-Tian-Shansky, M.A.: Dressing transformations and Poisson group actions. Publ. Res. Inst. Math. Sci. Kyoto **21**, 1237–1260 (1985)
26. Alekseev, A.Yu., Malkin, A.Z.: Symplectic structures associated to Lie-Poisson groups. Commun. Math. Phys. **162**, 147–174 (1994)
27. Zakharov, V.E., Shabat ,A.B.: Integration of nonlinear equations of mathematical physics by the method of inverse scattering. II. Funct. Anal. Appl. **13**, 166–174 (1979). [Funkt. Anal. Pril. 13N3, 13 (1979)]
28. Date, E., Jimbo, M., Kashiwara, M., Miwa, T.: Transformation groups for soliton equations. Physica **4D**, 343–365 (1982)
29. Date, E., Jimbo, M., Kashiwara, M., Miwa, T.: Transformation groups for soliton equations. Publ. RIMS Kyoto Univ. **18**, 1077 (1982)
30. Lu, J.-H., Weinstein, A.: Poisson lie groups, dressing transformations, and bruhat decompositions. J. Differ. Geom. **31**(2), 501–526 (1990)
31. Lu, J.-H.: Multiplicative and affine poisson structures on lie groups. PhD Thesis (1990)
32. Kosmann-Schwarzbach, Y.: Lie bialgebras, Poisson Lie Groups and Dressing Transformations. Integrability of Nonlinear Systems, 2 edn, Lecture Notes in Physics, Vol. 638, pp. 107–173 (2004)
33. Fock, V.V., Rosly, A.A.: Poisson structure on moduli of flat connections on Riemann surfaces and r matrix. Am. Math. Soc. Transl. **191**, 67–86 (1999)
34. Alekseev, A.Yu., Malkin, A.Z.: Symplectic structure of the moduli space of flat connection on a Riemann surface. Commun. Math. Phys. **169**, 99–120 (1995)
35. Arnold, V.I., Novikov, S.P. (Eds.): Dynamical Systems VII: Integrable Systems Nonholonomic Dynamical Systems. Springer (1993)
36. Adler, M., van Moerbeke, P.: The complex geometry of the Kowalewski-Painlevé analysis. Invent. Math. **97**(1), 3–51 (1989)
37. Adler, M., van Moerbeke, P., Vanhaeche, P.: Algebraic Integrability, Painlevé Geometry and Lie Algebras. Springer (2004)
38. Donagi, R.Y.: Seiberg-Witten integrable systems. alg-geom/9705010 (1997)
39. Fehér, L., Klimčík, C.: Poisson-Lie generalization of the Kazhdan-Kostant-Sternberg reduction. Lett. Math. Phys. **87**, 125–138 (2009)
40. Fehér, L., Klimčík, C.: Poisson-Lie interpretation of trigonometric Ruijsenaars duality. Commun. Math. Phys. **301**, 55–104 (2011)
41. Fehér, L., Kluck, T.J.: New compact forms of the trigonometric Ruijsenaars-Schneider system. Nucl. Phys. B **882**, 97–127 (2014)
42. Fehér, L., Görbe, T.F.: The full phase space of a model in the Calogero-Ruijsenaars family. J. Geom. Phys. **115**, 139–149 (2017)
43. Arutyunov, G.E., Frolov, S.A.: Quantum dynamical R-matrices and quantum frobenius group. Commun. Math. Phys. **191**, 15 (1998)
44. Ruijsenaars, S.N.M.: Complete integrability of relativistic Calogero-Moser systems and elliptic function identities. Commun. Math. Phys. **110**, 191 (1987)
45. Ruijsenaars, S.N.M.: Action-angle maps and scattering theory for some finite-dimensional integrable systems. I. The Pure Soliton Case. Commun. Math. Phys. **115**, 127–165 (1988)
46. Ruijsenaars, S.N.M.: Action-angle maps and scattering theory for some finite-dimensional integrable systems iii. Sutherland type systems and their duals. Publ. Res. Inst. Math. Sci. **31**(2), 247–353 (1995)
47. Fock, V., Gorsky, A., Nekrasov, N., Rubtsov, V.: Duality in integrable systems and gauge theories. JHEP **07**, 028 (2000)
48. Fehér, L., Klimčík, C.: On the duality between the hyperbolic Sutherland and the rational Ruijsenaars-Schneider models. J. Phys. A **42**, 185202 (2009)
49. Fehér, L., Ayadi, V.: Trigonometric Sutherland systems and their Ruijsenaars duals from symplectic reduction. J. Math. Phys. **51**, 103511 (2010)

50. Fehér, L., Klimčík, C.: Self-duality of the compactified Ruijsenaars-Schneider system from quasi-Hamiltonian reduction. Nucl. Phys. B **860**, 464–515 (2012)
51. Fehér, L., Marshall, I.: The action-angle dual of an integrable Hamiltonian system of Ruijsenaars-Schneider-van Diejen type. J. Phys. A **50**(31), 314004 (2017)
52. Babelon, O., Bernard, D.: The Sine-Gordon solitons as a N body problem. Phys. Lett. B **317**, 363–368 (1993)
53. Avan, J., Rollet, G.: The classical r matrix for the relativistic Ruijsenaars-Schneider system. Phys. Lett. A **212**, 50 (1996)
54. Suris, Y.B.: Why are the Ruijsenaars-Schneider and the Calogero-Moser hierarchies governed by the same *r* matrix? Phys. Lett. A **225**, 253–262 (1997)
55. Krichever, I., Zabrodin, A.: Spin generalization of the Ruijsenaars-Schneider model, non-Abelian 2-d Toda chain and representations of Sklyanin algebra. Russ. Math. Surv. **50**, 1101 (1995)
56. Arutyunov, G.E., Frolov, S.A.: On Hamiltonian structure of the spin Ruijsenaars-Schneider model. J. Phys. A **31**, 4203–4216 (1998)
57. Fehér, L.: Poisson-Lie analogues of spin Sutherland models. math-ph/1809.01529 (2018)
58. Chalykh, O., Fairon, M.: On the Hamiltonian formulation of the trigonometric spin Ruijsenaars-Schneider system. math-ph/1811.08727 (2018)
59. Ruijsenaars, S.N.M., Schneider, H.: A new class of integrable systems and its relation to solitons. Ann. Phys. **170**, 370–405 (1986)
60. Arutyunov, G.E., Chekhov, L., Frolov, S.A.: R-matrix quantization of the Elliptic Ruijsenaars-Schneider model. Commun. Math. Phys. **192**, 405–432 (1998)
61. Kac, V.G.: Infinite Dimensional Lie Algebras. University Press, Cambridge, UK (1990)
62. Pressley, A., Segal, G.: Loop Groups. Oxford mathematical monographs. Clarendon Press (1988)
63. Gorsky, A., Nekrasov, N.: Relativistic Calogero-Moser model as gauged WZW theory. Nucl. Phys. B **436**, 582–608 (1995)
64. Etingof, P.I., Frenkel, I.B.: Central extensions of current groups in two-dimensions. Commun. Math. Phys. **165**, 429–444 (1994)
65. Arutyunov, G.E., Frolov, S.A., Medvedev, P.B.: Elliptic Ruijsenaars-Schneider model from the cotangent bundle over the two-dimensional current group. J. Math. Phys. **38**, 5682–5689 (1997)
66. Sklyanin, E.K.: Dynamical r matrices for the elliptic Calogero-Moser model. Alg. Anal. **6**(2), 227–237 (1994). [St. Petersburg Math. J.6, 397 (1995)]
67. Nijhoff, F.W., Kuznetsov, V.B., Sklyanin, E.K., Ragnisco, O.: Dynamical r matrix for the elliptic Ruijsenaars-Schneider system. J. Phys. A **29**, L333–L340 (1996)
68. Arutyunov, G.E., Frolov, S.A., Medvedev, P.B.: Elliptic Ruijsenaars-Schneider model via the Poisson reduction of the affine Heisenberg double. J. Phys. A **30**, 5051–5063 (1997)

Chapter 3
Quantum-Mechanical Integrable Systems

The notion of complete integrability assumes central place in classical mechanics. Almost all great names in mechanics are associated with development of this notion and it is vividly discussed in textbooks. It is not so in quantum mechanics ...the quantization is rather an art than science. Thus the transfer of the features of classical mechanics into their quantum analogue is not automatic. Moreover, these features could be modified in this passage.

Ludwig Faddeev
What is complete integrability in quantum mechanics

In this chapter we deal with integrable models of quantum mechanics. After a brief introduction to the theory of quantum groups, which allows us to bring about some important tools and concepts like the deformation quantisation and the quantum Yang-Baxter equation, we show how to obtain the main algebraic structures related to the quantum RS models by using the reduction framework developed in the previous chapter. For the CMS and RS models with the discrete spectrum, we present the common eigenfunctions of commuting quantum integrals in terms of special families of orthogonal polynomials, thereby showing that for these quantum-mechanical models the spectral problem is fully solved.

3.1 Quantum Mechanics and Integrability

We open our discussion of quantum integrability by studying the case of finite-dimensional quantum integrable models. These are quantum-mechanical models in the usual sense of quantum mechanics: their unitary time evolution is described by

© Springer Nature Switzerland AG 2019
G. Arutyunov, *Elements of Classical and Quantum Integrable Systems*,
UNITEXT for Physics, https://doi.org/10.1007/978-3-030-24198-8_3

a multi-body wave function subjected to the Schrödinger equation. The Schrödinger equation is driven by the quantum hamiltonian $\widehat{H}$—a specially-chosen operator on the Hilbert space which represents a quantum-mechanical analogue of the classical hamiltonian H. In general, constructing a quantum model from its classical counterpart is a non-trivial problem, known under the common name "quantisation", that relies on a set of consistent axioms and that typically does not have a unique mathematical solution. Physically, the correctness of a well-defined quantum model is verified by comparing its theoretical predictions for observable quantities with results of an experiment.

Recall that in classical mechanics observables are real functions on a phase space $\mathcal{P}$ forming a linear space $\mathcal{F}(\mathcal{P})$. A quantisation map $Q_\hbar : \mathcal{F}(\mathcal{P}) \to \mathcal{A}$ puts in one-to-one correspondence to a classical observable f an operator $\widehat{f}$ from the set $\mathcal{A}$ of self-adjoint operators acting on some Hilbert space. This map depends on a parameter $\hbar$, the Planck constant, and is required to satisfy the following relations

$$\lim_{\hbar \to 0} \tfrac{1}{2} Q_\hbar^{-1}\big(Q_\hbar(f)Q_\hbar(h) + Q_\hbar(h)Q_\hbar(f)\big) = fh \qquad (3.1)$$

and

$$\lim_{\hbar \to 0} Q_\hbar^{-1}\big(\{Q_\hbar(f), Q_\hbar(h)\}_\hbar\big) = \{f, h\}, \qquad (3.2)$$

for any bounded classical observables $f, h \in \mathcal{F}(\mathcal{P})$, see e.g. [1]. Here $\{\cdot, \cdot\}_\hbar$ is a quantum Poisson bracket and it is related to the commutator as

$$\{\widehat{f}, \widehat{h}\}_\hbar = \frac{i}{\hbar}[\widehat{f}, \widehat{h}].$$

In the canonical quantisation procedure the quantisation map is first constructed for the phase space coordinates (p_i, q_i) with the canonical Poisson bracket $\{p_i, q_j\} = \delta_{ij}$. The corresponding operators $\widehat{p}_i, \widehat{q}_i$ are subject of the *Heisenberg commutation relations*

$$[\widehat{q}_i, \widehat{q}_j] = 0, \quad [\widehat{p}_i, \widehat{p}_j] = 0, \quad [\widehat{p}_i, \widehat{q}_j] = -i\hbar\,\delta_{ij}. \qquad (3.3)$$

The symbols $\widehat{q}_i$ and $\widehat{p}_i$ define a non-commutative algebra known as the Heisenberg algebra. By the Stone-von Neumann theorem this algebra has a unique (up to unitary equivalence) unitary irreducible representation.

The quantisation map $Q_\hbar$ is not unique. Even within the canonical procedure the non-uniqueness of this map shows up in the ordering problem that arises when trying to define it for an arbitrary classical observable $f(p, q)$: while classically p_i and q_i commute, this is not so for their quantum counterparts and one needs to specify the order in which the operators $\widehat{p}_i, \widehat{q}_i$ enter the expression for the corresponding classical observable $\widehat{f}$. For instance, for the well-known *Weyl* quantisation prescription, to get a quantum operator corresponding to a polynomial function of p_i, q_i, one postulates to symmetries products of non-commutative factors $\widehat{p}_i, \widehat{q}_i$. More generally, the question is, what is the Schrödinger operator of a quantum-mechanical system as a priori there are many (infinite number of) operators that correspond to the one and the same classical hamiltonian.

Coming back to integrable systems, we recall that Liouville integrability implies the existence of an involutive set of independent functions $H_k(p, q)$ which include the hamiltonian and which number is equal to half of the dimension of the phase space. The classical functions H_k pair-wise commute with respect to the Poisson bracket and it is natural to require their quantum counterparts to do so with respect to the commutator. Thus, one defines a quantum integrable system as a quantisation of a classical one with a complete involutive set H_k, if there exists a quantisation map such that the corresponding quantum hamiltonians $\widehat{H}_k$ all commute and, therefore, have a common spectrum of eigenvalues.

The above definition of a quantum integrable model, by no means, promises that our intuition gained from dealing with classical aspects of integrability can be judiciously extended to the corresponding quantum theory. To give an example, a functional independence on the classical H_k does not imply an algebraic independence of $\widehat{H}_k$. Indeed, consider, for instance, two commuting hermitian operators A and B with a simple spectrum acting on a finite-dimensional Hilbert space of dimension d and a scalar product $\langle \cdot , \cdot \rangle$. The corresponding spectral decomposition reads

$$A = \sum_{j=1}^{d} a_j P_j , \quad B = \sum_{j=1}^{d} b_j P_j ,$$

where P_j are the projector operators $P_j = \langle e_j, \cdot \rangle e_j$ and e_j is an orthonormal basis of common eigenvectors

$$A e_j = a_j e_j , \quad B e_j = b_j e_j .$$

The projector operator P_j can be represented as $P_j = \prod_{k \neq j}^{d} \frac{B - b_k}{b_j - b_k}$ and, therefore,

$$A = \sum_{j=1}^{d} a_j \prod_{k \neq j}^{d} \frac{B - b_k}{b_j - b_k} = \sum_{k=1}^{d} c_k B^{k-1} ,$$

where the numerical coefficients c_k are unambiguously determined in terms of non-degenerate eigenvalues a_i and b_i. This shows that, in general, any two commuting operators A and B are algebraically dependent. On the other hand, there are at most d linearly independent operators constituting a complete set of commuting operators, the latter can be taken either as projectors P_j or as powers of B. From this point of view, especially without referring to an underlying classical system, quantum integrability does not mean anything special—for a generic operator $\widehat{H}$ considered as the quantum hamiltonian, there always exists a complete set of commuting operators given, for instance, by powers of $\widehat{H}$. It is clear that without specifying additional features of commuting quantities in quantum theory, or without transferring attention to other manifestations of integrable behaviour, one can say little about the phenomenon of quantum integrability [2]. We will return to these issues later restricting ourselves here to the claim that the full power of integrability encoded in a large commutative operator algebra of quantum hamiltonians shows up in the possibility to find their common spectrum by means of the so-called Bethe Ansatz technique and various incarnations thereof, including the Quantum Inverse Scattering Method.

As we will see, quantum integrable systems give rise to a number of very rich algebraic structures, even in the finite-dimensional setting. Some of the models possess hidden symmetries, such as Yangians and quantum groups. Integrability properties are controlled by the *quantum Yang-Baxter equation* for the quantum R-matrix, the latter can be viewed as a "quantum" analogue of the r-matrix satisfying the classical Yang-Baxter equation. The quantum Yang-Baxter equation is a universal integrability structure that underlies the construction of non-commutative operator algebras having sufficiently large commutative subalgebras and it serves as a consistency condition for factorisation of the scattering process in a sequence of two-body events.

In this section we illustrate the main concepts of quantum integrability on the example of the CMS and RS models. We argue that in some sense these models can be naturally constructed by adopting a "quantum" analogue of the reduction procedure that has been used to obtain these models in the classical setting. This will also allow us to define an operator-valued Lax matrix that will play an essential role in constructing a family of commuting operators that render the corresponding model quantum integrable.

As has been explained in introductory remarks to Sect. 2.2, in this book we are mainly dealing with algebraic integrable systems for which the phase space variables are holomorphic. Quantisation of such systems will also be understood in the holomorphic sense. In particular, the Planck constant $\hbar$ is a complex number and the Heisenberg commutation relations (3.3) are replaced by their holomorphic analogue

$$[\widehat{q}_i, \widehat{q}_j] = 0, \quad [\widehat{p}_i, \widehat{p}_j] = 0, \quad [\widehat{p}_i, \widehat{q}_j] = \hbar\, \delta_{ij}, \qquad (3.4)$$

thus, giving up a possible realisation of $\widehat{p}_i, \widehat{q}_i$ by self-adjoint operators. As in the classical theory, proper reality and hermiticity conditions[1] can be imposed at the last stage, when the quantum model has already been constructed and quantum integrability has been revealed in the holomorphic setting.

The route we undertake to treat the quantum CMS and RS models relies heavily on the R-matrix formalism, the latter being the backbone of the Quantum Inverse Scattering Method [3, 4]. Perhaps, a natural way to introduce this formalism is to start from the discussion of quantum groups—algebraic objects that can be regarded as the result of the "quantisation" of Poisson-Lie groups.

---

[1]Together with $\hbar \to -i\hbar$, $\hbar \in \mathbb{R}$ for the physical case of self-adjoint $\widehat{p}$ and $\widehat{q}$.

3.2 Quantum Groups

Here we give a condensed introduction into the theory of quantum groups that is needed for our further purposes. For more details and history we refer the reader to the seminal papers [5–7] and other sources [8, 9].

In classical mechanics physical observables that are functions on the phase space constitute a commutative algebra with respect to a point-wise multiplication. Quantisation consists in replacing this commutative algebra with a non-commutative algebra of operators on a Hilbert space. From the point of view of deformation theory, the process of quantisation can be viewed as a deformation of a commutative algebra into a non-commutative one continuously depending on a parameter $\hbar$. When it is a Lie group that plays the role of the classical phase space, the corresponding algebra of functions $\mathcal{F}(G)$ carries the structure of a *Hopf algebra*, and the latter contains all the information on the algebraic structure of the Lie group. In this spirit, quantum groups are defined as non-commutative Hopf algebras that are deformations of the Hopf algebra structure on a commutative algebra of functions on a Poisson-Lie group [5]. Thus, a Hopf algebra is a fundamental concept for this theory and we recall the basic definitions.

Hopf algebras. An associative unital algebra $\mathcal{A}$ with a unit $\mathbb{1}$ and a product $\cdot :$ $\mathcal{A} \otimes \mathcal{A} \to \mathcal{A}$ is called a Hopf algebra over a field $\mathbb{C}$, if it is equipped with a coproduct $\Delta : \mathcal{A} \to \mathcal{A} \otimes \mathcal{A}$ (algebra homomorphism), a counit $\varepsilon : \mathcal{A} \to \mathbb{C}$ (algebra homomorphism) and an antipode $S : \mathcal{A} \to \mathcal{A}$ (algebra anti-homomorphism), satisfying the following axioms

$$(\Delta \otimes id)\Delta(a) = (id \otimes \Delta)\Delta(a), \quad (\varepsilon \otimes id)\Delta(a) = (id \otimes \varepsilon)\Delta(a) = a,$$
$$\cdot(S \otimes id)\Delta(a) = \cdot(id \otimes S)\Delta(a) = \mathbb{1}\varepsilon(a), \quad \Delta(S(a)) = \sigma(S \otimes S)\Delta(a), \quad (3.5)$$
$$\varepsilon(S(a)) = \varepsilon(a), \quad \Delta(\mathbb{1}) = \mathbb{1} \otimes \mathbb{1}, \quad S(\mathbb{1}) = \mathbb{1}, \quad \varepsilon(\mathbb{1}) = 1,$$

where σ is the twist map $\sigma(a \otimes b) = (b \otimes a)$ and $a \in \mathcal{A}$. Note, that a linear space supplied with only Δ and ε is called a coalgebra, and an algebra that is also a coalgebra is called a bialgebra.

Let G be a Lie group with Lie algebra $\mathfrak{g}$. A classical example of a Hopf algebra is given by the commutative algebra $\mathcal{F}(G)$ of functions on G. The corresponding operations Δ, ε and S are given by

$$\Delta f(g_1, g_2) = f(g_1 g_2), \quad (\varepsilon f)(g) = f(e), \quad (Sf)(g) = f(g^{-1}),$$

where e is the group unity and $\mathcal{F}(G) \otimes \mathcal{F}(G)$ is identified under a suitable topology with $\mathcal{F}(G \times G)$. If G is a Poisson-Lie group, then the Poisson-Lie condition (2.76) can be expressed via the coproduct as

$$\Delta\{f, h\}_G = \{\Delta(f), \Delta(h)\}_{G \times G}. \quad (3.6)$$

The antipode coincides with the inversion map i and it is anti-Poisson (2.87).

Poisson-Hopf algebras. In fact, relation (3.6) motivates a general definition of a Poisson-Hopf algebra. A Hopf algebra $\mathcal{A}$ supplied with the Poisson bracket $\{\cdot, \cdot\} : \mathcal{A} \times \mathcal{A} \to \mathcal{A}$ compatible with Δ in the sense

$$\Delta\{f, h\}_{\mathcal{A}} = \{\Delta(f), \Delta(h)\}_{\mathcal{A} \times \mathcal{A}} \tag{3.7}$$

Accordingly, $\mathcal{F}(G)$ is the Poisson-Hopf algebra.

Formal quantum group. Let $V = \mathbb{C}^N$ and consider a non-degenerate matrix $R \in \mathrm{Mat}_{N^2}(\mathbb{C})$ acting on $V \otimes V$ and satisfying the *quantum Yang-Baxter equation*

$$R_{12} R_{13} R_{23} = R_{23} R_{13} R_{12}, \tag{3.8}$$

where, in accordance with our conventions, the lower indices denote an embedding of the matrix R into endomorphisms of $V^{\otimes 3}$. Introducing the matrix $\mathcal{R} = C\hat{R}$, where C is the split Casimir (2.71) which acts on $V \otimes V$ as the permutation matrix, the quantum Yang-Baxter equation can be put in the form of the *braid relation*

$$\hat{R}_{12} \hat{R}_{23} \hat{R}_{12} = \hat{R}_{23} \hat{R}_{12} \hat{R}_{23}. \tag{3.9}$$

Let $\mathcal{A}$ be a free associative algebra over $\mathbb{C}$ generated by $1, t_{ij}, i, j = 1 \ldots, N$ modulo the following relations

$$R_{12} T_1 T_2 = T_2 T_1 R_{12}, \tag{3.10}$$

where $T = (t_{ij})_{i,j=1}^{N}$ and $T_1 = T \otimes \mathbb{1}$, $T_2 = \mathbb{1} \otimes T$ are $N^2 \times N^2$-matrices with entries in $\mathcal{A}$. According to [7], the algebra $\mathcal{A}$ is called an algebra of functions on the formal quantum group associated to the matrix R, or, simply, the *formal quantum group*. If R is the identity matrix, the algebra $\mathcal{A}$ is commutative.

The formal quantum group is a bialgebra (a Hopf algebra) with the coproduct

$$\Delta(t_{ij}) = \sum_{k=1}^{N} t_{ik} \otimes t_{kj},$$

and the counit $\epsilon(t_{ij}) = \delta_{ij}, i, j = 1, \ldots, N$. In matrix form the action of the coproduct on generators reads as $\Delta(T) = T \otimes T$. The statement that Δ is a Lie algebra homomorphism reduces to the equation

$$R_{12}(T_1 T_2 \otimes T_1 T_2) = (T_2 T_1 \otimes T_2 T_1) R_{12},$$

which is satisfied as a consequence of (3.10).

Quasi-classics and Poisson-Lie groups. A relationship between quantum groups and Poisson-Lie groups can be naturally explained in the framework of deformation quantisation [10–12].

Let $\mathcal{A}_0$ be a commutative Poisson algebra over $\mathbb{C}$ and let $\{\cdot, \cdot\}$ be its Poisson bracket. A non-commutative algebra $\mathcal{A}$ is a deformation quantisation of $\mathcal{A}_0$ over the ring $\mathbb{C}[[\hbar]]$, where $\hbar$ is a formal parameter, if $\mathcal{A}_0 \simeq \mathcal{A}/\hbar\mathcal{A}$ and the following two conditions are satisfied

$$\pi(a \star b) = \pi(a) \cdot \pi(b), \tag{3.11}$$

$$\{\pi(a), \pi(b)\} = \pi([a, b]/\hbar) \tag{3.12}$$

for all $a, b \in \mathcal{A}$. Here we denoted a non-commutative product in $\mathcal{A}$ by $\star$ to distinguish it from the commutative product $\cdot$ in $\mathcal{A}_0$ and introduced the canonical projection π

$$\pi : \mathcal{A} \to \mathcal{A}/\hbar\mathcal{A} \simeq \mathcal{A}_0.$$

A few comments are in order. We can think of elements a from $\mathcal{A}$ as formal power series in $\hbar$ with coefficients in $\mathcal{A}_0$, that is

$$a = \sum_{k=0}^{\infty} a_k \hbar^k.$$

Factoring over $\hbar\mathcal{A}$ means taking the classical limit $\hbar \to 0$, so that $\pi(a) = a_0$. According to the deformation theory, we can write

$$a \star b = \sum_{k=0}^{\infty} f_k(a, b)\hbar^k, \tag{3.13}$$

where the coefficients $f_k : \mathcal{A} \times \mathcal{A} \to \mathcal{A}$ should comply with the requirement of associativity of the $\star$-product. In the limit $\hbar \to 0$, the product $a \star b$ reduces to $f_0(a, b)$ which is, due to (3.11), equal to the commutative product $a_0 \cdot b_0$. Note that the commutator in (3.12) is defined as $[a, b] = a \star b - b \star a$. Commutativity of $\mathcal{A}_0$ guarantees that in the classical limit the expansion of $[a, b]$ starts for any $a, b \in \mathcal{A}$ at least at order $\hbar$, rendering the condition (3.12) consistent.

If $\mathcal{A}_0$ is a Poisson-Hopf algebra with coproduct Δ_0, then its deformation quantisation is defined as a Hopf algebra $\mathcal{A}$ with the coproduct Δ, where in addition to conditions (3.11) and (3.12) we require the fulfilment of the following relation

$$(\pi \otimes \pi) \circ \Delta = \Delta_0 \circ \pi. \tag{3.14}$$

Now let R be a solution of (3.8) which depends on a parameter $\hbar \in \mathbb{C}$ and in the limit $\hbar \to 0$ has an expansion

$$R = 1 + \hbar r + o(\hbar)\,. \tag{3.15}$$

Substituting this expansion into (3.8), at order $\hbar^2$ we find for r the classical Yang-Baxter equation (2.112). This shows that the quantum Yang-Baxter equation (3.8) can be regarded as a deformation (quantisation) of the classical one and justifies for R the name *quantum R-matrix*. To give an example, consider the following solution of (3.8)

$$R = \sum_{i \neq j}^{N} E_{ii} \otimes E_{jj} + e^{\hbar/2} \sum_{i=1}^{N} E_{ii} \otimes E_{ii} + (e^{\hbar/2} - e^{-\hbar/2}) \sum_{i>j}^{N} E_{ij} \otimes E_{ji}\,. \tag{3.16}$$

Using this R one can construct two more solutions $R_{\pm}$ of the quantum Yang-Baxter equation

$$R_{+12} = R_{21}\,, \qquad R_{-12} = R_{12}^{-1}\,. \tag{3.17}$$

These solutions are, therefore, related as

$$R_{+21} R_{-12} = \mathbb{1}\,. \tag{3.18}$$

Yet another relation is

$$R_{+} - R_{-} = (e^{\hbar/2} - e^{-\hbar/2})\,C\,, \tag{3.19}$$

where C is the split Casimir. In the limit $\hbar \to 0$ the matrices $R_{\pm}$ have an expansion

$$R_{\pm} = 1 + \hbar r_{\pm} + o(\hbar)\,, \tag{3.20}$$

where $r_{\pm}$ are the classical r-matrices (2.351). Further, we point out that $\hat{R}_{\pm} = C R_{\pm}$ satisfy the *Hecke condition*

$$\hat{R}_{\pm}^2 \mp (e^{\hbar/2} - e^{-\hbar/2})\hat{R}_{\pm} - \mathbb{1} = (\hat{R}_{\pm} - e^{\pm\hbar/2}\mathbb{1})(\hat{R}_{\pm} + e^{\mp\hbar/2}\mathbb{1}) = 0\,. \tag{3.21}$$

Consider now a quantum group $\mathcal{A}$ defined by some quantum R-matrix with the quasi-classical expansion (3.15). Let G be a Poisson-Lie group with the Sklyanin bracket

$$\{g_1, g_2\} = [r, g_1 g_2]\,, \tag{3.22}$$

where r is the same as in (3.15). Then, the Hopf algebra $\mathcal{A}$ is a deformation quantisation of the commutative Poisson-Hopf algebra $\mathcal{A}_0 = \mathcal{F}(G)$. Indeed, for generators t_{ij} of $\mathcal{A}$ we set $\pi(t_{ij}) = g_{ij}$. By expanding R in (3.10), we will get

$$[t_{ij}, t_{kl}] = t_{ij} \star t_{kl} - t_{kl} \star t_{ij} = \hbar(r_{im,kn} t_{mj} \star t_{nl} - t_{im} \star t_{kn} r_{mj,nl}) + o(\hbar)\,, \tag{3.23}$$

where again we used $\star$ for the product in $\mathcal{A}$. In the limit $\hbar \to 0$ the commutator $[t_{ij}, t_{kl}]$ vanishes and we restore the commutative algebra of coordinate functions g_{ij}. Further, it follows from (3.23) that the quantity

$$\pi\big((t_{ij} \star t_{kl} - t_{kl} \star t_{ij})/\hbar\big) = r_{im,kn}\pi(t_{mj})\pi(t_{nl}) - \pi(t_{im})\pi(t_{kn})r_{mj,nl} = [r, g_1 g_2]_{ij,kl}$$

coincides with the right hand side of the Sklyanin bracket $\{g_{ij}, g_{kl}\}$, therefore, showing the fulfilment of (3.12). Finally, relation (3.14) is also satisfied.

Quantum Yang-Baxter equation. Since a quantum group $\mathcal{A}$ is an associative but non-commutative algebra, one can ask whether it admits a basis of ordered monomials constructed from generators t_{ij}, i.e. if $\mathcal{A}$ satisfies the Poincaré-Birkhoff-Witt property. The significance of this property is that no new relations between cubic and higher monomials are introduced by the associative product in $\mathcal{A}$. For instance, consider a cubic monomial

$$T_1 T_2 T_3 = (T_1 T_2)T_3 = R_{12}^{-1} T_2 T_1 R_{12} T_3 = R_{12}^{-1} T_2 T_1 T_3 R_{12} =$$
$$= R_{12}^{-1} R_{13}^{-1} T_2 T_3 T_1 R_{13} R_{12} = R_{12}^{-1} R_{13}^{-1} R_{23}^{-1} T_3 T_2 T_1 R_{23} R_{13} R_{12} ,$$

where at each step we have used relations (3.10). By these manipulations we have reordered the monomial $T_1 T_2 T_3$ into $T_3 T_2 T_1$. However, we can reach the same answer by changing the order in which the factors are permuted

$$T_1 T_2 T_3 = T_1 (T_2 T_3) = R_{23}^{-1} T_1 T_3 T_2 R_{23} = R_{23}^{-1} R_{13}^{-1} T_3 T_1 T_2 R_{13} R_{12} =$$
$$= R_{12}^{-1} R_{13}^{-1} T_2 T_3 T_1 R_{13} R_{12} = R_{23}^{-1} R_{13}^{-1} R_{12}^{-1} T_3 T_2 T_1 R_{12} R_{13} R_{23} .$$

Would R be arbitrary, uniqueness of $T_1 T_2 T_3$ for an associative product would require an imposition of a new cubic relation

$$R_{12}^{-1} R_{13}^{-1} R_{23}^{-1} T_3 T_2 T_1 R_{23} R_{13} R_{12} - R_{23}^{-1} R_{13}^{-1} R_{12}^{-1} T_3 T_2 T_1 R_{12} R_{13} R_{23} = 0 ,$$

and it is not guaranteed that this relation together with the original quadratic one will be enough not to introduce quartic relations and so on. From the perspective of quantisation this would imply a physically awkward situation that a supply of quantum observables does not match its classical counterpart. Indeed, in the worst scenario, there could be so many algebraic relations between the generators that they would not have any non-trivial representation. However, if R is a solution of the quantum Yang-Baxter equation, the cubic relation above is satisfied automatically and no new relations are generated.

Quantum affine group. Replace in the construction of the formal quantum group the finite-dimensional vector space V by an infinite-dimensional $\mathbb{Z}$-grade vector space $W = \oplus_{n\in\mathbb{Z}}\lambda^n V$, where λ is a formal variable playing the role of the spectral parameter. Define the quantum R-matrix as an element $R \in \mathrm{Mat}(W^{\otimes 2}, \mathbb{C})$ satisfying the quantum Yang-Baxter equations. If we identify R with the matrix-valued function $R(\lambda, \mu) \in V \otimes V$, then the quantum Yang-Baxter equation reads as

$$R_{12}(\lambda, \mu) R_{13}(\lambda, \tau) R_{23}(\mu, \tau) = R_{23}(\mu, \tau) R_{13}(\lambda, \tau) R_{12}(\lambda, \mu) . \tag{3.24}$$

A simple solution of (3.24) is constructed as

$$R(\lambda, \mu) = \frac{\lambda R_+ - \mu R_-}{\lambda - \mu} , \tag{3.25}$$

where $R_\pm$ are defined in (3.16), (3.17). In fact, this R-matrix depends on the ratio λ/μ and satisfies[2]

$$R_{12}(\lambda, \mu) R_{21}(\mu, \lambda) = \mathbb{1} . \tag{3.26}$$

Having a solution of (3.24), one can define a quantum affine group as generated by the coefficients of the formal Lauren series $T(\lambda) = \sum_{m \in \mathbb{Z}} T_m \lambda^m$, modulo the relations

$$R_{12}(\lambda, \mu) T_1(\lambda) T_2(\mu) = T_2(\mu) T_1(\lambda) R_{12}(\lambda, \mu) . \tag{3.27}$$

More information on the infinite-dimensional quantum groups and quantum algebras can be found in [14].

3.3 Quantisation of RS Models

Here, using an algebraic realisation of the geometric idea of reduction, we obtain a formal quantisation of the classical RS and CMS models. Our considerations will be done in the R-matrix formalism exemplified in the previous section for quantum groups. Working with algebraic relations described in terms of various R-matrices gives a simple way to check their consistency and, most importantly, allows for an elementary proof of kinematical integrability[3] of the corresponding quantum models. Our treatment is based on [15].

3.3.1 Rational RS Model

We start with introducing the concept of the quantum cotangent bundle and further describe its algebraic relations in terms of variables Q, T, P, U, very similar to the classical case of T^*G. Then we construct the quantum L-operator—a quantum analogue of the Lax matrix (2.307), and obtain the permutation relations between its

---

[2]Upon identification $\lambda = e^u$ and $\mu = e^v$ this R-matrix is referred to as the standard trigonometric solution of the quantum Yang-Baxter equation [13].

[3]This approach deals with a formal construction of commuting integrals and it does not address the next level question about the Hilbert space integrability, that is whether the formal eigenfunctions belong to the Hilbert space of a quantum-mechanical system with a self-adjoint hamiltonian.

entries. This L-operator gives rise to a family of commuting operators, quantising the classical integrals of motion.

Quantum cotangent bundle. Consider a unital associative algebra $\mathcal{B}_\hbar$ over $\mathbb{C}$ generated by entries of two $N \times N$ matrices ℓ and g modulo the following relations[4]

$$
\begin{aligned}
[\ell_1, \ell_2] &= \tfrac{\hbar}{2}[C, \ell_1 - \ell_2], \\
[\ell_1, g_2] &= \hbar g_2 C, \\
[g_1, g_2] &= 0,
\end{aligned}
\tag{3.28}
$$

where $\hbar$ is a quantisation parameter and C is the split Casimir. We call $\mathcal{B}_\hbar$ an algebra of function on the quantum cotangent bundle associated with the group $G = \mathrm{GL}_N(\mathbb{C})$. Relations (3.28) define a deformation quantisation of the Poisson structure (2.65) in the sense of conditions (3.11) and (3.12).

In a similar spirit we can straightforwardly quantise the Poisson relations (2.272), (2.276), (2.279) and (2.278). To this end, let us introduce a free associative algebra $\mathcal{A}_\hbar$ generated by entries of the matrices T, U, T^{-1}, U^{-1} and Q, P modulo the following (non-trivial) relations

$$
T_1 T_2 = T_2 T_1 R_{12}(q), \quad U_1 U_2 = U_2 U_1 R_{12}^{-1}(q),
\tag{3.29}
$$
$$
P_2 T_1 = T_1 \bar{R}_{12}(q) P_2, \quad P_2 U_1 = U_1 \bar{R}_{12}(q) P_2,
\tag{3.30}
$$

and

$$
[P_1, Q_2] = \hbar P_1 \overline{C}_{12},
\tag{3.31}
$$

where $\overline{C}$ is given by (2.309). Here P and Q are diagonal with entries $Q_{ij} = q_i \delta_{ij}$ and $P_{ij} = e^{p_i} \delta_{ij}$, such that (3.31) is equivalent to the Heisenberg commutation relations $[p_i, q_j] = \hbar \delta_{ij}$. In the following we use the standard coordinate representation where q_j is realised by the operator of multiplication and $p_j = \hbar \frac{\partial}{\partial q_j}$. The inverse elements (antipods) T^{-1}, U^{-1} are assumed to satisfy the standard antipodal relations

$$
T_{ik} T_{kj}^{-1} = \delta_{ij} = T_{ik}^{-1} T_{kj}, \quad U_{ik} U_{kj}^{-1} = \delta_{ij} = U_{ik}^{-1} U_{kj},
\tag{3.32}
$$

and algebraic relations involving them follow compatibility of (3.29), (3.30) with (3.32), for instance,

$$
T_1^{-1} T_2 = T_2 R_{12}^{-1}(q) T_1^{-1}.
\tag{3.33}
$$

---

[4]Discussing quantum theory, we will use the same notations for algebra generators as for the corresponding coordinate functions in the classical theory.

The quantities $R(q)$ and $\bar{R}(q)$ featuring in (3.29) and (3.30) are quantum dynamical R-matrices for which we have to require the following behaviour close to $\hbar = 0$:

$$R(q) = 1 + \hbar r(q) + o(\hbar), \quad \bar{R}(q) = 1 + \hbar \bar{r}(q) + o(\hbar), \tag{3.34}$$

where the classical dynamical r-matrices $r(q)$ and $\bar{r}(q)$ are given by (2.273) and (2.280), respectively. The quantum R-matrices must satisfy a number of equations to guarantee the absence of additional higher-order relations beyond those given by (3.29), (3.30) and (3.31). For $R(q)$ we find the standard quantum Yang-Baxter equation

$$R_{12}(q)R_{13}(q)R_{23}(q) = R_{23}(q)R_{13}(q)R_{12}(q) \tag{3.35}$$

together with

$$R_{12}(q)R_{21}(q) = 1. \tag{3.36}$$

In addition, there are two more equations involving $\bar{R}(q)$

$$R_{12}(q)\bar{R}_{13}(q)\bar{R}_{23}(q) = \bar{R}_{23}(q)\bar{R}_{13}(q)P_3 R_{12}(q)P_3^{-1}, \tag{3.37}$$

$$\bar{R}_{12}(q)P_2\bar{R}_{13}(q)P_2^{-1} = \bar{R}_{13}(q)P_3\bar{R}_{12}(q)P_3^{-1}. \tag{3.38}$$

Let us demonstrate how one gets, for instance, (3.37). Consider a monomial $P_3 T_1 T_2$ and bring it to the form $T_2 T_1 P_3$ by using the relations (3.29) and (3.30). We can do it in two different ways

$$P_3 T_1 T_2 = T_1 \bar{R}_{13} P_3 T_2 = T_1 \bar{R}_{13} T_2 \bar{R}_{23} P_3 = T_2 T_1 P_3 \cdot P_3^{-1} R_{12} \bar{R}_{13} \bar{R}_{23} P_3,$$

$$P_3 T_1 T_2 = P_3 T_2 T_1 R_{12} = T_2 \bar{R}_{23} P_3 T_1 R_{12} = T_2 T_1 P_3 \cdot P_3^{-1} \bar{R}_{23} \bar{R}_{13} P_3 R_{12}.$$

Thus, Eq. (3.37) is a sufficient condition for the right hand sides of these expressions to coincide. Analogously, commuting $P_2 P_3 = P_3 P_2$ through T_1 and requiring the absence of new cubic relations, we obtain (3.38).

Before resuming the discussion of the quantum cotangent bundle, let us find $R(q)$ and $\bar{R}(q)$ that solve (3.35)–(3.38) and have the quasi-classical expansion (3.34). First, a relevant solution of (3.35) can be straightforwardly found if one notes that $r^2 = 0$. It is known that if $r \in \mathrm{Mat}_N(\mathbb{C}) \otimes \mathrm{Mat}_N(\mathbb{C})$ satisfies $r^3 = 0$ and solves the classical Yang-Baxter equation, then $R = e^r$ is a solution of the quantum Yang-Baxter equation [14]. Therefore,

$$R(q) = 1 + \hbar r(q) = 1 + \sum_{i \neq j}^{N} \frac{\hbar}{q_{ij}} f_{ij} \otimes f_{ji} \tag{3.39}$$

gives a desired solution of (3.35) and (3.36).

The solution of Eq. (3.37) can be found if we assume that $\bar{R}$ has the same matrix structure as $\bar{r}$ does:

$$\bar{R}(q) = 1 + \hbar \sum_{i \neq j}^{N} \bar{r}_{ij}(\hbar, q) f_{ij} \otimes E_{jj} \tag{3.40}$$

Then the following $\bar{R}$-matrix is a solution of Eqs. (3.37) and (3.38)

$$\bar{R}(q) = 1 + \sum_{i \neq j}^{N} \frac{\hbar}{q_{ij} - \hbar} f_{ij} \otimes E_{jj} , \tag{3.41}$$

as can be verified by an explicit calculation. This matrix has the inverse

$$\bar{R}^{-1}(q) = 1 - \sum_{i \neq j}^{N} \frac{\hbar}{q_{ij}} f_{ij} \otimes E_{jj}.$$

Now we return to our discussion of the quantum cotangent bundle and introduce the following (matrix) combinations of generators $\ell = TQT^{-1}$ and $g = UP^{-1}T^{-1}$. Formally, these combinations look the same as the factorised expressions (2.265), (2.277) for coordinate functions ℓ and g on T^*G. However, this time the entries of (T, U, P, Q) are elements of the non-commutative algebra (3.29)–(3.31) and, therefore, the order in which they appear in ℓ and g matters. Our immediate goal is to show that ℓ and g constructed in this way yield the commutation relations (3.28). From (3.29)–(3.31) we get

$$[\ell_1, \ell_2] = T_2 T_1 (R_{12} Q_1 R_{21} Q_2 - Q_2 R_{12} Q_1 R_{21}) T_1^{-1} T_2^{-1} ,$$
$$[\ell_1, g_2] = g_2 T_2 T_1 (\bar{R}_{12}(Q_1 + \hbar \bar{C}) \bar{R}_{12}^{-1} R_{12} - R_{12} Q_1) T_2^{-1} T_1^{-1} , \tag{3.42}$$
$$[g_1, g_2] = U_2 U_1 (R_{12}^{-1} P_1^{-1} \bar{R}_{21}^{-1} P_2^{-1} \bar{R}_{12}^{-1} R_{12} - P_2^{-1} \bar{R}_{12}^{-1} P_1^{-1} \bar{R}_{21}^{-1}) T_2^{-1} T_1^{-1} .$$

Using the explicit solution of R and $\bar{R}$, one can verify that these matrices satisfy the following identities

$$R_{12} Q_1 R_{21} Q_2 - Q_2 R_{12} Q_1 R_{21} = \hbar (C_{12} Q_1 R_{21} - R_{12} Q_1 C_{12}) ,$$
$$\bar{R}_{12}(Q_1 + \hbar \bar{C}_{12}) \bar{R}_{12}^{-1} R_{12} - R_{12} Q_1 = \hbar C_{12} , \tag{3.43}$$
$$R_{12}^{-1} P_1^{-1} \bar{R}_{21}^{-1} P_2^{-1} \bar{R}_{12}^{-1} R_{12} - P_2^{-1} \bar{R}_{12}^{-1} P_1^{-1} \bar{R}_{21}^{-1} = 0 .$$

Using (3.36) the last relation can be equivalently written as

$$P_2 \bar{R}_{21} P_1 R_{12} P_2^{-1} \bar{R}_{12}^{-1} P_1^{-1} = \bar{R}_{12}^{-1} R_{12} \bar{R}_{21} . \tag{3.44}$$

It remains to note that upon substituting (3.43) in (3.42), we recover the algebraic relations (3.28) of $\mathcal{B}_\hbar$. In other words, the algebra of functions on the quantum cotangent bundle is a closed subalgebra of $\mathcal{A}_\hbar$, its consistency is guaranteed by consistency of the algebraic relations of $\mathcal{A}_\hbar$.

Let us also show that the elements $\mathrm{Tr}\ell^k$, $k \in \mathbb{N}_+$, are expressed via the coordinates q_i only. Since the entries of Q and T commute, we have

$$\mathrm{Tr}\ell^k = \mathrm{Tr}(TQ^kT^{-1}) = \sum_{i,j=1}^{N} T_{ij}T_{ji}^{-1}q_j^k . \tag{3.45}$$

To proceed, we need to evaluate the sum

$$\sum_{i=1}^{N} T_{ij}T_{ji}^{-1} .$$

This can be done with the following trick. The first relation in (3.29) implies that $T_2^{-1}T_1 = T_1R_{12}(q)T_2^{-1}$. Applying to this equation the transposition in the first space followed by the multiplication of both sides of the resulting expression with $(R_{12}^{t_1}(q))^{-1}$ yields the following answer

$$T_1^tT_2^{-1} = (R_{12}^{t_1}(q))^{-1}T_2^{-1}T_1^t$$

or, in components,

$$T_{ij}T_{kl}^{-1} = (R_{12}^{t_1}(q))_{ja,kb}^{-1}T_{ai}^{-1}T_{lb} .$$

Putting here $k = j$, $l = i$, and using the fact that $T_{ai}^{-1}T_{ib} = \delta_{ab}$, we find for any j

$$\sum_{i=1}^{N} T_{ij}T_{ji}^{-1} = \sum_{i=1}^{N}(R_{12}^{t_1}(q))_{ji,ji}^{-1} = \prod_{i\neq j} \frac{q_{ij} - \hbar}{q_{ij}} , \tag{3.46}$$

where we have used the explicit form (3.39) of the R-matrix $R_{12}(q)$. Thus, $\mathrm{Tr}\ell^k$ is the following function of q

$$\mathrm{Tr}\ell^k = \sum_{j=1}^{N} q_j^k \prod_{i\neq j} \frac{q_{ij} - \hbar}{q_{ij}} .$$

The expression on the right hand side can be re-expanded via products of $\mathrm{Tr}Q^k$ with various positive integers k. In particular, we find that

$$\mathrm{Tr}\ell = \mathrm{Tr}\, Q + \frac{N(N-1)}{2}\hbar .$$

Quantum L-operator and integrals of motion. In full analogue with the classical case, we introduce the following matrix L

$$L = T^{-1}gT. \qquad (3.47)$$

We call this matrix L with coefficients in $\mathcal{A}_\hbar$ the *quantum Lax operator*, or simply, the Lax operator (L-operator), and regard it as a quantum analogue of the classical Lax matrix (2.307). By using (3.29)–(3.31) and (3.43), we first find

$$[Q_1, L_2] = \hbar L_2 \overline{C}_{12}, \qquad (3.48)$$

$$T_1 R_{12} L_2 = L_2 T_1 \bar{R}_{12}, \qquad (3.49)$$

and then establish the algebraic relations between the components of the Lax operator

$$R_{12} L_2 \bar{R}_{12}^{-1} L_1 = L_1 \bar{R}_{21}^{-1} L_2 \bar{R}_{12}^{-1} R_{12} \bar{R}_{21}. \qquad (3.50)$$

Note that in deriving this formula we also used (3.44). At this point it is natural to introduce the matrix

$$\underline{R}_{12}(q) = \bar{R}_{12}(q)^{-1} R_{12}(q) \bar{R}_{21}(q), \qquad (3.51)$$

so that (3.50) will acquire the form

$$R_{12} L_2 \bar{R}_{12}^{-1} L_1 = L_1 \bar{R}_{21}^{-1} L_2 \underline{R}_{12}. \qquad (3.52)$$

Obviously, the relation (3.52) is a quantum version of the Poisson algebra (2.308) and $\underline{R}$ is a quantum R-matrix corresponding to $\underline{r}$, that is $\underline{R} = \mathbb{1} + \hbar \underline{r}$, where $\underline{r}$ is given by (2.318). In fact, $\underline{R}$ satisfies the so-called quantum Gervais-Neveu-Felder equation [16, 17], see also [18] where it played an important role in quantisation of the CMS model with internal (spin) degrees of freedom. The Gervais-Neveu-Felder equation is dynamical, in the sense that it involves the momentum operator, and it reads

$$\underline{R}_{12}(q) P_2^{-1} \underline{R}_{13}(q) P_2 \underline{R}_{23}(q) = P_1^{-1} \underline{R}_{23}(q) P_1 \underline{R}_{13}(q) P_3^{-1} \underline{R}_{12}(q) P_3. \qquad (3.53)$$

Evidently, this equation is a quantum version of the classical relation (2.319). From (3.51) it is also clear that the matrix $\bar{R}$ plays a role of the *Drinfeld twist* F_{12}. For the non-dynamical R-matrices, the Drinfeld twist transforms a solution R_{12} of the quantum Yang-Baxter equation into another solution: $R_{12} \to F_{12}^{-1} R_{12} F_{21}$.

As is clear from (3.28), the quantities $I_k = \text{Tr} \, g^k$ form a set of mutually commuting operators. We are going to show that I_k can be solely expressed in terms of L and Q and, for this reason, can be interpreted as quantum integrals of motion for the rational RS model. To this end, invoking the definition of L, we rewrite I_k as

$$I_k = \text{Tr} g^k = \text{Tr} \, T L^k T^{-1} = \text{Tr}_{12} C_{12} T_1 L_1^k T_2^{-1} = \text{Tr}_{12} C_{12}^{t_2} T_1 L_1^k \, {}^t T_2^{-1}.$$

Here and below $t_1(t_2)$ denotes the matrix transposition in the (first) second factor of the tensor product. For the transpose of T^{-1} we use a special notation $\overset{t}{T}{}^{-1}$ to stress that in $\mathscr{A}_\hbar$ taking inverse and transposition are two non-commuting operations. For instance, applying transposition to (3.32), we will get

$$\sum_{k=1}^{N} \overset{t}{T}_{ki}\, \overset{t}{T}{}^{-1}{}_{jk} = \delta_{ij}\,, \qquad (3.54)$$

where to uniformize the notation we denote $\overset{t}{T} \equiv T^t$. To be able to identify the inverse of the transpose of T with $\overset{t}{T}{}^{-1}$, one needs to exchange the position of generators in the left hand side of (3.54) to restore the standard flow of the summation index. However, these generators do not commute which makes such an identification impossible.

To proceed, we first obtain from (3.49) the following relation between L and T^{-1}

$$L_1 \bar{R}_{21}^{-1} T_2^{-1} = R_{12} T_2^{-1} L_1\,. \qquad (3.55)$$

Applying to this relation the transposition in the second matrix space, we get

$$L_1 \overset{t}{T}_2^{-1} \left(\bar{R}_{21}^{-1}\right)^{t_2} = T_2^{-1} R_{12}^{t_2} L_1\,. \qquad (3.56)$$

Taking into account that $\left(\bar{R}_{21}^{-1}\right)^{t_2} = (\bar{R}_{21}^{t_2})^{-1}$, the last relation can be equivalently written as

$$L_1 \overset{t}{T}_2^{-1} = \overset{t}{T}_2^{-1} R_{12}^{t_2} L_1 \bar{R}_{21}^{t_2}\,. \qquad (3.57)$$

To avoid possible misinterpretation, we emphasise that t_2 refers to the transposition of the second matrix space in the tensor product, so that the explicit expressions for the matrices entering in (3.57) are

$$R_{12}^{t_2} = \mathbb{1} + \sum_{i\neq j}^{N} \frac{\hbar}{q_{ij}} (E_{ii} - E_{ij}) \otimes (E_{jj} - E_{ij})\,,$$

$$\bar{R}_{21}^{t_2} = \mathbb{1} - \sum_{i\neq j}^{N} \frac{\hbar}{q_{ij} + \hbar} E_{ii} \otimes (E_{jj} - E_{ij})\,. \qquad (3.58)$$

Now applying relation (3.57) successively, we bring I_k to the form

$$I_k = \mathrm{Tr}_{12} C_{12}^{t_2}\, T_1\, \overset{t}{T}_2^{-1}\, R_{12}^{t_2} L_1 \bar{R}_{21}^{t_2} \cdots R_{12}^{t_2} L_1 \bar{R}_{21}^{t_2}\,.$$

Further, Eq. (3.59) yields

$$T_2^{-1} T_1 = T_1 R_{12} T_2^{-1},$$ (3.59)

which upon transposing the second matrix space results into[5]

$$T_2^{t^{-1}} T_1 = T_1 T_2^{t^{-1}} R_{12}^{t_2}.$$ (3.60)

Exchanging T_1 and $T_2^{t^{-1}}$ with the help of the above relation, we obtain

$$I_k = \mathrm{Tr}_{12} C_{12}^{t_2} T_2^{t^{-1}} T_1 L_1 \bar{R}_{21}^{t_2} \cdots R_{12}^{t_2} L_1 \bar{R}_{21}^{t_2}.$$

Next, we consider a matrix element

$$(C_{12}^{t_2} T_2^{t^{-1}} T_1)_{ij,kl} = \sum_{a,b=1}^{N} (E_{ab})_{im} \otimes (E_{ab} T^{t^{-1}})_{kl} T_{mj} = \delta_{ik} T_{ml}^{t^{-1}} T_{mj} = \delta_{ik}\delta_{lj} = (C_{12}^{t_2})_{ij,kl},$$

in accord with the definition (3.32) of the inverse. In other words,

$$C_{12}^{t_2} T_2^{t^{-1}} T_1 = C_{12}^{t_2}.$$ (3.61)

Taking into account that $\bar{R}_{21}^{t_2} C_{12}^{t_2} = C_{12}^{t_2}$, we finally arrive at

$$I_k = \mathrm{Tr}_{12} C_{12}^{t_2} L_1 \bar{R}_{21}^{t_2} R_{12}^{t_2} L_1 \bar{R}_{21}^{t_2} R_{12}^{t_2} L_1 \cdots L_1 \bar{R}_{21}^{t_2} R_{12}^{t_2} L_1.$$ (3.62)

It is natural to regard this expression as the "quantum trace" of the operator L^n. We further note that $C_{12}^{t_2}$ featuring in this expression is a matrix of rank 1, it has a single non-zero eigenvalue equal to N.

The matrix that enters in between two L-operators in (3.62) is computed with the help of (3.58) and it has the following explicit form

$$\bar{R}_{21}^{t_2} R_{12}^{t_2} = \mathbb{1} - \sum_{i \neq j} \frac{\hbar}{q_{ij} + \hbar} E_{ij} \otimes (E_{jj} - E_{ij}).$$ (3.63)

Using the commutation relation (3.48), we can work out an explicit form of I_k. For instance, for the first three operators we find[6]

---

[5] Once again, we point out that we first write the algebraic relations for inverse elements and then take their transposition, but not vice versa.

[6] In the formulae below sums over each summation index run from 1 to N with the account of the indicated inequalities.

$$I_1 = \sum_j L_{jj},$$

$$I_2 = \sum_j L_{jj}^2 - \sum_{i \neq j} \frac{\hbar}{q_{ij}}(L_{ji} - L_{ii})L_{jj},$$

$$I_3 = \sum_j L_{jj}^3 - \sum_{i \neq j} \frac{\hbar}{q_{ij}}(L_{ji} - L_{ii})L_{jj}^2 - \sum_k \sum_{i \neq j}(L_{jk} - L_{ik})\frac{\hbar}{q_{ij}}L_{ki}L_{jj}$$

$$+ \sum_{\substack{i \neq j \\ k \neq j}} \frac{\hbar}{q_{ij}}(L_{ji} - L_{ii})\frac{\hbar}{q_{kj}}L_{jk}L_{jj} - \sum_{\substack{i \neq j \\ k \neq j}} \frac{\hbar}{q_{ij}}(L_{ji} - L_{ii})\frac{\hbar}{q_{jk}}L_{jj}L_{kk}.$$

(3.64)

Quantum Frobenius group. Just as in the classical case, the quantum L-operator has the form $L = WP^{-1}$, where $W = T^{-1}U$ satisfies the defining relations of the *quantum Frobenius group*

$$R_{12}W_2W_1 = W_1W_2R_{12}.$$

(3.65)

The algebra (3.48), (3.52) rewritten in terms of Q, P and W is given by (3.31), (3.65) and by the relation

$$\bar{R}_{12}P_2W_1 = W_1\bar{R}_{12}P_2.$$

(3.66)

This shows that the representation theory for L essentially reduces to the one for the quantum Frobenius group.

It is known that the algebra (3.65) admits a family of mutually commuting operators given by [19]

$$J_k = \mathrm{Tr}_{1...k}\left[\hat{R}_{12}\hat{R}_{23}\ldots\hat{R}_{k-1,k}W_1\ldots W_k\right],$$

where $\hat{R}_{ij} = R_{ij}C_{ij}$. Now we demonstrate that J_k commutes with P. For the sake of clarity we do it for $k = 3$. The fist step consists in pulling P_4 through $\hat{R}_{12}\hat{R}_{23}W_1W_2W_3$. To perform this operation, we note that Eq. (3.37) written in terms of $\hat{R}$ implies the relation

$$P_3\hat{R}_{12}P_3^{-1} = \bar{R}_{13}^{-1}\bar{R}_{23}^{-1}\hat{R}_{12}\bar{R}_{23}\bar{R}_{13}.$$

With the help of this relation together with (3.66), we get

$$P_4\hat{R}_{12}\hat{R}_{23}W_1W_2W_3 =$$
$$= \bar{R}_{14}^{-1}\bar{R}_{24}^{-1}\hat{R}_{12}\bar{R}_{24}\bar{R}_{14}\bar{R}_{24}^{-1}\bar{R}_{34}^{-1}\hat{R}_{23}\bar{R}_{34}\bar{R}_{24}\bar{R}_{14}^{-1}W_1\bar{R}_{14}\bar{R}_{24}^{-1}W_2\bar{R}_{24}\bar{R}_{34}^{-1}W_3\bar{R}_{34}P_4.$$

The second step consists in taking into account that $\bar{R}_{ij}$ is diagonal in the second space, so that any two $\bar{R}_{ij}$ and $\bar{R}_{kj}$ commute for $i \neq k$. This gives

$$P_4 \hat{R}_{12} \hat{R}_{23} W_1 W_2 W_3 = \bar{R}_{14}^{-1} \bar{R}_{24}^{-1} \bar{R}_{34}^{-1} \hat{R}_{12} \hat{R}_{23} W_1 W_2 W_3 \bar{R}_{14} \bar{R}_{24} \bar{R}_{34} \, P_4 \, .$$

Taking the trace of the above expression in the first, second and third spaces, and once again relying on diagonality of $\bar{R}_{ij}$ in the second space, one gets the desired property.

Representation for the quantum L-operator. Now we construct a representation for the quantum L-operator that will correspond to the classical Lax matrix on the reduced phase space. We start with identifying a relevant representation of the algebra (3.65) and (3.66). Namely, we prove that the W-operator given by (2.291) realises this algebra with

$$P_j = e^{\hbar \frac{\partial}{\partial q_j}} \, .$$

Since W depends on the coordinates q_i only, the commutator $[W_1, W_2]$ must vanish. Thus, the following relation has to be valid: $[r_{12}(q), W_1 W_2] = 0$. Substituting the explicit form of $r_{12}(q)$ we have

$$
\begin{aligned}
[r_{12}(q), W_1 W_2]_{kl,mn} = W_{kl} &\left(\frac{1}{q_{km}} W_{mn} - \frac{1}{q_{km}} W_{kn} - \frac{1}{q_{ln}} W_{mn} \right) \\
- W_{ml} &\left(\frac{1}{q_{km}} W_{mn} - \frac{1}{q_{km}} W_{kn} - \frac{1}{q_{ln}} W_{kn} \right) \quad (3.67) \\
+ \delta_{ln} &\sum_{j \neq l} \frac{1}{q_{lj}} (W_{kl} W_{mj} - W_{kj} W_{ml}).
\end{aligned}
$$

First we note that for $l \neq n$ the first line in (3.67) cancels the second one. Substituting $W_{ij} = \frac{\gamma}{\gamma + q_{ij}} b_j$ into (3.67), we find

$$
\begin{aligned}
\frac{\gamma}{\gamma + q_{kl}} &\left(\frac{1}{q_{km}} \frac{\gamma}{\gamma + q_{mn}} - \frac{1}{q_{km}} \frac{\gamma}{\gamma + q_{kn}} - \frac{1}{q_{ln}} \frac{\gamma}{\gamma + q_{mn}} \right) - \\
\frac{\gamma}{\gamma + q_{ml}} &\left(\frac{1}{q_{km}} \frac{\gamma}{\gamma + q_{mn}} - \frac{1}{q_{km}} \frac{\gamma}{\gamma + q_{kn}} - \frac{1}{q_{ln}} \frac{\gamma}{\gamma + q_{kn}} \right) = 0.
\end{aligned}
$$

In the case $l = n$ the right hand side of (3.67) reduces to

$$
\begin{aligned}
-\frac{1}{q_{km}} (W_{kl} - W_{ml})^2 + \sum_{j \neq l} \frac{1}{q_{lj}} (W_{kl} W_{mj} - W_{kj} W_{ml}) = \\
= -\frac{\gamma^2 q_{km}}{(\gamma + q_{km})^2 (\gamma + q_{ml})^2} b_l^2 + \sum_{j \neq l} \frac{1}{q_{lj}} \left(\frac{\gamma}{\gamma + q_{kl}} \frac{\gamma}{\gamma + q_{mj}} - \frac{\gamma}{\gamma + q_{kj}} \frac{\gamma}{\gamma + q_{ml}} \right) b_j b_l \\
= \frac{q_{mk}}{\gamma + q_{ml}} W_{kl} \sum_j \frac{W_{mj}}{\gamma + q_{kj}} .
\end{aligned}
$$

Thus, one has to show that the sum

$$S = \sum_j \frac{W_{mj}}{\gamma + q_{kj}}$$

vanishes. To this end, we consider the following integral

$$I = \frac{1}{2\pi i} \oint \frac{dz}{q_k - z + \gamma} \frac{\prod\limits_{a \neq m} (q_a - z + \gamma)}{\prod\limits_a (q_a - z)},$$

where the integration contour is taken around infinity. Since the integrand is non-singular at $z \to \infty$, we get $I = 0$. On the other hand, summing up the residues one finds

$$I = \sum_j \frac{1}{\gamma + q_{kj}} \frac{\prod\limits_{a \neq m} (q_{aj} + \gamma)}{\prod\limits_{a \neq j} q_{aj}} = S,$$

where we have taken into account formula (2.291) for W.

Now we turn to Eq. (3.66). Explicitly it reads as

$$P_j[W_{kl}, P_j^{-1}] = \left(\frac{\hbar}{q_{lj} - \hbar} - \frac{\hbar}{q_{kj}} - \frac{\hbar^2}{q_{kj}(q_{lj} - \hbar)} \right) W_{kl} + \left(\frac{\hbar^2}{q_{kj}(q_{lj} - \hbar)} + \frac{\hbar}{q_{kj}} \right) W_{jl}$$

$$+ \delta_{jl} \sum_{i \neq j} \left(\left(\frac{\hbar^2}{q_{kj}(q_{ij} - \hbar)} - \frac{\hbar}{q_{ij} - \hbar} \right) W_{ki} - \frac{\hbar^2}{q_{kj}(q_{ij} - \hbar)} W_{li} \right). \quad (3.68)$$

For the sake of shortness, in (3.68) we adopt a convention that if in some denominator q_{ij} becomes zero when $i = j$, the corresponding fraction is also regarded as zero. Thus, Eq. (3.68) is equivalent to the following system of equations

$$P_j[W_{kl}, P_j^{-1}] = \frac{\hbar}{q_{kj}(q_{lj} - \hbar)} (q_{kl} W_{kl} - q_{jl} W_{jl}), \quad \text{for } k \neq l \neq j;$$

$$P_j[W_{jl}, P_j^{-1}] = \frac{\hbar}{(q_{lj} - \hbar)} W_{jl}, \quad \text{for } j \neq l;$$

$$P_k[W_{kk}, P_k^{-1}] = -\sum_{i \neq k} \frac{\hbar}{q_{ik} - \hbar} W_{ki};$$

$$P_j[W_{kj}, P_j^{-1}] = \frac{\hbar}{q_{kj}} (W_{jj} - W_{kj}) + \frac{\hbar(\hbar - q_{kj})}{q_{kj}} \sum_{i \neq j} \frac{1}{q_{ij} - \hbar} W_{ki} \quad (3.69)$$

$$- \frac{\hbar^2}{q_{kj}} \sum_{i \neq j} \frac{1}{q_{ij} - \hbar} W_{ji}, \quad \text{for } k \neq j.$$

In the sequel we shall give an explicit proof only for the latter case since the other three cases are treated quite analogously. The left hand side of (3.69) is

$$P_j[W_{kj}, P_j^{-1}] = P_j \frac{\prod\limits_{a\neq k}(q_{aj}+\gamma)}{\prod\limits_{a\neq j} q_{aj}} P_j^{-1} - W_{kj} = \gamma \frac{\prod\limits_{\substack{a\neq k \\ a\neq j}}(q_{aj}+\gamma-\hbar)}{\prod\limits_{a\neq j}(q_{aj}-\hbar)} - W_{kj}. \quad (3.70)$$

As to the right hand side, one needs to calculate the sum $\sum\limits_{i\neq j} \frac{1}{q_{ij}-\hbar} W_{ki}$. For this purpose we evaluate the following integral with the integration contour around infinity

$$I = \frac{1}{2\pi i} \oint \frac{dz}{z-q_j-\hbar} \frac{\prod\limits_{a\neq k}(q_a - z + \gamma)}{\prod\limits_{a}(q_a - z)}.$$

The regularity of the integrand at $z \to \infty$ gives $I = 0$. On the other hand, summing up the residues one finds

$$I = -\frac{1}{\hbar} \frac{\prod\limits_{a\neq k}(q_{aj}+\gamma-\hbar)}{\prod\limits_{a\neq j}(q_{aj}-\hbar)} - \sum\limits_{i\neq j} \frac{1}{q_{ij}-\hbar} W_{ki} + \frac{1}{\hbar} W_{kj}$$

from here one deduces the desired sums

$$\sum\limits_{i\neq j} \frac{1}{q_{ij}-\hbar} W_{ki} = -\frac{1}{\hbar} \frac{\prod_{a\neq k}(q_{aj}+\gamma-\hbar)}{\prod_{a\neq j}(q_{aj}-\hbar)} + \frac{1}{\hbar} W_{kj},$$

$$\sum\limits_{i\neq j} \frac{1}{q_{ij}-\hbar} W_{ji} = -\frac{1}{\hbar} \frac{\prod_{a\neq j}(q_{aj}+\gamma-\hbar)}{\prod_{a\neq j}(q_{aj}-\hbar)} + \frac{1}{\hbar} W_{jj}.$$

Now substituting these sums in the right hand side of (3.69), one gets (3.70). This completes the proof of relation (3.69).

Finally, we can conclude from our treatment that the L-operator

$$L = \sum\limits_{ij} \frac{\gamma}{q_{ij}+\gamma} b_j e^{-\hbar\frac{\partial}{\partial q_j}} E_{ij}, \quad b_j = \prod\limits_{a\neq j} \frac{q_{aj}+\gamma}{q_{aj}}. \quad (3.71)$$

gives a representation of the algebra (3.52).

Simplifying quantum integrals. In the sequel it will be convenient to work with the shift operator[7]

$$\mathsf{T}_j = e^{-\hbar\frac{\partial}{\partial q_j}}. \quad (3.72)$$

---

[7]In fact, $\mathsf{T}_j = P_j^{-1}$, we use T_j to emphasise that we will be representing L and integrals as concrete operators.

On smooth functions $f(q_1, \ldots, q_N)$ it acts as

$$(\mathsf{T}_j f)(q_1, \ldots, q_N) = f(q_1, \ldots, q_j - \hbar, \ldots q_N).$$

In particular, we have the commutation relations

$$\mathsf{T}_i b_j = b_j \frac{q_{ij} + \gamma - \hbar}{q_{ij} + \gamma} \frac{q_{ij}}{q_{ij} - \hbar} \mathsf{T}_i \quad \text{for } i \neq j$$

and

$$\mathsf{T}_i b_j = \left(\prod_{a \neq j} \frac{q_{aj} + \gamma + \hbar}{q_{aj} + \hbar} \right) \mathsf{T}_i \quad \text{for } i = j.$$

These relations can be used to bring the integrals I_k given by (3.62) to the ordered form where all momentum operators stand on the right of the coordinates. Moreover, since the I_k form a ring, by taking proper linear combinations of products of I_k one can find a simpler basis of independent integrals S_k. Defining

$$b_{jk} = \frac{q_{kj} + \gamma}{q_{kj}}, \quad k \neq j,$$

this new basis is constructed as follows

$$S_1 = I_1 = \sum_j \left(\prod_{j \neq k} b_{jk} \right) \mathsf{T}_j$$

$$S_2 = \tfrac{1}{2} I_1^2 - \tfrac{1}{2} I_2 = \sum_{i<j} \left(\prod_{a \neq i,j} b_{ia} b_{ja} \right) \mathsf{T}_i \mathsf{T}_j$$

$$S_3 = \tfrac{1}{6} I_1^3 - \tfrac{1}{2} I_1 I_2 + \tfrac{1}{3} I_3 = \sum_{i<j<k} \left(\prod_{a \neq i,j,k} b_{ia} b_{ja} b_{ka} \right) \mathsf{T}_i \mathsf{T}_j \mathsf{T}_k.$$

Continuing this consideration further, one arrives at the following formula for S_k for arbitrary k, namely,

$$S_k = \sum_{\substack{J \subset \{1,\ldots,n\} \\ |J|=k}} \left(\prod_{\substack{j \in J \\ a \notin J}} b_{ja} \right) \left(\prod_{j \in J} \mathsf{T}_j \right), \tag{3.73}$$

where $|J|$ is cardinality of a subset J. The difference operators S_k represent the rational limit of the *Macdonald operators* [20] to be discussed later. One can further recognise that the relation between I_k and S_k is given by the well-known determinant formula

$$\mathcal{S}_k = \frac{1}{k!} \begin{vmatrix} I_1 & k-1 & 0 & \cdots & 0 \\ I_2 & I_1 & k-2 & \cdots & 0 \\ \vdots & \vdots & & \cdots & \vdots \\ I_{k-1} & I_{k-2} & \cdot & \cdots & 1 \\ I_k & I_{k-1} & \cdot & \cdots & I_1 \end{vmatrix}, \tag{3.74}$$

very similar to the one that relates two bases of symmetric functions—one is given by power sums and the other by elementary symmetric functions, see formulae (3.178) and the discussion surrounding it. Formula (3.74) suggests that the integrals $\mathcal{S}_k$ should follow from a sort of determinant expansion of the Lax operator L. Indeed, one can verify that $\mathcal{S}_k$ have the following generating function

$$: \det(L - \zeta \mathbb{1}) := \sum_{k=0}^{N} (-\zeta)^{N-k} \mathcal{S}_k, \quad \mathcal{S}_0 = 1. \tag{3.75}$$

Here ζ is a formal parameter, L is the Lax operator (3.71). Under the sign $::$ of normal ordering the operators T_j and q_j are considered as commuting and upon algebraic evaluation of the determinant all T_j are brought to the right. In the classical theory the normal ordering is omitted, and the corresponding generating function yields classical integrals of motion that are nothing else but the spectral invariants of the Lax matrix.

Quantum Lax representation. Undertaking an effort to develop quantum theory in close parallel to its classical counterpart, one can ask the question about the existence of a *quantum Lax Representation*. In general, having a Lax pair in the classical theory does not imply that a similar concept might exists in the corresponding quantum theory. For a given hamiltonian H and a quantum L-operator L, the quantum Lax representation (quantum Lax pair) means the existence of a matrix M with operator-valued entries such that the following relation is satisfied

$$[H, L_{ij}] = [M, L]_{ij}, \quad \forall i, j. \tag{3.76}$$

Here on the left hand side one has a commutator of two operators, H and L_{ij}, while on the right hand side one has the standard matrix commutator of two matrices, the entries of which are, in general, non-commutative operators. Even if such a representation exists, its usefulness is questionable, because it is a priori unclear how to apply it, for instance, to exhibit a commutative family of quantum integrals of motion.[8]

Specifically, it turns out that for the CMS and RS models a representation of the type (3.76) always exists, as a natural consequence of the quantum reduction procedure. Moreover, for the CMS models it allows for an elementary construction of quantum integrals, as will be discussed later. Here we will derive the corresponding

---

[8]For instance, the trace of the commutator of two matrices might not vanish if the entries of these matrices do not commute.

representation for the rational RS model and, although, it will not be useful for building up quantum integrals, it will allow us to understand why the CMS models behave differently in this respect. Additionally, the derivation itself provides an interesting application of the quantum reduction procedure and gives an example of a practical computation in quantum theory.

For concreteness, let us take the quantum hamiltonian to be $H = \mathrm{Tr}\, g = \mathrm{Tr} L$. The quantum Lax representation immediately follows from the form (3.47) of the L-operator and the fact that the entries of g commute between themselves, cf. the last line in (3.28). We have

$$[H, L] = [H, T^{-1}gT] = [M, L], \quad M = -T^{-1}[H, T] = H - T^{-1}HT. \quad (3.77)$$

To find an explicit form of M, all what we need is to commute $\mathrm{Tr}\, g$ through T. We, therefore, consider

$$\mathrm{Tr}_1 g_1 T_2 = \mathrm{Tr}_1 U_1 P_1^{-1} T_1^{-1} T_2 = \mathrm{Tr}_1 U_1 P_1^{-1} T_2 R_{21} T_1^{-1}$$
$$= \mathrm{Tr}_1 U_1 T_2 P_1^{-1} \bar{R}_{21}^{-1} R_{21} T_1^{-1} = T_2 \mathrm{Tr}_1 (U_1 P_1^{-1} \bar{R}_{21}^{-1} R_{21} T_1^{-1}),$$

where we commuted the matrix T_2 from the right of $\mathrm{Tr}_1 g_1$ to the left by using the relations (3.29) and (3.30). Thus, we are led to analyse

$$\mathscr{Z} \equiv \mathrm{Tr}_1 (U_1 P_1^{-1} \bar{R}_{21}^{-1} R_{21} T_1^{-1}) = \mathrm{Tr}_{13}(C_{13} U_1 P_1^{-1} T_3^{-1} \bar{R}_{21}^{-1} R_{21}), \quad (3.78)$$

where we have used commutativity of T with q_i. Our next goal is to commute T_3^{-1} to the left and try to combine the variables U, P, T into the L-operator (3.47). Applying the same transposition trick as was used for deriving the quantum integrals, we rewrite (3.78) as

$$\mathscr{Z} = \mathrm{Tr}_{13}(C_{13}^{t_3} U_1 P_1^{-1} \overset{t}{T_3^{-1}} \bar{R}_{21}^{-1} R_{21}). \quad (3.79)$$

From (3.30) we find

$$P_1^{-1} \overset{t}{T_3^{-1}} = \overset{t}{T_3^{-1}} P_1^{-1} ((\bar{R}_{31}^{-1})^{t_3})^{-1}.$$

Therefore,

$$\mathscr{Z} = \mathrm{Tr}_{13}\left(C_{13}^{t_3} \overset{t}{T_3^{-1}} U_1 P_1^{-1} ((\bar{R}_{31}^{-1})^{t_3})^{-1} \bar{R}_{21}^{-1} R_{21}\right)$$
$$= \mathrm{Tr}_{13}\left(C_{13}^{t_3} \overset{t}{T_3^{-1}} T_1 T_1^{-1} U_1 P_1^{-1} ((\bar{R}_{31}^{-1})^{t_3})^{-1} \bar{R}_{21}^{-1} R_{21}\right).$$

Recalling now the formula (3.61), we arrive at

$$\mathscr{Z} = \mathrm{Tr}_{13}\left(C_{13}^{t_3} L_1 ((\bar{R}_{31}^{-1})^{t_3})^{-1} \bar{R}_{21}^{-1} R_{21}\right) = \mathrm{Tr}_{13}\left(C_{13} L_1 (((\bar{R}_{31}^{-1})^{t_3})^{-1})^{t_3} \bar{R}_{21}^{-1} R_{21}\right).$$

A computation which uses an explicit form (3.41) of $\bar{R}_{13}$ shows that $\left(\left((\bar{R}_{31}^{-1})^{t_3}\right)^{-1}\right)^{t_3} = \bar{R}_{31}$, giving thereby

$$\mathcal{Z} = \mathrm{Tr}_{13}(C_{13}L_1\bar{R}_{31}\bar{R}_{21}^{-1}R_{21}) = \mathrm{Tr}_{13}(L_3 C_{13}\bar{R}_{31}\bar{R}_{21}^{-1}R_{21}).$$

As the next step, we compute

$$\mathrm{Tr}_3(C_{13}L_1\bar{R}_{31}) = L + \sum_{i \neq j}^{N}(L_{ii} - L_{ji})\frac{\hbar}{q_{ij} + \hbar}E_{ji}, \tag{3.80}$$

and

$$\bar{R}_{21}^{-1}R_{21} = \mathbb{1} - \sum_{i \neq j}^{N}\frac{\hbar}{q_{ij}}E_{ji} \otimes f_{ij} - \sum_{i \neq j}^{N}\frac{\hbar^2}{q_{ij}^2}f_{ji} \otimes f_{ij}. \tag{3.81}$$

Note that in (3.80) the position of the entries of L matters, because L_{ij} contains momentum operator that acts on q. For $\mathcal{Z}$ we have now

$$\mathcal{Z} = \mathrm{Tr}_1\left(L \otimes \mathbb{1} + \sum_{i \neq j}^{N}(L_{ii} - L_{ji})\frac{\hbar}{q_{ij} + \hbar}E_{ji} \otimes \mathbb{1}\right) \times$$

$$\times \left(\mathbb{1} - \sum_{i \neq j}^{N}\frac{\hbar}{q_{ij}}E_{ji} \otimes f_{ij} - \sum_{i \neq j}^{N}\frac{\hbar^2}{q_{ij}^2}f_{ji} \otimes f_{ij}\right) = (\mathrm{Tr}L)\,\mathbb{1} - \sum_{i \neq j}^{N}L_{ij}\frac{\hbar}{q_{ij}}f_{ij},$$

where upon opening the brackets and taking the trace in the first space most of the terms cancel leaving behind a simple expression. Substituting this result into the expression for M, we get

$$M = \sum_{i \neq j}^{N}L_{ij}\frac{\hbar}{q_{ij}}f_{ij} = \hbar\gamma \sum_{i \neq j}^{N}\frac{b_j}{(q_{ij} + \gamma)(q_{ij} + \hbar)}e^{-\hbar\frac{\partial}{\partial q_j}}(E_{ii} - E_{ij}), \tag{3.82}$$

where we brought the exponent of the momentum operator to the right. We observe that the quantum M-matrix appears to be an ordered version of the classical one, see (2.328) and, as its classical cousin, it takes values in the Frobenius Lie algebra. It is now relatively simple to check by explicit computation that the L-operator (3.71) and the M-matrix (3.82) obey the relation (3.76) with $H = \mathrm{Tr}L$. Finally, we note that Eq. (3.82) can be alternatively derived from the L-operator algebra (3.52). We return to the quantum Lax representation later when discussing the corresponding construction for the CMS models.

Limit to the rational CMS model. Let us briefly discuss the limit to the rational CMS model. Perform the canonical transformation $q_i \to q_i/\kappa$, $p_i \to \kappa p_i$ and then consider a limit $\kappa \to 0$. In this limit the L-operator of the rational RS model behaves as $L \to \mathbb{1} + \kappa \mathscr{L}^{(r)} + o(\kappa)$. Here ℓ^r is the L-operator of the rational CMS model

$$\mathscr{L}^{(r)} = -\hbar \sum_{i=1}^{N} E_{ii} \frac{\partial}{\partial q_i} + \gamma \sum_{i \neq j}^{N} \frac{1}{q_{ij}} E_{ij} \,. \tag{3.83}$$

From Eqs. (3.48), (3.52) we obtain algebraic relations satisfied by the L-operator of this model

$$[Q_1, \mathscr{L}_2^{(r)}] = \hbar \overline{C}_{12},$$
$$[\mathscr{L}_1^{(r)}, \mathscr{L}_2^{(r)}] = \hbar[r_{12} - \bar{r}_{12}, \mathscr{L}_1^{(r)}] - \hbar[r_{21} - \bar{r}_{21}, \mathscr{L}_2^{(r)}] + \hbar^2[r_{12} - \bar{r}_{12}, r_{21} - \bar{r}_{21}],$$

where the matrices r and $\bar{r}$ are given by (2.273) and (2.280). The last formula can be written in the following elegant form

$$[\mathscr{L}_1^{(r)} + \hbar(r_{21} - \bar{r}_{21}), \mathscr{L}_2^{(r)} + \hbar(r_{12} - \bar{r}_{12})] = 0 \,, \tag{3.84}$$

where the difference of r-matrices entering here is

$$r_{12} - \bar{r}_{12} = -\sum_{i \neq j}^{N} \frac{1}{q_{ij}} (E_{ii} - E_{ij}) \otimes E_{ji} \,. \tag{3.85}$$

The discussion of the CMS limit brings out another important point, namely, it reveals the limited applicability of the concept of a Lax pair in the quantum theory. Although we used the Lax operator to exhibit a family of commuting operators in the RS case, the latter become trivial in the RS limit. Indeed, assume that integrals I_n have the following perturbative expansion when $\kappa \to 0$

$$I_n = \text{const} + \kappa I_n^{(1)} + \kappa^2 I_n^{(2)} + \kappa^3 I_n^{(3)} + \cdots \,.$$

For the expansion of the commutator we will have

$$[I_n, I_m] = \kappa^2 [I_n^{(1)}, I_m^{(1)}] + \kappa^3 \left([I_n^{(2)}, I_m^{(1)}] + [I_n^{(1)}, I_m^{(2)}]\right) + \cdots \,.$$

Commutativity of I_n implies that the coefficient in front of each power of κ in the right hand side here must separately vanish, i.e. we obtain, as the consequence, a set of relations

$$[I_n^{(1)}, I_m^{(1)}] = 0 \,,$$
$$[I_n^{(2)}, I_m^{(1)}] + [I_n^{(1)}, I_m^{(2)}] = 0$$

and so on. Here only the first term $[I_n^{(1)}, I_m^{(1)}]$ looks as the commutativity condition for the operator family $I_n^{(1)}$. On the other hand, all $I_n^{(1)}$ collapse to one and the same expression

$$I_n = n\mathrm{Tr}\,\mathscr{L}^{(r)} = -\hbar \sum_{i=1}^{N} \frac{\partial}{\partial q_i}\,, \tag{3.86}$$

which is nothing else but the total momentum operator. Thus, our results for the commutative family of the RS model cannot be immediately extended to the CMS case and one has to find separate means to exhibit quantum integrability of the latter model. For instance, one can departure from the commuting integrals of the hyperbolic CMS model which remain non-trivial upon taking the rational limit. This will be done in the next sub-section.

3.3.2 Hyperbolic CMS Model

Recall that classically the hyperbolic CMS model was obtained by choosing a dual parametrisation of T^*G based on representation (2.331) of the group element g. In the quantum case we assume for the variable $g \in \mathcal{B}_\hbar$ the same representation (2.331) in terms of matrices T and Q, the entries of which generate an extension of the algebra $\mathcal{B}_\hbar$. We quantise the Poisson brackets (2.332) by postulating the following commutation relations

$$\begin{aligned} [T_1, \ell_2] &= \hbar T_1 T_2 s_{12} T_2^{-1}, \\ [Q_1, \ell_2] &= -\hbar Q_1 T_2 \overline{C}_{12} T_2^{-1}, \end{aligned} \tag{3.87}$$

while the entries of T and Q form a commutative algebra between themselves. Here s_{12} is the same as in (2.333). One can check that the compatibility of these relations with (3.28) follows from the identity (2.334).

Formally, we define the quantum L-operator $\mathscr{L}^{(h)}$ by the same formula (2.335). By using (3.28) and (3.87), we derive the commutation relations for the quantum L-operator

$$\begin{aligned} \left[Q_1, \mathscr{L}_2^{(h)}\right] &= \hbar Q_1 \overline{C}_{12}\,, \\ \left[T_1, \mathscr{L}_2^{(h)}\right] &= -\hbar T_1 s_{12}. \end{aligned}$$

and

$$\left[\mathscr{L}_1^{(h)}, \mathscr{L}_2^{(h)}\right] = \hbar[r_{12}^{(h)}, \mathscr{L}_1^{(h)}] - \hbar[r_{21}^{(h)}, \mathscr{L}_2^{(h)}] + \hbar^2[r_{12}^{(h)}, r_{21}^{(h)}], \tag{3.88}$$

where $r_{12}^{(h)}$ is given by (2.338). Note that this relation can be rewritten in a form similar to (3.84)

$$[\mathscr{L}_1^{(h)} + \hbar r_{21}^{(h)}, \mathscr{L}_2^{(h)} + \hbar r_{12}^{(h)}] = 0\,. \tag{3.89}$$

Finally, one can verify that the following quantum version

$$\mathcal{L}^{(h)} = -\hbar \sum_{i=1}^{N} E_{ii} \frac{\partial}{\partial q_i} + \gamma \sum_{i \neq j}^{N} \frac{Q_j}{Q_{ij}} E_{ij} \tag{3.90}$$

of the classical Lax matrix (2.342) realises the representation of the algebra (3.89).

To complete our discussion, let us show the existence of N mutually commuting operators in the algebra generated by the entries of the L-operator and the coordinates Q. Obviously, $I_k = \mathrm{Tr}\,(-\ell)^k$ mutually commute.[9] Applying the technique used in the previous section to derive the quantum integrals of motion, we find that the I_k can be expressed as the following combination of $\mathcal{L}^{(h)}$ and Q:

$$I_k = \mathrm{Tr}\,(-\ell)^k = \mathrm{Tr}\,T\,(\mathcal{L}^{(h)})^k T^{-1} = \mathrm{Tr}_{12} C_{12}^{t_2}(\mathcal{L}_1^{(h)} - \hbar s_{21}^{t_2})^k , \tag{3.91}$$

where

$$s_{21}^{t_2} \equiv (s_{21})^{t_2} = (s^{t_1})_{21} = \sum_{i \neq j}^{N} \frac{Q_j}{Q_{ij}} E_{ij} \otimes (E_{jj} - E_{ij}) .$$

As is clear from (3.91), the ratio $I_k/\hbar^k$ depends on the unique combination

$$\beta = \frac{\gamma}{\hbar} , \tag{3.92}$$

implying that common eigenfunctions of I_k will depend on the single parameter β. To give an impression of the explicit form of the integrals, we worked out the first three I_k. Introducing the concise notation $\partial_i \equiv \frac{\partial}{\partial q_i}$, these integrals read

$$I_1/\hbar = -\sum_i \partial_i - \frac{N(N-1)}{2} ,$$

$$I_2/\hbar^2 = \sum_i \partial_i^2 - \beta(\beta-1) \sum_{i \neq j} \frac{Q_i Q_j}{Q_{ij}^2} + (N-1) \sum_i \partial_i + \frac{N(N-1)(N-2)}{6} ,$$

$$I_3/\hbar^3 = -\sum_i \partial_i^3 + 3\beta(\beta-1) \sum_{i \neq j} \frac{Q_i Q_j}{Q_{ij}^2} \partial_i - \frac{1}{2} \left(\sum_{i \neq j} \partial_i \partial_j + \beta(\beta-1) \sum_{i \neq j} \frac{Q_i Q_j}{Q_{ij}^2} \right)$$

$$- (N-1) \left(\sum_i \partial_i^2 - \beta(\beta-1) \sum_{i \neq j} \frac{Q_i Q_j}{Q_{ij}^2} \right)$$

$$- \frac{(N-1)(N-2)}{2} \sum_i \partial_i - \frac{N(N-1)(N-2)(N-3)}{24} .$$

---

[9] We recall that I_k generate the center of the algebra spanned by ℓ.

Each I_k is a partial differential operator of order k, it contains admixtures of I_m with $m < k$ as well as a constant term. In this respect the basis of integrals produced through the reduction procedure is reducible. Omitting these admixtures and making proper linear combinations of resulting expressions, we can construct a simpler and irreducible basis. Application of this simplification procedure yields the integrals

$$H_1/\hbar = \sum_i \partial_i \, ,$$

$$H_2/\hbar^2 = \sum_i \partial_i^2 - \beta(\beta - 1) \sum_{i \neq j} \frac{Q_i Q_j}{Q_{ij}^2} \, ,$$

$$H_3/\hbar^3 = \sum_i \partial_i^3 - 3\beta(\beta - 1) \sum_{i \neq j} \frac{Q_i Q_j}{Q_{ij}^2} \partial_i$$

$$(3.93)$$

and so on. As $\gamma \to 0$ the integrals behave as $H_k \sim \sum_j \partial_j^k$, which gives a reason to call this basis the *power sum basis*.

As we will see in Chap. 4, the integrals H_k have the distinguished feature of being local. This means that in the asymptotic limit of large time, these integrals take an additive form with respect to the number of particles

$$H_k \sim \sum_j p_j^k \, ,$$

where p_j is an asymptotic momentum of j's particle. According to our discussion of the definition of quantum integrability in Sect. 3.1, exhibiting a commutative family of operators is not exciting by itself, but the fact that these operators are *local* is truly remarkable. For instance, powers of the hamiltonian, although commuting, are not local operators. Thus, the existence of local commuting operators adds to the notion of quantum integrability.

Departing from the power sum basis, one can construct another, *Hénon basis,*[10] where the quantum integral D_k is given by the following determinant formula

$$D_k = \frac{1}{k!} \begin{vmatrix} H_1 & 1 & 0 \cdots & 0 \\ H_2 & H_1 & 2 \cdots & 0 \\ \vdots & \vdots & \cdots & \vdots \\ H_{k-1} & H_{k-2} & \cdots & k-1 \\ H_k & H_{k-1} & \cdots & H_1 \end{vmatrix} \, .$$

$$(3.94)$$

Thus, similarly to (3.74), the relation between H_k and D_k is the same as between power sums and elementary symmetric functions, see the second formula in (3.178). Explicitly, for the first three integrals we have

---

[10] A similar basis was found by Hénon for the periodic Toda chain [21].

$$D_1 = \hbar \sum_i \partial_i \,,$$

$$D_2 = \frac{\hbar^2}{2!} \sum_{ij}{}' \left(\partial_i \partial_j + \beta(\beta - 1) \frac{Q_i Q_j}{Q_{ij}^2} \right),$$

(3.95)

$$D_3 = \frac{\hbar^3}{3!} \sum_{ijk}{}' \left(\partial_i \partial_j \partial_k + 3\beta(\beta - 1) \frac{Q_i Q_j}{Q_{ij}^2} \partial_k \right),$$

where Σ' means that no two summation indices coincide. In the limit $\gamma \to 0$ the integrals D_k are nothing but the elementary symmetric functions of derivatives and, for this reason, the Hénon basis can be also called *symmetric*. A welcome feature of this basis is that the ordering of the coordinate and momentum operators within any individual sum is not important. In fact, there is a closed formula for D_k that generalises (3.95) for arbitrary positive integers k [22],

$$D_k = \hbar^k \sum_{j=0}^{[k/2]} \frac{\beta^j (\beta - 1)^j}{2^j \, j! (k - 2j)!} \sum_{i_1,\ldots,i_k}{}' \frac{Q_{i_1} Q_{i_2}}{Q_{i_1 i_2}^2} \cdots \frac{Q_{i_{2j-1}} Q_{i_{2j}}}{Q_{i_{2j-1} i_{2j}}^2} \partial_{i_{2j+1}} \partial_{i_{2j+2}} \cdots \partial_{i_k} \,. \quad (3.96)$$

In applications the following combination

$$H = \tfrac{1}{2} H_2 = \tfrac{1}{2} (D_1^2 - 2D_2) = \frac{1}{2} \hbar^2 \sum_{i=1}^N \partial_i^2 - \frac{1}{2} \gamma(\gamma - \hbar) \sum_{i \neq j}^N \frac{Q_i Q_j}{Q_{ij}^2} \quad (3.97)$$

is typically taken as the quantum hamiltonian of the hyperbolic CMS model. It should be emphasised that the classical coupling constant γ^2 gets replaced in quantum theory by $g = \gamma(\gamma - \hbar)$.

Rational CMS model and quantum Lax representation. The quantum rational CMS model can be obtained from its hyperbolic counterpart by rescaling

$$q_i \to \kappa q_i \,, \quad \gamma \to \kappa \gamma \,, \quad \hbar \to \kappa \hbar \quad (3.98)$$

with further sending κ to zero. In this limit (3.90) turns into the quantum L-operator (3.83) of the rational CMS model (3.83):

$$\mathscr{L} = -\hbar \sum_{i=1}^N E_{ii} \frac{\partial}{\partial q_i} + \gamma \sum_{i \neq j}^N \frac{1}{q_{ij}} E_{ij} \,, \quad (3.99)$$

where to simply our further discussion we set $\mathscr{L} \equiv \mathscr{L}^{(r)}$. The hamiltonian defining the quantum rational CMS model is taken to be

$$H = \frac{1}{2}\hbar^2 \sum_{i=1}^{N} \frac{\partial^2}{\partial q_i^2} - \frac{1}{2}\gamma(\gamma - \hbar) \sum_{i \neq j}^{N} \frac{1}{q_{ij}^2} . \tag{3.100}$$

In the scaling limit (3.98) the integrals I_k remain non-trivial and take the form

$$I_k = \mathrm{Tr}_{12} C_{12}^{t_2} (\mathscr{L}_1 - \hbar s_{21}^{t_2})^k , \tag{3.101}$$

where $s_{21}^{t_2}$ reads as

$$s_{21}^{t_2} = \sum_{i \neq j}^{N} \frac{1}{q_{ij}} E_{ij} \otimes (E_{jj} - E_{ij}) .$$

It appears that formula (3.101) can be substantially simplified and directly related to the construction by Ujino, Hikami and Wadati [23], where integrals are obtained from the corresponding quantum Lax representation. This representation, see (3.76), can be constructed following the same steps as were done for the rational RS model. We leave this construction as an exercise to the reader and present here the final result. For the rational CMS model defined by (3.99) and (3.100) the quantum Lax representation holds with the matrix M given by

$$M = \hbar\gamma \sum_{i \neq j}^{N} \frac{1}{q_{ij}^2} (E_{ii} - E_{ij}) . \tag{3.102}$$

This matrix has a special property, namely, it belongs to the Lie algebra of the stabiliser F_e, as it satisfies the relations

$$\sum_{j=1}^{N} M_{ij} = 0 , \qquad \sum_{j=1}^{N} M_{ji} = 0 , \quad \forall i . \tag{3.103}$$

Now we are ready to explain the relationship between the quantum Lax representation and integrals (3.101). Writing

$$(\mathscr{L}_1 - \hbar s_{21}^{t_2})^k = \sum_{ijkl} h_{ij,kl}^{(k)} E_{ij} \otimes E_{kl} ,$$

it follows from (3.101) that

$$I_k = \sum_{ij} h_{ij,ij}^{(k)} .$$

For $k = 1$ it is straightforward to find[11] that $I_1 = \mathrm{Tr}\,\ell$. For I_2 we get

$$I_2 = \mathrm{Tr}\mathcal{L}^2 + \sum_{i \neq j} \mathcal{L}_{ii} \frac{\hbar}{q_{ij}} - \sum_{i \neq j} \mathcal{L}_{ji} \frac{\hbar}{q_{ij}} + \sum_{i \neq j} \frac{\hbar}{q_{ij}} \mathcal{L}_{jj} + \sum_{ij} (s_{21}^{t_2})_{ij,ij}^2 .$$

To proceed, we need the commutator

$$[\mathcal{L}_{ii}, \mathcal{L}_{ij}] = -\hbar \frac{\partial}{\partial q_i} \frac{\gamma}{q_{ij}} = \frac{\hbar\gamma}{q_{ij}^2} = \hbar \frac{\mathcal{L}_{ij}}{q_{ij}} , \qquad (3.104)$$

valid for $i \neq j$. Taking into account that

$$\sum_{ij} (s_{21}^{t_2})_{ij,ij}^2 = -\sum_{i \neq j} \frac{\hbar^2}{q_{ij}^2} , \qquad (3.105)$$

the expression for I_2 becomes

$$I_2 = \sum_{ik} \mathcal{L}_{ik} \mathcal{L}_{ki} + \frac{\hbar}{\gamma} \sum_{i \neq j} [\mathcal{L}_{ii}, \mathcal{L}_{ij}] + \sum_{i \neq j} \frac{\hbar \mathcal{L}_{ij}}{q_{ij}} + \sum_{i \neq j} \frac{\hbar}{q_{ij}} (\mathcal{L}_{ii} + \mathcal{L}_{jj}) - \sum_{i \neq j} \frac{\hbar^2}{q_{ij}^2} \mathbb{1} .$$

Now we see that due to (3.104) and (3.105) the second sum cancels against the last one and the fourth sum vanishes for symmetry reasons. Thus, I_2 simplifies to

$$I_2 = \sum_{ik} \mathcal{L}_{ik} \mathcal{L}_{ki} + \sum_{i \neq j} [\mathcal{L}_{ii}, \mathcal{L}_{ij}] ,$$

where we used (3.104). Adding to I_2 the vanishing term $\sum_{i \neq j} \mathcal{L}_{ij} (\mathcal{L}_{ii} + \mathcal{L}_{jj})$, we get

$$I_2 = \sum_{ik} \mathcal{L}_{ik} \mathcal{L}_{ki} + \sum_{i \neq j} \mathcal{L}_{ii} \mathcal{L}_{ij} + \sum_{i \neq j} \mathcal{L}_{ij} \mathcal{L}_{jj} .$$

This can be compared to the expression

$$\sum_{ij} \mathcal{L}_{ij}^2 = \sum_{ijk} \mathcal{L}_{ik} \mathcal{L}_{kj} = \sum_{ik} \mathcal{L}_{ik} \mathcal{L}_{ki} + \sum_{i \neq j} \mathcal{L}_{ii} \mathcal{L}_{ij} + \sum_{i \neq j} \mathcal{L}_{ij} \mathcal{L}_{jj} + \sum_{i \neq k \neq j} \mathcal{L}_{ik} \mathcal{L}_{kj} .$$

Since the last term here vanishes, we conclude that[12] $I_2 = \sum_{ij} \mathcal{L}_{ij}^2$. Working out the cases of higher $k = 3, 4, \ldots$, one can verify the validity of the general expression

---

[11] Note that $\mathrm{Tr}\, C_{12}^{t_2} s_{21}^{t_2} = 0$.

[12] Explicitly, $I_2 = 2H$, where H is given by (3.100).

$$I_k = \sum_{ij} \mathcal{L}_{ij}^k = \mathbf{e}^t \mathcal{L}^k \mathbf{e} \equiv \mathrm{T}_\Sigma(\mathcal{L}^k)\,,$$

in the notation of [24]. According to [23], the fact that all I_k are quantum integrals of motion follows immediately from the quantum Lax representation (3.76). Indeed,

$$[H, \mathrm{T}_\Sigma(\mathcal{L}^k)] = \sum_{ij}[M, \mathcal{L}^k]_{ij} = \sum_{ijn}(M_{in}\mathcal{L}_{nj}^k - \mathcal{L}_{in}^k M_{nj}) = 0\,,$$

because M satisfies (3.103). In our approach the crucial relations (3.103) have a clear geometric origin in the hamiltonian reduction procedure.

Quantum Lax and integrals for the hyperbolic CMS model. At this point we recall that for the classical hyperbolic CMS model we have found the Lax representation (2.334), where the matrix M does satisfy relations (3.103). It is then straightforward to see that picking up the hamiltonian

$$H = \frac{1}{2}\hbar^2 \sum_{i=1}^{N} \frac{\partial^2}{\partial q_i^2} - \frac{1}{2}\gamma(\gamma - \hbar) \sum_{i \neq j}^{N} \frac{Q_i Q_j}{Q_{ij}^2}\,, \tag{3.106}$$

the corresponding quantum model admits the Lax representation (3.76), where $L = \mathcal{L}^{(h)}$ is (3.90) and the matrix M is given by

$$M = \hbar\gamma \sum_{i \neq j}^{N} \frac{Q_i Q_j}{Q_{ij}^2}(E_{ii} - E_{ij})\,. \tag{3.107}$$

Up to an overall constant $\hbar$, this is the same matrix (2.347) as in the classical theory. In the rational limit it turns into M-matrix (3.102) of the rational CMS model. As follows from our previous discussion, due to the same properties (3.103), the quantum integrals of the hyperbolic CMS model can also be constructed as [23]

$$J_k = \mathrm{T}_\Sigma(\mathcal{L}^{(h)})^k\,. \tag{3.108}$$

Because of slightly different properties of the L-operators and the matrices $s_{21}^{t_2}$ in the rational and trigonometric cases, the integrals I_k introduced by the formula (3.91) do not coincide with J_k but can be re-expanded via those. Moreover, recalling the issue of non-uniqueness of the L-operator, we can alternatively consider another $\mathcal{L}^{(h)}$, namely,

$$\mathcal{L}^{(h)} = -\hbar \sum_{i=1}^{N} E_{ii} \frac{\partial}{\partial q_i} + \frac{\gamma}{2} \sum_{i \neq j}^{N} \cot(\tfrac{1}{2} q_{ij}) E_{ij}\,. \tag{3.109}$$

This L-operator has the same Lax representation (3.76) with the same M-matrix (3.107). It will generate another set of quantum integrals through the same formula (3.108).

Concerning the difference between the quantum Lax representation for CMS and RS models, in a nutshell, the one for RS models has M-matrix that depends on momentum and does not belong to the Lie algebra of the stabiliser group of the moment map. This makes it inefficient for exhibiting quantum integrability of the corresponding models.

3.3.3   Hyperbolic RS Model

Quantum Heisenberg double. Recall that at the classical level we obtained the hyperbolic RS model by means of the Poisson reduction of the Heisenberg double. It is therefore natural to start with the quantum analogue of the Heisenberg double. The Poisson algebra (2.221) can be straightforwardly quantised in the same spirit of deformation theory. We thus introduce an associative unital algebra $D_\hbar$ generated by the entries of matrices x, y modulo the relations [25]

$$R_+ x_1 x_2 = x_2 x_1 R_-^{-1}, \quad R_+ x_1 y_2 = y_2 x_1 R_+^{-1},$$
$$R_+ y_1 y_2 = y_2 y_1 R_-^{-1}, \quad R_- y_1 x_2 = x_2 y_1 R_-^{-1}. \tag{3.110}$$

Here $R_\pm$ are solutions of the quantum Yang-Baxter equation given by (3.16) and (3.17). Of course, one can use different system of generators to describe the quantum Heisenberg double. In particular, in terms of (A, B) generators the algebraic relations of the double read as

$$R_-^{-1} A_2 R_+ A_1 = A_1 R_-^{-1} A_2 R_+,$$
$$R_-^{-1} B_2 R_+ A_1 = A_1 R_-^{-1} B_2 R_-,$$
$$R_+^{-1} A_2 R_+ B_1 = B_1 R_-^{-1} A_2 R_+, \tag{3.111}$$
$$R_-^{-1} B_2 R_+ B_1 = B_1 R_-^{-1} B_2 R_+,$$

and they can be regarded as the quantisation of the Poisson relations (2.260).

The first (last) line in (3.111) is a set of defining relations for the corresponding subalgebra that describes quantisation of the Semenov-Tian-Shansky bracket, the latter has a set of Casimir functions generated by $C_k = \mathrm{Tr} A^k$. In the quantum case an analogue $\mathrm{Tr} A^k$ can be defined by means of the quantum trace formula

$$C_k = \mathrm{Tr}_q A^k \equiv \mathrm{Tr}(D A^k), \quad q \equiv e^{-\hbar},$$

where D is a diagonal matrix $D = \mathrm{diag}(q, q^2, \ldots, q^n)$. The elements C_k are central in the subalgebra generated by A. Indeed, by successively using the permutation relations for A, one gets

$$A_2 R_+ A_1^k R_+^{-1} = R_- A_1^k R_-^{-1} A_2 \, .$$

We then multiply both sides of this relation by D_1 and take the trace in the first matrix space

$$A_2 \text{Tr}_1 \left(D_1 R_+ A_1^k R_+^{-1} \right) = \text{Tr}_1 \left(D_1 R_- A_1^k R_-^{-1} \right) A_2 \, .$$

It remains to notice that $\text{Tr}_1 \left(D_1 R_+ A_1^k R_+^{-1} \right) = \text{Tr}_1 \left(D_1 R_- A_1^k R_-^{-1} \right) = \text{Tr}_q A^k \cdot 1$, so that

$$A \, \text{Tr}_q A^k = \text{Tr}_q A^k \, A \, , \tag{3.112}$$

i.e. $\text{Tr}_q A^k$ is central in the subalgebra generated by A. Analogously, the $I_k = \text{Tr}_q B^k$ are central in the algebra generated by B and, in particular, the I_k form a commutative family.

In principle, we can start with (3.111) and develop a proper parametrisation of the (A, B) generators suitable for reduction. It is an interesting path that should lead to understanding how to implement the Dirac constraints at the quantum level. We will leave this to the reader and note here an existence of a short cut to the algebra of the quantum L-operator.

Quantum R-matrices and the L-operator. An alternative route to the quantum R-matrices and to the corresponding L-operator algebra is based on the observation that in the classical theory, the Poisson brackets between the entries of the Lax matrix have the same structure (2.308) for both rational and hyperbolic cases. As a consequence, equations satisfied by the classical rational and hyperbolic r-matrices are also the same. This should also be applied to the equations obeyed by the corresponding quantum R-matrices. We thus assume that the matrices $R(Q)$ and $\bar{R}(Q)$ for the hyperbolic RS model satisfies the system of Eqs. (3.35), (3.37) and (3.38) and have the standard semi-classical limit where they match the classical r-matrices (2.370). Our considerations here follow [26].

In fact, it is not difficult to guess a proper solution for these R-matrices basing on the analogy with the rational case. For $R(Q)$ we can take[13]

$$R(Q) = \exp \hbar r(Q) \, . \tag{3.113}$$

To simplify the presentation, below we omit the collective variable Q in the notation of the R-matrices, since it is clear that from now on all the discussion involves only the dynamical R-matrices. We also adopt the notation $R_+ \equiv R(Q)$. Since the classical r-matrix satisfies the property $r^2 = -r$, the exponential in (3.113) can be easily evaluated and we find

---

[13]In order not to overload the presentation, in the following we take the deformation parameter $\varkappa = 1$, so that $Q_i = \exp q_i$. One can always restore $\varkappa$ in the final formulae by rescaling $q_i \to \varkappa q_i$, $\hbar \to \varkappa \hbar$.

$$R_+ = 1 + (1-q) \sum_{i \neq j}^{N} \left(\frac{Q_j}{Q_{ij}} E_{ii} - \frac{Q_i}{Q_{ij}} E_{ij} \right) \otimes (E_{jj} - E_{ji}). \qquad (3.114)$$

A direct check shows that (3.114) is a solution of (3.35).

In comparison to the rational model, a new feature is that there exists yet another solution R_- of the Yang-Baxter equation, namely,

$$R_- = 1 - (1-q^{-1}) \sum_{i \neq j}^{N} (E_{ii} - E_{ij}) \otimes \left(\frac{Q_i}{Q_{ij}} E_{jj} - \frac{Q_j}{Q_{ij}} E_{ji} \right). \qquad (3.115)$$

These solutions are related as

$$R_{+21} R_{-12} = 1, \qquad (3.116)$$

i.e. precisely in the same way as their non-dynamical counterparts, cf. (3.18). Further, the matrices $R_\pm$ satisfy the following analog of Eq. (3.19)

$$R_+ - q R_- = (1-q)C. \qquad (3.117)$$

They are also of Hecke type and the matrices $\hat{R}_\pm = C R_\pm$ have the following property

$$\left(\hat{R}_\pm - 1 \right) \left(\hat{R}_\pm + q^{\pm 1} 1 \right) = 0. \qquad (3.118)$$

Concerning the generalisation of Eq. (3.37) to the hyperbolic case, we can imagine two different versions - one involving R_+ and another R_-, that is,

$$R_{\pm 12} \bar{R}_{13} \bar{R}_{23} = \bar{R}_{23} \bar{R}_{13} P_3 R_{\pm 12} P_3^{-1}, \qquad (3.119)$$

In this relation and below the variables (Q_i, P_i) satisfy the commutation relations of the Weyl algebra

$$Q_i Q_j = Q_j Q_i, \qquad P_i P_j = P_j P_i, \qquad P_i Q_j = e^{\hbar \delta_{ij}} Q_j P_i.$$

It appears that there exist a unique matrix $\bar{R}$ which satisfies both these equations. It is given by

$$\bar{R} = 1 - \sum_{i \neq j}^{N} \frac{q Q_i - Q_i}{q Q_i - Q_j} (E_{ii} - E_{ij}) \otimes E_{jj}. \qquad (3.120)$$

and its inverse is

$$\bar{R}^{-1} = \mathbb{1} - (1-q) \sum_{i \neq j}^{N} \frac{Q_i}{Q_{ij}} (E_{ii} - E_{ij}) \otimes E_{jj} \,. \tag{3.121}$$

The matrix (3.120) also obeys (3.38),

$$\bar{R}_{12} P_2 \bar{R}_{13} P_2^{-1} = \bar{R}_{13} P_3 \bar{R}_{12} P_3^{-1} \,. \tag{3.122}$$

Using the definition (3.51), we further get

$$\underline{R}_+ = \mathbb{1} + (1-q) \sum_{i \neq j}^{N} \frac{Q_i}{Q_{ij}} (E_{ij} \otimes E_{ji} - E_{ii} \otimes E_{jj}) \,,$$

$$\underline{R}_- = \mathbb{1} - (1-q^{-1}) \sum_{i \neq j}^{N} \frac{Q_j}{Q_{ij}} (E_{ij} \otimes E_{ji} - E_{ii} \otimes E_{jj}) \,. \tag{3.123}$$

These matrices satisfy the Gervais-Neveu-Felder equation

$$\underline{R}_{\pm 12} P_2^{-1} \underline{R}_{\pm 13} P_2 \underline{R}_{\pm 23} = P_1^{-1} \underline{R}_{\pm 23} P_1 \underline{R}_{\pm 13} P_3^{-1} \underline{R}_{\pm 12} P_3 \,. \tag{3.124}$$

and are related to each other as

$$\underline{R}_{+21} \underline{R}_{-12} = \mathbb{1} \,. \tag{3.125}$$

They also have another important property, usually refereed to as the *zero weight condition*,

$$[P_1 P_2, \underline{R}_\pm] = 0 \,. \tag{3.126}$$

Finally, the quantum L-operator is literally the same as its classical counterpart (2.363), of course, with the natural replacement of p_i by the corresponding derivative

$$L = \sum_{i,j=1}^{N} \frac{Q_i - t Q_j}{Q_i - t Q_j} \, b_j \mathsf{T}_j E_{ij} \,, \quad b_j \equiv \prod_{a \neq j}^{N} \frac{t Q_j - Q_a}{Q_j - Q_a} \,, \tag{3.127}$$

where $t = e^{-\gamma}$. It is a straightforward exercise to check that this L-operator satisfies the algebraic relations

$$R_{+12} L_2 \bar{R}_{12}^{-1} L_1 = L_1 \bar{R}_{21}^{-1} L_2 \underline{R}_{+12} \,,$$

$$R_{-12} L_2 \bar{R}_{12}^{-1} L_1 = L_1 \bar{R}_{21}^{-1} L_2 \underline{R}_{-12} \,. \tag{3.128}$$

with the R-matrices given by (3.114), (3.115), (3.120) and (3.123). The consistency of these relations follow from (3.116) and (3.125).

Concerning commuting integrals, the Heisenberg double has a natural commutative family $I_k = \mathrm{Tr}_q B^k$. It is not clear, however, how these integrals can be expressed via L, because we are lacking an analogue of the quantum factorisation formula $B = TLT^{-1}$, where T and L would be a subject of well-defined algebraic relations. Rather, what we could do is to conjecture the same formula (3.62) for quantum integrals, where now the R-matrices are those of the hyperbolic model. Interestingly, the existence of two R-matrices, $R_\pm$, gives rise to two families of commuting integrals $I_k^\pm$ [26]. These families look the same as in the rational case (3.62), except for R_{12} one takes R_{+12} for the I_k^+ family and R_{-12} for I_k^- family, respectively. Thus,

$$I_k^\pm = \mathrm{Tr}_{12}\left(C_{12}^{t_2} L_1 \bar{R}_{21}^{t_2} R_{\pm 12}^{t_2} L_1 \ldots L_1 \bar{R}_{21}^{t_2} R_{\pm 12}^{t_2} L_1\right),\tag{3.129}$$

where k is a number of L_1's on the right hand side. Commutativity of $I_k^\pm$ can be then verified by direct computation.

Analogously to the rational case, there exists a simpler basis of integrals $\{S_k\}$, where

$$S_k = t^{\frac{1}{2}k(k-1)} \sum_{\substack{J\subset\{1,\ldots,n\} \\ |J|=k}} \prod_{\substack{i\in J \\ j\notin J}} \frac{tQ_i - Q_j}{Q_i - Q_j} \prod_{i\in J} T_i.\tag{3.130}$$

The commuting integrals S_k are known as *Macdonald operators* and they have the generating function (3.75) where L is given by (3.127). Also, there is an explicit formula between the families $\{I_k^\pm\}$ and S_k. To present it, we need a notion of a q-number $[k]_q$ associated to an integer k

$$[k]_q = \sum_{n=0}^{k-1} q^n = \frac{1-q^k}{1-q},\tag{3.131}$$

so that $[k]_1 = k$, which corresponds to the limit $\hbar \to 0$. Then S_k is expressed via I_m^+ or I_m^- as

$$S_k = \frac{1}{[k!]_{q^{\pm 1}}} \begin{vmatrix} I_1^\pm & [k-1]_{q^{\pm 1}} & 0 & \cdots & 0 \\ I_2^\pm & I_1^\pm & [k-2]_{q^{\pm 1}} & \cdots & 0 \\ \vdots & \vdots & & \cdots & \vdots \\ I_{k-1}^\pm & I_{k-2}^\pm & & \cdots & [1]_{q^{\pm 1}} \\ I_k^\pm & I_{k-1}^\pm & & \cdots & I_1^\pm \end{vmatrix}.\tag{3.132}$$

These determinant formulae replace (3.74) in the hyperbolic case. They can be inverted to express each integral $I_k^\pm$ as the determinant of a $k \times k$ matrix depending on S_j, namely,

$$
I_k^\pm = \begin{vmatrix} S_1 & 1 & 0 & \cdots & 0 \\ [2]_{q^{\pm1}} S_2 & S_1 & 1 & 0 & \cdots \\ \vdots & \vdots & \cdots & \cdots & 1 \\ [k]_{q^{\pm1}} S_k & S_{k-1} & S_{k-2} & \cdots & S_1 \end{vmatrix}. \tag{3.133}
$$

Spectral parameter and quantum L-operator. The quantum L-operator depending on the spectral parameter is naturally introduced as a normal ordered version of its classical counterpart

$$
L(\lambda) = \frac{1-t}{\lambda} \sum_{i,j=1}^{N} \frac{\lambda Q_i - e^{-\hbar/2} t Q_j}{Q_i - t Q_j} b_j \top_j E_{jj} = L - \frac{t e^{\hbar/2}}{\lambda} Q^{-1} L Q, \tag{3.134}
$$

where b_j are the same as in (3.127). This L-operator satisfies the following quadratic relation

$$
R_{12}(\lambda, \mu) L_2(\mu) \bar{R}_{12}^{-1}(\lambda) L_1(\lambda) = L_1(\lambda) \bar{R}_{21}^{-1}(\mu) L_2(\mu) \underline{R}_{12}(\lambda, \mu), \tag{3.135}
$$

where

$$
\underline{R}_{12}(\lambda, \mu) = \bar{R}_{12}^{-1}(\lambda) R_{12}(\lambda, \mu) \bar{R}_{21}(\mu). \tag{3.136}
$$

In (3.135) the quantum R-matrices are

$$
R(\lambda, \mu) = \frac{\lambda e^{\hbar/2} R_+ - \mu e^{-\hbar/2} R_-}{\lambda - \mu} - \frac{e^{\hbar/2} - e^{-\hbar/2}}{e^{\hbar/2}\lambda - 1} X_{12} + \frac{e^{\hbar/2} - e^{-\hbar/2}}{e^{-\hbar/2}\mu - 1} X_{21}.
$$
$$
\bar{R}(\lambda) = \bar{R} - \frac{e^\hbar - 1}{e^{\hbar/2}\lambda - 1} X_{12}. \tag{3.137}
$$

Here R_+ and R_- are the solutions (3.114) and (3.115) of the quantum Yang-Baxter equation, $\bar{R}$ is (3.120) and we have introduced the matrix $X \equiv X_{12}$,

$$
X = \sum_{i,j=1}^{N} E_{ij} \otimes E_{jj}. \tag{3.138}
$$

This matrix satisfies a number of simple relations with $\bar{R}$ and $R_\pm$, which are

$$
\bar{R} X = X \bar{R} \tag{3.139}
$$

and

$$
\begin{aligned}
R_- X_{12} &= R_- X_{12}, & R_- X_{21} - X_{21} R_- &= (1 - q^{-1})(X_{12} - X_{21}), \\
R_+ X_{21} &= X_{21} R_+, & R_+ X_{12} - X_{12} R_+ &= -(1 - q)(X_{12} - X_{21}).
\end{aligned} \tag{3.140}
$$

We also present the formula for the inverse of $\bar{R}(\lambda)$

$$\bar{R}(\lambda)^{-1} = \bar{R}^{-1} + \frac{e^{\hbar} - 1}{e^{\hbar/2}\lambda - e^{\hbar}} X_{12}. \tag{3.141}$$

With the help of this formula and (3.137) one can show that (3.136) boils down to

$$\underline{R}_{12}(\lambda, \mu) = \frac{\lambda e^{\hbar/2} \underline{R}_+ - \mu e^{-\hbar/2} \underline{R}_-}{\lambda - \mu}, \tag{3.142}$$

where $\underline{R}_\pm$ are the same as given by (3.123). We note also the relation

$$R_{12}(\lambda, \mu) R_{21}(\mu, \lambda) = \underline{R}_{12}(\lambda, \mu)\underline{R}_{21}(\mu, \lambda) = \frac{(e^{\hbar/2}\lambda - e^{-\hbar/2}\mu)(e^{-\hbar/2}\lambda - e^{\hbar/2}\mu)}{(\lambda - \mu)^2}\mathbb{1}. \tag{3.143}$$

Finally, in addition to (3.136) there is one more relation between $R(\lambda, \mu)$ and $\underline{R}(\lambda, \mu)$, namely,

$$\underline{R}_{12}(\lambda, \mu) = P_1^{-1}\bar{R}_{21}(\mu)P_1 R_{12}(\lambda, \mu)P_2^{-1}\bar{R}_{12}^{-1}(\lambda)P_2. \tag{3.144}$$

An interesting observation is that the combination

$$R^{\mathrm{YB}}(\lambda, \mu) = \frac{\lambda e^{\hbar/2} R_+ - \mu e^{-\hbar/2} R_-}{\lambda - \mu}$$

solves the standard quantum Yang-Baxter equation with the spectral parameter (3.24). However, the full R-matrix in (3.137) differs from R^{YB} by the terms that violate scale invariance. As a result, this matrix obeys the shifted version of the quantum Yang-Baxter equation, namely,

$$R_{12}(\lambda, \mu) R_{13}(q\lambda, q\tau) R_{23}(\mu, \tau) = R_{23}(q\mu, q\tau) R_{13}(\lambda, \tau) R_{12}(q\lambda, q\mu). \tag{3.145}$$

In addition, there are two more equations—the one involving both R and $\bar{R}$, and the other involving $\bar{R}$ only,

$$R_{12}(\lambda, \mu)\bar{R}_{13}(q\lambda)\bar{R}_{23}(\mu) = \bar{R}_{23}(q\mu)\bar{R}_{13}(\lambda)P_3 R_{12}(q\lambda, q\mu)P_3^{-1}, \tag{3.146}$$

$$\bar{R}_{12}(\lambda)P_2\bar{R}_{13}(q\lambda)P_2^{-1} = \bar{R}_{13}(\lambda)P_3\bar{R}_{12}(q\lambda)P_3^{-1}. \tag{3.147}$$

It is immediately recognisable that Eqs. (3.145), (3.146) and (3.147) is a quantum analogue (quantisation) of the classical Eqs. (2.399), (2.401) and (2.402), respectively. In the semi-classical expansion

$$R(\lambda, \mu) = \mathbb{1} + \hbar r(\lambda, \mu) + o(\hbar), \quad \bar{R}(\lambda) = \mathbb{1} + \hbar r(\lambda) + o(\hbar) \tag{3.148}$$

the matrices (3.137) yield

$$r_{12}(\lambda, \mu) = \frac{\lambda r_{12} + \mu r_{21}}{\lambda - \mu} + \frac{\sigma_{12}}{\lambda - 1} - \frac{\sigma_{21}}{\mu - 1}$$
$$+ \left(\frac{1}{2} \frac{\lambda + \mu}{\lambda - \mu} - \frac{1}{\lambda - 1} + \frac{1}{\mu - 1} \right) \mathbb{1} \otimes \mathbb{1},$$
$$\bar{r}_{12}(\lambda) = \bar{r}_{12} + \frac{\sigma_{12}}{\lambda - 1} - \frac{\mathbb{1} \otimes \mathbb{1}}{\lambda - 1},$$

which is different from the canonical classical r-matrices (2.397) by allowed symmetry shifts. Thus, (3.137) should be regarded as a quantisation of the classical r-matrices satisfying the shifted Yang-Baxter equation. In this respect it is interesting to point out that the corresponding quantisation of the r-matrices solving the usual CYBE remains known at the time of writing this book. For further studies of some algebraic structures related to the shifted Yang-Baxter equation, see [27].

Finally, the algebra (3.135) should be completed by the following additional relations encoding the commutation properties of L with Q

$$L_1 Q_2 = Q_2 L_1 \Omega_{12}, \quad Q_1^{-1} L_2 = L_2 Q_1^{-1} \Omega_{12}, \tag{3.149}$$

where $\Omega_{12} = \mathbb{1} - (1 - q)\bar{C}_{12}$ and $\bar{C}_{12}$ is given by (2.309).

Now we derive a couple of important consequences of the algebraic relation (3.135). Namely, we establish the quantum Lax representation, similar to the and in the rational case, and also prove the commutativity of the operators $\mathrm{Tr} L(\lambda)$ for different values of the spectral parameter.

Following considerations of the dynamics in the classical theory, we take $H = \lim_{\lambda \to \infty} \mathrm{Tr} L(\lambda)$ as the hamiltonian. From (3.135) we get

$$\mathrm{Tr}_1 \left[R_{21}(\mu, \lambda) L_1(\lambda) \bar{R}_{21}^{-1}(\mu) \right] L_2(\mu) = L_2(\mu) \, \mathrm{Tr}_1 \left[\bar{R}_{12}^{-1}(\lambda) L_1(\lambda) \underline{R}_{21}(\mu, \lambda) \right], \tag{3.150}$$

where (3.142) was used. A tedious but straightforward computation reveals that the traces on the left and the right hand side of the last expression are equal and that, for instance,

$$e^{\hbar/2} \mathrm{Tr}_1 \left[\bar{R}_{12}^{-1}(\lambda) L_1(\lambda) \underline{R}_{21}(\mu, \lambda) \right] = \mathrm{Tr} L(\lambda) \, \mathbb{1} - M(\lambda, \mu), \tag{3.151}$$

where

$$M(\lambda, \mu) = (e^{\hbar} - 1) \frac{\lambda}{\lambda - \mu} \frac{\mu - e^{-\hbar/2}}{\lambda - e^{\hbar/2}} L(\lambda)$$
$$+ \frac{e^{\hbar} - 1}{\lambda - e^{\hbar/2}} \sum_{i \neq j}^{N} \frac{\lambda e^{-\hbar} Q_j - e^{-\hbar/2} Q_i}{Q_i - e^{-\hbar} Q_j} L_{ij}(\lambda)(E_{ii} - E_{ij}). \tag{3.152}$$

Thus, Eq. (3.150) turns into

$$\text{Tr}L(\lambda)L(\mu) - L(\mu)\text{Tr}L(\lambda) = [M(\lambda, \mu), L(\mu)]. \qquad (3.153)$$

From (3.152) we, therefore, derive the quantum-mechanical operator M

$$M = \lim_{\lambda \to \infty} M(\lambda, \mu) = (e^{\hbar} - 1) \sum_{i \neq j}^{N} \frac{e^{-\hbar}Q_j}{Q_i - e^{-\hbar}Q_j} L_{ij}(E_{ii} - E_{ij})$$

$$= (e^{\hbar} - 1) \sum_{i \neq j}^{N} L_{ij} \frac{Q_j}{Q_{ij}} (E_{ii} - E_{ij}), \qquad (3.154)$$

where in the last expression we commuted the entries of L_{ij} to the left so that it formally coincides with its classical counterpart (2.391). In the limit $\lambda \to \infty$, (3.153) becomes the quantum Lax equation. Note that in the derivation of the equation we did not use a concrete form of L except that it factorises as $L = WP^{-1}$, where W is a function of coordinates only.

Taking the trace of (3.153), one gets

$$\text{Tr}L(\lambda)\text{Tr}L(\mu) - \text{Tr}L(\mu)\text{Tr}L(\lambda) = \text{Tr}[M(\lambda, \mu), L(\mu)]. \qquad (3.155)$$

A priory the trace of the commutator on the right hand side might not be equal to zero, because it involves the matrices with operator-valued entries. An involved calculation that uses representation (3.134) shows that it nevertheless vanishes[14], identically for λ and μ. Fortunately, there is a simple and transparent way to show the commutativity of traces of the Lax operator, which directly relies on the algebraic relations (3.150), thus bypassing the construction of the quantum Lax pair [28]. Indeed, let us multiply both sides of (3.135) with $P_2^{-1}\bar{R}_{12}(\lambda)P_2R_{12}^{-1}(\lambda, \mu)$ and take the trace with respect to both spaces, 1 and 2. We get

$$\text{Tr}_{12}\left[P_2^{-1}\bar{R}_{12}(\lambda)P_2L_2(\mu)\bar{R}_{12}^{-1}(\lambda)L_1(\lambda) \right] =$$
$$\text{Tr}_{12}\left[P_2^{-1}\bar{R}_{12}(\lambda)P_2R_{12}^{-1}(\lambda, \mu)L_1(\lambda)\bar{R}_{21}^{-1}(\mu)L_2(\mu)\underline{R}_{12}(\lambda, \mu) \right].$$

From (3.144) we have

$$P_2^{-1}\bar{R}_{12}(\lambda)P_2R_{12}^{-1}(\lambda, \mu) = \underline{R}_{12}^{-1}(\lambda, \mu)P_1^{-1}\bar{R}_{21}(\mu)P_1,$$

so that the right hans side of the above equation can be transformed as

---

[14]For this result to hold, the presence in (3.152) of the first term proportional to $L(\lambda)$ is of crucial importance.

$$\mathrm{Tr}_{12}\Big[P_2^{-1} \bar{R}_{12}(\lambda) P_2 L_2(\mu) \bar{R}_{12}^{-1}(\lambda) L_1(\lambda) \Big] =$$
$$\mathrm{Tr}_{12}\Big[\underline{R}_{12}^{-1}(\lambda, \mu) P_1^{-1} \bar{R}_{21}(\mu) P_1 L_1(\lambda) \bar{R}_{21}^{-1}(\mu) L_2(\mu) \underline{R}_{12}(\lambda, \mu) \Big].$$
$$\tag{3.156}$$

Further progress is based on the fact that the matrix $\bar{R}_{12}(\lambda)$ and $\bar{R}_{12}^{-1}(\lambda)$ is diagonal in the second space. We represent it in the factorised form

$$\bar{R}_{12}(\lambda) = \sum_{j=1}^{N} G_j(\lambda) \otimes E_{jj}, \tag{3.157}$$

see (3.137), (3.120) and (3.138). Therefore,

$$P_2^{-1} \bar{R}_{12}(\lambda) P_2 = \sum_{j=1}^{N} P_j^{-1} G_j(\lambda) P_j \otimes E_{jj}. \tag{3.158}$$

Although this expression involves the shift operator, it commutes with any function of coordinates q_j, because when pushed through (3.158), this function will undergo the shifts of q_j in opposite directions which compensate each other. Similarly,

$$\bar{R}_{12}^{-1}(\lambda) = \sum_{j=1}^{N} G_j(\lambda)^{-1} \otimes E_{jj} = \sum_{j=1}^{N} (\mathbb{1} \otimes E_{jj})(G_j(\lambda)^{-1} \otimes \mathbb{1}).$$

Consider first the left hans side of (3.156)

$$\mathrm{Tr}_{12}\Big[\sum_{j=1}^{N} \sum_{k=1}^{N} (P_j^{-1} G_j(\lambda) P_j \otimes E_{jj} L(\mu) E_{kk})(G_k(\lambda)^{-1} \otimes \mathbb{1}) L_1(\lambda) \Big].$$

Using the cyclic property of the trace in the second space, this expression is equivalent to

$$\mathrm{Tr}_{12}\Big[\sum_{j=1}^{N} \sum_{k=1}^{N} (P_j^{-1} G_j(\lambda) P_j \otimes L(\mu) E_{jj} E_{kk})(G_k(\lambda)^{-1} \otimes \mathbb{1}) L_1(\lambda) \Big].$$

Taking into account that $L = W P^{-1}$ and commutativity of $P_j^{-1} G_j(\lambda) P_j$ with any function of coordinates, we arrive at

$$\mathrm{Tr}_{12}\Big[\sum_{j=1}^{N} (\mathbb{1} \otimes W(\mu))(P_j^{-1} G_j(\lambda) P_j \otimes P_j^{-1} E_{jj}) \bar{R}_{12}^{-1}(\lambda) L_1(\lambda) \Big] = \mathrm{Tr} L(\mu) \mathrm{Tr} L(\lambda).$$

Now we look at the right hand side of (3.156). Using the cyclic property of the trace, the matrix $\underline{R}_{12}(\lambda, \mu)$ can be moved to the left where it cancels with its inverse. This manipulation is allowed because $L_1(\lambda)$ and $L_2(\mu)$ produce together a factor $P_1^{-1}P_2^{-1}$ with which $\underline{R}_{12}(\lambda, \mu)$ commutes due to the zero weigh condition (3.126). Also, the individual entries of $\underline{R}_{12}(\lambda, \mu)$ are freely moved through $P_1^{-1}\bar{R}_{21}(\mu)P_1$, because of the diagonal structure of the latter matrix in the first matrix space, analogous to the similar property of (3.158). Then, to eliminate $\bar{R}_{21}(\mu)$, one employs the same procedure as was used for the left hand side of (3.156) and the final result is $\mathrm{Tr}L(\lambda)\mathrm{Tr}L(\mu)$. This proves the commutativity of traces of the Lax matrix for different values of the spectral parameter.

We finally remark that writing the analogue of (3.75) with spectral parameter dependent Lax operator [29, 30]

$$: \det(L(\lambda) - \zeta \mathbb{1}) := \sum_{k=0}^{N}(-\zeta)^{N-k}S_k(\lambda), \qquad (3.159)$$

the quantities $S_k(\lambda)$ are commuting integrals and they are related to Macdonald operators (3.130) by a simple coupling- and spectral parameter-dependant rescaling

$$S_k(\lambda) = \lambda^{-k}(\lambda - t^k e^{-\hbar/2})(\lambda - e^{-\hbar/2})^{k-1}S_k.$$

3.4 Spectral Problem

Determination of the spectrum of the Schrödinger operator is one of the main problems of quantum mechanics. The specific of an integrable model is that the hamiltonian defining the corresponding Schrödinger operator is a member of a commuting family of differential operators, so that one can search for a basis of common eigenfunctions for all of them. Among various approaches to the spectral problem in the context of the CMS and RS models, the following two are the most advanced—the first is rooted in the theory of multi-variable orthogonal polynomials, and the second is based on a special ansatz for the asymptotic wave function. In particular, the second approach, known as Bethe Ansatz, is applicable to a large variety of integrable models that support scattering, and we address it later. Here we concentrate on the method of orthogonal polynomials, suitable for the case of a purely discrete spectrum. The corresponding models include, for instance, the rational CMS model with a harmonic oscillator potential, the trigonometric CMS and RS models. In this section we discuss these three models and the construction of their spectrum in terms of the corresponding orthogonal polynomials, although in a brief manner, as the corresponding subject is well treated in the literature, see e.g. [31–33].

3.4.1 Rational CMS Model and Hermite Polynomials

The hamiltonian for the rational CMS model in a confining harmonic potential reads as

$$H = -\frac{\hbar^2}{2} \sum_{i=1}^{N} \frac{\partial^2}{\partial q_i^2} + \frac{\omega^2}{2} \sum_{i=1}^{N} q_i^2 + \frac{g}{2} \sum_{i \neq j}^{N} \frac{1}{q_{ij}^2}, \tag{3.160}$$

where $g = \gamma(\gamma - \hbar)$ is an effective coupling constant.[15] This hamiltonian is obtained from (3.100) by performing an analytic continuation $q_i \to iq_i$ and by adding the harmonic potential. The presence of the harmonic potential does not change the integrability status, the model is still integrable as a classical theory; one can give the corresponding Lax representation and solve the equations of motion. Here we will be interested in the spectrum of the quantum model rather than in its classical integrable properties.

To present the main idea of a polynomial construction of the spectrum, it is convenient to start from the one-particle case, where the inverse square potential is absent. Following the standard quantum-mechanical treatment of the harmonic oscillator, we introduce the creation and annihilation operators

$$a^\dagger = \frac{1}{\sqrt{2}}\left(-\hbar \frac{\partial}{\partial q} + \omega q\right), \quad a = \frac{1}{\sqrt{2}}\left(\hbar \frac{\partial}{\partial q} + \omega q\right), \tag{3.161}$$

acting in the Hilbert space of real square-integrable functions. The creation and annihilation operators are conjugate to each other. In terms of these operators the hamiltonian takes a factorised form

$$H = a^\dagger a + E_0, \tag{3.162}$$

where E_0 is the ground state energy $E_0 = \frac{1}{2}\hbar\omega$. The ground state wave function $\Delta(q) = e^{-\frac{\omega q^2}{2\hbar}}$ is annihilated by a, it is symmetric under $q \to -q$ and has no zeros. Conjugating H with the ground state wave function and subtracting E_0 we obtain a new operator

$$\mathcal{H} = \Delta^{-1}(H - E_0)\Delta = -\frac{1}{2}\hbar^2 \frac{\partial^2}{\partial q^2} + \hbar\omega q \frac{\partial}{\partial q}. \tag{3.163}$$

---

[15]Note that $g > 0$ corresponds to the repulsive case, while $g < 0$ to the attractive one, the latter cannot be realised in classical theory as it leads to a collapse. Quantum-mechanically, for the inverse-square potential negative values of the coupling constant $-\frac{\hbar^2}{4} < g < 0$ are also allowed, see [34]. Since g is parametrised as $g = \gamma(\gamma - \hbar)$, it automatically satisfies $g > -\frac{\hbar^2}{4}$ for any real $\gamma \neq \frac{\hbar}{2}$.

Finding the eigenstates and eigenvalues of $\mathcal{H}$ is a text book problem and it can be solved in a variety of ways. Here we present a solution that is suitable for further generalisations to the multi-particle case.

A characteristic feature of $\mathcal{H}$ is that it acts in the space spanned by monomials

$$\{1, q, q^2, \ldots, q^n\}$$

as a triangular matrix, which, in fact, reduces the problem of finding the spectrum to the problem of diagonalising triangular matrices. Indeed, introduce a polynomial of degree n

$$h_n = \sum_{k=0}^{n} c_k q^{n-k}, \quad c_0 \neq 0. \tag{3.164}$$

Then

$$\mathcal{H} h_n = -\frac{1}{2}\hbar^2 \sum_{k=0}^{n-2} c_k(n-k)(n-k-1)q^{n-k-2} + \hbar\omega \sum_{k=0}^{n-1} c_k(n-k)q^{n-k}.$$

The requirement h_n to be an eigenfunction yields the corresponding eigenvalue $E_n = \hbar\omega n$ together with the recurrence relation for the coefficients c_k:

$$c_k = -\frac{\hbar}{2\omega}\frac{(n-k+1)(n-k+2)}{k}c_{k-2}. \tag{3.165}$$

This relation implies that all the odd coefficients are zero and all the even ones are determined in terms of c_0

$$c_k = \left(-\frac{\hbar}{\omega}\right)^{k/2}\frac{n!}{(n-k)!}\frac{c_0}{2^k\Gamma(k/2+1)}. \tag{3.166}$$

Picking up for c_0 a convenient value $c_0 = 2^n(\omega/\hbar)^{n/2}$ and evaluating the sum (3.164), one finds that h_n is nothing else but the Hermite polynomial H_n

$$h_n = H_n\left(\sqrt{\omega/\hbar}\,q\right). \tag{3.167}$$

In this way we constructed the eigenfunctions $\psi_n = \Delta H_n\left(\sqrt{\omega/\hbar}\,q\right)$ and the corresponding eigenvalues $E_n = \hbar\omega(n + \frac{1}{2})$ of the original hamiltonian, which is, of course, the well-known result for the harmonic oscillator.

Quite remarkably, it appears that the general case can be treated in a similar fashion. We introduce the following Hermitian conjugate operators[16] [35]

---

[16]These are not creation and annihilation operators because they do not have the proper commutation relations.

$$\mathcal{A}_i = \hbar \frac{\partial}{\partial q_i} + \omega q_i - \sum_{j \neq i}^{N} \frac{\gamma}{q_{ij}}, \qquad \mathcal{A}_i^\dagger = -\hbar \frac{\partial}{\partial q_i} + \omega q_i - \sum_{j \neq i}^{N} \frac{\gamma}{q_{ij}}. \quad (3.168)$$

The hamiltonian (3.160) factorises in a manner similar to (3.162)

$$H = \frac{1}{2} \sum_i \mathcal{A}_i^\dagger \mathcal{A}_i + E_0 , \qquad (3.169)$$

where

$$E_0 = \tfrac{1}{2} N \omega \big(\hbar + (N-1)\gamma \big) .$$

Since $\sum_i \mathcal{A}_i^\dagger \mathcal{A}_i$ is a non-negative Hermitian operator, E_0 coincides with the ground state energy. The ground state wave function is determined by a set of equations

$$\mathcal{A}_j \Delta = 0 , \qquad j = 1, \ldots, N ,$$

and its is given by the Laughlin's type wave function

$$\Delta = \prod_{i=1}^{N} e^{-\frac{\omega q_i^2}{2\hbar}} \prod_{i<j}^{N} (q_i - q_j)^\beta , \qquad (3.170)$$

where β was introduced in (3.92). At the points where the coordinates of any two particles coincide the ground state wave function and the corresponding probability current vanish. This reflects the fact that the inverse-square potential acts in a repulsive manner forbidding the particles to overtake each other. We can therefore restrict our attention to the region of the configuration space where particle coordinates are ordered, for instance, as

$$q_1 \geq q_2 \geq \ldots \geq q_N . \qquad (3.171)$$

Once eigenfunctions are determined for q_i in this region, their extension to other regions can be made by taking into account the particle statistics, the latter is determined by the symmetry properties of the ground state wave function [31]. For particle orderings other than (3.171) one can take the wave function ψ to obey

$$\psi(\sigma q) = \eta_\sigma \psi(q) \qquad (3.172)$$

where σ is an arbitrary permutation and q is in the region (3.171). The statistical factor η_σ is equal to unity for bosons and to the parity of the permutation if the particles are fermions. As is clear from (3.170), the case $\gamma = 0$ corresponds to free bosons, while $\gamma = \hbar$ describes free fermions. One can also consider the Boltzmann statistics where all the particles are distinguishable. In this case ψ corresponds to one of the $N!$ different states, each of them is characterised by the wave function which

vanishes for all particle orderings except the one featuring on the left hand side of
(3.172), where it coincides with the right hand side of (3.172).

To consider excited states, we perform the same trick of conjugating H with the
ground state wave function. This defines a new operator $\mathcal{H} = \Delta^{-1}(H - E_0)\Delta$ that
reads as

$$\mathcal{H} = \sum_{i=1}^{N} \left(-\frac{1}{2}\hbar^2 \frac{\partial}{\partial q_i^2} + \hbar\omega q_i \frac{\partial}{\partial q_i} \right) - \hbar\gamma \sum_{i<j}^{N} \frac{1}{q_{ij}} \left(\frac{\partial}{\partial q_i} - \frac{\partial}{\partial q_j} \right). \quad (3.173)$$

Let $\lambda \equiv [\lambda_1, \lambda_2, \lambda_3, \ldots]$ be a *partition* that is a sequence of non-negative integers
λ_i (dominant weights) in decreasing order

$$\lambda_1 \geq \lambda_2 \geq \lambda_3 \geq \ldots \quad (3.174)$$

and such that it contains only a finite number of non-zero terms. Two partitions that
differ only by strings of zeros at the end are considered to be identical. Non-zero
λ_i are called *parts*. The number of parts is the *length* $l(\lambda)$ of λ and the sum of all
parts is the *weight* $|\lambda| = \sum_i \lambda_i$. Partitions can be conveniently identified with Young
diagram, see Fig. 3.1.

Consider a monomial

$$q^\lambda \equiv q_1^{\lambda_1} q_2^{\lambda_2} \ldots q_N^{\lambda_N}. \quad (3.175)$$

The operator $\mathcal{H}$ does not preserve the space spanned by such multi-variable monomials because of the presence in (3.173) of the non-polynomial term $1/q_{ij}$. However,
this operator does preserve the space of symmetric polynomials. To understand this
point [36], we first assume that $\lambda_i \neq \lambda_j$ and, setting $\partial_i \equiv \frac{\partial}{\partial q_i}$, consider

Fig. 3.1 Young diagram

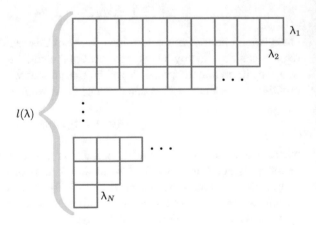

$$\frac{1}{q_i - q_j}(\partial_i - \partial_j)(q_i^{\lambda_i} q_j^{\lambda_j} + q_i^{\lambda_j} q_j^{\lambda_i}) =$$

$$= \lambda_i q_i^{\lambda_j} q_j^{\lambda_j} \frac{q_i^{\lambda_i - \lambda_j - 1} - q_j^{\lambda_i - \lambda_j - 1}}{q_i - q_j} - \lambda_j q_i^{\lambda_j - 1} q_j^{\lambda_j - 1} \frac{q_i^{\lambda_i - \lambda_j + 1} - q_j^{\lambda_i - \lambda_j + 1}}{q_i - q_j}$$

$$= \lambda_i \sum_{k=1}^{\lambda_i - \lambda_j - 1} q_i^{\lambda_i - k - 1} q_j^{\lambda_j + k - 1} - \lambda_j \sum_{k=0}^{\lambda_i - \lambda_j} q_i^{\lambda_i - k - 1} q_j^{\lambda_j + k - 1}$$

$$= (\lambda_i - \lambda_j) \sum_{k=1}^{\lambda_i - \lambda_j - 1} q_i^{\lambda_i - k - 1} q_j^{\lambda_j + k - 1} - \lambda_j (q_i^{\lambda_i - 1} q_j^{\lambda_j - 1} + q_i^{\lambda_j - 1} q_j^{\lambda_i - 1}).$$

The last formula remains also valid for the case $\lambda_i = \lambda_j$. Thus, acting with the symmetric operator $\frac{1}{q_{ij}}(\partial_i - \partial_j)$ on a symmetric polynomial of two variables gives a symmetric polynomial.

In the following we need some notions from the theory of symmetric functions and polynomials[17] [20]. We start with recalling two natural bases on the space of symmetric functions. The first one is given in terms of power-sum symmetric functions p_λ defined as

$$p_\lambda = p_{\lambda_1} p_{\lambda_2} \cdots ,$$

where

$$p_i = \sum_k q_k^i .$$

The second basis is given in terms of monomial symmetric functions $m_\lambda \equiv m_{\lambda_1, \dots, \lambda_N}$ defined as the sum of all distinct monomials that can be obtained from q^λ by permutations of q_i's

$$m_\lambda = \sum_{\sigma \in \mathfrak{S}_N} q_{\sigma(1)}^{\lambda_1} q_{\sigma(2)}^{\lambda_2} \cdots q_{\sigma(N)}^{\lambda_N} , \tag{3.176}$$

where $\mathfrak{S}_N$ is the symmetric group. If the number of indeterminates is less than the number of parts, the corresponding function m_λ is equal to zero. One also considers elementary symmetric functions that are

$$e_k = \sum_{i_1 < \dots < i_k} q_{i_1} q_{i_2} \cdots q_{i_k} = m_{1^k} . \tag{3.177}$$

We point out the relations between power-sum and elementary symmetric functions given by the following *determinant formulae*

---

[17]Generalising the notion of symmetric polynomial to an infinite countable set of variables, one normally speaks about symmetric functions.

$$p_k = \begin{vmatrix} e_1 & 1 & 0 & \cdots & 0 \\ 2e_2 & e_1 & 1 & \cdots & 0 \\ \vdots & \vdots & & \cdots & \vdots \\ (k-1)e_{k-1} & e_{k-2} & \cdots & \cdots & 1 \\ ke_k & e_{k-1} & \cdots & \cdots & e_1 \end{vmatrix} , \quad e_k = \frac{1}{k!} \begin{vmatrix} p_1 & 1 & 0 & \cdots & 0 \\ p_2 & p_1 & 2 & \cdots & 0 \\ \vdots & \vdots & & \cdots & \vdots \\ p_{k-1} & p_{k-2} & \cdots & \cdots & k-1 \\ p_k & p_{k-1} & \cdots & \cdots & p_1 \end{vmatrix} . \tag{3.178}$$

Then, we need the notion of a dominance partial order defined on a set of partitions which is

$$\mu \le \lambda \quad \text{if and only if} \quad \mu_1 + \ldots + \mu_j \le \lambda_1 + \ldots + \lambda_j , \quad \text{for all } j = 1, \ldots, N .$$

This order is only partial because not all the partitions can be compared in this way.

Now, acting with $\mathcal{H}$ on m_λ, we find

$$\mathcal{H} m_\lambda = \hbar\omega |\lambda| m_\lambda - \frac{1}{2}\hbar^2 \sum_{i=1}^{N} \lambda_i (\lambda_i - 1) \, m_{\lambda_1,\ldots,\lambda_i-2,\ldots,\lambda_n} + \hbar\gamma \sum_{i<j}^{N} \lambda_j \, m_{\lambda_1,\ldots,\lambda_i-1,\ldots,\lambda_j-1,\ldots,\lambda_n}$$

$$- \hbar\gamma \sum_{i<j} \lambda_{ij} \left(\sum_{k=1}^{\theta_{ij}} m_{\lambda_1,\ldots,\lambda_i-k-1,\ldots,\lambda_j+k-1,\ldots,\lambda_n} - \frac{\delta_{\lambda_{ij},2\mathbb{Z}}}{2} m_{\lambda_1,\ldots,\lambda_i-\theta_{ij}-1,\ldots,\lambda_j+\theta_{ij}-1,\ldots,\lambda_n} \right) .$$

Here θ_{ij} is the greatest integer less or equal to $\frac{1}{2}\lambda_{ij}$ and $\lambda_{ij} = \lambda_i - \lambda_j$. Some of the terms m_μ appearing on the right hand side do not look as corresponding to partitions because they break the ordering pattern (3.174). However, each of these terms can be identified in a unique way with a partition by properly permuting its parts.

Thus, one can recognise that the differential operator (3.173) transforms m_λ into a linear combination of m_μ with $\mu \le \lambda$, i.e. it acts in the space of symmetric polynomials $\mu \le \lambda$ as a triangular matrix

$$\mathcal{H} m_\lambda = \hbar\omega |\lambda| m_\lambda + \sum_{\mu<\lambda} c_{\lambda\mu} m_\mu .$$

Therefore, the operator $\mathcal{H}$ can be diagonalised by using $m_\lambda + \sum_{\mu<\lambda} u_{\lambda\mu} m_\mu$ as an ansatz for the corresponding eigenfunction. For the coefficients $u_{\lambda\mu}$ one obtains a recurrence relation which is, however, much more complicated to solve than in the one-particle case [36]. Anyway, the corresponding solutions yield symmetric polynomials known as *generalised Hermite polynomials*. These polynomials provide a complete basis in the space of symmetric polynomials that can be made orthogonal with respect to the following scalar product [32]

$$(f_1, f_2) = \int_{\mathbb{R}^n} \mathrm{d}q f_1(q) f_2(q) \prod_{i=1}^{N} e^{-\frac{\omega}{\hbar}q_i^2} \prod_{i \ne j} |q_i - q_j|^\beta . \tag{3.179}$$

Table 3.1 The first seven orthogonal polynomials corresponding to the first four eigenvalues

Eigenvalue	Diagram	Generalised Hermite polynomial
0	[0]	1
$\hbar\omega$	[1]	m_1
$2\hbar\omega$	[1, 1]	$m_{1,1} + \frac{\gamma}{2\omega}\frac{N(N-1)}{2}$
	[2]	$m_2 + \frac{2\beta}{\beta+1}m_{1,1} - \frac{\hbar}{2\omega}\frac{N(\beta N+1)}{\beta+1}$
$3\hbar\omega$	[1, 1, 1]	$m_{1,1,1} + \frac{\gamma}{2\omega}\frac{(N-1)(N-2)}{2}m_1$
	[2, 1]	$m_{2,1} + \frac{6\beta}{2\beta+1}m_{1,1,1} +$ $\frac{\hbar}{2\omega}\frac{(N-1)(\beta N+1)(\beta-1)}{2\beta+1}m_1$
	[3]	$m_3 + \frac{3\beta}{\beta+2}m_{2,1} + \frac{6\beta^2}{(\beta+1)(\beta+2)}m_{1,1,1} -$ $\frac{3\hbar}{2\omega}\frac{(\beta N+1)(\beta N+2)}{(\beta+1)(\beta+2)}m_1$

Taking into account the expression for E_0, we get the formula for the energy of a state corresponding to a given Young diagram

$$E_\lambda = \hbar\omega\sum_{j=1}^{N}\lambda_j + \frac{1}{2}N\hbar\omega + \frac{N(N-1)}{2}\gamma\omega. \qquad (3.180)$$

Obviously, the spectrum appears highly degenerate: states with different Young diagrams but with the same $|\lambda|$ have the same energy. Since the application of the Gram-Schmidt procedure is technically problematic, one can build up an orthogonal basis of eigenstates of the hamiltonian by requiring these eigenstates to also diagonalise higher commuting integrals. As an example,[18] in Table 3.1 we give the first seven polynomials corresponding to the first three eigenvalues of $\mathcal{H}$. These polynomials are orthogonal with respect to the scalar product (3.179). Further, as was discussed in Sect. 1.1.3, degeneracy of the spectrum signals the presence of a non-abelian symmetry. Indeed, it appears that the rational CMS model is invariant with respect to the so-called W_N-algebra [38]. This explains the high degeneracy of the spectrum and renders the rational CMS model superintegrable.

Quasi-particles and quasi-momentum. Formula (3.180) can be presented in a manner that reveals another important point about integrable systems. Introducing the new variables

$$k_j = \lambda_{N+1-j} + (j-1)\beta, \quad j = 1, \dots, N,$$

the spectrum of energies reads as the one for N uncoupled harmonic oscillators

---

[18]The example was adopted from [37], see also [24].

$$E = \frac{1}{2}\hbar\omega N + \hbar\omega \sum_{i=1}^{N} k_j \,. \tag{3.181}$$

The only difference is that for oscillators k_j are non-negative integers, while in our case they are positive numbers satisfying the restriction

$$k_{j+1} - k_j \geq \beta \,.$$

This restriction can be viewed as a *generalised exclusion principle*: momenta of different particles are at least separated by the distance β from each other, for free bosons this distance is zero and for free fermions it is one.

A lesson to learn from this observation is that, being written in proper variables, the energy of an interacting system looks like the sum of energies of individual "free" particles. It is this additivity of the energy spectrum that appears to be characteristic for integrable behavior. In this context, an integrable model can be given a suggestive interpretation in terms of *quasi-particles*, the latter carry *quasi(pseudo)-momentum* k_j and essentially behave themselves as free. As we will see, the dispersion relation $e(k)$, which expresses the energy e of a single quasi–particle excitation via its quasi-momentum k, does depend on the nature of the interactions and varies from model to model, but once found for a given model, the spectrum of the latter is obtained as a simple sum of quasi-particle energies $e(k_j)$. In general, the notion of a quasi-particle is an important theoretical concept which use goes far beyond integrable systems [39]. One way to determine the quasi-momenta for an integrable system, at least in some asymptotic regimes, is provided by the Bethe Ansatz—a universal integrability tool, which will be explained in Chap. 5.

Concluding the discussion of the rational CMS model in the harmonic potential, we mention that it is also possible to describe its spectrum in terms of the Fock space construction based on creation- and annihilation-like operators, see [40], or by using the formalism of Dunkle operators [40, 41].

3.4.2 Trigonometric CMS Model and Jack Polynomials

The trigonometric CMS model can be viewed as a finite-size version of the rational CMS model. Its consideration goes much in parallel to the rational case. The hamiltonian is obtained from (3.106) by analytically continuing the coordinates $q_i \to iq_i$. We also introduce an explicit dependence on the parameter ℓ which controls the size of the system. This gives

$$H = -\frac{\hbar^2}{2} \sum_{i=1}^{N} \frac{\partial^2}{\partial q_i^2} + \frac{g}{2} \sum_{i\neq j}^{N} \frac{1}{4\ell^2 \sin^2(q_{ij}/2\ell)} \,. \tag{3.182}$$

or

$$H = -\frac{\hbar^2}{2} \sum_{i=1}^{N} \frac{\partial^2}{\partial q_i^2} - \frac{\gamma(\gamma - \hbar)}{2\ell^2} \sum_{i \neq j}^{N} \frac{Q_i Q_j}{Q_{ij}^2} . \tag{3.183}$$

The variables $Q_i = e^{iq_i/\ell}$ are thus periodic functions of q_i with the period $2\pi\ell$. Commuting integrals are given by the analytically continued S_k from (3.96), in particular, the total momentum is

$$P = S_1 = -i\hbar \sum_{i=1}^{N} \frac{\partial}{\partial q_i} .$$

Up to a phase that can be adjusted to describe given particle statistics, the ground state wave-function $\triangle$ is a Jastrow wave function[19]

$$\triangle = \prod_{i<j}^{N} (Q_i - Q_j)^{\beta} \prod_{i=1}^{N} Q_i^{-\frac{1}{2}(N-1)\beta} , \tag{3.184}$$

where we recall that $\beta = \frac{\gamma}{\hbar}$ and

$$E_0 = \frac{\gamma^2}{\ell^2} \frac{N(N^2 - 1)}{24} \tag{3.185}$$

is the ground state energy. As in the rational case, the phase of $\triangle$ controls the particle statistics. Formula (3.185) follows from representing the hamiltonian in the form (3.169), where

$$\mathcal{A}_i = \hbar \frac{\partial}{\partial q_i} - \frac{\gamma}{2\ell} \sum_{j \neq i}^{N} \cot \left(\frac{q_{ij}}{2\ell}\right), \qquad \mathcal{A}_i^{\dagger} = -\hbar \frac{\partial}{\partial q_i} - \frac{\gamma}{2\ell} \sum_{j \neq i}^{N} \cot \left(\frac{q_{ij}}{2\ell}\right).$$

We note that the ground state energy admits the thermodynamic limit where the length $L = 2\pi\ell$ of the box (period) and the particle number N both tend to infinity, such that the particle density $\mathcal{D} = N/L$ remains finite in this limit. From (3.185) for the density $\mathscr{E}_0$ of the ground state energy of the we find in the thermodynamics limit

$$\mathscr{E}_0 = \lim_{N,L\to\infty} \frac{E_0}{\ell} = \frac{(\pi\gamma)^2}{6} \mathcal{D}^3 . \tag{3.186}$$

---

[19] A type of trial wave function used in the theory of quantum liquids. We also point out that

$$\prod_{1 \leqslant i < j \leqslant j} (Q_i - Q_j) = \det \left[Q_i^{N-j}\right]_{i,j=1}^{N} = \sum_{\sigma \in \mathfrak{S}_N} \text{sign}(\sigma) \prod_{j=1}^{N} Q_{\sigma(j)}^{N-j} .$$

is the so-called Vandermonde determinant.

Similarly to the rational case, the spectrum of excited states of the trigonometric model is constructed by conjugating H with the ground state wave-function and subtracting the ground state energy. This defines a new hamiltonian operator $\mathcal{H} \equiv \frac{2\ell^2}{\hbar^2}(\Delta^{-1} H \Delta - E_0)$ that reads as

$$
\mathcal{H} = \sum_{i=1}^{N} \left(Q_i \frac{\partial}{\partial Q_i} \right)^2 + \beta \sum_{i<j}^{N} \frac{Q_i + Q_j}{Q_i - Q_j} \left(Q_i \frac{\partial}{\partial Q_i} - Q_j \frac{\partial}{\partial Q_j} \right).
$$

This operator is invariant under the action of the symmetric group $\mathfrak{S}_N$ and its spectrum will be spanned by the multivariable symmetric polynomials. Indeed, one can verify that acting on the monomial symmetric functions

$$
\boldsymbol{m}_\lambda = \sum_{\sigma \in \mathfrak{S}_N} Q_{\sigma(1)}^{\lambda_1} Q_{\sigma(2)}^{\lambda_2} \cdots Q_{\sigma(N)}^{\lambda_N} \tag{3.187}
$$

associated to the Young diagram λ, the operator $\mathcal{H}$ yields

$$
\mathcal{H} \boldsymbol{m}_\lambda = \varepsilon_\lambda \boldsymbol{m}_\lambda + \sum_{\mu < \lambda} c_{\lambda\mu} \boldsymbol{m}_\mu \,,
$$

where

$$
\varepsilon_\lambda = \sum_j \lambda_j^2 + \beta \sum_{i<j} (\lambda_i - \lambda_j) \,. \tag{3.188}
$$

Thus, $\mathcal{H}$ is triangular in a symmetric monomial basis and its eigenfunctions can be searched as linear combinations of $\boldsymbol{m}_\lambda$ with all subordinate $\boldsymbol{m}_\mu$. This leads to the construction of the spectrum in terms of symmetric Jack polynomials.

Jack polynomials. For a given partition λ denote by m_1 multiplicity of its part 1, by m_2 multiplicity of its part 2, etc. Let us associate to λ the following number

$$
z_\lambda = 1^{m_1} m_1! \, 2^{m_2} m_2! \dots . \tag{3.189}
$$

With this notation we define on the space of symmetric functions the following scalar product

$$
\langle p_\lambda, p_\mu \rangle_\beta = \delta_{\lambda,\mu} z_\lambda \beta^{-\ell(\lambda)} \,. \tag{3.190}
$$

The Jack polynomials $J_\lambda(Q; \beta)$ are uniquely defined by the following conditions[20] [20, 43]

---

[20] Our notation $J_\lambda(Q_i; \beta)$ is a bit different from the one adopted in the mathematical literature, see e.g. [42], where one defines Jack polynomials as $J_\lambda(Q_i; \alpha)$ with α related to our β as $\alpha = 1/\beta$.

Table 3.2 The first seven Jack polynomials and the corresponding eigenvalues of $\mathcal{H}$ for $N = 3$. The coefficients of the linear combinations of monomial functions depend rationally on β and do not depend on the particle number N. Eigenvalues do depend on N

$N = 3$ eigenvalue	Diagram	Jack polynomial $J_\lambda(Q; \beta)$
0	[0]	1
$2\beta + 1$	[1]	m_1
$2\beta + 2$	[1, 1]	$m_{1,1}$
$4\beta + 4$	[2]	$m_2 + \frac{2\beta}{\beta+1} m_{1,1}$
3	[1, 1, 1]	$m_{1,1,1}$
$4\beta + 5$	[2, 1]	$m_{2,1} + \frac{6\beta}{2\beta+1} m_{1,1,1}$
$6\beta + 9$	[3]	$m_3 + \frac{3\beta}{\beta+2} m_{2,1} +$ $\frac{6\beta^2}{(\beta+1)(\beta+2)} m_{1,1,1}$

$$J_\lambda(Q_i; \beta) = m_\lambda + \sum_{\mu < \lambda} u_{\mu\lambda}(\beta) m_\mu \,,$$

$$\langle J_\lambda, J_\mu \rangle_\beta = 0 \quad \text{if} \quad \lambda \neq \mu \,, \tag{3.191}$$

where the coefficients $u_{\mu\lambda}(\beta)$ are rational functions of β. If $l(\lambda) > N$, then $J_\lambda(Q, \beta) = 0$. For $\beta = 1$ the Jack polynomials coincide with Schur polynomials.

The Jack polynomials are also orthogonal with respect to the scalar product

$$(f_1, f_2) = \frac{1}{(2\pi i)^N} \oint_{|Q_1|=1} \frac{dQ_1}{Q_1} \cdots \oint_{|Q_N|=1} \frac{dQ_N}{Q_N} f_1(Q) f_2(\bar{Q}) \prod_{i<j}^{N} |Q_i - Q_j|^{2\beta} \,. \tag{3.192}$$

In fact, the scalar product (3.192) is proportional to (3.190). Among other properties of the Jack polynomials the following relation is noteworthy

$$\left(\prod_i Q_i \right) J_\lambda(Q; \beta) = J_{\lambda+1}(Q; \beta) \,, \quad \lambda + 1 \equiv [\lambda_1 + 1, \lambda_2 + 1, \ldots] \,. \tag{3.193}$$

A crucial point is that the Jack polynomials defined by the conditions (3.191) are the eigenstates of $\mathcal{P}$ and $\mathcal{H}$

$$\mathcal{P} J_\lambda(Q; \beta) = p_\lambda J_\lambda(Q; \beta) \,,$$

$$\mathcal{H} J_\lambda(Q; \beta) = \varepsilon_\lambda J_\lambda(Q; \beta) \,,$$

where eigenvalues ε_λ are given by (3.188) and

$$p_\lambda = \sum_{j=1}^{N} \lambda_j \,. \tag{3.194}$$

Here $\mathscr{P} = \frac{\ell}{\hbar} \Delta^{-1} P \Delta = \frac{\ell}{\hbar} P$ is the operator of total momentum because P annihilates the ground state wave function. The action of higher commuting integrals on Jack polynomials will be discussed later.

A simple and explicit formula for Jack polynomials is currently unknown. To give the reader an impression on the structure of J_λ, in the Table 3.2 we list the first seven polynomials together with the corresponding eigenvalues of the operator $\mathscr{H}$. We can now present a complete formula for an arbitrary eigenfunction of H

$$\psi_{\lambda,s}(Q) = \left(\prod_{i=1}^{N} Q_i\right)^{s-(N-1)\beta/2} \prod_{i<j}^{N}(Q_i - Q_j)^\beta J_\lambda(Q; \beta), \tag{3.195}$$

Here s is an arbitrary real number and $l(\lambda) \leq N - 1$. The overall prefactor depending on s can be understood as an application of an arbitrary Galilean boost, see e.g. [33]. To avoid overcounting, the number of parts of λ was restricted to be strictly smaller than the number of particles. Indeed, due to relation (3.193), one can always adjust the parameter s of the Galilean boost in such a way that $\lambda_N = 0$.

Quasi-momentum. Now we comment on the quasi-particle interpretation of the spectrum. Taking into account the expression for the ground state energy, we obtain the following formula for the spectrum of the hamiltonian (3.183)

$$E = \frac{\hbar^2}{2\ell^2} \sum_{j=1}^{N} \left(\lambda_j + \beta(N + 1 - 2j)\right)\lambda_j + \frac{\gamma^2}{\ell^2} \frac{N(N^2 - 1)}{24}. \tag{3.196}$$

Introducing this time the dimensionful quasi-momenta[21]

$$p_j = \frac{\hbar}{\ell}\left(\lambda_{N+1-j} + \left(j - \tfrac{N+1}{2}\right)\beta\right), \quad j = 1, \ldots, N, \tag{3.197}$$

the formulas for the energy and total momentum can be cast in the form

$$E = \frac{1}{2} \sum_{j=1}^{N} p_j^2, \quad P = \sum_{j=1}^{N} p_j, \tag{3.198}$$

where they obviously coincide with the sums of energies and momenta of free non-relativistic particles with momenta p_j. For p_j we find a similar selection rule to the rational case, $p_{j+1} - p_j \geq \gamma/\ell$. As in the rational case, the spectrum is degenerate and the eigenstates of the hamiltonian are separated by the values of higher commuting charges that we now discuss.

Sekiguchi operators. Recall that the higher commuting integrals of the model are given by the trigonometric version of (3.96). Conjugating this integrals with the ground state wave function, we get

---

[21]Quasi-momenta p_j have the physical dimension of momentum.

$$\mathcal{D}_k = \Delta^{-1} D_k \Delta . \tag{3.199}$$

These new mutually commuting differential operators are known as *Sekiguchi operators*. Originally, they were introduced in [44], see also [45], as a one-parametric deformation of the generators of the algebra of the invariant differential operators on symmetric spaces.

In the following it is convenient to introduce the following generating functions for D_k and $\mathcal{D}_k$

$$D(\zeta) = \sum_{k=0}^{N} \zeta^{N-k} (-1)^k S_k , \qquad \mathcal{D}(\zeta) = \sum_{k=0}^{N} \zeta^{N-k} (-1)^k \mathcal{S}_k , \tag{3.200}$$

so that

$$\mathcal{D}(\zeta) = \Delta^{-1} D(\zeta) \Delta .$$

According to [44], the function $\mathcal{D}(\zeta)$ has the following explicit form

$$\mathcal{D}(\zeta) = \frac{1}{\displaystyle\prod_{i<j} Q_{ij}} \sum_{\sigma \in \mathfrak{S}_N} (-1)^\sigma \prod_{i=1}^{N} Q_i^{N-\sigma(i)} \left(\zeta + \frac{\gamma}{\ell}(\sigma(i) - \tfrac{N+1}{2}) + i\hbar \frac{\partial}{\partial q_i} \right), \tag{3.201}$$

where $\mathfrak{S}_N$ is the symmetric group. The Jack polynomials diagonalise not only the momentum and the hamiltonian, but also all the Sekiguchi operators, namely,

$$\mathcal{D}(\zeta) \cdot J_\lambda(Q; \beta) = \mathcal{Q}(\lambda) \cdot J_\lambda(Q; \beta) ,$$

where the eigenvalue is

$$\mathcal{Q}(\zeta) = \prod_{j=1}^{N} \left(\zeta - \frac{\hbar}{\ell}(\lambda_j - (j - \tfrac{N+1}{2})\beta) \right) .$$

Upon shifting the index $j \to N + 1 - j$, the expression for $\mathcal{Q}(\zeta)$ can be rewritten in terms of the quasi-momenta (3.197) as

$$\mathcal{Q}(\zeta) = \prod_{j=1}^{N} (\zeta - p_j) . \tag{3.202}$$

The polynomial $\mathcal{Q}(\zeta)$ is an example of a *Baxter polynomial*. Its roots p_j coincide with the allowed values of quasi-momentum.

3.4.3 *Trigonometric RS Model and Macdonald Polynomials*

The quantum RS model [46] is described by the following hamiltonian H, the momentum operator P

$$H = \tfrac{1}{2}(S_1 + S_{-1}), \quad P = \tfrac{1}{2}(S_{-1} - S_1),\qquad(3.203)$$

and by the following set of N independent mutually commuting quantum integrals

$$S_{\pm k} = \sum_{\substack{J \subset \{1,\dots,N\} \\ |J|=k}} \prod_{\substack{i \in J \\ j \notin J}} h^{\pm}(q_i - q_j) \left(\prod_{l \in J} \mathsf{T}_l^{\pm} \right) \prod_{\substack{i \in J \\ j \notin J}} h^{\pm}(q_j - q_i).\qquad(3.204)$$

Here $k = 1, \dots, N$ and the set of integrals with positive index, equivalently, the set of integrals with negative index is complete. For symmetry reasons we keep both. The summation index J with $|J| = k$ runs over all subsets of the set $\{1, \dots, N\}$ that have the cardinality equal to k. Further, for the trigonometric variant of the hyperbolic model the functions $h^{\pm}$ are

$$h^{\pm}(q_i - q_j) = \left[\frac{\sin \frac{1}{2\ell}(q_i - q_j \pm i\gamma)}{\sin \frac{1}{2\ell}(q_i - q_j)} \right]^{1/2} = \left[t^{\mp 1/2} \frac{t^{\pm 1} Q_i - Q_j}{Q_i - Q_j} \right]^{1/2},\quad(3.205)$$

where $Q_i = e^{iq_i/\ell}$ and $t = e^{-\gamma/\ell}$ is a deformation parameter depending on γ. The shift operators $\mathsf{T}_k^{\pm} = e^{\pm i \hbar \partial_k}$ acts on functions $f(Q_1, \dots, Q_N)$ as

$$(\mathsf{T}_k^{\pm} f)(Q_1, \dots, Q_N) = f(Q_1, \dots Q_{k-1}, q^{\pm 1} Q_k, Q_{k+1}, \dots, Q_N),\qquad(3.206)$$

where we introduced the quantum deformation parameter $q = e^{-\hbar/\ell}$, $\hbar \in \mathbb{R}_+$. The model describes N relativistic particles (in the sense discussed in Sect. 1.3.2) of mass $m = 1$ moving on a circle of length $2\pi\ell$. For the reader's convenience we recall that to restore the actual physical units, one needs to make the substitutions

$$\hbar \to \frac{\hbar}{mc}, \quad \gamma \to \frac{\gamma}{mc},\qquad(3.207)$$

where m is the mass of a particle and c is the speed of light. Also, since the S_k are dimensionless, the right hand side of the expressions (3.203) is written for H/mc^2 and P/mc, respectively. In the following, however, we proceed to work in our standard conventions where $m = 1$ and $c = 1$.

Commuting all the shift operators to the right,[22] we get for $S_{\pm k}$ the following expressions

$$S_{\pm k} = t^{\mp \frac{1}{2}k(N-k)} \sum_{\substack{J \subset \{1,\dots,N\} \\ |J|=k}} \prod_{\substack{i \in J \\ j \notin J}} \left(\frac{t^{\pm 1}Q_i - Q_j}{Q_i - Q_j} \frac{t^{\pm 1}Q_j - q^{\pm 1}Q_i}{Q_j - q^{\pm 1}Q_i} \right)^{1/2} \left(\prod_{l \in J} T_l^{\pm} \right). \quad (3.208)$$

These operators involve square roots and for this reason are not simple to analyse. However, a similarity transformation[23] with the ground state wave function, the latter is given by the following infinite product [47]

$$\Delta = \left(\prod_{i \neq j} \prod_{k=0}^{\infty} \frac{Q_i - q^k Q_j}{Q_i - tq^k Q_j} \right)^{1/2}, \quad (3.209)$$

turn (3.208) into Macdonald operators.

To prove this statement, let us fix a subset $J \subset \{1, \dots, N\}$ and compute

$$T_l^{\pm} \cdot \Delta = \left(\prod_{j \neq l} \frac{q^{\pm 1}Q_l - Q_j}{q^{\pm 1}Q_l - t^{\pm 1}Q_j} \frac{t^{\pm 1}Q_l - Q_j}{Q_l - Q_j} \right)^{1/2} \Delta \cdot T_l^{\pm}, \quad (3.210)$$

for $l \in J$. Let us now apply to this expression T_m, where $m \in J$ different from l. We get

$$T_m^{\pm} T_l^{\pm} \cdot \Delta = \left(\frac{Q_l - Q_m}{Q_l - t^{\pm 1}Q_m} \frac{t^{\pm 1}Q_l - q^{\pm 1}Q_m}{Q_l - q^{\pm 1}Q_m} \right)^{1/2} \times$$

$$\times \left(\prod_{\substack{j \neq l \\ j \neq m}} \frac{q^{\pm 1}Q_l - Q_j}{q^{\pm 1}Q_l - t^{\pm 1}Q_j} \frac{t^{\pm 1}Q_l - Q_j}{Q_l - Q_j} \right)^{1/2} \left(\prod_{j \neq m} \frac{q^{\pm 1}Q_m - Q_j}{q^{\pm 1}Q_m - t^{\pm 1}Q_j} \frac{t^{\pm 1}Q_m - Q_j}{Q_m - Q_j} \right)^{1/2} \Delta \cdot T_m^{\pm} T_l^{\pm},$$

where we have used (3.210) to commute $T_m^{\pm}$ through Δ. Singling out from the last product the term with $j = l$, we see that it cancels against the term in the first line in the right hand side of the above expression. Thus, we can write

$$T_m^{\pm} T_l^{\pm} \cdot \Delta = \left(\prod_{\substack{j \neq l \\ j \neq m}} \frac{q^{\pm 1}Q_l - Q_j}{q^{\pm 1}Q_l - t^{\pm 1}Q_j} \frac{t^{\pm 1}Q_l - Q_j}{Q_l - Q_j} \cdot \frac{q^{\pm 1}Q_m - Q_j}{q^{\pm 1}Q_m - t^{\pm 1}Q_j} \frac{t^{\pm 1}Q_m - Q_j}{Q_m - Q_j} \right)^{1/2} \Delta \cdot T_m^{\pm} T_l^{\pm}.$$

Proceeding in the same manner, we arrive at

---

[22] At this point we would like to invoke analogy with the formula (3.130) obtained by the reduction procedure. All shift operators there appear on the right.

[23] In classical theory this is a canonical transformation.

$$\left(\prod_{l\in J} \mathsf{T}_l\right)\cdot\mathbf{\Delta} = \left(\prod_{\substack{i\in J\\ j\notin J}} \frac{q^{\pm 1}Q_i - Q_j}{q^{\pm 1}Q_i - t^{\pm 1}Q_j}\frac{t^{\pm 1}Q_i - Q_j}{Q_i - Q_j}\right)^{1/2}\mathbf{\Delta}\cdot\left(\prod_{l\in J}\mathsf{T}_l\right).$$

Thus, conjugating the quantum integrals $S_{\pm k}$ with $\mathbf{\Delta}$, we find

$$\mathbf{\Delta}^{-1}S_{\pm k}\,\mathbf{\Delta} = t^{\mp\frac{1}{2}k(N-k)}\sum_{\substack{J\subset\{1,\dots,N\}\\ |J|=k}}\prod_{\substack{i\in J\\ j\notin J}}\frac{t^{\pm 1}Q_i - Q_j}{Q_i - Q_j}\left(\prod_{l\in J}\mathsf{T}_l^{\pm}\right). \tag{3.211}$$

Now, define the Macdonald operators as the two-parametric finite-difference operators

$$\mathcal{S}_k(q,t) = t^{\frac{1}{2}k(k-1)}\sum_{\substack{J\subset\{1,\dots,N\}\\ |J|=k}}\prod_{\substack{i\in J\\ j\notin J}}\frac{tQ_i - Q_j}{Q_i - Q_j}\prod_{i\in J}\mathsf{T}_i\,, \tag{3.212}$$

where $k = 1,\dots,N$. Here the dependence on q on the right hand side is implicit in $\mathsf{T}_i \equiv \mathsf{T}_i^+$ through its dependence on $\hbar = -\ell\log q$. We thus see that

$$\mathbf{\Delta}^{-1}S_{\pm k}\,\mathbf{\Delta} = t^{\mp\frac{1}{2}k(N-1)}\mathcal{S}_k(q^{\pm 1}, t^{\pm 1})\,. \tag{3.213}$$

This completes the proof.

It should be noted that formula (3.212) for Macdonald operators looks literally the same as formula (3.130) we obtained through the reduction procedure. The only difference is that in (3.212) the variables Q_i are on a circle, while in (3.130) they are on a line.

As in the previous cases, one can show [20] that the action of the Macdonald operators on the monomial symmetric functions (3.187) is triangular. Simultaneous eigenstates of commuting $\mathcal{S}_k$ are given by Macdonald polynomials.

Macdonald polynomials. Introduce the following q, t-Hall scalar product on the space of symmetric functions

$$\langle p_\lambda, p_\mu\rangle_{q,t} = \delta_{\lambda,\mu}\, z_\lambda\prod_{i=1}^{l(\lambda)}\frac{1 - q^{\lambda_i}}{1 - t^{\lambda_i}}\,, \tag{3.214}$$

where z_λ is defined in (3.189). In order for this scalar product to be strictly positive one has to assume that $0 < q, t < 1$ which in the present physical context implies $\hbar > 0$ and $\gamma > 0$. The *Macdonald polynomials* $P_\lambda(Q; q, t)$ constitute a unique family of symmetric polynomials such that [20]

Table 3.3 The first seven Macdonald polynomials for $N = 3$ and the corresponding eigenvalues of $\mathcal{S}_1$

$N = 3$ eigenvalue	Diagram	Macdonald polynomial $P_\lambda(Q; q, t)$
$1 + t + t^2$	[0]	1
$1 + t + qt^2$	[1]	m_1
$1 + qt + qt^2$	[1, 1]	$m_{1,1}$
$1 + t + q^2 t^2$	[2]	$m_2 + \frac{(1-t)(1+q)}{1-qt} m_{1,1}$
$q + qt + qt^2$	[1, 1, 1]	$m_{1,1,1}$
$1 + qt + q^2 t^2$	[2, 1]	$m_{2,1} + \frac{(1-t)(2+q+t+2qt)}{1-qt^2} m_{1,1,1}$
$1 + t + q^3 t^2$	[3]	$m_3 + \frac{(1-t)(1+q+q^2)}{1-q^2 t} m_{2,1} +$ $\frac{(1-t)^2(1+q)(1+q+q^2)}{(1-qt)(1-q^2 t)} m_{1,1,1}$

$$P_\lambda(Q; q, t) = m_\lambda + \sum_{\mu < \lambda} u_{\mu\lambda}(q, t) m_\mu \,,$$

$$\langle P_\lambda, P_\mu \rangle_{q,t} = 0 \quad \text{if} \quad \lambda \neq \mu \,, \tag{3.215}$$

where the coefficients $u_{\mu\lambda}(q, t)$ are rational functions of q and t. Importantly, the Macdonald polynomials are simultaneous eigenfunctions of the operators $\mathcal{S}_k$:

$$\mathcal{S}_k P_\lambda(Q; q, t) = h_k P_\lambda(Q; q, t) \,, \tag{3.216}$$

with eigenvalues

$$h_k = \sum_{j_1 < \ldots < j_k} \mu_{j_1} \ldots \mu_{j_k} \,, \quad \mu_j = q^{\lambda_j} t^{N-j} \,. \tag{3.217}$$

We also point out the quantum-mechanical scalar product

$$(f_1, f_2) = \frac{1}{(2\pi i)^N} \oint_{|Q_1|=1} \frac{dQ_1}{Q_1} \cdots \oint_{|Q_N|=1} \frac{dQ_N}{Q_N} f_1(Q) f_2(\overline{Q}) \mathbf{\Delta}^2(Q) \,, \tag{3.218}$$

where $\mathbf{\Delta}(Q)$ is the ground state wave function (3.209). Macdonald polynomials are pairwise orthogonal also with respect to this scalar product. Since $t, q \in \mathbb{R}$, the ground state wave function is real $\overline{\mathbf{\Delta}(Q)} = \mathbf{\Delta}(1/Q) = \mathbf{\Delta}(Q)$.

Quasi-momentum. The considerations above show that the hamiltonian and the momentum operator have a common basis of eigenstates given by

$$\psi_\lambda(Q) = \mathbf{\Delta} P_\lambda(Q, q, t) \,. \tag{3.219}$$

To find the corresponding eigenvalues, we evaluate

$$P\psi_\lambda(Q) = \tfrac{1}{2}(S_{-1}\boldsymbol{\Delta} - S_1\boldsymbol{\Delta})P_\lambda$$

$$= \tfrac{1}{2}\Big(t^{1/2(N-1)}\sum_{j=1}^{N}q^{-\lambda_j}t^{-N+j} - t^{-1/2(N-1)}\sum_{j=1}^{N}q^{\lambda_j}t^{N-j}\Big)\psi_\lambda(Q)$$

and

$$E\psi_\lambda(Q) = \tfrac{1}{2}(S_{-1}\boldsymbol{\Delta} + S_1\boldsymbol{\Delta})P_\lambda$$

$$= \tfrac{1}{2}\Big(t^{1/2(N-1)}\sum_{j=1}^{N}q^{-\lambda_j}t^{-N+j} + t^{-1/2(N-1)}\sum_{j=1}^{N}q^{\lambda_j}t^{N-j}\Big)\psi_\lambda(Q)\,.$$

Substituting here $q = e^{-\hbar/\ell}$, $t = e^{-\gamma/\ell}$ and introducing the quasi-momentum p_j by the same formula (3.197), we obtain the eigenvalues for the energy and momentum

$$E = \sum_{j=1}^{N}\cosh p_j\,,\qquad P = \sum_{j=1}^{N}\sinh p_j\,. \tag{3.220}$$

Upon restoration of mass m and the speed of light c according to (3.207), it becomes clear from these formulae that the quantity $\vartheta_j = p_j/mc$ plays the role of the relativistic rapidity. The model itself can be interpreted as an ideal gas of "relativistic" quasi-particles with rapidities ϑ_j satisfying the selection rule $\vartheta_{j+1} - \vartheta_j \geq \frac{\hbar}{mc\ell}\beta = \frac{\gamma}{mc\ell}$. For the ground state all $\lambda_j = 0$ and one finds $P_0 = 0$, as well as

$$E_0 = \frac{\sinh\frac{\gamma}{2\ell}N}{\sinh\frac{\gamma}{2\ell}} = \frac{\sinh\pi\gamma\frac{N}{L}}{\sinh\frac{\pi\gamma}{L}}\,, \tag{3.221}$$

where $L = 2\pi\ell$. In the thermodynamic limit we find the energy density of the ground state to be

$$\mathcal{E}_0 = \lim_{N,L\to\infty}\frac{E_0}{L} = \lim_{N,L\to\infty}\frac{\sinh\pi\gamma\frac{N}{L}}{L\sinh\frac{\pi\gamma}{L}} = \frac{\sinh\pi\gamma\mathcal{D}}{\pi\gamma}\,, \tag{3.222}$$

where $\mathcal{D} = \lim_{N,L\to\infty} N/L$ is the particle density.

Introducing a generating function for the RS integrals, see (3.75),

$$S(\zeta) = \sum_{k=0}^{n}(-\zeta)^{N-k}S_k\,,\qquad S_0 = 1\,, \tag{3.223}$$

we evaluate its action on ψ_λ with the following result

$$S(\zeta)\cdot\psi_\lambda(Q) = Q(\zeta)\cdot\psi_\lambda(Q)\,,$$

where the eigenvalue

$$Q(\zeta) = \sum_{k=0}^{N} (-\zeta)^{N-k} h_k t^{-1/2k(N-1)} \tag{3.224}$$

admits the following remarkable factorisation

$$Q(\zeta) = \prod_{j=1}^{N} (e^{-p_j} - \zeta). \tag{3.225}$$

This formula can be viewed as a trigonometric deformation of the Baxter polynomial (3.202).

Connection to Jack polynomials and Sekiguchi operators. The trigonometric RS model has the trigonometric CMS model as its limiting case. Therefore, some relation between commutative families and eigenfunctions in respective theories should arise in this limit.

This relation with Jack polynomials arises as follows. Let us rescale

$$\hbar \rightarrow \kappa \hbar, \quad \gamma \rightarrow \kappa \gamma$$

and consider the limit $\kappa \rightarrow 0$. In this limit

$$\frac{1-t^m}{1-q^m} = \frac{1-e^{-\kappa m \gamma}}{1-e^{-\kappa m \hbar}} \rightarrow \frac{\gamma}{\hbar} = \beta, \quad \forall m \in \mathbb{Z}_+, \tag{3.226}$$

and, therefore, the scalar product (3.214) turns into (3.190) used to define the Jack polynomials. In particular, one can also observe that the seven Macdonald polynomials from Table 3.3 turn in this limit into the corresponding Jack polynomials from Table 3.2.

Further, comparison of (3.225) with (3.202) gives an idea on how to connect the commutative family (3.208) of the RS model with (3.96) of the CMS one. In the limit $\kappa \rightarrow 0$ one has

$$e^{-\kappa p_j} - \zeta = 1 - \kappa p_j - \zeta + \mathcal{O}(\kappa^2).$$

Thus, if we replace in $Q(\zeta)$ the variable $\zeta \rightarrow 1 - \kappa \zeta$ and divide it by κ^N, then in the limit $\kappa \rightarrow 0$ we obtain $\mathcal{Q}(\zeta)$ of (3.96). Applying the same procedure to the generating function $S(\zeta)$ yields

$$\lim_{\kappa \to 0} \frac{1}{\kappa^N} \sum_{k=0}^{N} (\kappa\zeta - 1)^{N-k} S_k = \sum_{k=0}^{N} \sum_{m=0}^{N-k} C_{N-k}^{m} \zeta^{N-k-m} \lim_{\kappa \to 0} \frac{(-1)^m S_k}{\kappa^{k+m}}$$

$$= \sum_{k=0}^{N} \zeta^{N-k} (-1)^k \lim_{\kappa \to 0} \kappa^{-k} \sum_{l=0}^{k} C_{N-l}^{k-l} (-1)^l S_l \,,$$

where C_r^m are binomial coefficients. Comparing the last formula with (3.200), we obtain the relation between the commutative families of the CMS and RS models [46]

$$D_k = \lim_{\kappa \to 0} \sum_{l=0}^{k} C_{N-l}^{N-k} (-1)^l \frac{S_l}{\kappa^k} \,, \tag{3.227}$$

where $C_{N-l}^{k-l} = C_{N-l}^{N-k}$ was used.

Let us also show that in the limit $\kappa \to 0$ the ground state wave function of the RS model tends to the one of its CMS counterpart. We have

$$\log \mathbf{\Delta} = \frac{1}{2} \sum_{i \neq j} \sum_{k=0}^{\infty} \Big(\log(1 - q^k Q_j/Q_i) - \log(1 - tq^k Q_j/Q_i) \Big)$$

$$= \frac{1}{2} \sum_{i \neq j} \sum_{k=0}^{\infty} \sum_{m=1}^{\infty} \frac{1}{m} (t^m - 1) q^{km} \Big(\frac{Q_j}{Q_i} \Big)^m = -\frac{1}{2} \sum_{i \neq j} \sum_{m=1}^{\infty} \frac{1}{m} \frac{1 - t^m}{1 - q^m} \Big(\frac{Q_j}{Q_i} \Big)^m.$$

In the limit $\kappa \to 0$ we have (3.226) and, therefore,

$$\lim_{\kappa \to 0} \log \mathbf{\Delta} = -\frac{\beta}{2} \sum_{i \neq j} \sum_{m=1}^{\infty} \frac{1}{m} \Big(\frac{Q_j}{Q_i} \Big)^m = \frac{\beta}{2} \sum_{i \neq j} \log \Big(1 - \frac{Q_j}{Q_i} \Big).$$

This formula implies that $\lim_{\kappa \to 0} \mathbf{\Delta} = \triangle$, where $\triangle$ is the expression (3.184). Owing to this relation between $\triangle$ and $\mathbf{\Delta}$, we deduce from (3.227) that

$$\mathcal{D}_k = \triangle^{-1} D_k \triangle = \lim_{\kappa \to 0} \sum_{l=0}^{k} C_{N-l}^{N-k} (-1)^l \frac{\triangle^{-1} S_l \triangle}{\kappa^k} \,.$$

We thus observe that the Sekiguchi differential operators can be naturally viewed as the limiting case of the finite-difference Macdonald operators

$$\mathcal{D}_k = \lim_{\kappa \to 0} \sum_{l=0}^{k} C_{N-l}^{N-k} (-1)^l \frac{S_l}{\kappa^k} \,, \tag{3.228}$$

because $t^{-\frac{1}{2}k(N-1)} \to 1$ as $\kappa \to 0$.

Further developments. The approach to the spectrum in terms of multivariable symmetric polynomials that we outlined above on the concrete examples of the rational CMS and trigonometric CMS and RS models provides a basis for investigation of dynamical correlation functions and thermodynamic properties of these systems, see, for instance, [48, 49]. Further important developments that we do not touch upon here include the relation of orthogonal polynomials with Virasoro and W-algebras, and q-deformations thereof [50], the role these polynomials play in the theory of double affine Hecke algebras [51–53], and also generalisations to the theory of associated non-symmetric polynomials [32].

Finally, we point out that at the classical level the CMS and RS models can be solved by means of the method of *separation of variables* [54] and at the quantum level this leads to the construction of common eigenfunctions of commuting (Macdonald) operators in the factorised form of the product of functions of one variable, and gives new relations between special functions [55]. The integrable N-body Schrödinger operators from the CMS and RS families were investigated by various means in [56], where different representations for their eigenfunctions were found.

References

1. Takhtajan, L.A.: *Quantum Mechanics for Mathematicians*, vol. 95. AMS (2008)
2. Weigert, S.: The problem of quantum integrability. Phys. D **56**, 107–119 (1992)
3. Sklyanin, E.K.: Quantum version of the method of inverse scattering problem. J. Sov. Math. **19**, 1546–1596 (1982). [Zap. Nauchn. Semin. 95, 55 (1980)]
4. Sklyanin, E.K.: Some algebraic structures connected with the Yang-Baxter equation. Funct. Anal. Appl. **16**, 263–270 (1982). [Funkt. Anal. Pril. 16N4, 27 (1982)]
5. Drinfel'd, V.G.: Quantum groups. In: Proceedings of the International Congress of Mathematicians, pp. 798–820 (1987)
6. Jimbo, M.: A q-difference analogue of $U(J)$ and the Yang-Baxter equation. Lett. Math. Phys **10**, 63–69 (1985)
7. Faddeev, L.D., Reshetikhin, N.Y., Takhtajan, L.A.: Quantization of lie groups and lie algebras. Leningrad Math. J. **1**, 193–225 (1990). [Alg. Anal. 1, no. 1, 178 (1989)]
8. Majid, S.: Quasitriangular Hopf Algebras and Yang-Baxter equations. Int. J. Mod. Phys. A **5**, 1–91 (1990)
9. Kassel, C.: Quantum Groups (Graduate Text in Mathematics, 155). Springer, New York (1995)
10. Moyal, J.E.: Quantum mechanics as a statistical theory. Proc. Camb. Phil. Soc. **45**, 99–124 (1949)
11. Bayen, F., Flato, M., Fronsdal, C., Lichnerowicz, A., Sternheimer, D.: Deformation theory and quantization. 1. Deformations of symplectic structures. Annal. Phys. **111**, 61 (1978)
12. Bayen, F., Flato, M., Fronsdal, C., Lichnerowicz, A., Sternheimer, D.: Deformation theory and quantization. 2. Physical applications. Annal. Phys., **111**, 111 (1978)
13. Jimbo, M.: A q-difference analog of U(g) and the Yang-Baxter equation. Lett. Math. Phys. **10**, 63–69 (1985)
14. Chari, V., Pressley, A.: A Guide to Quantum Groups. University Press, Cambridge (1994)
15. Arutyunov, G.E., Frolov, S.A.: Quantum dynamical R-matrices and quantum frobenius group. Commun. Math. Phys. **191**, 15 (1998)
16. Gervais, J.-L., Neveu, A.: Novel triangle relation and absence of tachyons in Liouville string field theory. Nucl. Phys. B **238**, 125–141 (1984)

17. G. Felder. Conformal field theory and integrable systems associated to elliptic curves. *hep-th/9407154* (1994)
18. Avan, J., Babelon, O., Billey, E.: The Gervais-Neveu-Felder equation and the quantum Calogero-Moser systems. Commun. Math. Phys. **178**, 281–300 (1996)
19. Maillet, J.M.: Lax equations and quantum groups. Phys. Lett. B **245**, 480–486 (1990)
20. Macdonald, I.G.: Symmetric functions and Hall polynomials. Clarendon Press, New York; Oxford University Press, Oxford (1995)
21. Hénon, M.: Integrals of the Toda lattice. Phys. Rev. B. **9**, 1921–1923 (1974)
22. Ochiai, H., Oshima, T., Sekiguchi, H.: Commuting families of symmetric differential operators. Proc. Jpn. Acad. Ser. A Math. Sci. **70**(2), 62–66 (1994)
23. Ujino, H., Hikami, K., Wadati, M.: Integrability of the quantum Calogero-Moser model. J. Phys. Soc. Jpn. **61**(10), 3425–3427 (1992)
24. Ujino, H.: Algebraic study on the quantum Calogero model. Ph.D. thesis, University of Tokyo, 61, No. 10, 1–112 (1996)
25. Semenov-Tian-Shansky, M.A.: Poisson Lie groups, quantum duality principle, and the quantum double. Theor. Math. Phys. **93**, 1292–1307 (1992). [Teor. Mat. Fiz. 93N2, 302 (1992)]
26. Arutyunov, G.E., Klabbers, R., Olivucci, E.: Quantum trace formulae for the integrals of the hyperbolic Ruijsenaars-Schneider model. *hep-th/1902.06755* (2019)
27. Sechin, I., Zotov, A.: Associative Yang-Baxter equation for quantum (semi-)dynamical R-matrices. J. Math. Phys. **57**(5), 053505 (2016)
28. Arutyunov, G.E., Chekhov, L., Frolov, S.A.: R-matrix quantization of the Elliptic Ruijsenaars-Schneider model. Commun. Math. Phys. **192**, 405–432 (1998)
29. Hasegawa, K.: Ruijsenaars' commuting difference operators as commuting transfer matrices. Comm. Math. Phys. **187**, 289–325 (1997)
30. Antonov, A., Hasegawa, K., Zabrodin, A.: On trigonometric intertwining vectors and nondynamical R matrix for the Ruijsenaars model. Nucl. Phys. B **503**, 747–770 (1997)
31. Calogero, F.: Solution of the one-dimensional N body problems with quadratic and/or inversely quadratic pair potentials. J. Math. Phys. **12**, 419–436 (1971)
32. Baker, T.H., Forrester, P.J.: The Calogero-Sutherland model and generalized classical polynomials. Commun. Math. Phys. **188**, 175–216 (1997)
33. Lapointe, L., Vinet, L.: Exact operator solution of the Calogero-Sutherland model. Commun. Math. Phys. **178**, 425–452 (1996)
34. Landau, L.D., Lifshitz, E.M.: *Quantum Mechanics Non-Relativistic Theory*, vol. 3, 3 edn. Butterworth-Heinemann (1981)
35. Perelomov, A.M.: Algebraical approach to the solution of one-dimensional model of n interacting particles. Teor. Mat. Fiz. **6**, 364–391 (1971)
36. Hallnas, M., Langmann, E.: Explicit formulas for the eigenfunctions of the N-body Calogero model. J. Phys. A **39**, 3511 (2006)
37. Ujino, H., Wadati, M.: The Calogero model: integrable structure and orthogonal basis. In: van Diejen, J.F., Vinet, L. (eds.) Calogero-Moser-Sutherland Models, pp. 521–537. CRM Series in Mathematical Physics. Springer, New York (2000)
38. Ujino, H., Wadati, M.: The quantum Calogero model and the W-algebra. J. Phys. Soc. Jpn. **63**(10), 3585–3597 (1994)
39. Landau, L.D., Lifshitz, E.M., Pitaevskij, L.P.: *Statistical Physics: Part 2 : Theory of Condensed State*, 2 edn. Oxford (1980)
40. Polychronakos, A.P.: Exchange operator formalism for integrable systems of particles. Phys. Rev. Lett. **69**, 703–705 (1992)
41. Dunkle, C.F.: Differential-difference operators associated to reflection group. Tran. Am. Math. Soc. **311**, 167–183 (1989)
42. Vilenkin, N.J., Klimyk, A.U.: *Representation of Lie Groups and Special Functions. Recent Advances*, 497 p. Kluwer Academic Publishers (1995)
43. Stanley, R.P.: Some combinatorial properties of Jack symmetric functions. Adv. Math. **77**, 76–115 (1989)

44. Sekiguchi, J.: Zonal spherical functions on some symmetric spaces. Publ. Res. Inst. Math. Sci. **12**, 455–459 (1977)
45. Debiard, A.: Polynômes de Tchébychev et de Jacobi dans un espace euclidien de dimension p. C. R. Acad. Sc. Paris **I**(296), 529–532 (1983)
46. Ruijsenaars, S.N.M.: Complete integrability of relativistic Calogero-Moser systems and elliptic function identities. Commun. Math. Phys. **110**, 191 (1987)
47. van Diejen, J.F.: On the diagonalization of difference Calogero-Sutherland systems. CRM Proceedings and Lecture Notes, pp. 1–10 (1995)
48. Lesage, F., Pasquier, V., Serban, D.: Dynamical correlation functions in the Calogero-Sutherland model. Nucl. Phys. B **435**, 585–603 (1995)
49. Konno, H.: Dynamical correlation functions and finite size scaling in Ruijsenaars-Schneider model. Nucl. Phys. B **473**, 579–600 (1996)
50. Awata, H., Odake, S., Shiraishi, J.: Integral representations of the Macdonald symmetric functions. Commun. Math. Phys. **179**, 647–666 (1996)
51. Cherednik, I.: A unification of Knizhnik-Zamolodchikov and Dunkl operators via affine Hecke algebras. I. Invent. Math. **106**, 411–431 (1991)
52. Cherednik, I.: Double Affine Hecke Algebras and Macdonald's Conjectures. Ann. Math. **95**, 191–216 (1995)
53. Cherednik, I.: Lectures on Knizhnik-Zamolodchikov equations and Hecke algebras. MSJ Memoirs, pp. 1–96 (1998)
54. Kuznetsov, V., Nijhoff, F., Sklyanin, E.: Separation of variables for the Ruijsenaars system. Comm. Math. Phys. **189**, 855–877 (1997)
55. Kuznetsov, V.B., Sklyanin, E.K.: Separation of variables for A2 Ruijsenaars model and new integral representation for A2 Macdonald polynomials. J. Phys. A **29**, 2779–2804 (1996)
56. Felder, G., Varchenko, A.: Three formulas for Eigen functions of integrable Schrodinger operators. *hep-th/9511120* (1995)

Chapter 4
Factorised Scattering Theory

> *...unlike the classical case, to know a system is integrable buys us absolutely nothing.*
> *... any reasonable definition of quantum integrability should imply that the system scatters without diffraction. We know what diffractionless scattering means, while we do not know what quantum integrability means. ... Therefore, for the quantum system that supports scattering, we shall take quantum integrability to mean scattering without diffraction.*
>
> Bill Sutherland
> Beautiful models

In this chapter we discuss implications of integrability for systems that support scattering. These are the systems that fly apart into individual constituents as time $t \to \pm\infty$. The constituents ultimately emerge to an observer as free particles or bound states thereof. We show that integrability leads to the conservation of asymptotic momenta of particles, meaning that the scattering process is non-diffractive. We then define the notion of the classical phase shift and show that the multi-body phase shift acquired in the scattering process factorises into the sum of the two-body shifts. Quantum-mechanically, integrability results into the special form of the asymptotic wave function, known as the Bethe wave function. Most importantly, the multi-body S-matrix factorises into the product of the two-body ones, corresponding to subsequent two-body collision events. The consistency condition for this factorisation is the Yang-Baxter equation satisfied by two-body S-matrix. We present the two-body S-matrices for the main models discussed in the book and explain the relation between the Bethe wave function and particle statistics.

© Springer Nature Switzerland AG 2019
G. Arutyunov, *Elements of Classical and Quantum Integrable Systems*,
UNITEXT for Physics, https://doi.org/10.1007/978-3-030-24198-8_4

4.1 Classical Scattering in Integrable Models

It is natural to start from the picture of scattering in classical mechanics. Here we deal with well-defined classical trajectories and the simplest situation corresponds to a single non-relativistic particle scattering elastically off a fixed (heavy) target modelled by a potential with a finite interaction range. Asymptotic trajectories describing the motion of a particle far away from an interaction region are straight lines.[1] Restraining to a one-dimensional situation, this means that before the collision, as $t \to -\infty$, the actual orbit asymptotes to a free orbit

$$q(t) \to q_{\text{in}}(t) \equiv q^- + v^- t, \quad t \to -\infty, \tag{4.1}$$

for some fixed (q^-, v^-). Analogously, after the collision,

$$q(t) \to q_{\text{out}}(t) \equiv q^+ + v^+ t, \quad t \to +\infty. \tag{4.2}$$

The scattering process is completely characterised by $q_{\text{in}}(t)$ and $q_{\text{out}}(t)$, the incoming and outgoing asymptotic orbits. The pairs $(q^\pm, v^\pm)$ represent the classical scattering data and a transformation from (q^-, v^-) to (q^+, v^+) defines a classical scattering operator, also known as *classical S-matrix*. Scattering of several particles interacting with each other via an admissible pair-wise potential is considered in a similar manner. Every particle trajectory has well-defined in- and out-asymptotics and scattering results to a transformation of incoming into outgoing scattering data.

To understand what is special about scattering processes in an integrable model, we consider the example of N particles of equal mass $m = 1$ governed by an integrable Hamiltonian

$$H = \frac{1}{2} \sum_{j=1}^{N} p_j^2 + \sum_{i<j}^{N} v(q_{ij}). \tag{4.3}$$

We assume that the potential v is symmetric: $v(q) = v(-q)$, repulsive and impenetrable, and falls off sufficiently rapidly with the distance between particles, to guarantee the existence of an asymptotic region. Concretely, one can think about the rational or hyperbolic CMS models, which potentials satisfy the above-mentioned conditions. Classical integrability for these models follows from the existence of the Lax representation with the matrix L given by

$$L = \sum_{j=1}^{N} p_j E_{jj} + \sum_{i<j}^{N} u(q_{ij}) E_{ij} \tag{4.4}$$

with an appropriate function $u(q)$.

----

[1]We exclude bounded orbits which might exist for the case of attractive potentials.

Non-diffractive scattering. Since the potential is repulsive and impenetrable, particles cannot overtake each other and one can label them according to the order

$$q_1(t) < q_2(t) < \cdots < q_N(t), \quad \forall t, \tag{4.5}$$

see Fig. 4.1. For $t \to -\infty$ the asymptotic condition has the form

$$q_i(t) = p_i^- t + q_i^- + o(1), \quad p_i^- \equiv p_i(-\infty), \tag{4.6}$$

and it is described by $2N$ numbers (p_i^-, q_i^-). Analogously, for $t \to +\infty$,

$$q_i(t) = p_i^+ t + q_i^+ + o(1), \quad p_i^+ \equiv p_i(+\infty). \tag{4.7}$$

From (4.5) that is valid for any t, in particular, for the asymptotic conditions (4.6) and (4.7), we deduce that[2]

$$p_1^- > p_2^- > \cdots > p_N^-,$$
$$p_1^+ < p_2^+ < \cdots < p_N^+.$$

Because of the asymptotic conditions, $|q_i - q_j| = \mathcal{O}(t)$ as $t \to \pm\infty$, the Lax matrix has the asymptotic limits $L(\pm\infty)$, where it becomes diagonal; the diagonal supports the set of eigenvalues λ_j of L. These eigenvalues are integrals of motion, and if we set $p_j^- = \lambda_j$ then,

$$\lambda_1 > \lambda_2 > \cdots > \lambda_N.$$

Obviously, the same order of eigenvalues must be found at $t = +\infty$, which is only possible if $L(+\infty)$ has the same eigenvalues as $L(-\infty)$ but in the reversed order:

Fig. 4.1 The picture of scattering in one dimension. Integrability implies that scattering is non-diffractive

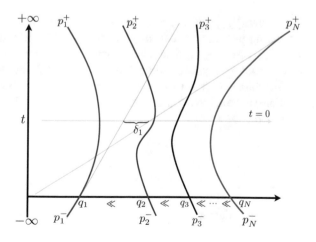

---

[2]We omit certain subtleties which are discussed in [1].

$$L(-\infty) = \begin{pmatrix} p_1^- & & \\ & \ddots & \\ & & p_N^- \end{pmatrix}, \quad L(+\infty) = \begin{pmatrix} p_N^+ & & \\ & \ddots & \\ & & p_1^+ \end{pmatrix}.$$

This implies in turn a very simple relation between the scattering data:

$$p_{N+1-j}^+ = p_j^- . \tag{4.8}$$

Thus, the set of incoming asymptotic momenta $\{p_i^-\}$ coincides with the set of outgoing ones $\{p_i^+\}$. This central result is usually referred to as *conservation of asymptotic momenta* and the corresponding scattering process is described as *non-diffractive* [2–4]. Note that (4.8) is independent of the value of the coupling constant.

Classical phase shift. Due to the coincidence of the sets of incoming and outgoing momenta, we can reinterpret the scattering picture in a different way. Namely, we can associate to each particle a unique asymptotic momentum and assume that the order of particles is the same as that of their momenta. In particular, before the scattering the fastest particle is the most left one, and after the scattering it reappears on the right of all the others, as if interactions would be completely absent. This is the so-called *transmission* representation of scattering in comparison to the *reflection* representation we started with. We have to say more on these interpretations when it comes to the discussion of quantum scattering.

From the transmission point of view, individual particles always keep their asymptotic momenta, while scattering shows up in the discontinuity δ_j of the asymptotic coordinates

$$\begin{aligned} \delta_j &= q_{N+1-j}^+ - q_j^- \\ &= \lim_{t \to +\infty} \left(q_{N+1-j}(t) - q_j(-t) - 2p_j^- t \right). \end{aligned} \tag{4.9}$$

The quantity δ_j, also known as the *classical phase shift*, completely characterises the scattering process: it shows how much the jth particle has advanced in comparison to a freely moving particle with momentum p_j. Our next goal is to understand how to compute δ_j, especially for an integrable model with the Hamiltonian (4.3).

We start with the two-body problem. For the two-particle case the equations of motion can be solved in quadratures by making use of the conservation laws of energy and momentum

$$\frac{p_1^2}{2} + \frac{p_2^2}{2} + v(q_{12}) = \frac{(p_1^-)^2}{2} + \frac{(p_2^-)^2}{2} ,$$
$$p_1 + p_2 = p_1^- + p_2^- .$$

Indeed, introducing $k = p_1^- - p_2^- > 0$, from these equations we get

$$p_{1,2} = \frac{1}{2}(p_1^- + p_2^-) \pm \frac{1}{2}\sqrt{k^2 - 4v(q_{12})}. \tag{4.10}$$

To correctly associate the particle labels to the signs on the right hand side of the last formula, we recall that for $t \to -\infty$ the potential $v(q_{12})$ vanishes and from equations above we restore the asymptotic conditions for particle momenta, we have chosen

$$p_1(t) = \frac{1}{2}(p_1^- + p_2^-) + \frac{1}{2}\sqrt{k^2 - 4v(q_{12})},$$
$$p_2(t) = \frac{1}{2}(p_1^- + p_2^-) - \frac{1}{2}\sqrt{k^2 - 4v(q_{12})}.$$
(4.11)

As time goes, p_1 decreases, while p_2 increases and at the value t_0 such that $k^2 = 4v(x_0)$, $x_0 \equiv x(t_0)$, the particle momenta become equal. The difference $x(t) = q_2(t) - q_1(t) > 0$ is governed by the equation

$$\dot{x}(t) = -\sqrt{k^2 - 4v(x(t))}.$$
(4.12)

As time grows starting from $-\infty$, the distance between particles diminishes and at $t = t_0$ it reaches its minimum $x = x_0$. By continuity, after passing the value t_0, p_1 and p_2 continue to decrease and increase, respectively, which for $t > t_0$ enforces the identification

$$p_1(t) = \frac{1}{2}(p_1^- + p_2^-) - \frac{1}{2}\sqrt{k^2 - 4v(q_{12})},$$
$$p_2(t) = \frac{1}{2}(p_1^- + p_2^-) + \frac{1}{2}\sqrt{k^2 - 4v(q_{12})}.$$

The distance between the particles starts to increase again according to

$$\dot{x}(t) = \sqrt{k^2 - 4v(x(t))}.$$
(4.13)

This discussion provides a qualitative picture of the dynamics, see Fig. 4.2. In particular, the center–of–mass undergoes free motion in accordance with the prescribed asymptotic behaviour

$$q_1 + q_2 = (p_1^- + p_2^-)t + q_1^- + q_2^- = (p_1^+ + p_2^+)t + q_1^+ + q_2^+,$$
(4.14)

which, together with (4.8), implies for the discontinuities (4.9) that $\delta_1 + \delta_2 = 0$.

Equations (4.12) and (4.13) are solved by quadrature

$$t = -\int^x \frac{dx}{\sqrt{k^2 - 4v(x)}}, \quad t \leqslant t_0,$$
$$t = \int^x \frac{dx}{\sqrt{k^2 - 4v(x)}}, \quad t \geqslant t_0.$$
(4.15)

Before we proceed with a general $v(q)$, let us consider the two concrete examples of the rational and hyperbolic CMS models corresponding to the potentials

Fig. 4.2 Time evolution of
particle momenta in the
two-body problem

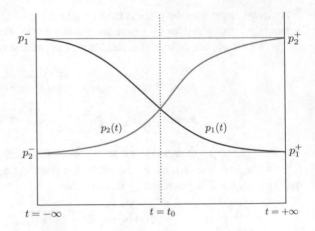

$$v(q) = \frac{\gamma^2}{q^2} \quad \text{and} \quad v(q) = \frac{\gamma^2}{4\ell^2 \sinh^2 \frac{q}{2\ell}}.$$

In the latter case the parameter ℓ controls the interaction range and we set $\ell = 1$ to
simplify considerations. For these models, performing integration in the first equation
of (4.15) yields

$$t = -\frac{1}{k}\sqrt{x^2 - \frac{4\gamma^2}{k^2}} + \frac{a}{k} \qquad \text{(rat)},$$

$$t = -\frac{2}{k}\log\left(\cosh\frac{x}{2} + \sinh\frac{x}{2}\sqrt{1 - \frac{\gamma^2}{k^2 \sinh^2\frac{x}{2}}}\right) + \frac{a}{k} \quad \text{(hyper)},$$

where in both cases a is an integration constant. This constant is found by matching
the above formulae with the asymptotic expansion

$$x(t) = -kt + (q_2^- - q_1^-) + a_1/t + a_2/t^2 + \cdots,$$

for $t \to -\infty$. In both cases we find $a = q_2^- - q_1^-$. For $t > t_0$ the solution of the
second equation in (4.15) is

$$t = \frac{1}{k}\sqrt{x^2 - \frac{4\gamma^2}{k^2}} - \frac{b}{k} \qquad \text{(rat)},$$

$$t = \frac{2}{k}\log\left(\cosh\frac{x}{2} + \sinh\frac{x}{2}\sqrt{1 - \frac{\gamma^2}{k^2 \sinh^2\frac{x}{2}}}\right) - \frac{b}{k} \quad \text{(hyper)}.$$

(4.16)

The integration constant b is found from the condition that at $t = t_0$ corresponding
to $x = x_0$, solutions obtained for $t \leqslant t_0$ and $t \geqslant t_0$ must coincide. This gives $b = -a$
for the rational case and $b = -a + 4\log\cosh\frac{x_0}{2}$ for the hyperbolic one. Matching

now the asymptotics

$$x(t) = kt + (q_2^+ - q_1^+) + b_1/t + b_2/t^2 + \cdots$$

for $t \to \infty$ with that of (4.16), one gets

$$q_2^+ - q_1^+ = b = -a = q_1^- - q_2^- \quad \text{(rat)},$$
$$q_2^+ - q_1^+ = b = -a + 4 \log \cosh \tfrac{x_0}{2} = q_1^- - q_2^- + 4 \log \cosh \tfrac{x_0}{2} \quad \text{(hyper)}.$$

Combining these formulae with (4.14), yields

$$\delta_1 = 0 = \delta_2 \quad \text{(rat)},$$

$$\delta_1 = \log \cosh^2 \tfrac{x_0}{2} = \log \left(1 + \frac{\gamma^2}{k^2}\right) = -\delta_2 \quad \text{(hyper)},$$

(4.17)

where we used the fact that x_0 is defined by $\sinh^2 \tfrac{x_0}{2} = \gamma^2/k^2$. In particular, if we restore the physical dimensions and the dependence on ℓ, the answer for the hyperbolic case $\delta(k) \equiv \delta_1$ reads as

$$\delta(k) = \ell \log \left(1 + \frac{\gamma^2}{k^2 \ell^2}\right).$$

(4.18)

These examples make clear how to treat the case of a generic potential $v(x)$. A general solution of (4.15) can be written as

$$t = -\int_{x_0}^x \frac{dx}{\sqrt{k^2 - 4v(x)}} + t_0, \quad t \leqslant t_0,$$

$$t = \int_{x_0}^x \frac{dx}{\sqrt{k^2 - 4v(x)}} + t_0, \quad t \geqslant t_0,$$

(4.19)

where t_0 is chosen such that for $x_0 = x(t_0)$ one has $4v(x_0) = k^2$. When $x \to \infty$, due to the rapid decrease of the potential at infinity, the first formula in (4.19) admits a well-defined asymptotic expansion

$$t = -\int_{x_0}^x dx \left(\frac{1}{k} + \frac{2}{k^3} v(x) + \frac{6}{k^5} v(x)^2 + \cdots\right) + t_0$$

$$= -\left(\frac{1}{k}(x - x_0) + \frac{2}{k^3} V(x)\Big|_{x_0}^x + \cdots\right) + t_0,$$

where $V'(x) = v(x)$. Assuming $V(x) \to 0$ for $x \to \infty$ and similar behaviour for other terms in the above expansion, we see that $-x/k$ gives the leading term of the asymptotics $t \to -\infty$, while the constant term equal to $q_2^- - q_1^-$ should be identified with

$$q_2^- - q_1^- = kt_0 + x_0 + k\left(\frac{2}{k^3}V(x_0) + \cdots\right)$$

$$= kt_0 + x_0 - k\int_{x_0}^{\infty} dx\left(\frac{1}{\sqrt{k^2 - 4v(x)}} - \frac{1}{k}\right).$$

This expresses the constant t_0 in terms of the asymptotic data and the potential. Analogously, considering the asymptotics of the second formula in (4.19) for $x \to \infty$, we find

$$q_2^+ - q_1^+ = -kt_0 + x_0 - k\int_{x_0}^{\infty} dx\left(\frac{1}{\sqrt{k^2 - 4v(x)}} - \frac{1}{k}\right)$$

$$= q_1^- - q_2^- + 2x_0 - 2\int_{x_0}^{\infty} dx\left(\frac{k}{\sqrt{k^2 - 4v(x)}} - 1\right).$$

This ultimately gives [5] the coordinate discontinuity $\delta_1 \equiv \delta(k)$ for a generic potential subject to the requirements formulated above

$$\delta(k) = x_0(k) - \int_{x_0(k)}^{\infty} dx\left(\frac{1}{\sqrt{1 - \frac{4v(x)}{k^2}}} - 1\right), \tag{4.20}$$

where $x_0 = x_0(k)$ is found from $4v(x_0) = k^2$. Obviously, the discontinuity depends only on the difference k of asymptotic momenta but not on the asymptotic coordinates q_i^-. From (4.20) we also read that $\delta(k)$ is an even function of k. Now we can turn to the case of many particles.

Factorisation of the classical S-matrix. First we recall the notion of the scattering matrix in classical theory. Let $f(p, q)$ be an observable defined on a phase space $\mathcal{P} = \mathbb{R}^{2N}$ with the canonical bracket (1.9). Given $f(p, q)$ at some moment of time, say at $t = 0$, we can find its value $f(p, q, t) \equiv f(p(t), q(t))$ at any moment t through the Hamiltonian equations. Assuming the time-dependence is continuous, we may expand $f(p, q, t)$ in powers of t

$$f(p, q, t) = f + \frac{t}{1!}\dot{f} + \frac{t^2}{2!}\ddot{f} + \frac{t^3}{3!}\dddot{f} + \cdots$$

Using the Hamiltonian equations this can be rewritten as

$$f(t) = f + \frac{t}{1!}\{H, f\} + \frac{1}{2!}t^2\{H\{H, f\}\} + \frac{t^3}{3!}\{H\{H\{H, f\}\}\} + \cdots$$

Formally, this series can be viewed as an action on f of a certain evolution operator U_t

$$f(p, q, t) = e^{t\{H, \cdot\}} \circ f(p, q) \equiv (U_t f)(p, q).$$

The evolution operator U_t defines a one-parametric continuous group of canonical transformations:

$$U_t(\{f, h\}) = \{U_t(f), U_t(h)\}.$$

In the context of scattering theory we assume the existence of in- and out-asymptotics, namely,

$$(p_i(t), q_i(t)) \rightarrow (p_i^{\pm}, p_i^{\pm} t + q_i^{\pm}), \quad t \rightarrow \pm\infty, \tag{4.21}$$

where $(p_i^{\pm}, q_i^{\pm})$ are coordinates on the asymptotic phase spaces

$$\begin{aligned}
\mathscr{P}^- &= \{(p^-, q^-) \in \mathbb{R}^{2N}, \quad p_1^- > p_2^- > \cdots > p_N^-\}, \\
\mathscr{P}^+ &= \{(p^+, q^+) \in \mathbb{R}^{2N}, \quad p_1^+ < p_2^+ < \cdots < p_N^+\}.
\end{aligned}$$

Since the time evolution preserves the Poisson brackets between canonical variables, the asymptotic data (p_i^-, q_i^-) and (q_i^+, p_i^+) also form canonical pairs, i.e. the asymptotic spaces are symplectic. In terms of evolution maps, Eq. (4.21) can be written as

$$e^{t\{H, \cdot\}} \circ (p_i, q_i) \rightarrow e^{t\{H_0^{\pm}, \cdot\}} \circ (p_i^{\pm}, q_i^{\pm}) \quad t \rightarrow \pm\infty, \tag{4.22}$$

where $H_0^{\pm}$ are the free Hamiltonians constructed from the asymptotic momenta

$$H_0^{\pm} = \frac{1}{2} \sum_{j=1}^{N} (p_j^{\pm})^2. \tag{4.23}$$

This allows one to define the classical analogue of the quantum-mechanical Møller operators (the wave maps) $\Omega_{\pm} : \mathscr{P}^{\mp} \rightarrow \mathscr{P}$,[3]

$$\Omega_{\pm} = \lim_{t \rightarrow \mp\infty} e^{-t\{H, \cdot\}} \circ e^{t\{H_0^{\mp}, \cdot\}} \tag{4.24}$$

The wave maps are canonical and they are used to construct the classical S-matrix

$$S_{\text{class}} = \Omega_-^{-1} \Omega_+ : \mathscr{P}^- \rightarrow \mathscr{P}^+, \tag{4.25}$$

which is also a canonical transformation.

Coming back to the problem of finding the phase shifts δ_j for many-body scattering, we note that this can be done by relying on canonicity of the asymptotic phase spaces and the relations (4.8). Indeed, we have

$$\delta_{ij} = \{p_{N+1-i}^+, q_{N+1-j}^+\} = \{p_i^-, q_j^- + \delta^j\} = \delta_{ij} + \frac{\partial \delta_j}{\partial q_i^-}.$$

---

[3] We define $\Omega_{\pm}$ following the tradition of the quantum scattering theory, see e.g. [6].

This shows that $\partial \delta_j / \partial q_i^- = 0$, that is δ_j *depends on* p^- *only* and, therefore, it can be immediately found by arranging the asymptotic data $q_1^-, \ldots, q_N^-$, such that collisions take place pairwise, with asymptotically large times in between of any two subsequent collisions. This shows that the multi-body phase shift can be found by simply summing up the two-body phase shifts arising from collisions of j's particle with the rest[4]

$$\delta_j = q_{N-j+1}^+ - q_j^- = \sum_{k>j} \delta(p_j^- - p_k^-) - \sum_{k<j} \delta(p_j^- - p_k^-) . \tag{4.26}$$

This strikingly simple answer is, of course, a consequence of the existence of a complete set of integrals of motion responsible for (4.8). There are no separate three- and higher-body events and the multi-particle scattering process is completely characterised by the two-body phase shift. A direct argument supporting this statement is based on using evolution equations produced by higher commuting integrals to rearrange the scattering process into a sequence of two-body events [5].

Finally, to determine the classical S-matrix for the scattering problem at hand, we consider the generating function $\Phi(q^-, p^+)$ of the canonical transformation $(q_i^-, p_i^-) \rightarrow (q_i^+, p_i^+)$

$$p_i^- = \frac{\partial \Phi(q^-, p^+)}{\partial q_i^-} = p_{N+1-i}^+ , \quad q_i^+ = \frac{\partial \Phi(q^-, p^+)}{\partial p_i^+} = q_{N+1-i} + \delta_{N+1-i} , \tag{4.27}$$

where, according to (4.26),

$$\delta_{N+1-i} = \sum_{k>N+1-i} \delta(p_{N+1-i}^- - p_k^-) - \sum_{k<N+1-i} \delta(p_{N+1-i}^- - p_k^-)$$
$$= \sum_{k<j} \delta(p_i^+ - p_k^+) - \sum_{k>i} \delta(p_i^+ - p_k^+) .$$

Integrating (4.27), we find

$$\Phi(q^-, p^+) = \sum_{i=1}^N q_i^- p_{N+1-i}^+ + \sum_{i<j}^N \theta(p_i^+ - p_j^+) , \tag{4.28}$$

where $\theta(k)$ is an integrated phase shift

$$\theta(k) = kx_0(k) - k \int_{x_0(k)}^\infty dx \left(\sqrt{1 - \frac{4v(x)}{k^2}} - 1 \right) , \quad \frac{\partial \theta}{\partial k} = \delta(k) . \tag{4.29}$$

---

[4]In the S-matrix picture where particles are labelled according to their conserved asymptotic momenta, initially j's particle has $j - 1$ particles on its left and $N - j$ on its right, and, after all the collisions, it will have $N - j$ particles on its left and $j - 1$ on its right.

Obviously, $\theta(k)$ is an odd function of k. Were the theory free, the relation between $\mathscr{P}^+$ and $\mathscr{P}^-$ would reduce to relabelling of particles described by the generating function

$$\Phi_0(q^-, p^+) = \sum_{i=1}^{N} q_i^- \, p_{N+1-i}^+ \, . \tag{4.30}$$

Thus, the non-trivial part of the generating function can be identified with the classical S-matrix that is, therefore, is given by

$$S_{\text{cl}} = \sum_{i<j}^{N} \theta(p_i^+ - p_j^+) \, . \tag{4.31}$$

This quantity has the physical dimension of the action: $[S_{\text{cl}}] = [\hbar]$ and, from the point of view of the correspondence between classical and quantum mechanics, can be thought of as the leading term in the semi-classical expansion of the phase θ of the quantum-mechanical wave function $\Psi = a e^{\frac{i}{\hbar}\theta}$ in powers of $\hbar$ [7].

A fundamental fact about this classical S-matrix is that it has a factorised structure, i.e. it is written as the sum of two-body integrated phase shifts. This is a direct consequence of integrability for scattering theory and, as we will see, it will persist in the quantum case as well.

Classical phase shift for RS models. We conclude this section by pointing out that our discussion of the factorised scattering theory can be also applied to the RS models. These models have with the following Hamiltonian H and the total momentum P

$$H = \sum_{i=1}^{N} \cosh p_i \prod_{j\neq i}^{N} v(q_{ij}) \, , \quad P = \sum_{i=1}^{N} \sinh p_i \prod_{j\neq i}^{N} v(q_{ij}) \, .$$

Here the function $v(q)$ tends to 1 as the distance between particles tends to infinity. One can show that in the limits $t \to \pm\infty$, H, P, as well as higher commuting integrals are functions of the asymptotic momenta, albeit more complicated than in the case of the CMS models. As an exercise, we evaluate the phase shift for the hyperbolic RS model, for which

$$v(q) = \sqrt{1 + \frac{\sin^2 \frac{1}{2}\gamma}{\sinh^2 \frac{1}{2}q}} \, .$$

For the case of two particles we have

$$\dot{q}_1 = v(x) \sinh p_1 \, , \quad \dot{q}_2 = v(x) \sinh p_2 \, , \quad x = q_2 - q_1 > 0 \, ,$$

and the asymptotic conditions are

$$q_{1,2} \rightarrow \sinh p_{1,2}^{\pm} t + q_{1,2}^{\pm}, \quad t \rightarrow \pm\infty. \tag{4.32}$$

The total momentum, $P = \dot{q}_1 + \dot{q}_2$, is conserved and, therefore, the sum $q_1 + q_2$ enjoys free motion implying $p_1^- = p_2^+$ and $q_1^+ + q_2^+ = q_1^- + q_2^-$. For the variable x that describes relative motion one gets

$$\dot{x} = \pm H \sqrt{1 - \frac{4v^2(x)}{H^2 - P^2}}. \tag{4.33}$$

Here one has to take minus for $t \leqslant t_0$ and plus for $t \geqslant t_0$, where t_0 corresponds to $x_0 = x(t)$ such that $4v^2(x_0) = H^2 - P^2$. Introducing $k = p_1^- - p_2^-$ and taking into account the expressions for H and P in terms of asymptotic momenta

$$\begin{aligned} H &= \cosh p_1^- + \cosh p_2^-, \\ P &= \sinh p_1^- + \sinh p_2^-, \end{aligned} \tag{4.34}$$

we find $H^2 - P^2 = 4\cosh^2 \frac{k}{2}$. The Eq. (4.33) can be integrated

$$t = \mp \frac{2}{H \tanh \frac{k}{2}} \log \left(\cosh \frac{x}{2} + \sinh \frac{x}{2} \sqrt{1 - \frac{\sin^2 \frac{\gamma}{2}}{\sinh^2 \frac{k}{2} \sinh^2 \frac{x}{2}}} \right) \pm \frac{a^{\mp}}{H \tanh \frac{k}{2}},$$

where $a^{\pm}$ are integration constants. It also follows from (4.34) that

$$H \tanh \frac{k}{2} = \sinh p_1^- - \sinh p_2^- = \sinh p_2^+ - \sinh p_1^+. \tag{4.35}$$

Matching the two different expressions for t at $t = t_0$, we find

$$a^+ = 2 \log \cosh^2 \frac{x_0}{2} - a^-,$$

while the comparison with the asymptotics (4.32) gives $a^+ = q_2^+ - q_1^+$ and $a^- = q_2^- - q_1^-$. Putting everything together, we obtain the phase shift for the hyperbolic RS model [8]

$$\delta(k) = \log \left(1 + \frac{\sin^2 \frac{\gamma}{2}}{\sinh^2 \frac{k}{2}} \right). \tag{4.36}$$

Restoring the physical units and the dependence on the interaction length ℓ is done by the substitutions $k \rightarrow k/\mu$, $\gamma \rightarrow \gamma/\mu\ell$ and $\delta \rightarrow \ell\delta$, where $\mu = mc$. Needless to say, our previous results for the CMS models (4.17) follow from this formula as limiting cases.

4.2 Quantum-Mechanical Scattering and Integrability

The scattering problem in quantum mechanics deals with the continuous spectrum of the Schrödinger operator. We will deal with the time-independent scattering theory and refer the reader to Appendix E for the time-dependent interpretation of the corresponding results.

4.2.1 Two-Body Scattering

Consider the two-body Schrödinger equation

$$-\frac{\hbar^2}{2}\left(\frac{\partial^2}{\partial q_1^2} + \frac{\partial^2}{\partial q_2^2}\right)\Psi + v(q_1 - q_2)\Psi = E\Psi. \tag{4.37}$$

Here q_1 and q_2 are associated to the wave function $\Psi(q_1, q_2)$ and they are not ordered. The potential $v(x)$, where $x = q_2 - q_1$ is the relative coordinate, satisfies the standard assumptions of the previous section. To simplify the presentation, in the following we set $\hbar = 1$.

The Schrödinger equation (4.37) is separated in the center-of-mass frame so that the wave function is

$$\Psi(q_1, q_2) = e^{iK(q_1+q_2)/2}\psi(x), \tag{4.38}$$

where K is the value of the total momentum and ψ solves

$$-\psi''(x) + v(x)\psi(x) = \tfrac{1}{4}k^2\psi(x). \tag{4.39}$$

The energy is $E = (K^2 + k^2)/4$. Equation (4.39) defines the standard quantum-mechanical scattering problem in one dimension. Its spectrum is continuous and doubly degenerate, i.e. for given E there are two linearly independent solutions $\psi_{1,2}(k, x)$ bounded on the whole real line. Below we recall the construction of the corresponding scattering matrix.

Far away from the potential region, Eq. (4.39) is well approximated by

$$-\psi''(x) = \tfrac{1}{4}k^2\psi(x), \tag{4.40}$$

which has two independent solutions $e^{\pm ikx/2}$, $k > 0$. It is convenient to introduce asymptotic momenta p_1 and p_2 such that

$$K = p_1 + p_2, \quad k = p_1 - p_2 > 0.$$

In terms of these momenta the energy reads as $E = \tfrac{1}{2}p_1^2 + \tfrac{1}{2}p_2^2$.

Table 4.1 Asymptotics of two independent solutions ψ_1 and ψ_2 of the Schrödinger equation. The scattering process is characterised by complex amplitudes A, B, C, D

Solution	$x \to +\infty$	$x \to -\infty$
$\psi_1(k, x)$	$e^{-ikx/2} + A(k)e^{ikx/2}$	$B(k)e^{-ikx/2}$
$\psi_2(k, x)$	$D(k)e^{ikx/2}$	$e^{ikx/2} + C(k)e^{-ikx/2}$

Independent solutions $\psi_{1,2}(k, x)$ of (4.39) are usually chosen to have asymptotics presented in Table 4.1. These are called *scattering solutions*. The existence of scattering solutions is based on the following argument [9]: a general solution of (4.39) is constructed as a linear combination of any two linearly independent solutions φ_1 and φ_2. One of the coefficients in this linear combination can always be chosen in such a way that the corresponding solution does not have an exponent $e^{ikx/2}$ in the asymptotic expansion as $x \to +\infty$. The coefficient 1 in front of $e^{-ikx/2}$ in the expansion $x \to +\infty$ is then obtained by fixing the overall normalisation. The function ψ_2 is constructed through $\varphi_{1,2}$ in a similar way. One can further show that the scattering solutions are normalised as [9]

$$\int_{-\infty}^{\infty} \overline{\psi_1(k_1, x)}\, \psi_2(k_2, x)\mathrm{d}x = 0\,, \quad \int_{-\infty}^{\infty} \overline{\psi_{1,2}(k_1, x)}\, \psi_{1,2}(k_2, x)\mathrm{d}x = 2\pi\delta(k_1 - k_2)\,.$$

From the asymptotics of $\psi_{1,2}$, by using (4.38), the asymptotics of the corresponding full two-particle wave functions $\Psi_{1,2}$ follows, see Table 4.2.

The Wronskian $W(\varphi_1, \varphi_2) = \varphi_1\varphi_2' - \varphi_2\varphi_1'$ is x-independent and different from zero if these solutions are linearly independent. In particular, computing $W(\psi_1, \psi_2)$ in the two different asymptotic regimes $x \to \pm\infty$, we get that $D = B$. Further, if φ is a solution then its complex conjugate $\bar{\varphi}$ is also a solution. We have four solutions $\psi_1, \bar{\psi}_1, \psi_2, \bar{\psi}_2$ that give rise to 3 new independent Wronskians:

$$W(\psi_1, \bar{\psi}_1)\,, \quad W(\psi_2, \bar{\psi}_2)\,, \quad W(\psi_1, \bar{\psi}_2)\,.$$

Matching their values in the different asymptotic regimes, we get three more relations for the amplitudes

$$|A|^2 + |B|^2 = 1\,, \quad |C|^2 + |D|^2 = 1\,, \quad A\bar{D} + B\bar{C} = 0\,. \tag{4.41}$$

Table 4.2 Asymptotics behaviour of the full wave functions Ψ_1 and Ψ_2 corresponding to ψ_1 and ψ_2, respectively

Solution for Ψ	$q_1 \ll q_2$	$q_1 \gg q_2$
$\Psi_1(q_1, q_2)$	$e^{ip_1q_1+ip_2q_2} + A(k)e^{ip_2q_1+ip_1q_2}$	$B(k)e^{ip_1q_1+ip_2q_2}$
$\Psi_2(q_1, q_2)$	$B(k)e^{ip_2q_1+ip_1q_2}$	$e^{ip_2q_1+ip_1q_2} + C(k)e^{ip_1q_1+ip_2q_2}$

Introduce the following symmetric matrix

$$S = \begin{pmatrix} A & B \\ B & C \end{pmatrix}. \tag{4.42}$$

Due to (4.41), this matrix is unitary, $S^\dagger S = \mathbb{1}$. Matrix S is called the *scattering matrix* or simply the S-matrix. It is the main object of one-dimensional scattering theory.

Note that the equality $D = B$ follows from the invariance of the Schödinger equation under time reversal. For the special case of even functions $v(x)$: $v(x) = v(-x)$, solution $\psi_1(x)$ should turn into $\psi_2(x)$ under the parity transformation $x \to -x$, which implies that $C(k) = A(k)$ for this case. In the following we assume that $v(x)$ is even. The S-matrix eigenvalues $A \pm B$ are pure phases which we set to parametrise as

$$A \pm B = -e^{-\frac{i}{\hbar}\theta_\pm}, \tag{4.43}$$

where the dependence on $\hbar$ reminds that $\theta_\pm$ have the physical dimension of the classical action.

To recall the reader the physical interpretation of A and B as *reflection* and *transmission coefficients*, in Appendix E we briefly describe the time-dependent scattering theory. Finally, we note that the practical evaluation of the S-matrix consists in finding $\psi_{1,2}(k, x)$ exactly or by using the WKB approach.

Representations of the asymptotic wave function. Obviously, an arbitrary solution $\Psi(q_1, q_2)$ of the Schrödinger equation can be constructed by making an arbitrary linear combination of scattering solutions. The asymptotics of $\Psi(q_1, q_2)$ are, therefore,

$$\Psi(q_1, q_2) \to \begin{cases} c_1 e^{ip_1 q_1 + ip_2 q_2} + (c_1 A + c_2 B)e^{ip_2 q_1 + ip_1 q_2}, & q_1 \ll q_2, \\ c_2 e^{ip_2 q_1 + ip_1 q_2} + (c_2 A + c_1 B)e^{ip_1 q_1 + ip_2 q_2}, & q_1 \gg q_2, \end{cases} \tag{4.44}$$

where c_1 and c_2 are arbitrary complex numbers. In particular, considering the sum or difference of the scattering solutions, we can build up symmetric $\Psi_+(q_1, q_2)$ and anti-symmetric $\Psi_-(q_1, q_2)$ wave functions describing bosons and spinless fermions, respectively. The asymptotics of these functions are shown in Table 4.3.

In the following it is convenient to call the asymptotic regions $q_1 \ll q_2$ and $q_1 \gg q_2$ *sectors* and to parametrise the asymptotic wave function (4.44) with four complex coefficients $\mathcal{A}(\sigma|\tau)$ called *amplitudes* as

$$\Psi(q_1, q_2) \to \begin{cases} \mathcal{A}(12|12)e^{ip_1 q_1 + ip_2 q_2} + \mathcal{A}(12|21)e^{ip_2 q_1 + ip_1 q_2}, & q_1 \ll q_2, \\ \mathcal{A}(21|12)e^{ip_2 q_1 + ip_1 q_2} + \mathcal{A}(21|21)e^{ip_1 q_1 + ip_2 q_2}, & q_1 \gg q_2. \end{cases} \tag{4.45}$$

The labels σ and τ are given by permutations of the set $\{12\}$, in particular (12) corresponds to the identity. Parametrising the amplitude $\mathcal{A}(\sigma|\tau)$, these labels describe the position of coordinates and momentum variables in the exponential obtained from

Table 4.3 Asymptotic behaviour of the symmetric Ψ_+ and anti-symmetric Ψ_- wave functions. The S-matrix eigenvalues equal to $A \pm B$ are pure phases

Ψ	$q_1 \ll q_2$	$q_1 \gg q_2$
$\Psi_+(q_1, q_2)$	$e^{ip_1q_1+ip_2q_2} + (A+B)e^{ip_2q_1+ip_1q_2}$	$e^{ip_2q_1+ip_1q_2} + (A+B)e^{ip_1q_1+ip_2q_2}$
$\Psi_-(q_1, q_2)$	$e^{ip_1q_1+ip_2q_2} + (A-B)e^{ip_2q_1+ip_1q_2}$	$-\left[e^{ip_2q_1+ip_1q_2} + (A-B)e^{ip_1q_1+ip_2q_2}\right]$

$e^{ip_1q_1+ip_2q_2}$ by applying the corresponding permutation σ of coordinates and τ of momenta. With these conventions, σ simultaneously labels the sectors, to be exact, 12 corresponds to $q_1 \ll q_2$ and 21 to $q_1 \gg q_2$. Comparing (4.45) with (4.44), we get $c_1 = \mathcal{A}(12|12)$ and $c_2 = \mathcal{A}(21|12)$, together with the following relation between amplitudes

$$\mathcal{A}(12|21) = A\,\mathcal{A}(12|12) + B\,\mathcal{A}(21|12)\,,$$
$$\mathcal{A}(21|21) = B\,\mathcal{A}(12|12) + A\,\mathcal{A}(21|12)\,. \tag{4.46}$$

This shows that from the four amplitudes two are independent and the other two are determined via those and elements of S. Depending on which amplitudes are considered as independent, we obtain various representations for the scattering operator. Below we indicate the three most important ones [4].

(1) *Reflection-diagonal representation* is obtained by simply rewriting formulae (4.46) in the matrix form

$$\begin{pmatrix} \mathcal{A}(12|21) \\ \mathcal{A}(21|21) \end{pmatrix} = S \begin{pmatrix} \mathcal{A}(12|12) \\ \mathcal{A}(21|12) \end{pmatrix}, \tag{4.47}$$

where S is given by (4.42). Acting on the column made of the amplitudes with fixed momenta it produces a column of amplitudes with momenta permuted. If the potential $v(x)$ is impenetrable, there is no transmission and S becomes diagonal in this representation. In general, S has two eigenvectors

$$S \begin{pmatrix} 1 \\ 1 \end{pmatrix} = (A+B) \begin{pmatrix} 1 \\ 1 \end{pmatrix}, \quad S \begin{pmatrix} 1 \\ -1 \end{pmatrix} = (A-B) \begin{pmatrix} 1 \\ -1 \end{pmatrix}, \tag{4.48}$$

with eigenvalues $A \pm B$ giving the asymptotics of the symmetric and anti-symmetric wave functions.

(2) *Transmission-diagonal representation* follows by changing the order of amplitudes in the column featuring in the left hand side of the reflection-diagonal representation above

$$\begin{pmatrix} \mathcal{A}(21|21) \\ \mathcal{A}(12|21) \end{pmatrix} = \begin{pmatrix} B & A \\ A & B \end{pmatrix} \begin{pmatrix} \mathcal{A}(12|12) \\ \mathcal{A}(21|12) \end{pmatrix}. \tag{4.49}$$

The matrix acting on the column of amplitudes becomes diagonal for reflection-less potentials.

(3) *Transfer-matrix representation* arises when we relate amplitudes in different sectors, namely,

$$\begin{pmatrix} \mathcal{A}(21|12) \\ \mathcal{A}(21|21) \end{pmatrix} = \begin{pmatrix} -\frac{A}{B} & \frac{1}{B} \\ \frac{B^2-A^2}{B} & \frac{A}{B} \end{pmatrix} \begin{pmatrix} \mathcal{A}(12|12) \\ \mathcal{A}(12|21) \end{pmatrix}. \tag{4.50}$$

At this point we recall the relation $S(k)S(-k) = \mathbb{1}$, where S is given by (4.42). In components this relation is equivalent to

$$A(k)A(-k) + B(k)B(-k) = 1, \quad A(k)B(-k) + A(-k)B(k) = 0, \tag{4.51}$$

the latter allows one to express

$$B^2(k) - A^2(k) = \frac{B(k)}{B(-k)}, \quad \frac{A(k)}{B(k)} = -\frac{A(-k)}{B(-k)} \tag{4.52}$$

and, therefore, to rewrite formula (4.50) in the form [4]

$$\begin{pmatrix} \mathcal{A}(21|12) \\ \mathcal{A}(21|21) \end{pmatrix} = \frac{1}{B(k)B(-k)} \begin{pmatrix} -A(k)B(-k) & B(-k) \\ B(k) & -A(-k)B(k) \end{pmatrix} \begin{pmatrix} \mathcal{A}(12|12) \\ \mathcal{A}(12|21) \end{pmatrix},$$

where due to relations (4.51) the matrix

$$T = \frac{1}{B(k)B(-k)} \begin{pmatrix} -A(k)B(-k) & B(-k) \\ B(k) & -A(-k)B(k) \end{pmatrix} \tag{4.53}$$

has the properties

$$\operatorname{Tr} T = 0, \quad T^2 = \mathbb{1}. \tag{4.54}$$

The transfer matrix T transfers a collection of amplitudes from one sector into another one.

Before we proceed with multi-body scattering, in the next subsection we explain how to practically compute two-body S-matrices in some integrable models. This will also give a good feeling of what kind of results (special functions) one could expect there. The reader who is more interested in the Factorised Scattering Theory can move directly to Sect. 4.3.

4.2.2 Examples of Two-Body S-Matrices

Our examples include here the S-matrices for the delta-interaction model, the hyperbolic CMS and RS models and the rational degeneration thereof.

Delta-interaction model. For this model Eq. (4.39) reads as

$$-\psi''(x) + \varkappa\delta(x)\psi(x) = \frac{k^2}{4}\psi(x). \tag{4.55}$$

Here $\varkappa$ has the dimension of inverse length, for $\varkappa > 0$ the potential is repulsive, while for $\varkappa < 0$ it is attractive. Only the case of bosons is interesting, because for fermions the odd wave function vanishes at the origin and the potential causes no effect on the otherwise free motion. As a result, $A - B = -1$, that is $\theta_- = 0$, see Table 4.3.

As to a symmetric wave function describing bosons, we first integrate both sides of Eq. (4.55) over a small segment $[-\epsilon, \epsilon]$ and then send $\epsilon \to 0$. Due to continuity of the wave function, we get an equation for the discontinuity of its derivative at the origin

$$\psi'(+0) - \psi'(-0) = \varkappa\psi(0), \tag{4.56}$$

which is equivalent to $2\psi'(+0) = \varkappa\psi(0)$ because the wave function is symmetric, $\psi(-x) = \psi(x)$. For $x > 0$ the motion is free and the wave function is

$$\psi(x) = e^{-ikx/2} + (A + B)e^{ikx/2}. \tag{4.57}$$

Using this $\psi(x)$, from the discontinuity equation we find

$$A + B = -e^{-i\theta_+} = \frac{k - i\varkappa}{k + i\varkappa}, \tag{4.58}$$

which yields in turn

$$\theta_+(k) = -2\arctan\frac{k}{\varkappa}. \tag{4.59}$$

Combining this result with $A - B = -1$, we separate the reflection and transmission amplitudes

$$A = -\frac{i\varkappa}{k + i\varkappa}, \quad B = \frac{k}{k + i\varkappa}. \tag{4.60}$$

The *Tonks-Girardeau limit* $\varkappa \to \infty$ corresponds to the case of impenetrable bosons because in this limit the transmission coefficient B vanishes, while the reflection coefficient A tends to -1.

Further, we observe that in the attractive case $\varkappa < 0$, the S-matrix (4.60) has a pole on the positive imaginary axis at $k = i|\varkappa|$. For k around this pole the wave function (4.57) is dominated by the second term $\psi(x) \approx (A + B)e^{-|\varkappa|x/2}$. This exponentially suppressed wave function corresponds to a bound state with energy $E = -\varkappa^2/4$.[5]

Hyperbolic CMS model. Equation (4.39) takes the form

$$- \psi''(x) + \frac{\gamma(\gamma - 1)}{4\ell^2 \sinh^2 \frac{x}{2\ell}} \psi(x) = \frac{k^2}{4} \psi(x) , \qquad (4.61)$$

where $\hbar = m = c = 1$. In the following we also set $\ell = 1$. For $0 < \gamma < 1$ the potential is negative (attractive) and for $\gamma > 1$ it is positive (repulsive). It is impenetrable implying vanishing of the wave-function at the origin. In the case of an anti-symmetric wave function we, as usual, describe *fermions*, while for a symmetric function we deal with *hard-core bosons*. Since the potential is impenetrable, there is no transmission, such that there is only one non-trivial reflection amplitude A.

To solve (4.61), we introduce a new variable

$$z = 1/(1 - e^x) .$$

For $0 < x < \infty$, the variable z ranges as $-\infty < z < 0$ and, therefore, for these values of x we have

$$\sinh \frac{x}{2} = \frac{1}{2\sqrt{-z}\sqrt{1 - z}} .$$

Analogously, for $-\infty < x < 0$ one has $1 < z < \infty$ and, the corresponding relation reads as

$$\sinh \frac{x}{2} = -\frac{1}{2\sqrt{z}\sqrt{z - 1}} .$$

For definiteness we restrict ourselves to the interval $0 < x < \infty$. For $x < 0$ the wave function is defined $\psi(-|x|) = \pm\psi(|x|)$, where the plus sign is for bosons and the minus for fermions.

Performing now the substitution

$$\psi(x) = (2 \sinh \tfrac{x}{2})^{-ik} f(z) = (-z)^{ik/2}(1 - z)^{ik/2} f(z) = (z(z - 1))^{ik/2} f(z) , \quad (4.62)$$

one obtains the equation

$$z(1 - z) f''(z) + (ik + 1 - 2(ik + 1)z) f'(z) - (ik + \gamma)(ik + 1 - \gamma) f(z) = 0 .$$

This has the form of the standard hypergeometric equation

---

[5] Another way to argue is to change the normalisation of the wave function (4.57) to get $\psi(x) = (A + B)^{-1}e^{-ikx/2} - e^{ikx/2}$. A pole of $A + B$ is the same as a zero of $(A + B)^{-1}$ and for $k = i|\varkappa|$ we recover a normalisable wave function $e^{-|\varkappa|x/2}$.

$$z(1-z)u''(z) + (c - (a+b+1)z)u'(z) - abu(z) = 0$$

with

$$a = ik + \gamma, \quad b = ik + 1 - \gamma, \quad c = ik + 1.$$

The two independent solutions are given in terms of the hypergeometric function $_2F_1(a, b, c; z)$

$$
\begin{aligned}
f_1(z) &= {}_2F_1(\gamma + ik, 1 - \gamma + ik, 1 + ik; z), \\
f_2(z) &= (-z)^{-ik}{}_2F_1(\gamma, 1 - \gamma, 1 - ik; z).
\end{aligned}
\tag{4.63}
$$

Recall that $_2F_1(z)$ is a single-valued analytic function in the complex z-plane with a cut along $(1, +\infty)$. We note also that

$$f_1(z) = (1-z)^{-ik}{}_2F_1(\gamma, 1 - \gamma, 1 + ik; z).$$

For $\gamma = 0$ and $\gamma = 1$ the potential vanishes, i.e. we deal with free theory, where solutions (4.63) reduce to $f_1(z) = (1-z)^{-ik}$ and $f_2 = (-z)^{-ik}$. According to (4.62), the corresponding $\psi(x)$ boils down to free waves $e^{\pm ikx/2}$, as expected.

For $x \to +0$, we have $z \approx -1/x \to -\infty$. Taking into account the asymptotic behaviour of the hypergeometric function in this limit

$$
{}_2F_1(a, b, c; z) \approx \frac{\Gamma(c)\Gamma(b-a)}{\Gamma(b)\Gamma(c-a)}(-z)^{-a} + \frac{\Gamma(c)\Gamma(a-b)}{\Gamma(a)\Gamma(c-b)}(-z)^{-b}, \tag{4.64}
$$

we obtain the asymptotics of our solutions for large negative z

$$
\begin{aligned}
f_1(z) &\approx \frac{\Gamma(ik+1)\Gamma(1-2\gamma)}{\Gamma(ik+1-\gamma)\Gamma(1-\gamma)}(-z)^{-ik-\gamma} + \frac{\Gamma(ik+1)\Gamma(2\gamma-1)}{\Gamma(ik+\gamma)\Gamma(\gamma)}(-z)^{\gamma-1-ik}, \\
f_2(z) &\approx \frac{\Gamma(1-ik)\Gamma(1-2\gamma)}{\Gamma(1-\gamma)\Gamma(1-\gamma-ik)}(-z)^{-ik-\gamma} + \frac{\Gamma(1-ik)\Gamma(2\gamma-1)}{\Gamma(\gamma)\Gamma(\gamma-ik)}(-z)^{\gamma-1-ik}.
\end{aligned}
$$

The second term in the asymptotic expressions for both f_1 and f_2 diverges for $\gamma > 1$. For $\gamma < 1$ we deal with an attractive potential that is singular. A proper treatment consists in replacing the potential near the origin by any regular potential. Then, removing the regularisation singles out a unique solution, as was shown for $1/x^2$ potential in [7, 10]. In our present case, this unique solution is the one which does not contain the term $(-z)^{\gamma-1}$ in the asymptotic expansion around infinity, and it is given by the following linear combination

$$f(z) = f_1(z) - \frac{\Gamma(1+ik)\Gamma(\gamma-ik)}{\Gamma(1-ik)\Gamma(\gamma+ik)}f_2(z). \tag{4.65}$$

Now we look at the behaviour of this solution for $x \to +\infty$, corresponding to $z \approx -e^{-x} \to -0$.[6] Restoring from (4.62) the corresponding asymptotics of $\psi(x)$

$$\psi(x) \approx e^{-ikx/2} - \frac{\Gamma(1+ik)\Gamma(\gamma-ik)}{\Gamma(1-ik)\Gamma(\gamma+ik)} e^{ikx/2}, \qquad (4.66)$$

we find the reflection amplitude $A = -e^{-i\theta}$ with the phase [4]

$$\theta = i \log\left[\frac{\Gamma(1+ik)\Gamma(\gamma-ik)}{\Gamma(1-ik)\Gamma(\gamma+ik)}\right]. \qquad (4.67)$$

In the limiting cases, for $\gamma = 0$ (free bosons) this formula gives $\theta = -\pi$ and for $\gamma = 1$ (free fermions or hard-core bosons) it yields $\theta = 0$. The amplitude $A(k)$, originally derived for $k > 0$, is in fact an analytic function of k without poles in the upper-half plane (no bound states).

It is instructive to restore in (4.67) the physical units and the length parameter ℓ. A simple dimensional analysis shows that this is done by performing in (4.67) the substitutions $\gamma \to \gamma/\hbar$, $k \to k\ell/\hbar$ and $\theta \to \theta/\hbar$. This yields

$$\theta = i\hbar \log\left[\frac{\Gamma\left(1+i\frac{k\ell}{\hbar}\right)\Gamma\left(\frac{\gamma}{\hbar}-i\frac{k\ell}{\hbar}\right)}{\Gamma\left(1-i\frac{k\ell}{\hbar}\right)\Gamma\left(\frac{\gamma}{\hbar}+i\frac{k\ell}{\hbar}\right)}\right], \qquad (4.68)$$

where k has the physical dimension of momentum. Further, we note the following integral representation[7]

$$\frac{\Gamma\left(1+i\frac{k\ell}{\hbar}\right)\Gamma\left(\frac{\gamma}{\hbar}-i\frac{k\ell}{\hbar}\right)}{\Gamma\left(1-i\frac{k\ell}{\hbar}\right)\Gamma\left(\frac{\gamma}{\hbar}+i\frac{k\ell}{\hbar}\right)} = \exp\left[2i\int_0^\infty \frac{dt}{t} \frac{\sinh t(1-\frac{\gamma}{\hbar})\,e^{-\frac{\gamma}{\hbar}t}}{\sinh t} \sin 2\frac{k\ell}{\hbar}t\right]. \qquad (4.69)$$

To exhibit an analytic structure of the S-matrix (4.69), we employ the Weierstrass representation for the gamma function

$$\Gamma(z) = \frac{e^{-\gamma_{\text{EM}}}}{z} \prod_{n=1}^\infty \left(1+\frac{z}{n}\right)^{-1} e^{z/n}, \qquad (4.70)$$

valid in the whole complex plane except the non-positive integers. Here γ_{EM} is the Euler-Mascheroni constant. We then obtain the following representation

---

[6]The asymptotics $x \to -\infty$ would correspond to $z \to 1$, but this does not concern us here.

[7]It stems from the following formula, see e.g. Appendix A of [11],

$$\frac{\Gamma(s+(1+\sigma+\tau)/2)\Gamma(-s+(1-\sigma+\tau)/2)}{\Gamma(-s+(1+\sigma+\tau)/2)\Gamma(s+(1-\sigma+\tau)/2)} = \exp\left[2\int_0^\infty \frac{dt}{t} \frac{\sinh \sigma t \sinh 2st}{\sinh t} e^{-\tau t}\right]$$

valid for $s \in i\mathbb{R}$ and $\text{Re}\,t - |\text{Re}\lambda| > -1$.

$$\frac{\Gamma\left(1 + i\frac{k\ell}{\hbar}\right)\Gamma\left(\frac{\gamma}{\hbar} - i\frac{k\ell}{\hbar}\right)}{\Gamma\left(1 - i\frac{k\ell}{\hbar}\right)\Gamma\left(\frac{\gamma}{\hbar} + i\frac{k\ell}{\hbar}\right)} = -\prod_{n=0}^{\infty} \frac{(n - i\frac{k\ell}{\hbar})(n + \frac{\gamma}{\hbar} + i\frac{k\ell}{\hbar})}{(n + i\frac{k\ell}{\hbar})(n + \frac{\gamma}{\hbar} - i\frac{k\ell}{\hbar})}. \tag{4.71}$$

In the upper-half plane of complex k this expression exhibits an infinite number of kinematic poles (independent of γ) at $k = in\hbar/\ell$ and an infinite number of zeros at $k = i(n\hbar + \gamma)/\ell$, where in both cases $n = 0, 1, \ldots, \infty$.

We conclude by considering the limiting cases. First, the rational degeneration is obtained taking the dimensionless parameter $k\ell/\hbar$ to be large in comparison to $\gamma/\hbar$, which corresponds to $\ell \to \infty$. In this limit (4.68) gives

$$\theta = \pi(\gamma - \hbar)\operatorname{sign}(k), \tag{4.72}$$

For $k > 0$ the derivative $\delta(k) = \theta'(k)$ vanishes, as for the corresponding classical case, c.f. the first formula in (4.17).

Concerning the classical limit of (4.68), it is clear that it corresponds to taking the "action" $k\ell$ and the coupling γ to be large in comparison to $\hbar$. Taking the limit, we find

$$\theta \approx k\ell \log\left(1 + \frac{\gamma^2}{k^2\ell^2}\right) + 2\gamma \arctan\frac{k\ell}{\gamma} - \frac{\pi\hbar}{2}\operatorname{sign}(k).$$

The first two terms here represent the integrated classical phase shift corresponding to $\delta(k)$ in (4.18). The last term goes beyond the classical result, i.e. it is a correction, but we keep it to numerically match the asymptotic expression for $e^{-i/\hbar\theta}$ with the actual result in terms of gamma functions, as the reader can verify.

Hyperbolic RS model. Determining the scattering matrix for the hyperbolic RS model is rather non-trivial, mainly, because the Schrödinger differential equation is now replaced by a complicated analytic difference equation that brings along many new and subtle analytic questions. Construction of the solution basis relies on some asymptoticity assumptions and entails the use of the quantum Ruijsenaars duality, the classical counterpart of the duality that was mentioned in Sect. 2.2.1. Leaving most of this fascinating subject to the reader, here we only formulate the difference equation and present the result on the scattering matrix due to Ruijsenaars [8, 11, 12].

Recall that the Hamiltonian and momentum operators of the RS model are given by (3.203), where $S_{\pm 1}$ belong to a commutative family of difference operators (3.204), where for the hyperbolic case the functions $h_{\pm}$ are

$$h_{\pm}(q) = \left[\frac{\sinh\frac{1}{2\ell}(q \pm i\gamma/\mu)}{\sinh\frac{1}{2\ell}q}\right]^{1/2}, \quad h_+(-q) = h_-(q). \tag{4.73}$$

and the shift operators $T_k^{\pm} = e^{\pm i\frac{\hbar}{\mu}\frac{\partial}{\partial q_k}}$ act on functions $f(q_1, \ldots, q_N)$ as

$$(T_k^{\pm} f)(q_1, \ldots, q_N) = f(q_1, \ldots q_{k-1}, q_k \pm i\tfrac{\hbar}{\mu}, q_{k+1}, \ldots, q_N).$$

For the present discussion we prefer to explicitly keep all the physical parameters, in particular, the parameter $\mu \equiv mc$. Note the periodicity property $h_{\pm}(q + 2\pi i \ell) = h_{\pm}(q)$.

For the two-body problem we are interested in here, the operators $S_{\pm 1}$ are

$$S_{+1} = h_-(x) T_1^+ h_+(x) + h_+(x) T_2^+ h_-(x),$$
$$S_{-1} = h_+(x) T_1^- h_-(x) + h_-(x) T_2^- h_+(x),$$

where, as usual, we introduced $x = q_2 - q_1$ and $X = q_1 + q_2$. Since

$$\frac{\partial}{\partial q_1} = \frac{\partial}{\partial X} - \frac{\partial}{\partial x}, \quad \frac{\partial}{\partial q_2} = \frac{\partial}{\partial X} + \frac{\partial}{\partial x},$$

the operators $S_{\pm 1}$ are

$$S_{\pm 1} = \mathcal{H} \, e^{\pm i \frac{\hbar}{\mu} \frac{\partial}{\partial X}},$$

where we have introduced the reduced Hamiltonian

$$\mathcal{H} = h_-(x) e^{-i\frac{\hbar}{\mu}\frac{\partial}{\partial x}} h_+(x) + h_+(x) e^{i\frac{\hbar}{\mu}\frac{\partial}{\partial x}} h_-(x). \tag{4.74}$$

For the wave function $\Psi(q_1, q_2)$ that diagonalises the operators $S_{\pm 1}$ and, therefore, the Hamiltonian and the total momentum, we naturally assume the factorised form (4.38), $\Psi(q_1, q_2) = e^{\frac{i}{\hbar} K X/2} \psi(x)$. Then, $S_{\pm 1}\Psi = \varkappa e^{\mp K/2\mu}\Psi$, where $\varkappa$ is an eigenvalue of $\mathcal{H}$

$$\mathcal{H}\psi(x) = \varkappa \psi(x). \tag{4.75}$$

In the following it is convenient to use instead of K and $\varkappa$, the new variables p_1 and p_2, so that
$$K = p_1 + p_2, \quad \varkappa = 2\cosh \tfrac{k}{2\mu}.$$

where, as usual, $k = p_1 - p_2$. The variables, p_1 and p_2 will play the role of the asymptotic momenta. The corresponding rescaled (dimensionless) eigenvalues of energy and momentum then read

$$E = 2\cosh \tfrac{k}{2\mu} \cosh \tfrac{K}{2\mu}, \quad P = 2\cosh \tfrac{k}{2\mu} \sinh \tfrac{K}{2\mu}. \tag{4.76}$$

In this way the spectral problem is reduced to a linear second-order analytic difference equation [13]

$$G_+(x)\psi\big(x + i\tfrac{\hbar}{\mu}\big) + G_-(x)\psi\big(x - i\tfrac{\hbar}{\mu}\big) = 2\cosh \tfrac{k}{2\mu} \psi(x) \tag{4.77}$$

with coefficients

$$G_\pm(x) = h_\pm(x)h_\mp(x \pm i\hbar/\mu) = \left[\frac{\sinh \frac{1}{2\ell}(x \pm i\frac{\gamma}{\mu})\, \sinh \frac{1}{2\ell}(x \mp i\frac{\gamma}{\mu} \pm i\frac{\hbar}{\mu})}{\sinh \frac{1}{2\ell}x\; \sinh \frac{1}{2\ell}(x \pm i\frac{\hbar}{\mu})} \right]^{\frac{1}{2}}. \quad (4.78)$$

In comparison to a second-order differential equation where the parameter space of solutions is two-dimensional, the space of solutions of (4.77) is infinite-dimensional. Indeed, multiplying a solution by any periodic function with period $i\hbar/\mu$ gives another solution. Evidently, special properties of solutions should now play an important role in singling them out.

Concerning specific features of Eq. (4.77), we note the following: the coefficients $G_\pm$ are periodic functions with the period $2\pi i\ell$. There are two distinguished cases $\gamma = 0$ and $\gamma = \hbar$. For these cases $G_\pm(x) = 1$ and Eq. (4.77) becomes free, quite similar to the Schrödinger differential equation for the CMS model. In the free case, Eq. (4.77) is solved by the Fourier transform and the fundamental solutions are $e^{\pm ikx/(2\mu\hbar)}$. Also, for large x, $G_\pm(x) \to 1$, and we are interested in solutions with the same scattering asymptotics $e^{\pm ikx/(2\mu\hbar)}$. Equation (4.77), its solutions, and, therefore, the scattering matrix, are controlled by three distinct dimensionless parameters for which we can take, for instance, $\gamma/\hbar$, $k\ell/\hbar$ and $\hbar/\mu\ell$.

Discussion of the scattering solutions of (4.77) is beyond the scope of this book. We, therefore, restrict ourselves to presenting the resulting scattering matrix which, according to [11], admits the following integral representation

$$e^{-\frac{i}{\hbar}\theta(k)} = \exp\left[2i \int_0^\infty \frac{dt}{t} \frac{\sinh t(1 - \frac{\gamma}{\hbar})\, \sinh t(2\pi\frac{\mu\ell}{\hbar} - \frac{\gamma}{\hbar})}{\sinh t\, \sinh 2\pi\frac{\mu\ell}{\hbar}t} \sin 2\frac{k\ell}{\hbar}t \right]. \quad (4.79)$$

This formula can be viewed as a deformation of (4.69) depending on an additional parameter $\hbar/\mu\ell$. Indeed, in the non-relativistic limit $c \to \infty$, corresponding to $\frac{\mu\ell}{\hbar} \to \infty$, Eq. (4.79) goes into the integral representation (4.69) for the scattering amplitude of the hyperbolic CMS model. Further, performing in (4.79) a change of the integration variable $t \to \frac{\hbar}{\mu\ell}t$, for the derivative of the phase we can write

$$\theta'(k) = 4\ell x \int_0^\infty dt \frac{\sinh t(\frac{\gamma}{\mu\ell} - x)\, \sinh t(2\pi - \frac{\gamma}{\mu\ell})}{\sinh xt\, \sinh 2\pi t} \cos \frac{2k}{\mu}t, \quad x \equiv \frac{\hbar}{\mu\ell}. \quad (4.80)$$

The classical limit $\hbar \to 0$ corresponds to $x \to 0$. In this limit the integral above yields[8]

$$\lim_{\hbar \to 0} \theta'(k) = 4\ell \int_0^\infty \frac{dt}{t} \frac{\sinh \frac{\gamma}{\mu\ell}t\, \sinh t(2\pi - \frac{\gamma}{\mu\ell})}{\sinh 2\pi t} \cos \frac{2k}{\mu}t = \ell \log\left[1 + \frac{\sin^2 \frac{\gamma}{2\mu\ell}}{\sinh^2 \frac{k}{2\mu}} \right],$$

thereby reproducing the classical result (4.36).

---

[8]Consult, for instance, formula (A.44) in [11].

As was shown in [11], in the rational limit $\ell \to \infty$ formula (4.79) yields the same result as for the rational CMS model, where the phase is given by (4.72).

Among other results, we point out a representation of the scattering phase in terms of an infinite product [11]

$$\theta(k) = i\hbar \log \left[\frac{\Gamma\left(1 + i\frac{k\ell}{\hbar}\right)\Gamma\left(\frac{\gamma}{\hbar} - i\frac{k\ell}{\hbar}\right)}{\Gamma\left(1 - i\frac{k\ell}{\hbar}\right)\Gamma\left(\frac{\gamma}{\hbar} + i\frac{k\ell}{\hbar}\right)} \prod_{n=1}^{\infty} \gamma_n \right], \tag{4.81}$$

where

$$\gamma_n = \frac{\Gamma\left(1 + i\frac{k\ell}{\hbar} + 2\pi n\frac{\mu\ell}{\hbar}\right)\Gamma\left(i\frac{k\ell}{\hbar} + 2\pi n\frac{\mu\ell}{\hbar}\right)}{\Gamma\left(1 - i\frac{k\ell}{\hbar} + 2\pi n\frac{\mu\ell}{\hbar}\right)\Gamma\left(-i\frac{k\ell}{\hbar} + 2\pi n\frac{\mu\ell}{\hbar}\right)} \frac{\Gamma\left(\frac{\gamma}{\hbar} - i\frac{k\ell}{\hbar} + 2\pi n\frac{\mu\ell}{\hbar}\right)\Gamma\left(1 - i\frac{k\ell}{\hbar} - \frac{\gamma}{\hbar} + 2\pi n\frac{\mu\ell}{\hbar}\right)}{\Gamma\left(\frac{\gamma}{\hbar} + i\frac{k\ell}{\hbar} + 2\pi n\frac{\mu\ell}{\hbar}\right)\Gamma\left(1 + i\frac{k\ell}{\hbar} - \frac{\gamma}{\hbar} + 2\pi n\frac{\mu\ell}{\hbar}\right)}.$$

For a generic value of γ a representation for (4.79) in terms of elementary functions is unknown. Remarkably, however, for γ parametrised by two integers (M, S) as

$$\gamma = (M + 1)\hbar - 2\pi\mu\ell S, \quad M \geqslant 1, \ S \geqslant 0, \tag{4.82}$$

the expression (4.79) turns into [11]

$$e^{-\frac{i}{\hbar}\theta} = (-1)^{M+S} \prod_{m=1}^{M} \frac{\sinh \frac{\hbar}{2\mu\ell}\left(\frac{k\ell}{\hbar} + im\right)}{\sinh \frac{\hbar}{2\mu\ell}\left(\frac{k\ell}{\hbar} - im\right)} \prod_{s=1}^{S} \frac{\sinh 2\pi\left(\frac{k\ell}{\hbar} + i\pi\frac{\mu\ell}{\hbar}s\right)}{\sinh 2\pi\left(\frac{k\ell}{\hbar} - i\pi\frac{\mu\ell}{\hbar}s\right)}. \tag{4.83}$$

For $S = 0$ the coupling γ is integer-valued in units of $\hbar$. If a pair (M, S) runs over all the values allowed in (4.82), the point set of γ-values is dense.

4.3 Factorised Scattering

As is already clear from our consideration of two-particle scattering, permutations provide an efficient and physically appealing way to parametrise partial amplitudes of the wave function. In this respect, passing to multi-particle scattering brings into light the symmetric group $\mathfrak{S}_N$. As a preparation for the discussion of the Bethe wave function in the next section, we find it convenient to start by recalling some standard facts about $\mathfrak{S}_N$.

4.3.1 Symmetric Group $\mathfrak{S}_N$

The symmetric group $\mathfrak{S}_N$ is defined as the set of all one-to-one mappings of the set of numbers $\{1, \ldots, N\}$ to itself. Every element $\sigma \in \mathfrak{S}_N$, called permutation, can be written in two-line notation as

$$\sigma = \begin{pmatrix} 1 & 2 & \cdots & N \\ \sigma(1) & \sigma(2) & & \sigma(N) \end{pmatrix},$$

meaning that σ maps 1 to $\sigma(1)$, 2 to $\sigma(2)$ and so on. Since in two-line notation the top line is fixed, one can drop it obtaining a one-line notation. The product $\sigma\tau$ of two permutations σ and τ is constructed as follows. In the τ-string one takes an integer standing in the position $\sigma(j)$ and moves it in the position j of the product $\sigma\tau$. For instance, given two permutations $\sigma = (132)$ and $\tau = (213)$ in $\mathfrak{S}_3$, written in one-line notation, one has for their product[9]

$$(132)(213) = (231).$$

In the two-line notation the same product looks like

$$\begin{pmatrix} 1 & 2 & 3 \\ 1 & 3 & 2 \end{pmatrix} \begin{pmatrix} 1 & 2 & 3 \\ 2 & 1 & 3 \end{pmatrix} = \begin{pmatrix} 1 & 2 & 3 \\ 2 & 3 & 1 \end{pmatrix}.$$

As the flow of indices shows, this result means that $(\sigma\tau)(j) = \tau(\sigma(j))$, that is it corresponds to the application to the index j of σ followed by τ.

Next, we need a transposition α_{ij} that interchanges the positions of i and j and leaves all the other elements unchanged

$$\alpha_{ij} \equiv (i|j) = \begin{pmatrix} 1 & & i & & j & & N \\ 1 & \cdots & j & \cdots & i & \cdots & N \end{pmatrix}.$$

Multiplication of σ by α_{ij} from the right exchanges the positions of $\sigma(i)$ and $\sigma(j)$. Thus,

$$\alpha_{ij}\sigma = (\sigma(1), \ldots, \sigma(j), \ldots, \sigma(i), \ldots, \sigma(N)).$$

Transpositions satisfy $\alpha_{ij}^2 = e$, where e is the identity, and the relations

$$\alpha_{kj}\alpha_{ik} = \alpha_{ik}\alpha_{ij}, \quad i \neq j. \tag{4.84}$$

Every permutation can be expressed as a product of transpositions, albeit not in a unique way. For a given permutation $\sigma \in \mathfrak{S}_N$ the number of transpositions in its decomposition is always either even or odd. This allows to define the *sign function*, also called *parity*, on $\mathfrak{S}_N$: $\text{sign}(\sigma) = 1$, if σ is given by an even number of transpositions, and $\text{sign}(\sigma) = -1$, if the corresponding number is odd. The sign function has the properties

$$\text{sign}(\sigma\tau) = \text{sign}(\sigma)\text{sign}(\tau), \quad \text{sign}(\sigma^{-1}) = \text{sign}(\sigma).$$

---

[9]To display a permutation σ in the one-line notation, we confine the corresponding sequence of numbers from the set $\{1, 2, \ldots, N\}$ within the brackets $(\ldots)$. This notation should not be confused with the one used to represent σ via its *cycles*. In this book we never use cycle notation.

Importantly, the group $\mathfrak{S}_N$ is generated by simple transpositions $\alpha_j \equiv \alpha_{jj+1}$, $i = 1, \ldots, N - 1$, subject to the Coxeter relations

$$
\begin{aligned}
\alpha_j^2 &= e\,, & 1 &\leqslant j \leqslant N - 1\,, \\
\alpha_j \alpha_{j+1} \alpha_j &= \alpha_{j+1} \alpha_j \alpha_{j+1}\,, & 1 &\leqslant j \leqslant N - 2\,, \quad (4.85) \\
\alpha_i \alpha_j &= \alpha_j \alpha_i\,, & 1 &\leqslant i, j \leqslant N - 1 \text{ and } |i - j| \geqslant 2\,.
\end{aligned}
$$

It is useful to have in mind that the invariance of the euclidean scalar product under the action of $\sigma, \tau \in \mathfrak{S}_N$ implies that

$$
\sum_j q_{\sigma(j)} p_{\tau(j)} = \sum_j q_j p_{(\sigma^{-1}\tau)(j)} = \sum_j q_{(\tau^{-1}\sigma)(j)} p_j\,.
$$

Let $\mathrm{Fun}(\mathfrak{S}_N)$ be the algebra of functions on $\mathfrak{S}_N$. The left π and the right π' regular representations of $\mathfrak{S}_N$ are defined as

$$
\pi(\sigma_0)\mathcal{A}(\sigma) = \mathcal{A}(\sigma_0^{-1}\sigma)\,, \tag{4.86}
$$

$$
\pi'(\sigma_0)\mathcal{A}(\sigma) = \mathcal{A}(\sigma\sigma_0)\,, \tag{4.87}
$$

for $\mathcal{A} \in \mathrm{Fun}(\mathfrak{S}_N)$. The left and right regular representations are equivalent: $\pi'\mathfrak{I} = \mathfrak{I}\pi$, where an intertwining operator $\mathfrak{I}$ acts as $(\mathfrak{I}\mathcal{A})(\sigma) = \mathcal{A}(\sigma^{-1})$.

The left (right) regular representation is decomposed into a sum of irreducible representations according to

$$
\pi \simeq \bigoplus_\lambda \dim \pi_\lambda \cdot \pi_\lambda\,, \tag{4.88}
$$

where π_λ is the irreducible representation of $\mathfrak{S}_N$ corresponding to a partition λ of N and $\dim \pi_\lambda$ is the dimension of this representation. The sum runs over all partitions of N. According to (4.88), the multiplicity with which a representation π_λ appears in the decomposition of the regular representation is equal to $\dim \pi_\lambda$. This dimension is given by the determinant formula

$$
\dim \pi_\lambda = N! \det_{l \times l} \frac{1}{(\lambda_i - i + j)!}\,, \tag{4.89}
$$

where $i, j = 1, \ldots, l$ and $\lambda = [\lambda_1, \lambda_2, \ldots, \lambda_l]$ is the associated partition or Young diagram.

Example. As an illustrative example for later use, we consider the decomposition (4.88) for $\mathfrak{S}_3$. There are 6 permutations which we enumerate as

$$
\sigma_1 = (123) = e\,, \ \sigma_2 = (213)\,, \ \sigma_3 = (231)\,, \ \sigma_4 = (321)\,, \ \sigma_5 = (312)\,, \ \sigma_6 = (132)\,. \tag{4.90}
$$

The corresponding multiplication table $\sigma_i \sigma_j$, where i and j enumerate its rows and columns, respectively, looks as

	σ_1	σ_2	σ_3	σ_4	σ_5	σ_6
σ_1	σ_1	σ_2	σ_3	σ_4	σ_5	σ_6
σ_2	σ_2	σ_1	σ_4	σ_3	σ_6	σ_5
σ_3	σ_3	σ_6	σ_5	σ_2	σ_1	σ_4
σ_4	σ_4	σ_5	σ_6	σ_1	σ_2	σ_3
σ_5	σ_5	σ_4	σ_1	σ_6	σ_3	σ_2
σ_6	σ_6	σ_3	σ_2	σ_5	σ_4	σ_1

In the basis $\mathbf{g}_i \equiv \mathcal{A}(\sigma_i)$ the representation π is realised by the following 6×6 real orthogonal matrices

$$
\pi(\sigma_1) = \mathbb{1}, \qquad
\pi(\sigma_2) = \begin{pmatrix} 0&1&0&0&0&0 \\ 1&0&0&0&0&0 \\ 0&0&0&1&0&0 \\ 0&0&1&0&0&0 \\ 0&0&0&0&0&1 \\ 0&0&0&0&1&0 \end{pmatrix}, \quad
\pi(\sigma_3) = \begin{pmatrix} 0&0&0&0&1&0 \\ 0&0&0&1&0&0 \\ 1&0&0&0&0&0 \\ 0&0&0&0&0&1 \\ 0&0&1&0&0&0 \\ 0&1&0&0&0&0 \end{pmatrix},
$$

$$
\pi(\sigma_4) = \begin{pmatrix} 0&0&0&1&0&0 \\ 0&0&0&0&1&0 \\ 0&0&0&0&0&1 \\ 1&0&0&0&0&0 \\ 0&1&0&0&0&0 \\ 0&0&1&0&0&0 \end{pmatrix}, \quad
\pi(\sigma_5) = \begin{pmatrix} 0&0&1&0&0&0 \\ 0&0&0&0&0&1 \\ 0&0&0&0&1&0 \\ 0&1&0&0&0&0 \\ 1&0&0&0&0&0 \\ 0&0&0&1&0&0 \end{pmatrix}, \quad
\pi(\sigma_6) = \begin{pmatrix} 0&0&0&0&0&1 \\ 0&0&1&0&0&0 \\ 0&1&0&0&0&0 \\ 0&0&0&0&1&0 \\ 0&0&0&1&0&0 \\ 1&0&0&0&0&0 \end{pmatrix},
$$

and, hence, this representation is unitary. Transpositions are σ_2, σ_4 and σ_6. They are realised by symmetric matrices. In particular, the simple transpositions are $\alpha_1 = \sigma_2$ and $\alpha_2 = \sigma_6$. The operator $\mathfrak{I}$ that intertwines π' and π, $\mathfrak{I}\pi' = \pi\mathfrak{I}$, is

$$
\mathfrak{I} = \begin{pmatrix} 1&0&0&0&0&0 \\ 0&1&0&0&0&0 \\ 0&0&0&0&1&0 \\ 0&0&0&1&0&0 \\ 0&0&1&0&0&0 \\ 0&0&0&0&0&1 \end{pmatrix}, \quad \mathfrak{I}^2 = \mathbb{1}. \tag{4.91}
$$

The matrix realisation of π' on the same basis is

$$\pi'(\sigma_1) = \mathbb{1}, \qquad \pi'(\sigma_2) = \begin{pmatrix} 0\,1\,0\,0\,0\,0 \\ 1\,0\,0\,0\,0\,0 \\ 0\,0\,0\,0\,0\,1 \\ 0\,0\,0\,0\,1\,0 \\ 0\,0\,0\,1\,0\,0 \\ 0\,0\,1\,0\,0\,0 \end{pmatrix}, \quad \pi'(\sigma_3) = \begin{pmatrix} 0\,0\,1\,0\,0\,0 \\ 0\,0\,0\,1\,0\,0 \\ 0\,0\,0\,0\,1\,0 \\ 0\,0\,0\,0\,0\,1 \\ 1\,0\,0\,0\,0\,0 \\ 0\,1\,0\,0\,0\,0 \end{pmatrix},$$

$$\pi'(\sigma_4) = \begin{pmatrix} 0\,0\,0\,1\,0\,0 \\ 0\,0\,1\,0\,0\,0 \\ 0\,1\,0\,0\,0\,0 \\ 1\,0\,0\,0\,0\,0 \\ 0\,0\,0\,0\,0\,1 \\ 0\,0\,0\,0\,1\,0 \end{pmatrix}, \quad \pi'(\sigma_5) = \begin{pmatrix} 0\,0\,0\,0\,1\,0 \\ 0\,0\,0\,0\,0\,1 \\ 1\,0\,0\,0\,0\,0 \\ 0\,1\,0\,0\,0\,0 \\ 0\,0\,1\,0\,0\,0 \\ 0\,0\,0\,1\,0\,0 \end{pmatrix}, \quad \pi'(\sigma_6) = \begin{pmatrix} 0\,0\,0\,0\,0\,1 \\ 0\,0\,0\,0\,1\,0 \\ 0\,0\,0\,1\,0\,0 \\ 0\,0\,1\,0\,0\,0 \\ 0\,1\,0\,0\,0\,0 \\ 1\,0\,0\,0\,0\,0 \end{pmatrix},$$

The representation π has three irreducible components: $\lambda = [3]$, $\lambda = [2, 1]$ and $\lambda = [1, 1, 1]$. The first is a trivial (symmetric) representation, the last is a one-dimensional anti-symmetric one. The representation $[2, 1]$ is the two-dimensional standard (defining) representation and it occurs with multiplicity 2. The decomposition (4.88) is obtained by performing a similarity transformation $\pi \to T\pi T^{-1}$, where T is the following unitary matrix

$$T = \frac{1}{\sqrt{6}} \begin{pmatrix} 1 & 1 & 1 & 1 & 1 & 1 \\ 1 & 1 & e^{-\frac{2\pi i}{3}} & e^{\frac{2\pi i}{3}} & e^{\frac{2\pi i}{3}} & e^{-\frac{2\pi i}{3}} \\ 1 & 1 & e^{\frac{2\pi i}{3}} & e^{-\frac{2\pi i}{3}} & e^{-\frac{2\pi i}{3}} & e^{\frac{2\pi i}{3}} \\ -1 & 1 & -e^{-\frac{2\pi i}{3}} & e^{\frac{2\pi i}{3}} & -e^{\frac{2\pi i}{3}} & e^{-\frac{2\pi i}{3}} \\ 1 & -1 & e^{\frac{2\pi i}{3}} & -e^{-\frac{2\pi i}{3}} & e^{-\frac{2\pi i}{3}} & -e^{\frac{2\pi i}{3}} \\ 1 & -1 & 1 & -1 & 1 & -1 \end{pmatrix}.$$

Under the action of T the basis $\mathbf{g} = \{\mathbf{g}_i\}$ of π transforms into

$$T\mathbf{g} = \frac{1}{\sqrt{6}} \begin{pmatrix} \mathbf{g}_1 + \mathbf{g}_2 + \mathbf{g}_3 + \mathbf{g}_4 + \mathbf{g}_5 + \mathbf{g}_6 \\ \mathbf{g}_1 + \mathbf{g}_2 + \mathbf{g}^{-\frac{2\pi i}{3}}(\mathbf{g}_3 + \mathbf{g}_6) + e^{\frac{2\pi i}{3}}(\mathbf{g}_4 + \mathbf{g}_5) \\ \mathbf{g}_1 + \mathbf{g}_2 + e^{-\frac{2\pi i}{3}}(\mathbf{g}_4 + \mathbf{g}_5) + e^{\frac{2\pi i}{3}}(\mathbf{g}_3 + \mathbf{g}_6) \\ -\mathbf{g}_1 + \mathbf{g}_2 - e^{-\frac{2\pi i}{3}}(\mathbf{g}_3 - \mathbf{g}_6) + e^{\frac{2\pi i}{3}}(\mathbf{g}_4 - \mathbf{g}_5) \\ \mathbf{g}_1 - \mathbf{g}_2 - e^{-\frac{2\pi i}{3}}(\mathbf{g}_4 - \mathbf{g}_5) + e^{\frac{2\pi i}{3}}(\mathbf{g}_3 - \mathbf{g}_6) \\ \mathbf{g}_1 - \mathbf{g}_2 + \mathbf{g}_3 - \mathbf{g}_4 + \mathbf{g}_5 - \mathbf{g}_6 \end{pmatrix} \equiv \frac{1}{\sqrt{6}} \begin{pmatrix} v_1 \\ v_2 \\ v_3 \\ v_4 \\ v_5 \\ v_6 \end{pmatrix}.$$

In this basis the matrices of $T\pi T^{-1}$ take a block-diagonal form which corresponds to the decomposition

$$T\pi T^{-1} = \pi_{[3]} \oplus \pi_{[2,1]} \oplus \pi_{[2,1]} \oplus \pi_{[1,1,1]}. \tag{4.92}$$

In particular, v_1 is a projection on the invariant subspace of the trivial representation of $\pi_{[3]}$ and v_6 plays a similar role for the anti-symmetric one-dimensional representation $\pi_{[1,1,1]}$. Analogously, (v_2, v_3) and (v_4, v_5) are invariant subspaces for the two-dimensional representations $\pi_{[2,1]}$. This example shows a clear pattern of how the tensor product decomposition of the regular representation looks like. In particular, for the general case of $\mathfrak{S}_N$, a decomposition of π will always contain trivial and anti-symmetric representations.

4.3.2 Bethe Wave Function and S-Matrix

Now we come to multi-body scattering in quantum mechanics. In the coordinate representation a quantum-mechanical system is described by a multi-variable wave function $\Psi(q_1, \ldots, q_N)$. In the time-independent approach the wave function is a solution of the stationary Schrödinger equation ($\hbar = 1$)

$$-\frac{1}{2m} \sum_{i=1}^{N} \frac{\partial^2}{\partial q_i^2} \Psi(q_1, \ldots, q_N) + \sum_{i \neq j} v(q_i - q_j) \Psi(q_1, \ldots, q_N) = E\Psi(q_1, \ldots, q_N). \quad (4.93)$$

The potential v is translation invariant, so that the total momentum P is conserved, $[H, P] = 0$. For the case of two particles, the corresponding wave function is then searched as a common eigenstate of two commuting operators, H and P, and is naturally labeled by the asymptotic momenta p_1 and p_2. In the scattering process asymptotic momenta are conserved, i.e. incoming and outgoing plane waves are built on the one and the same set of asymptotic momenta.

For a generic potential there are no conservation laws beyond energy and momentum and, as a result, scattering is diffractive if more than two particles are involved. To make this evident, consider scattering for an initial state $|p_1, p_2, p_3\rangle$ corresponding to three particles with fixed asymptotic momenta p_1, p_2 and p_3. For large separation between the particles, this state is described by an incoming wave $\Psi_{\text{in}} \sim e^{i \sum_{i=1}^{3} p_i q_i}$ that solves the free Schrödinger equation. After scattering happens, one expects to find an outgoing wave Ψ_{out}, also given by a superposition of plane waves, albeit with all possible asymptotic momenta permitted by the conservation laws of energy and momentum

$$\Psi_{\text{out}} \sim \sum_{k_1, k_2, k_3} \mathcal{A}(k_1, k_2, k_3) e^{ik_1 q_1 + ik_2 q_2 + ik_3 q_3} \delta\Big(\sum_i k_i - P\Big) \delta\Big(\sum_i E(k_i) - E\Big).$$

where $E(k) = k^2/2m$. Obviously, the two conservation laws are not anymore enough to forbid a continuous distribution of momenta among scattering constituents, hence, diffraction and genuine three-body events may occur.

Conservation laws and Bethe wave function. Let us now assume that our quantum-mechanical model is integrable in the sense that there exists a family of N linearly independent, local in particle momenta, pairwise commuting operators I_m, $m = 1, \ldots, N$, with H and P included in this family. In this case we can search for the wave function as a common solution of N compatible eigenvalue problems

$$I_m \Psi(q_1, \ldots, q_N) = h_m \Psi(q_1, \ldots, q_N), \quad m = 1, \ldots, N. \tag{4.94}$$

Solutions to this system are thus labelled by the set $\{h_1, \ldots, h_N\}$ constituting the common spectrum of the commutative operator family. Further, we assume that I_m are deformations of the conservation laws $I_m^{(0)}(p_i)$ of free theory, the latter being symmetric functions of particle momenta p_i. For instance, for the hyperbolic CMS model I_m's are given by (3.93) or by (3.95) and $I_m^{(0)}$ is obtained from I_m by putting $\gamma = 0$. Our immediate goal is to show that the system (4.94) implies its scattering solutions to have a peculiar asymptotic form compatible with non-diffractive scattering. To simplify the discussion, we assume hereafter that particles are *distinguishable*.

Consider a special kinematic domain where particle coordinates are arranged as

$$q_1 < q_2 < \cdots < q_N. \tag{4.95}$$

Further, consider a special asymptotic regime in which distances between any two neighbouring particles become very large in comparison to the interaction range set by the potential. In this regime the system (4.94) turns into

$$I_m^{(0)} \Psi(q_1, \ldots, q_N) = h_m \Psi(q_1, \ldots, q_N), \quad m = 1, \ldots, N, \tag{4.96}$$

where $I_m^{(0)}$ are free conservation laws. For the corresponding asymptotic form of wave function we make the following ansatz

$$\Psi \sim e^{i p_1 q_1 + \cdots i p_N q_N}, \tag{4.97}$$

with numbers p_i called the asymptotic momenta. Substitution of (4.97) into (4.96) yields a system of equations for the asymptotic momenta

$$I_m^{(0)}(p_i) = h_m, \quad m = 1, \ldots, N. \tag{4.98}$$

Given a set of h_m, this system imposes very tight restrictions on possible values of the asymptotic momenta. Suppose we found a particular solution of (4.98) for which the individual momenta are enumerated according to the ordering pattern

$$p_1 > p_2 > \cdots > p_N. \tag{4.99}$$

Since $I_m^{(0)}$ are assumed to be symmetric functions of the p_i, it is plausible that all the other solution to (4.98) are simply obtained by permutations of the set (4.99). Thus, in the domain (4.95), called the *fundamental sector*, the asymptotic wave function is given by a superposition of plane waves constructed from the set of asymptotic momenta obeying (4.98). Explicitly,

$$\Psi(q_1, \ldots, q_N) = \sum_{\tau \in \mathfrak{S}_N} \mathcal{A}(\tau) e^{iq_1 p_{\tau(1)} + \cdots + iq_N p_{\tau(N)}} , \qquad (4.100)$$

where the sum runs over all permutations τ from the symmetric group $\mathfrak{S}_N$, which act on indices of the asymptotic momenta, the latter form an ordered set according to (4.99).

In general, the configuration space $\mathbb{R}^N$ can be divided into $N!$ disconnected domains, each domain corresponds to a certain ordering of coordinates

$$q_{\sigma(1)} < q_{\sigma(2)} < \cdots < q_{\sigma(N)} , \qquad (4.101)$$

where the latter are labelled by permutations $\sigma \in \mathfrak{S}_N$.

The domain (4.101) will be called σ-*sector*. In particular, $\sigma = e$ corresponds to the fundamental sector (4.95). In each σ-sector we zoom in on the asymptotic region where the difference between any two neighbouring coordinates is very large so that the contribution of the potential terms in (4.94) is negligibly small. Thus, in each σ-sector we will have one and the same asymptotic system (4.96) with the same kind of asymptotic solution (4.100). To deal with all sectors at once, we conveniently parametrise the asymptotic wave function as

$$\Psi(q_1, \ldots, q_N | \sigma) = \sum_{\tau \in \mathfrak{S}_N} \mathcal{A}(\sigma | \tau) e^{iq_{\sigma(1)} p_{\tau(1)} + \cdots + iq_{\sigma(N)} p_{\tau(N)}} , \qquad (4.102)$$

where the complex amplitudes $\mathcal{A}(\sigma | \tau)$ naturally form a $N! \times N!$ matrix depending on the particle momenta. Evidently, this formula can also be written as

$$\Psi(q_1, \ldots, q_N) = \sum_{\tau \in \mathfrak{S}_N} \mathcal{A}(\sigma | \sigma \tau) e^{iq_1 p_{\tau(1)} + \cdots + iq_N p_{\tau(N)}} . \qquad (4.103)$$

In this form the range of q_i is unrestricted. Which σ is picked up on the right hand side is determined from a given coordinate configuration $(q_1, \ldots, q_N)$ by ordering it as (4.101).

The expression (4.102) is the celebrated *Bethe wave function* that goes back to the *Bethe hypothesis* on the form of the wave function in the spin-wave problem [14]. It was introduced by Yang in his work [15] on the delta-interaction Bose gas. Formula (4.102) generalises two-particle expression (4.45) to the case of N particles. The variable σ indicates that coordinates are restricted to lie in the domain (4.101). Different domains contribute with different and a priori unrelated amplitudes.

It is important to realise that the Bethe form of the asymptotic wave function is only possible due to integrability. Far away from the boundaries of a given σ-sector the Schrödinger equation becomes free and has a general solution given by a super-position of free waves. The set of asymptotic momenta is, however, the one and the same for each sector and for any type of scattering wave (incoming, outgoing), because this set is uniquely determined from Eq. (4.98) driven by global spectral invariants h_m. In turn, the fact that any asymptotic wave is determined by the same set of momenta of the incoming wave, up to permutations, means that the scattering process is non-diffractive. Three- and higher-body events that would lead to a continuous redistribution of momenta are prohibited by a sufficiently large number of conservation laws. Further insight into the structure of (4.102) can be thus derived from a relatively simple picture of successive two-body scatterings of classical particles.

Transmission and reflection representations. The fact that we deal with distinguishable particles can be made explicit by indicating the nature of a particle's identity. This can be, for instance, colour, or any other quantum number, like spin or charge. For now we take *colour* as an additional (internal) quantum number, so that all particles have different colours. We assume for simplicity that under collisions no new colours can be created, i.e. when they collide particles either keep or interchange their colours.

A collision process in which particles keep the same order of colours along the line before and after the collision is called *reflection* (backward scattering), and a process in which they interchange this order is called *transmission* (forward scattering), see Figs. 4.3 and 4.5. Were the particles to have the same colour, we would not be able to distinguish between reflection and transmission.

Regardless of colour, we have to decide on how to associate initial asymptotic momenta to particles after their two-body collision. Two different assignments are possible and they give rise to the so-called *transmission* and *reflection representations* of scattering [4]. These are two alternative but equivalent ways to describe the scattering theory.

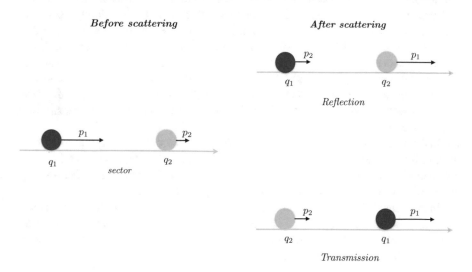

Fig. 4.3 Interpretation of quantum-mechanical scattering in the transmission representation. Under scattering particles always keep their momenta. The position of a particle is rigidly tight with the colour (q_1 is red, q_2 is yellow). Under reflection the original order of colours along the line pertains, while it gets interchanged under transmission

Fig. 4.4 Sectors for $N = 3$

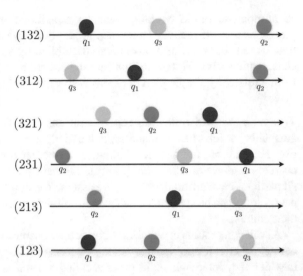

In the transmission representation particles are uniquely associated with their momenta, the latter are kept intact through collisions. If the i'th particle in an initial state has momentum p_i, after all possible collisions we have to identify the i'th particle as the one which has the same momentum p_i. In this representation the colour of a particle is fixed to its position on the line.[10] Formula (4.102) gives the Bethe wave function in the transmission representation, therefore, implying that we associate to each σ-sector a unique order of colours. Transition from sector to sector happens due to transmission. For instance, in Fig. 4.4 we pictured 6 sectors arising for the case of $N = 3$ particles. Each sector is associated to the corresponding permutation σ from the list (4.90) and it comes with a particular colour ordering. If interactions are completely absent, the incoming wave $\Psi_{\text{in}} \sim e^{i q_i p_i}$ propagates to all sectors without changing its amplitude, which corresponds to perfect transmission; from the viewpoint of (4.103) perfect transmission means that $\mathcal{A}(\sigma|\sigma\tau) = \delta_{\tau e}$. Note that the conventional definition of the S-matrix relies on the use of the transmission representation.

In the transmission representation (4.102) each σ-sector has a unique colour ordering, so that the notations of σ- and colour sector are in fact equivalent and one can use them interchangeably. The situation, however, is different for the reflection representation of the scattering process which we now describe.

In the reflection representation a collision is an event where particles always interchange their momenta (see Fig. 4.5). Because of this interchange, particles cannot overtake each other, i.e. in this representation scattering happens within the fundamental sector and the Bethe wave function is understood as

---

[10]Classically, particles are characterised by their momentum p, coordinate q and a colour. From a quantum-mechanical viewpoint, the q's control the sectors of the asymptotic wave function.

$$\Psi(q_1, \ldots, q_N | \sigma) = \sum_{\tau \in \mathfrak{S}_N} \mathcal{A}(\sigma | \tau) e^{iq_1 p_{\tau(1)} + \cdots + iq_N p_{\tau(N)}}, \qquad (4.104)$$

where q_i satisfy (4.95). As usual, under collisions colours can be kept (reflection) or interchanged (transmission). Permutations σ are now in one-to-one correspondence to colour orderings, which are now unrelated to coordinates q_i, the latter *always* lie within the fundamental sector and cannot be continued outside. In the reflection representation, Fig. 4.4 would look the same as it looks, except now, regardless of its colour, the most left particle on the line will always be labelled as q_1, the second to it as q_2 and so on. Elementary reflections are characterised by passing from one plane wave to another one with interchanged momenta of two particles, but in the same sector which signifies no colour change. Elementary transmissions are the same but they necessarily involve the change of the sector. Were the particles impenetrable, there would be no transmissions, and the amplitude in (4.104) would be $\mathcal{A}(\sigma | \tau) = \delta_{\sigma e}$. For the fundamental sector the transmission representation coincides with the reflection one.

Comparing our discussion here with the treatment of classical scattering in Sect. 4.1, we note that we started there with scattering in the reflection representation, which is, of course, physically very appealing if we deal with a repulsive potential. Specifying (4.8) to the two-particle case, we found that $p_1^+ = p_2^-$ and $p_2^+ = p_1^-$, which corresponds to the situation of interchanging the asymptotic momenta under scattering. To compute the classical phase shift, which is the same as the classical S-matrix, we then changed to the transmission representation, see the discussion around (4.9).

S-matrix. The S-matrix of the problem is an $N! \times N!$ matrix, the elements of which encode how one of the $N!$ initial configurations of particles on a line couples to each of the final $N!$ configurations. To obtain the whole S-matrix it is enough to consider one distinguished configuration as an initial state. For instance, if we set up an incoming wave in the fundamental sector

$$e^{ip_1 q_1 + ip_2 q_2 + \cdots + ip_N q_N}, \qquad (4.105)$$

where momenta satisfy (4.99) to guarantee that scattering happens, then in the σ-sector we register an outgoing wave

$$e^{ip_N q_{\sigma(1)} + ip_{N-1} q_{\sigma(2)} + \cdots + ip_1 q_{\sigma(N)}} \qquad (4.106)$$

with the amplitude given by the S-matrix element $S(\sigma | \varpi)$, where $\tau = \varpi$ is the reversed permutation

$$\varpi \equiv \begin{pmatrix} 1 & 2 & \cdots & N \\ N & N-1 & & 1 \end{pmatrix}. \qquad (4.107)$$

In particular, (4.106) for $\sigma = e$ is a reflected wave in the fundamental sector. The remaining elements of the $N! \times N!$ matrix are obtained by permutations of particles in the initial state.

Thus, finding the S-matrix requires an extrapolation of the wave function from one asymptotic sector to another through sectorial boundaries where particle interactions are essential and cannot be neglected. Solving the multi-body interacting problem is, in general, very complicated. However, one could notice the following: consider two sectors which differ only by a permutation of two neighbouring particles with coordinates q_i and q_j, so that in the first sector $q_i < q_j$ and $q_i > q_j$ in the second. Geometrically, these sectors are neighbours and have the hyperplane $q_i = q_j$ as a common boundary. Extrapolation of the wave function through this boundary can always be done in the asymptotic regime where all the other coordinates are kept far away from $q_i \approx q_j$ and from each other. Physically, this extrapolation corresponds to a two-body scattering event. Starting from any sector, one can obviously reach any other by passing through the adjacent sectorial boundaries, albeit not in a unique way. The sectors are thus connected by simple transpositions α_j, $j = 1, \ldots, N - 1$, the latter generate the symmetric group $\mathfrak{S}_N$.

Scattering operators in the reflection representation. The scattering process is described most elementary in the reflection representation. If we have two neighbouring particles at q_j and q_{j+1}, then under collision they interchange their momenta $p_{\tau(j)}$ and $p_{\tau(j+1)}$. If this collision is a pure reflection, then colours are preserved, if this is a pure transmission then also the colour sector changes as $\sigma \to \alpha_j \sigma$. This picture suggests that for these two pure processes the amplitudes must be related as

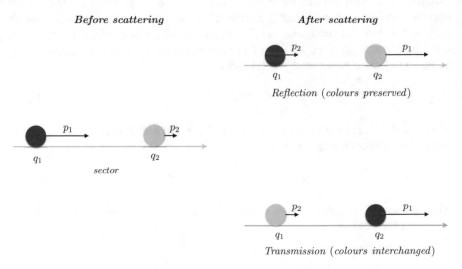

Fig. 4.5 Interpretation of quantum-mechanical scattering in reflection representation. Under scattering momenta are always interchanged. Since particles cannot overtake each other, they live in a single sector $q_1 < q_2$. Transmission corresponds to a scattering channel where particles exchange their colours. Under reflection colours are preserved

$$\mathcal{A}(\sigma|\alpha_j\tau) = A(p_{\tau(j)}, p_{\tau(j+1)})\mathcal{A}(\sigma|\tau)\,,$$
$$\mathcal{A}(\alpha_j\sigma|\alpha_j\tau) = B(p_{\tau(j)}, p_{\tau(j+1)})\mathcal{A}(\sigma|\tau)\,,$$

(4.108)

where A and B are reflection and transmission coefficients, respectively. They depend on the momenta of scattered particles. The second formula in (4.108) is equivalent to

$$\mathcal{A}(\sigma|\alpha_j\tau) = B(p_{\tau(j)}, p_{\tau(j+1)})\mathcal{A}(\alpha_j^{-1}\sigma|\tau) = B(p_{\tau(j)}, p_{\tau(j+1)})(\pi(\alpha_j)\mathcal{A})(\sigma|\tau)\,,$$

where we have written the final result via the action of the right regular representation π of $\mathfrak{S}_N$, confer (4.86). To combine two scattering processes into one formula, we introduce a column vector $\Phi(\tau)$ which comprises the amplitudes in all the sectors corresponding to the same permutation τ of momenta

$$\Phi(\tau) \equiv \{\mathcal{A}(\sigma|\tau),\ \sigma \in \mathfrak{S}_N\}\,.$$

(4.109)

Then (4.108) can be combined into a single formula

$$\Phi(\alpha_j\tau) = Y_j(p_{\tau(j)}, p_{\tau(j+1)})\Phi(\tau)\,,$$

(4.110)

where we introduced Yang's scattering operators Y_j [15]

$$Y_j(p_1, p_2) = A(p_1, p_2)\,\mathbb{1} + B(p_1, p_2)\,\pi(\alpha_j)\,,$$

(4.111)

where $j = 1, \ldots, N - 1$. Each Y_j is a $N! \times N!$ matrix which acts on the vector $\Phi(\tau)$. This matrix can be naturally viewed as a momentum-dependent connection on the symmetric group that defines the transport of the vector $\Phi(\tau)$ by a "discrete" amount α_j. As such, it must satisfy certain compatibility conditions that render the system of $(N - 1)(N!)^2$ Eq. (4.110) for $(N!)^2$ unknowns $\mathcal{A}(\sigma|\tau)$ consistent. Indeed, we have

$$\Phi(\alpha_j^2\tau) = Y_j\big(p_{(\alpha_j\tau)(j)}, p_{(\alpha_j\tau)(j+1)}\big)\Phi(\alpha_j\tau) = Y_j(p_{\tau(j+1)}, p_{\tau(j)})Y_j(p_{\tau(j)}, p_{\tau(j+1)})\Phi(\tau)\,,$$

where we have taken into account that according to our rules $(\alpha_j\tau)(j) = \tau(\alpha_j(j)) = \tau(j + 1)$ and $(\alpha_j\tau)(j + 1) = \tau(\alpha_j(j + 1)) = \tau(j)$. The defining relation $\alpha_j^2 = e$ then demands the fulfilment of the following relation

$$Y_j(p_1, p_2)Y_j(p_2, p_1) = \mathbb{1}\,.$$

(4.112)

Analogously, the second relation in (4.85) implies that

$$\Phi(\alpha_j\alpha_{j+1}\alpha_j) = \Phi(\alpha_{j+1}\alpha_j\alpha_{j+1})\,,$$

so that Y_j must satisfy

$$Y_j(p_2, p_3)Y_{j+1}(p_1, p_3)Y_j(p_1, p_2) = Y_{j+1}(p_1, p_2)Y_j(p_1, p_3)Y_{j+1}(p_2, p_3). \quad (4.113)$$

Finally, the third relation in (4.85) leads to

$$Y_i Y_j = Y_j Y_i \qquad (4.114)$$

for $|i - j| \geqslant 2$.

Provided the matrices Y_j satisfy the conditions above, the system (4.110) is consistent. Since α_j generate the whole $\mathfrak{S}_N$, the connection (4.110) transports the value of Φ at one point, for instance, at the identity e, to any other point of the group. The value $\Phi(e)$ is thus an initial condition for (4.110), it depends on $N!$ arbitrary parameters which are nothing else but the amplitudes $\mathcal{A}(\sigma|e)$ of purely incoming waves. Note that if the transmission is absent, i.e. $B = 0$, then each Y_j is the identity matrix times the reflection coefficient A, justifying the name *reflection* for the corresponding representation of the Bethe wave function. Thus, Y_j can be interpreted as the two-body scattering matrices in the reflection representation. In particular, for $N = 2$ there is only one transposition α represented by the matrix

$$\pi(\alpha) = \begin{pmatrix} 0 & 1 \\ 1 & 0 \end{pmatrix}, \qquad (4.115)$$

and, therefore, Y coincides with S-matrix (4.42).

Now we are ready to construct the full scattering matrix. The momenta of the incoming wave (4.99) are related to those of the outgoing wave by means of permutation (4.107) that acts as $\varpi(j) = N - j + 1$. Writing ϖ in two-line notation makes it obvious that it can be represented as the following product of transpositions

$$\varpi = (1\,|N)(2\,|N - 1)(3\,|N - 2)\dots. \qquad (4.116)$$

In turn, each of the transpositions entering this expression can be represented as a product of simple transpositions

$$(1|N) = (1|2)(2|3)(3|4)\dots(N - 2|N - 1)(N - 1|N)(N - 2|N - 1)\dots(3|4)(2|3)(1|2),$$
$$(2|N - 1) = (2|3)(3|4)\dots(N - 3|N - 2)(N - 2|N - 1)(N - 3|N - 2)\dots(3|4)(2|3),$$
$$(3|N - 2) = (3|4)\dots(N - 4|N - 3)(N - 3|N - 2)(N - 4|N - 3)\dots(3|4)$$

and so on. Successively multiplying these expressions and using the defining relations of $\mathfrak{S}_N$, one finds that ϖ reduces to

$$\varpi = (1|2) \cdot (2|3)(1|2) \cdot (3|4)(2|3)(1|2) \cdot \dots \cdot (N - 1|N)(N - 2|N - 1)\dots(1|2), \quad (4.117)$$

where to make the structure of ϖ more visible, we separated the groups of simple transpositions by an explicit multiplication sign. Taking this structure of ϖ into account, formula (4.110) yields

$$\Phi(\varpi) = Y_1(p_{N-1}, p_N)$$
$$\times Y_2(p_{N-2}, p_N)Y_1(p_{N-2}, p_{N-1})$$
$$\times Y_3(p_{N-3}, p_N)Y_2(p_{N-3}, p_{N-1})Y_1(p_{N-3}, p_{N-2}) \qquad (4.118)$$
$$\times \cdots$$
$$\times Y_N(p_1, p_N)Y_{N-1}(p_1, p_{N-1})\ldots Y_1(p_1, p_2)\ \Phi(e)\,.$$

Here the order of Y matrices follows the pattern of simple transpositions in (4.117). The arguments of Y's were determined according to the scattering history built in (4.117).

To illustrate the last point, consider an example of $N = 4$. As the first step of using the connection formula (4.110), we have

$$\Phi((1|2) \cdot (2|3)(1|2) \cdot (3|4)(2|3)(1|2)) = Y_1(p_{\varpi^{(1)}(1)}, p_{\varpi^{(1)}(2)})\Phi((2|3)(1|2) \cdot (3|4)(2|3)(1|2))\,.$$

Here $\varpi^{(1)}$ is the permutation

$$\varpi^{(1)} = (2|3)(1|2) \cdot (3|4)(2|3)(1|2) = (3421)\,,$$

so that $\varpi^{(1)}(1) = 3$ and $\varpi^{(1)}(2) = 4$. Therefore,

$$\Phi((1|2) \cdot (2|3)(1|2) \cdot (3|4)(2|3)(1|2)) =$$
$$= Y_1(p_3, p_4)Y_2(p_{\varpi^{(2)}(2)}, p_{\varpi^{(2)}(3)})\Phi((1|2) \cdot (3|4)(2|3)(1|2))\,,$$

where $\varpi^{(2)} = (1|2) \cdot (3|4)(2|3)(1|2) = (3241)$ yielding $\varpi^{(2)}(2) = 2$ and $\varpi^{(2)}(3) = 4$. Thus, we have

$$\Phi((1|2) \cdot (2|3)(1|2) \cdot (3|4)(2|3)(1|2)) = Y_1(p_3, p_4)Y_2(p_2, p_4)\Phi((1|2) \cdot (3|4)(2|3)(1|2))\,.$$

Continuing along the same lines, we will arrive at the final expression

$$\Phi((1|2) \cdot (2|3)(1|2) \cdot (3|4)(2|3)(1|2)) =$$
$$= Y_1(p_3, p_4)Y_2(p_2, p_4)Y_1(p_2, p_3)Y_3(p_1, p_4)Y_2(p_1, p_3)Y_1(p_1, p_2)\Phi(e)\,,$$

which is a specification of (4.118) to the four-particle case.

S-matrix in the transmission representation. More generally, we can associate the scattering operator of Yang with an arbitrary transposition α_{ij}, namely,

$$Y_{ij}(p_1, p_2) = A(p_1, p_2)\, \mathbb{1} + B(p_1, p_2)\, \pi(\alpha_{ij})\,,$$

so that Y_j's introduced in (4.111) are $Y_j \equiv Y_{jj+1}$. In turn, by using Y_{ij}, we define the following matrix

$$S_{ij}(p_1, p_2) \equiv \pi(\alpha_{ij})Y_{ij} = B(p_1, p_2)\mathbb{1} + A(p_1, p_2)\pi(\alpha_{ij})\,. \qquad (4.119)$$

Due to (4.84), we observe the following "braiding" property

$$S_{kj}\pi(\alpha_{ik}) = \pi(\alpha_{ik})S_{ij}\,. \tag{4.120}$$

Now we rewrite the main formula (4.118) via S_{ij} and, using (4.120), bring the answer to the following form

$$
\begin{aligned}
\Phi(\varpi) = {} & \pi(\varpi) \cdot S_{N-1\,N}(p_{N-1},\, p_N) \\
& \times\ S_{N-2\,N}(p_{N-2},\, p_N)S_{N-2\,N-1}(p_{N-2},\, p_{N-1}) \\
& \times\ S_{N-3\,N}(p_{N-3},\, p_N)S_{N-3\,N-1}(p_{N-3},\, p_{N-1})S_{N-3\,N-2}(p_{N-3},\, p_{N-2}) \\
& \times \ldots \\
& \times\ S_{1N}(p_1,\, p_N)S_{1\,N-1}(p_1,\, p_{N-1})\ldots S_{12}(p_1,\, p_2)\ \Phi(e)\,, \tag{4.121}
\end{aligned}
$$

where ϖ is the permutation (4.116). A welcome feature of this formula is that the index of each S-matrix perfectly matches with the index of the momenta on which this S-matrix depends. Thus, in the future we may not indicate the momentum dependence of S, as the latter can be unambiguously restored from the S-matrix subscript. Note that this is not true for the Y-representation (4.118). If we introduce

$$S \equiv S_{N-1\,N} \cdot S_{N-2\,N}S_{N-2\,N-1}\cdot \ldots \cdot S_{1N}S_{1\,N-1}\ldots S_{12}\,, \tag{4.122}$$

then, keeping in mind that $\varpi^2 = e$, we obtain

$$\pi(\varpi)\Phi(\varpi) = S\,\Phi(e)\,. \tag{4.123}$$

The expression on the left hand side is $\pi(\varpi)\Phi(\varpi) = \mathcal{A}(\varpi\sigma|\varpi)$. This is the amplitude of the outgoing wave in the $\varpi\sigma$-sector where

$$
\varpi\sigma = \begin{pmatrix} 1 & 2 & \cdots & N \\ N & N-1 & \cdots & 1 \end{pmatrix}\begin{pmatrix} 1 & 2 & \cdots & N \\ \sigma(1) & \sigma(2) & \cdots & \sigma(N) \end{pmatrix} = \begin{pmatrix} 1 & 2 & \cdots & N \\ \sigma(N) & \sigma(N-1) & \cdots & \sigma(1) \end{pmatrix}\,.
$$

It is clear that in this sector p_i couples to $q_{\sigma(i)}$, i.e. precisely in the same way as in the incoming wave in the σ-sector. The incoming wave in the σ-sector

$$e^{ip_1 x_{\sigma(1)}+ip_2 x_{\sigma(2)}+\cdots+ip_N x_{\sigma(N)}}$$

with an amplitude $\mathcal{A}(\sigma|e)$ is transmitted to the $\varpi\sigma$-sector

$$e^{ip_N x_{(\varpi\sigma)(1)}+ip_{N-1} x_{(\varpi\sigma)(2)}+\cdots+ip_1 x_{(\varpi\sigma)(N)}} = e^{ip_1 x_{\sigma(1)}+ip_2 x_{\sigma(2)}+\cdots+ip_N x_{\sigma(N)}}$$

with the amplitude $\mathcal{A}(\varpi\sigma|\varpi)$. Thus, S in (4.123) is nothing else but the S-matrix in the transmission representation. In particular, if the reflection coefficient $A = 0$ the S-matrix is diagonal, as can be seen from (4.119).

The structure of the scattering matrix encoded in (4.121) and (4.122) has also a very clear physical meaning: in the transmission picture the fastest particle with momentum p_1, which is the most left before scattering should undergo collisions with the remaining particles with momenta $p_2, p_3, \ldots, p_N$ to appear the most right after scattering. Every time it transfers through the i'th particle, its amplitude undergoes a change (a phase shift) by the corresponding two-body S-matrix $S_{1i}(p_1, p_i)$. After all these collisions the accumulated change of the amplitude is

$$S_{1N} S_{1\,N-1} \ldots S_{12} \, \Phi(e).$$

Then the p_2-particle, which is now the most left, goes to cross $p_3, p_4, \ldots, p_N$ and take its position in between p_N and p_1. This leads to further accumulation of successive amplitude changes and we get

$$S_{2N} S_{2\,N-1} \ldots S_{23} \cdot S_{1N} S_{1\,N-1} \ldots S_{12} \, \Phi(e).$$

Continuing in the same fashion, we find that after all particles crossed and reached the final order $p_N, p_{N-1}, \ldots, p_1$, the initial amplitude turns into $S\Phi(e)$, where S is exactly the S-matrix (4.121). Obviously, the number of two-body collisions that happens before the final configuration is reached is

$$N - 1 + N - 2 + N - 3 + \cdots + 1 = \frac{N(N-1)}{2}.$$

Hence, the N-body S-matrix factorises into the product of $N(N-1)/2$ two-body S-matrices. This factorised structure of the S-matrix is a consequence of a large number of conservation laws that prohibit diffraction and render the wave function in each asymptotic domain to be a superposition of a finite number of waves.

Another important observation about the factorised structure of the S-matrix is that the order in which $N(N-1)/2$ two-body collisions occur *does not matter*. This statement is a consequence of the consistency conditions obeyed by the two-body scattering matrix. These conditions can be immediately derived from those satisfied by the corresponding S-matrix in the reflection representation. Indeed, using the relation (4.119), the formulae (4.112) and (4.113) yield

$$S_{12}(p_1, p_2) S_{21}(p_2, p_1) = \mathbb{1} \tag{4.124}$$

and

$$S_{12}(p_1, p_2) S_{13}(p_1, p_3) S_{23}(p_2, p_3) = S_{23}(p_2, p_3) S_{13}(p_1, p_3) S_{12}(p_1, p_2), \tag{4.125}$$

respectively. In addition, it follows from (4.119) that $S_{ij} S_{kl} = S_{kl} S_{ij}$, if among the indices i, j, k and l there are no two coincident ones.

Relation (4.125) is the *Yang-Baxter equation* for the two-body S-matrix. Physically, it expresses the equivalence of two different ways to factorise a three-body

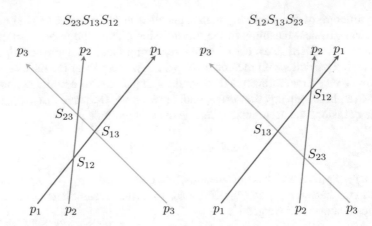

Fig. 4.6 Two topologically different three-body space-time diagrams and factorisation of the three-body S-matrix. The result of the three-body scattering process does not depend on the order in which two-body scattering events take place

S-matrix S_{123} into a product of two-body S-matrices, see Fig. 4.6. Thus, integrability implies consistent factorisation of scattering process and the corresponding S-matrix in a sequence of two-body events and S-matrices, giving rise to the notion of *Factorised Scattering Theory* [16]. This result is of fundamental nature, it reduces the problem of calculating the multi-body S-matrix in an integrable model to the one just for the two-body S-matrix, making the latter the main object of study, at least in the context of scattering theory.

Among further properties of the S-matrix, we point out that it is unitary and symmetric. Unitarity of (4.122) follows immediately if we require the two-body S-matrix to be unitary. Taking into account the unitarity of the representation π, the latter requirement reduces to the familiar conditions: $|A|^2 + |B|^2 = 1$ and $\bar{A}B + \bar{B}A = 0$, check (4.41). As to the symmetric property required by the time-reversal invariance of the interaction, using the fact that the two-body S-matrix is symmetric,[11] we have

$$S^t = S_{12} \ldots S_{1\,N-1}S_{1N} \cdot \ldots \cdot S_{N-2\,N-1}S_{N-2\,N} \cdot S_{N-1\,N}.$$

Now by successive use of (4.125) the right hand side of the last formula can be brought to the original form (4.122) that proves the relation $S^t = S$.

Historically, one of the first systems with a factorised behaviour was the delta-interaction model of repulsive (hard-core) bosons solved by Lieb and Liniger [17, 18]. Conditions for factorisation of the scattering matrix in this model have been first observed by McGuire in 1964 [2] and then by Yang for the multi-component case [15, 19]. Earlier considerations of the S-matrix factorisation in integrable quantum field theories include [20–22]. A bootstrap S-matrix approach to relativistic quantum

---

[11] Representation π is orthogonal and, since for permutation $\pi(\alpha_{ij})^2 = e$, one gets $\pi(\alpha_{ij}) = \pi(\alpha_{ij})^t$.

field theories based on axiomatisation of the Factorised Scattering Theory was put forward in [16].

4.3.3 Bethe Wave Function and Statistics

First we recall the standard treatment of a quantum-mechanical system of many particles with internal degrees of freedom. Suppose that the Hamiltonian of such a system does not involve terms acting on internal degrees of freedom. In this situation the wave function Ψ can be searched in a factorised form

$$\Psi(x_1, \ldots, x_N) = \Psi(q_1, \ldots, q_N)\chi(s_1, \ldots, s_N), \qquad (4.126)$$

where Ψ depends on the particle coordinates q_i, while χ – on variables s_i describing internal degrees of freedom (spin). The variables x_i stand for the pairs (q_i, s_i). The Schrödinger equation determines the coordinate wave function Ψ only, leaving χ arbitrary. Indistinguishable particles are either bosons or fermions. Accordingly, the wave function Ψ is symmetric or anti-symmetric, which can only be possible if there is a rigid correlation between symmetry properties of Ψ and χ with respect to simultaneous permutations of their respective arguments.

Assume that the Hamiltonian is invariant under permutations of particles, and, therefore, commutes with all operators representing permutations. Permutations, however, do not all commute with each other, and, as a consequence, can not be simultaneously brought to a diagonal form. This means that the spectrum of the Hamiltonian is degenerate, and, in general, there will be several solutions Ψ of the Schrödinger equation with the same energy transforming into each other under the action of the symmetric group. In other words, these solutions can be combined into an irreducible multiplet λ of $\mathfrak{S}_N$. Correspondingly, the wave function Ψ_λ is labelled by an index λ and is said to be of the symmetry type λ. Similarly, the action of $\mathfrak{S}_N$ on χ can also be decomposed into the sum of irreducible components.

Now, it is well known how to choose χ such that it can be combined with the coordinate wave function of a given symmetry type λ to produce symmetric or anti-symmetric Ψ. Namely, if particles are bosons, then the symmetry of Ψ and χ must be defined by the same Young diagram, and the full symmetric wave function is expressed via certain bilinear combinations of those. If particles are spin-$\frac{1}{2}$ fermions, then the full wave function is anti-symmetric and the Young diagrams of the coordinate and the spin wave functions must be conjugate, i.e. one is obtained from the other by replacing rows for columns and vice versa. This follows from the fact that if $\pi_\lambda(\sigma)$ is the representation of $\mathfrak{S}_N$ corresponding to the partition λ, then the representation corresponding to the conjugate (the same as "transposed") partition $\bar\lambda$ is $\pi_{\bar\lambda}(\sigma) = \text{sign}(\sigma)\pi_\lambda(\sigma)$, $\sigma \in \mathfrak{S}_N$. Hence,

$$\Psi = \sum_{\sigma \in \mathfrak{S}_N} \Psi_\lambda(q_{\sigma(1)}, \ldots, q_{\sigma(N)})\chi_{\bar\lambda}(s_{\sigma(1)}, \ldots, s_{\sigma(N)}). \qquad (4.127)$$

For the purpose of our present discussion, the most convenient way to describe representations λ (or $\bar{\lambda}$) of $\mathfrak{S}_N$ in the space of functions of N-variables is to use *Hund's method* [23]. According to this method, a function $\Psi_\lambda(q_1, \ldots q_N)$ has a definite symmetry type λ, i.e. it is *one of the basis functions* of a representation $\lambda = [\lambda_1, \lambda_2, \ldots, \lambda_l]$ of $\mathfrak{S}_N$, if

(1) it is anti-symmetric in a set of λ_1 arguments, anti-symmetric in another set of λ_2 arguments and so on,
(2) it satisfies the Fock symmetry conditions

$$\left[\mathbb{1} - \sum_{k \in \lambda_i} \alpha_{km} \right] \Psi = 0 \,, \tag{4.128}$$

where α_{km} is a transposition such that m is in λ_j for all choices of λ_i and λ_j with $\lambda_i \geqslant \lambda_j$.

One usually writes a function Ψ satisfying the first condition as

$$\Psi\big([q_1, \ldots, q_{\lambda_1}][q_{\lambda_1+1}, \ldots, q_{\lambda_1+\lambda_2}] \ldots [q_{N-\lambda_l-\lambda_{l-1}+1}, \ldots, q_{N-\lambda_l}][q_{N-\lambda_l+1}, \ldots, q_N]\big) \,,$$

meaning that it is separately anti-symmetric in the λ_1 variables $q_1, \ldots, q_{\lambda_1}$, in the λ_2 variables $q_{\lambda_1+1}, \ldots q_{\lambda_1+\lambda_2}$, and so forth.

Later, by using a different realisation of irreducible representations of $\mathfrak{S}_N$, we show that for electrons the spin wave functions are associated to the two-row Young diagrams $\bar{\lambda}$ depicted in Fig. 4.7. Any such diagram with N boxes has $N - M$ boxes in the first row and M boxes in the second row, where M takes values from 0 to the integer part of $N/2$. For a fixed M the (2)-representation associated with this diagram has spin $S = 1/2(N - 2M)$. Therefore, for electrons the corresponding coordinate wave function must transform in the representation λ depicted in Fig. 4.8 and, according to Hund's method, be of the type

$$\Psi\big([q_1, \ldots, q_{N-M}][q_{N-M+1}, \ldots, q_N]\big) \,, \tag{4.129}$$

and satisfy the Fock condition

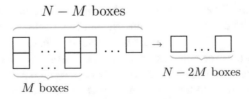

Fig. 4.7 Young diagram $\bar{\lambda} = [N - M, M]$ represents an irrep of $\mathfrak{S}_N$ under which the spin wave function of electrons with the total spin $S = 1/2(N - 2M)$ transforms

Fig. 4.8 Young diagram
$\lambda = [2^M, 1^{N-M}]$ for a
coordinate function of
electrons

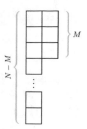

$$\left[\mathbb{1} - \sum_{i=1}^{N-M} \alpha_{ij} \right] \Psi = 0, \qquad j > N - M. \tag{4.130}$$

This completes our recollection of the standard treatment of multi-particle systems in quantum mechanics.

Coming back to the Bethe wave function, we would like to understand the conditions it must satisfy in order to be of a definite symmetry type. To this end, we consider this function in the transmission representation (4.102) and determine how it transforms under permutations of coordinates. For any particle configuration from the σ-sector the wave function is

$$\Psi(q_1, \ldots, q_N) = \sum_{\tau \in \mathfrak{S}_N} \mathcal{A}(\sigma|\tau) e^{iq_{\sigma(i)} P_{\tau(i)}}. \tag{4.131}$$

Consider $\Psi(q_{\varsigma(1)}, \ldots q_{\varsigma(N)})$, where ς is any permutation. Denote $q_i' \equiv q_{\varsigma(i)}$, so that $q_i = q_{\varsigma^{-1}(i)}'$. Replacing in the last equality the index i with $\sigma(i)$, we get $q_{\sigma(i)} = q_{\varsigma^{-1}(\sigma(i))}' = q_{\sigma\varsigma^{-1}(i)}'$. This means that the configuration of q_i' belongs to the $\sigma\varsigma^{-1}$-sector and we have

$$\Psi(q_{\varsigma(1)}, \ldots q_{\varsigma(N)}) = \sum_{\tau \in \mathfrak{S}_N} \mathcal{A}(\sigma\varsigma^{-1}|\tau) e^{iq_{\sigma\varsigma^{-1}(i)}' P_{\tau(i)}} = \sum_{\tau \in \mathfrak{S}_N} \mathcal{A}(\sigma\varsigma^{-1}|\tau) e^{iq_{\sigma(i)} P_{\tau(i)}}. \tag{4.132}$$

Comparison of (4.131) with (4.132) shows that the action of the symmetric group on the wave function by permuting its arguments induces the following action on the amplitudes

$$\mathcal{A}(\sigma|\tau) \to \mathcal{A}(\sigma\varsigma^{-1}|\tau), \tag{4.133}$$

or for the vector $\Phi(\tau)$

$$\Phi(\tau) \to \pi'(\varsigma^{-1})\Phi(\tau), \tag{4.134}$$

where this time π' is the *right* regular representation of $\mathfrak{S}_N$. It is now clear that if we want the Bethe wave function to be of the symmetry type λ, the vector $\Phi(\tau)$ for any τ must obey the following two conditions:

(1) The anti-symmetry requirement

$$\pi'(\alpha_i)\Phi(\tau) = -\Phi(\tau)\,, \tag{4.135}$$

where $i \in \{1, \ldots \lambda_1 - 1\} \cup \{\lambda_1 + 1, \ldots, \lambda_1 + \lambda_2 - 1\} \cup \ldots \cup \{N - \lambda_l + 1, \ldots, N - 1\}$.

(2) The Fock condition

$$\left[\mathbb{1} - \sum_{k \in \lambda_i} \pi'(\alpha_{km})\right]\Phi(\tau) = 0\,, \tag{4.136}$$

where m is in λ_j for all choices of λ_i and λ_j with $\lambda_i \geqslant \lambda_j$.

It is now time to recall that Φ transforms linearly under the left regular representation π of $\mathfrak{S}_N$. Because π' and π commute, if Φ satisfies the constraints (4.135) and (4.136), the vector $\pi(\sigma)\Phi$ will also satisfy them for any σ. In fact, these constraints project out in decomposition (4.88) all components but one that coincides with λ. Thus, we arrive at the important conclusion that the requirement for the Bethe wave function to have the symmetry type λ is equivalent for the amplitude vector Φ to transform in the irreducible representation λ of the symmetric group [15].

Example. To illustrate how the constraints (4.135) and (4.136) single out an irreducible component of π, we look at a simple example of $N = 3$ particles. The group $\mathfrak{S}_3$ and its representations π and π' were discussed in Sect. 4.3.1. Consider the diagram $\lambda = [2, 1]$ and fill it as

$$\begin{array}{|c|c|}\hline 1 & 3 \\\hline 2 \\\cline{1-1}\end{array}\,.$$

This gives one of two possible standard tableaux. There is one anti-symmetry condition on the six-dimensional vector Φ, namely,

$$\pi'(\sigma_2)\Phi = -\Phi\,, \tag{4.137}$$

where $\sigma_2 = \alpha_{12}$ is the second element from the list (4.90). The Fock condition is

$$\left[\mathbb{1} - \pi'(\sigma_4) - \pi'(\sigma_6)\right]\Phi = 0\,, \tag{4.138}$$

where σ_4 and σ_6 are the corresponding permutations from the same list. A solution of the first equation leaves 3 parameters undetermined, while a subsequent imposition of the second equation leaves a two-dimensional vector space

$$\Phi^t = (u_1 - u_2, -u_1 + u_2, -u_1, -u_2, u_2, u_1)\,, \quad u_1, u_2 \in \mathbb{C}\,.$$

It is not hard to see that this is a two-dimensional invariant subspace of the representation π. On this subspace π acts irreducibly and coincides, in fact, with one of the irreducible components $\pi_{[2,1]}$ in the decomposition (4.92).

Analogously, we can consider the second standard tableau

$$\begin{array}{|c|c|} \hline 1 & 2 \\ \hline 3 \\ \cline{1-1} \end{array}.$$

The conditions on Φ are

$$\pi'(\sigma_4)\Phi = -\Phi\,,$$
$$\left[\mathbb{1} - \pi'(\sigma_2) - \pi'(\sigma_6)\right]\Phi = 0\,.$$

Together they single out another two-dimensional invariant subspace

$$\Phi' = (-v_2, -v_1 - v_2, v_1 + v_2, v_2, -v_1, v_1)\,, \quad v_1, v_2 \in \mathbb{C}\,,$$

on which π acts irreducibly and coincides with another component $\pi_{[2,1]}$ in (4.92).

Further, there is an anti-symmetric representation

$$\begin{array}{|c|} \hline 1 \\ \hline 2 \\ \hline 3 \\ \hline \end{array}.$$

It is singled out by the conditions

$$\pi'(\sigma_2)\Phi = -\Phi\,, \quad \pi'(\sigma_6)\Phi = -\Phi\,,$$

that have the solution $\Phi' = w(1, -1, 1, -1, 1, -1)$. Finally, the trivial representation [3] completes the list of irreducible components appearing in (4.92) for the regular representation of $\mathfrak{S}_3$.

Compatibility of scattering with statistics. Recall that the action of scattering operators on amplitudes is realised through the *left* regular representation π. Taking into account that π and π' commute, we conclude that imposition of symmetry conditions on the wave function should also be compatible with scattering. More precisely, by construction the S-matrix is an element of the group algebra of $\mathfrak{S}_N$ evaluated in the left regular representation π. This representation is reducible and its decomposition into a sum of irreducibles is given by (4.88). Projecting the Bethe wave function on an irreducible component λ, we obtain the wave function with a type of symmetry described by the Young diagram λ. The corresponding S-matrix is still given by (4.122), where the two-body S-matrices are substituted with

$$S_{ij}(p_1, p_2) = B_\lambda(p_1, p_2)\mathbb{1} + A_\lambda(p_1, p_2)\pi_\lambda(\alpha_{ij})\,,$$

where the subscript λ of A and B is used to emphasise that these scattering coefficients can be, in fact, different for different representations. If $\lambda = [1^N]$ is the anti-symmetric representation, then $\pi_\lambda(\alpha_{ij}) = -1$ and $S = B - A$. If $\lambda = [N]$ is the symmetric representation, then $\pi_\lambda(\alpha_{ij}) = 1$ and $S = A + B$. In both cases the S-matrices are scalar.

Example. Consider the delta-interaction model for the case of fermions, see Sect. 4.2.2. The delta-function potential produce no effect and the two-body S-matrix is trivial, $S = B - A = 1$. Since we deal with fermions, the Bethe wave function must transform in the anti-symmetric representation for which $\pi_\lambda(\alpha_{ij}) = -1$. Therefore, from (4.110) we obtain the following equation

$$\Phi(\alpha_j\tau) = (A - B)\Phi(\tau) = -(B - A)\Phi(\tau) = -\Phi(\tau).$$

It follows from here that for an arbitrary ς the vector Φ satisfies $\Phi(\varsigma\tau) = \text{sign}(\varsigma)$ $\Phi(\tau)$, the latter equation is obviously solved as $\Phi(\tau) = \text{sign}(\tau)\Phi(e)$. Further, the requirement (4.135) of anti-symmetry of the wave function allows one to completely determine the amplitudes $\mathcal{A}(\sigma|\tau) = \text{sign}(\sigma^{-1}\tau)\mathcal{A}(e|e)$. Up to an overall normalisation factor $\mathcal{A}(e|e)$, the Bethe wave function (4.102) is then given by the Slater determinant

$$\Psi(q_1, \ldots, q_N) = \sum_{\tau \in \mathfrak{S}_N} \text{sign}(\sigma^{-1}\tau)e^{iq_{\sigma(i)}p_{\tau(i)}}$$

$$= \sum_{\tau \in \mathfrak{S}_N} \text{sign}(\tau)e^{iq_i p_{\tau(i)}} = \det_{N \times N}\left(e^{iq_i p_j}\right). \quad (4.139)$$

This completes our discussion of the fermionic delta-interaction model.

References

1. Moser, J.: Three integrable Hamiltonian systems connected with isospectral deformations. Adv. Math. **16**, 197–220 (1975)
2. McGuire, J.B.: Study of exactly soluble one-dimensional N-body problems. J. Math. Phys. **5**(5), 622–636 (1964)
3. Marchioro, C.: Solution of a three-body scattering problem in one dimension. J. Math. Phys. **11**, 2193–2196 (1970)
4. Sutherland, B.: Beautiful Models: 70 Years of Exactly Solved Quantum Many-body Problems, 381 p. World Scientific (2004)
5. Kulish, P.P.: Factorization of the classical and quantum S matrix and conservation laws. Theor. Math. Phys. **26**, 132 (1976) [Teor. Mat. Fiz. 26, 198 (1976)]
6. Taylor, J.R.: Scattering Theory: The quantum Theory on Nonrelativistic Collisions. Wiley, New York (1972)
7. Landau, L.D., Lifshitz, E.M.: Quantum Mechanics Non-Relativistic Theory, vol. 3, 3rd edn. Butterworth-Heinemann (1981)
8. Ruijsenaars, S.N.M.: Systems of calogero-moser type. In: Semenoff, G., Vinet, L. (eds.) Particles and Fields. CRM Series in Mathematical Physics, pp. 251–352. Springer, New York, NY (1999)
9. Faddeev, L.D., Yakubovskii, O.A.: Lectures on Quantum Mechanics for Mathematics Students. American Mathematical Society (2009)

10. Sutherland, B.: Quantum many body problem in one-dimension: ground state. J. Math. Phys. **12**, 246–250 (1971)
11. Ruijsenaars, S.N.M.: First order analytic difference equations and integrable quantum systems. J. Math. Phys. **38**, 1069–1146 (1997)
12. Ruijsenaars, S.N.M.: Finite-dimensional soliton systems. In: Kupershmidt, B. (ed.) Integrable and Superintegrable Systems. World Scientific, Singapore (1990)
13. Ruijsenaars, S.N.M.: Sine-gordon solitons vs. relativistic calogero-moser particles. In: Pakuliak, S., von Gehlen, G. (eds.) Integrable Structures of Exactly Solvable Two-Dimensional Models of Quantum Field Theory. NATO Science Series (Series II: Mathematics, Physics and Chemistry), vol. 35, pp. 251–352 (2001)
14. Bethe, H.: On the theory of metals, 1. Eigenvalues and eigenfunctions for the linear atomic chain. Z. Phys. **71**, 205–226 (1931)
15. Yang, C.N.: Some exact results for the many body problems in one dimension with repulsive delta function interaction. Phys. Rev. Lett. **19**, 1312–1314 (1967)
16. Zamolodchikov, A.B., Zamolodchikov, A.B.: Factorized S matrices in two-dimensions as the exact solutions of certain relativistic quantum field models. Ann. Phys. **120**, 253–291 (1979) [559 (1978)]
17. Lieb, E.H., Liniger, W.: Exact analysis of an interacting bose gas, 1. The General solution and the ground state. Phys. Rev. **130**, 1605–1616 (1963)
18. Lieb, E.H., Liniger, W.: Exact analysis of an interacting bose gas II. The excitation spectrum. Phys. Rev. **130**, 1616 (1963)
19. Yang, C.N.: S matrix for the one-dimensional N body problem with repulsive or attractive delta function interaction. Phys. Rev. **168**, 1920–1923 (1968)
20. Aref'eva, I., Korepin, V.: Scattering in two-dimensional model with Lagrangian $L = 1/\gamma[(1/2)(\partial_\mu u)^2 + m^2(\cos u - 1)]$. Pisma Zh. Eksp. Teor. Fiz. **20**, 680 (1974)
21. Vergeles, S.N., Gryanik, V.M.: Two-dimensional quantum field theories having exact solutions. Sov. J. Nucl. Phys. **23**, 704–709 (1976) [Yad. Fiz. 23, 1324 (1976)]
22. Iagolnitzer, D.: Factorization of the multiparticle S matrix in two-dimensional space-time models. Phys. Rev. D **18**, 1275 (1978)
23. Hamermesh, M.: Group Theory and Its Application to Physical Problems. Dover Publications Inc., New York (1989)

Chapter 5
Bethe Ansatz

A 1931 result that lay in obscurity for decades, Bethe's solution to a quantum mechanical model now finds its way into everything from superconductors to string theory.

Murray T. Batchelor
The Bethe ansatz after 75 years,
Physics Today, January 2007

In the previous chapter we have studied the consequences of integrability for the scattering theory. We saw that integrability implies a special form of the asymptotic wave function and factorisation of the multi-body scattering matrix. Our considerations were done for one-dimensional quantum-mechanical systems defined on an infinite line. To study thermodynamics, one needs, however, to confine the system in a finite volume. Physically, this can be realised in our one-dimensional context by putting the particles on a circle of length L. Mathematically, such a system is described by requiring the corresponding wave function to be periodic. This leads to the determination of the spectrum of a quantum integrable system in terms of solutions of the so-called Bethe or Bethe-Yang equations, at least in the asymptotic regime where the notion of the Bethe wave function applies. In this chapter we introduce and consider various forms of the Bethe-Yang equations including the elements of the Quantum Inverse Scattering Method.

5.1 Coordinate Bethe Ansatz

Nowadays various specifications of the general Bethe Ansatz technique are known, the basic ones include

© Springer Nature Switzerland AG 2019
G. Arutyunov, *Elements of Classical and Quantum Integrable Systems*,
UNITEXT for Physics, https://doi.org/10.1007/978-3-030-24198-8_5

- *Coordinate Bethe Ansatz.* This technique was originally introduced by H. Bethe to solve the Heisenberg model of magnetism [1].
- *Nested Bethe Ansatz.* This is a generalisation of the Bethe Ansatz to models with internal degrees of freedom, where scattering involves changes of the internal states of scatters. This problem was eventually solved by Yang [2] and Gaudin [3] by means of the technique known today as "Nested Bethe Ansatz".
- *Asymptotic Bethe Ansatz.* Many integrable systems in finite volume cannot be solved by the Bethe Ansatz methods. However, the Bethe Ansatz provides the leading finite-size correction to the wave function, energy levels, etc. for systems in infinite volumes. This technique was introduced and extensively studied by Sutherland [4]. Solutions of many integrable quantum field theories rely on the use of the asymptotic Bethe Ansatz.
- *Algebraic Bethe Ansatz.* It was realised afterwards that the Bethe Ansatz can be formulated in such a way that it can be understood as the quantum analogue of the classical inverse scattering method. "Algebraic Bethe Ansatz" is another name for the powerful technique known as the "Quantum Inverse Scattering Method". The main development of this approach is due to Faddeev and the Leningrad School [5–7].
- *Functional Bethe Ansatz.* The algebraic Bethe Ansatz is not the only approach to solve the spectral problems for models connected with the Yang-Baxter algebra. In fact, it can only be applied to models that admit a so-called (pseudo-) vacuum state. For models like the periodic Toda chain, which has the same R-matrix as the Heisenberg magnet (spin-$\frac{1}{2}$ chain), but no pseudo-vacuum, the algebraic Bethe Ansatz fails. Inspired by Gutzwiller's work on the periodic Toda chain [8], another powerful technique—the method of separation of variables—was devised by Sklyanin in the framework of the Quantum Inverse Scattering Method [9, 10]. This technique is also known as the "Functional Bethe Ansatz" and it is connected to Baxter's method based on the notion of Q-operator [11, 12].
- *Thermodynamic Bethe Ansatz.* This method allows to investigate the thermodynamic properties of integrable systems. It was introduced by Yang and Yang [13] and further applied to a variety of models, see [14–16].

In this book we restrict ourselves to a subset of the above techniques and will start our discussion from the Coordinate Bethe Ansatz.

5.1.1 Periodicity Condition for the Bethe Wave Function

Consider a system of N interacting particles confined in a one-dimensional box. The full description of this system requires an imposition of certain boundary conditions. In the following we choose *periodic boundary conditions*. These conditions relate the quantum-mechanical amplitude for finding a particle with given colour and momentum at one end of the box to the amplitude for finding a particle with the same colour and momentum at the other end. To be able to use the Bethe function formalism developed in the previous section, we assume that the size L of the box is large enough to fit in a kinematic configuration corresponding to asymptotically free particles.

Under these assumptions, let us consider the Bethe wave function Ψ in the transmission representation (4.102), which we write as

$$\Psi(q_1,\ldots,q_N) = \sum_{\sigma\in\mathfrak{S}_N}\sum_{\tau\in\mathfrak{S}_N} \mathcal{A}(\sigma|\tau) e^{i\sum_{i=1}^{N} q_{\sigma(i)} P_{\tau(i)}} \Theta\big(q_{\sigma(1)} < \cdots < q_{\sigma(N)}\big). \qquad (5.1)$$

Here we introduced a concise notation

$$\Theta\big(q_{\sigma(1)} < \cdots < q_{\sigma(N)}\big) = \prod_{i=1}^{N-1} \Theta\big(q_{\sigma(i+1)} - q_{\sigma(i)}\big),$$

where $\Theta(x)$ is the Heaviside Θ-function. Coordinates q_i can take any values but now within a segment of length L, that is $0 \leqslant q_i \leqslant L$ for $\forall i$. The periodic boundary conditions for Ψ mean that

$$\Psi(q_1,\ldots,q_j=0,\ldots,q_N) = \Psi(q_1,\ldots,q_j=L,\ldots,q_N), \quad \forall j. \qquad (5.2)$$

For the left hand side of this equation, denoted by LHS, we get

$$\text{LHS} = \sum_{\substack{\sigma\in\mathfrak{S}_N \\ \sigma(1)=j}}\sum_{\tau\in\mathfrak{S}_N} \mathcal{A}(\sigma|\tau) e^{i\sum_{k=2}^{N} q_{\sigma(k)} P_{\tau(k)}} \Theta\big(q_{\sigma(2)} < \cdots < q_{\sigma(N)}\big), \qquad (5.3)$$

while for the right hand side called RHS,

$$\text{RHS} = \sum_{\substack{\sigma\in\mathfrak{S}_N \\ \sigma(N)=j}}\sum_{\tau\in\mathfrak{S}_N} \mathcal{A}(\sigma|\tau) e^{iL P_{\tau(N)}} e^{i\sum_{k=1}^{N} q_{\sigma(k)} P_{\tau(k)}} \Theta\big(q_{\sigma(1)} < \cdots < q_{\sigma(N-1)}\big), \qquad (5.4)$$

where in the exponent we have now $q_{\sigma(N)} = q_j = 0$.

To compare (5.4) with (5.3), we introduce the following element $\xi \in \mathfrak{S}_N$

$$\xi = \alpha_{N-1}\ldots\alpha_1 = \begin{pmatrix} 1 & 2 & & N-1 & N \\ 2 & 3 & \cdots & N & 1 \end{pmatrix} \qquad (5.5)$$

and make a replacement of the summation variable σ as $\sigma \to \xi\sigma$. Since σ obeys $\sigma(1) = j$, then $\xi\sigma(N) = \sigma(\xi(N)) = \sigma(1) = j$, and we get

$$\text{RHS} = \sum_{\substack{\sigma \in \mathfrak{S}_N \\ \xi\sigma(N)=j}} \sum_{\tau \in \mathfrak{S}_N} \mathcal{A}(\xi\sigma|\tau) e^{iLp_{\tau(N)}} e^{i\sum_{k=1}^{N} q_{\xi\sigma(k)} p_{\tau(k)}} \Theta\big(q_{\xi\sigma(1)} < \cdots < q_{\xi\sigma(N-1)}\big)$$

$$= \sum_{\substack{\sigma \in \mathfrak{S}_N \\ \sigma(1)=j}} \sum_{\tau \in \mathfrak{S}_N} \mathcal{A}(\xi\sigma|\tau) e^{iLp_{\tau(N)}} e^{i\sum_{k=1}^{N} q_{\sigma(k)} P_{\xi^{-1}\tau(k)}} \Theta\big(q_{\sigma(2)} < \cdots < q_{\sigma(N)}\big).$$

Here the argument of the theta-function and the condition $\sigma(1) = j$ are precisely the same as in the expression LHS. Further, making a change of variable $\tau \to \xi\tau$, we arrive at

$$\text{RHS} = \sum_{\substack{\sigma \in \mathfrak{S}_N \\ \sigma(1)=j}} \sum_{\tau \in \mathfrak{S}_N} \mathcal{A}(\xi\sigma|\xi\tau) e^{iLp_{\xi\tau(N)}} e^{i\sum_{k=2}^{N} q_{\sigma(k)} p_{\tau(k)}} \Theta\big(q_{\sigma(2)} < \cdots < q_{\sigma(N)}\big), \qquad (5.6)$$

where we have taken into account that $q_{\sigma(1)} = q_j = 0$. Note also that $p_{\xi\tau(N)} = p_{\tau(\xi(N))} = p_{\tau(1)}$. Finally, comparison of (5.6) with LHS yields the following equations

$$\mathcal{A}(\sigma|\tau) = \mathcal{A}(\xi\sigma|\xi\tau) e^{iLp_{\tau(1)}}. \qquad (5.7)$$

This is a requirement on the coefficients of the asymptotic wave function in order for the latter to satisfy periodic boundary conditions. Using the left regular representation π of $\mathfrak{S}_N$, Eq. (5.7) can be written as conditions for the vector $\Phi(\tau)$ defined in (4.109):

$$\pi(\xi)\Phi(\tau) = e^{iLp_{\tau(1)}} \Phi(\xi\tau). \qquad (5.8)$$

To make further progress, we note that (5.8) must be satisfied for any momentum ordering τ, which means that, in order to proceed, we can make for τ a convenient choice. We take

$$\tau = \alpha_1 \ldots \alpha_{j-1} = \begin{pmatrix} 1 & 2 & 3 & \cdots & j & j+1 & \cdots & N \\ j & 1 & 2 & & j-1 & j+1 & & N \end{pmatrix},$$

so that $\xi\tau = \alpha_{N-1} \ldots \alpha_j$. With this choice $\tau(1) = j$ and (5.8) boils down to

$$\pi(\alpha_{N-1} \ldots \alpha_1)\Phi(\alpha_1 \ldots \alpha_{j-1}) = e^{iLp_j} \Phi(\alpha_{N-1} \ldots \alpha_j). \qquad (5.9)$$

Next, we evaluate both sides of the last expression with the help of connection formula (4.110) and get

$$\pi(\alpha_{j-1}) \ldots \pi(\alpha_1) Y_1(p_1, p_j) \ldots Y_{j-1}(p_{j-1}, p_j)\Phi(e)$$
$$= e^{iLp_j} \pi(\alpha_j) \ldots \pi(\alpha_{N-1})Y_{N-1}(p_j, p_N) \ldots Y_j(p_j, p_{j+1})\Phi(e).$$

Using the definition (4.119) of the two-body S-matrix, the last expression can be rewritten in the following elegant form [2]

$$S_{j+1\,j}S_{j+2\,j}\dots S_{Nj}\cdot S_{1j}\dots S_{j-1\,j}\,\Phi(e) = e^{iLp_j}\,\Phi(e)\,. \tag{5.10}$$

If we introduce the following matrix operators

$$T_j = S_{j+1\,j}S_{j+2\,j}\dots S_{Nj}\cdot S_{1j}\dots S_{j-1\,j}\,, \tag{5.11}$$

then (5.10) tells that $\Phi \equiv \Phi(e)$ is a common eigenvector of N matrix operators T_j

$$T_j\,\Phi = e^{iLp_j}\,\Phi\,, \tag{5.12}$$

where $j = 1, \dots, N$. Compatibility of these equations requires that matrices T_j for various j pair-wise commute. The important fact that they do so is a consequence of the condition (4.124) and the Yang-Baxter equation (4.125).[1] In the following we refer to (5.12) as the *(matrix) Bethe-Yang equations*. In the case of scalar S-matrices, where Φ is one-dimensional, the diagonalisation problem of T_j does not arise, and equations (5.12) are usually called Bethe equations.

As was argued in Sect. 4.3.3, demanding the Bethe wave function to be of the symmetry type λ implies that the vector Φ transforms in the same representation. Correspondingly, the scattering operators and the operators T_j are also restricted to λ. The problem now is to solve the system (5.12) for a given irreducible representation of $\mathfrak{S}_N$ and, subsequently, use this solution to reconstruct the corresponding Bethe wave function.

Twisted boundary conditions. More generally, one can impose on the Bethe wave function the *twisted boundary conditions*, namely,

$$\Psi(q_1, \dots, q_j + L, \dots, q_N) = e^{i\varphi}\Psi(q_1, \dots, q_j, \dots, q_N)\,, \quad \forall j\,. \tag{5.13}$$

In particular, the twist $\varphi = \pi$ corresponds to the anti-periodic wave function. Physically, twisted boundary conditions can be interpreted as a change of the phase of the wave function due to a magnetic flux passing through the ring of charged particles (the Aharonov-Bohm effect). Most easily this can be seen for the spin chain models where a contribution to the Hamiltonian arising due to interactions of spins with an external magnetic field can be gauged away by a non-unitary transformation

---

[1] As an example, consider the $N = 4$ case and the operators $T_2 = S_{32}S_{42}S_{12}$ and $T_3 = S_{43}S_{13}S_{23}$. We have

$$T_2 T_3 = S_{32}S_{42}S_{12}\cdot S_{43}S_{13}S_{23} = \underbrace{S_{32}S_{42}S_{43}}_{\text{YB}}\cdot\underbrace{S_{12}S_{13}S_{23}}_{\text{YB}} = S_{43}S_{42}\underbrace{S_{32}\cdot S_{23}}_{=1}S_{13}S_{12}$$

$$= S_{43}S_{42}\cdot S_{13}S_{12} = S_{43}S_{13}\cdot S_{42}S_{12} = S_{43}S_{13}\underbrace{S_{23}\cdot S_{32}}_{=1}S_{42}S_{12} = T_3 T_2\,.$$

that results in the new wave function obeying twisted boundary conditions. In the presence of the twisted boundary conditions the Bethe equations (5.13) modify as

$$e^{i\varphi} T_j \, \Phi = e^{iLp_j} \, \Phi \,, \tag{5.14}$$

Note that in general φ can depend on the number N.

The twist provides a very useful concept. Even if the original model was untwisted, introducing a twist allows one to trace a fate of an individual state, *i.e.* a solution of (5.14) as a function of φ and in some cases it serves as a regularisation parameter for states that otherwise lead to singular Bethe equations.

5.1.2 Examples

To shed more light on the Bethe-Yang equations (5.12), we consider a few concrete examples. The first example is provided by the delta-interaction model for three cases different by the nature of interacting particles.

(1) The case of spinless fermions corresponds to picking up the anti-symmetric representation, $\lambda = [1^N]$, so that the wave function is anti-symmetric. As was already mentioned, the S-matrix for this model is trivial, $S = 1$. As a result, equations (5.12) become the familiar quantisation condition $e^{ip_j L} = 1$ for momenta of free fermions put on a circle of length L. The corresponding wave function is given by the Slater determinant

$$\Psi_{\{p_j\}}(q_1, \dots, q_N) = \det \left(e^{iq_i p_j} \right), \tag{5.15}$$

confer (4.139). The wave function is characterised by a set of quantised momenta $\{p_j\}$, it is anti-symmetric under permutations of coordinates which reflects the Fermi statistics of particles. It is also anti-symmetric under permutations of momenta.

(2) The Lieb-Liniger model [17, 18] describes Bose gas with repulsive delta-function interaction (1.57). The corresponding wave function transforms in the symmetric representation, $\lambda = [N]$. The two-body S-matrix is scalar and reads as

$$S(p_1, p_2) = \frac{p_1 - p_2 - i\varkappa}{p_1 - p_2 + i\varkappa} \,, \tag{5.16}$$

where $\varkappa > 0$ is the coupling constant, see (4.60). Equation (5.12) reduce to

$$e^{ip_jL} = \prod_{\substack{k=1 \\ k\neq j}}^{N} S_{kj}(p_k, p_j) = \prod_{\substack{k=1 \\ k\neq j}}^{N} \frac{p_j - p_k + i\varkappa}{p_j - p_k - i\varkappa} = -\prod_{k=1}^{N} \frac{p_j - p_k + i\varkappa}{p_j - p_k - i\varkappa}, \quad (5.17)$$

This set of N equations determine the allowed values of the particle momenta in this model [17]. The vector Φ in (5.8) is one-dimensional, *i.e.* the amplitude $\mathcal{A}(\sigma|\tau)$ does not depend on σ. From (4.110) we then have

$$\mathcal{A}(\alpha_j\tau) = \frac{p_{\tau(j)} - p_{\tau(j+1)} - i\varkappa}{p_{\tau(j)} - p_{\tau(j+1)} + i\varkappa} \mathcal{A}(\tau).$$

The last equation has a unique, up to an overall normalisation, solution

$$\mathcal{A}(\tau) = -(-1)^{\mathrm{sign}(\tau)} \prod_{i<j}^{N} \left(p_{\tau(j)} - p_{\tau(i)} - i\varkappa \right)$$

$$= \prod_{i<j}^{N}(p_j - p_i) \prod_{i<j}^{N} \left(1 + \frac{i\varkappa}{p_{\tau(i)} - p_{\tau(j)}} \right).$$

Substituting this expression into (5.1), we get

$$\Psi(q_1, \ldots, q_N) = -\sum_{\sigma\in\mathfrak{S}_N} \sum_{\tau\in\mathfrak{S}_N} (-1)^{\mathrm{sign}(\tau)} e^{i\sum_{k=1}^{N} q_k p_{\tau(k)}}$$

$$\times \prod_{i<j}^{N} \left(p_{\tau(j)} - p_{\tau(i)} - i\varkappa \right) \prod_{i=1}^{N-1} \Theta\left(q_{\sigma(i+1)} - q_{\sigma(i)} \right).$$

Taking into account that $\mathrm{sign}(q) = \Theta(q) - \Theta(-q)$, this formula simplifies to

$$\Psi_{\{p_j\}}(q_1, \ldots, q_N) = \prod_{i<j}^{N}(p_j - p_i) \sum_{\tau\in\mathfrak{S}_N} e^{i\sum_{k=1}^{N} q_k p_{\tau(k)}} \prod_{i<j}^{N} \left[1 - \frac{i\varkappa\,\mathrm{sign}(q_i - q_j)}{p_{\tau(i)} - p_{\tau(j)}} \right].$$

The wave function is parametrised by a set of N momenta $\{p_j\}$ which was reflected in its notation above. The function is symmetric under permutations of coordinates and anti-symmetric under permutations of momenta, so that it vanishes if any two momenta coincide. The coordinate symmetry of the wave function is compatible with the boson statistics. In the momentum space the particles behave rather like fermions: each value of momentum can be occupied by at most one particle. To emphasise this behaviour, in the expression for Ψ we singled out the overall anti-symmetric factor $\prod_{i<j}(p_j - p_i)$, without which the wave function would be symmetric under permutations of momenta. Dropping this factor is not allowed, however, because this would lead to an ill-defined

wave function that would not be bounded on the whole $\mathbb{R}^N$ when two momenta coincide.

This completes our discussion of the Bethe wave function for this model. For the normalisation issue, orthogonality and completeness, see [3].

(3) Interacting particles with internal degrees of freedom, the representation λ for the wave function remains unspecified. Using the expression (4.60) for the reflection and transmission coefficients, we can write down for the two-body S-matrix the following expression [2]

$$S_{ij}(p_i, p_j) = \frac{(p_i - p_j)\mathbb{1} - i\varkappa\,\pi_\lambda(\alpha_{ij})}{p_i - p_j + i\varkappa}\,, \tag{5.18}$$

where $\pi_\lambda(\alpha_{ij})$ is the transposition α_{ij} evaluated in the representation λ. In this case T_j are matrices, and the non-trivial problem of their diagonalisation will be postponed till Sect. 5.2.

Second, we consider the rational CMS model for which the Bethe-Yang equations in the repulsive regime can be solved exactly. For this model the phase shift θ is given by (4.72). The S-matrix reduces to the reflection amplitude $S = A \cdot \pi_\lambda = -e^{-i\theta} \cdot \pi_\lambda$, where $\pi_\lambda = 1$ for bosons in the symmetric, $\lambda = [N]$, representation and $\pi_\lambda = -1$ for fermions in the anti-symmetric, $\lambda = [1^N]$, representation. Accordingly, the Bethe equations read as

$$e^{\frac{i}{\hbar}p_j L} = (-1)^{(N-1)\sigma} \prod_{k\neq j}^N e^{i\pi(\beta-1)\text{sign}(p_j-p_k)}\,, \qquad \beta = \frac{\gamma}{\hbar}\,, \tag{5.19}$$

where $\sigma = 0$ for fermions and $\sigma = 1$ for bosons and we restored the the Planck constant $\hbar$. Traditionally, one chooses to order momenta p_j in (5.19) as

$$p_1 < \cdots < p_N\,. \tag{5.20}$$

This is in contrary to the momentum ordering (4.99) that was chosen for the initial scattering states. However, the choice (5.20) does not influence our previous considerations, because the Bethe wave function (5.1) involves summation over all permutations of the momenta and, for this reason, does not depend on the initial momentum ordering. Equation (5.19) can be solved by passing to its logarithmic version

$$p_j = \frac{2\pi\hbar I_j}{L} + \frac{\pi\hbar(\beta - 1)}{L} \sum_{k\neq j}^N \text{sign}(p_j - p_k)\,, \quad j = 1, \ldots, N\,, \tag{5.21}$$

where the numbers $I_j = n_j + \frac{N-1}{2}\sigma$, $n_j \in \mathbb{Z}$, arise from the phase ambiguity. They are integers for bosons with N odd and fermions, and half-odd integers for bosons with N even. The ordering (5.20) implies that $p_k < p_j$ for $k = 1, \ldots, j - 1$, and

$p_k > p_j$ for $k = j + 1, \ldots, N$, one gets

$$p_j = \frac{2\pi \hbar I_j}{L} + \frac{2\pi \hbar (\beta - 1)}{L} \left(j - \frac{N+1}{2} \right), \tag{5.22}$$

We recall that the case $\beta = 1$ corresponds to free fermions or hard-core bosons. For this case $p_j = 2\pi \hbar I_j / L$ are the quantised momenta of particles, provided I_j are distinct (half)-integers satisfying

$$I_1 < \cdots < I_N. \tag{5.23}$$

Note that the ground state of a system with an even number of fermions is doubly degenerate and carries non-zero momentum.[2] Bosons and fermions with N odd can be treated on equal footing and for this case I_j are integers.

Introducing a reduced length parameter $\ell = L/2\pi$, we can write the formula (5.22) as

$$p_j = \frac{\hbar}{\ell} \left(I_j + \frac{N+1}{2} - j + \beta(j - \frac{N+1}{2}) \right). \tag{5.24}$$

A crucial observation is that this formula coincides with the expression for the quasi-momenta (3.197) of the trigonometric CMS model, provided we make the following identification of quantum numbers I_j with integers λ_k parametrising Young diagrams

$$\lambda_{N+1-j} = I_j + \frac{N+1}{2} - j.$$

As was found in Sect. 3.4.2, the exact ground state of the trigonometric model has all λ vanishing, which gives

$$I_j = j - \frac{N+1}{2}$$

and $p_j = \frac{\gamma}{\ell}(j - \frac{N+1}{2})$, N odd, for the ground state. The expressions for the total energy and momentum written in terms of p_j are the same for both models, see (3.198). In the trigonometric model p_j is a quasi-momentum and in the rational model it is an asymptotic momentum. Thus, *solutions of the Bethe Ansatz equations for the rational CMS model yield the exact spectrum of its trigonometric (finite-size) counterpart.* It is worth pointing out that while for the trigonometric model the length parameter $L = 2\pi\ell$ (the period) occurs directly in the hamiltonian, for the rational model it enters the spectrum only through the Bethe equations that follow from the periodicity condition imposed on the wave function.

---

[2] This could be understood from the fact that to minimise energy for N even, one of the momenta must take zero value, so that the odd number of remaining integers I_j is distributed asymmetrically and maximally close to zero in one of two possible ways.

A similar result holds for the rational RS model. For this model the two-body S-matrix coincide with the one for the rational CMS model (4.72). Thus, the Bethe equations have the same form as (5.19) and the corresponding asymptotic momenta coincide with (5.22). The dispersion relation in (3.220) is, however, different from its non-relativistic CMS counterpart and coincides with the one for the trigonometric RS model. Thus, *the Bethe Ansatz for the rational RS model yields the exact spectrum of its trigonometric cousin*, in full analogy, with the corresponding relationship between rational and trigonometric CMS models.

Finally, recalling the S-matrix (4.69) for the hyperbolic CMS model, we also present the Bethe equations for this model with periodic boundary conditions

$$e^{\frac{i}{\hbar}p_j L} = (-1)^{N+1} \prod_{k \neq j}^{N} \frac{\Gamma\big(1 - i\frac{\ell}{\hbar}(p_j - p_k)\big)\Gamma\big(\beta + i\frac{\ell}{\hbar}(p_j - p_k)\big)}{\Gamma\big(1 + i\frac{\ell}{\hbar}(p_j - p_k)\big)\Gamma\big(\beta - i\frac{\ell}{\hbar}(p_j - p_k)\big)} . \tag{5.25}$$

Here a priori no any special relation between the interaction length ℓ and the size of the circle L is assumed. According to the logic of our construction, the Bethe equation (5.25) must encode, at least asymptotically, the spectrum of the elliptic CMS model with the real period equal to L.

5.1.3 Spin Chains and Transfer Matrix

Any further success in solving the system (5.12) relies on finding an appropriate realisation of a representation λ of $\mathfrak{S}_N$, where the permutation operators $\pi_\lambda(\alpha_{ij})$ act in the simplest way.

At least a wide class of such convenient realisations is provided by *spin chains*. Let $V \simeq \mathbb{C}^n$ be an $n = 2s + 1$-dimensional vector space. The configuration space of a spin chain of spin s is given by the N-fold tensor product

$$\mathscr{H} = V_1 \otimes \cdots \otimes V_N \equiv V^{\otimes N}, \quad V_i \simeq V,$$

where N is the number of particles carrying spin. The group $\mathfrak{S}_N$ acts on this space by permutations, in particular, transpositions $(i|j)$ are realised by the matrices π_{ij}

$$\pi_{ij} = \sum_{\alpha,\beta=1}^{n} \mathbb{1} \otimes \cdots \otimes \underset{i}{E_{\alpha\beta}} \otimes \cdots \otimes \underset{j}{E_{\beta\alpha}} \otimes \cdots \otimes \mathbb{1} . \tag{5.26}$$

Here $E_{\alpha\beta}$, $\alpha, \beta = 1, \ldots, n$, are the standard matrix unities, so that (5.26) is nothing but the split Casimir (2.71) with the corresponding components of the tensor product being in i'th and j'th positions. In accordance with this realisation of permutations, the two-body S-matrices S_{ij} are defined as

$$S_{ij}(p_i, p_j) = B(p_i, p_j)\mathbb{1} + A(p_i, p_j)\pi_{ij}.$$

If momenta of two particles coincide, transmission cannot happen, *i.e.* $B(p, p) = 0$, while the conservation of probability $|A(p, p)|^2 = 1$ implies $A(p, p) = e^{i\varphi(p)}$ for some function $\varphi(p)$. It is then follows from (4.124) that $e^{2\varphi(p)} = 1$, *i.e.* $\varphi(p) = 0$ or $\varphi(p) = \pi$. Thus, we have two possibilities for the value of the S-matrix at coincident momenta

$$S_{ij}(p, p) = +\pi_{ij} \quad \text{or} \quad S_{ij}(p, p) = -\pi_{ij}. \tag{5.27}$$

In particular, for the Lieb-Liniger model the second possibility takes place, as is seen from the explicit form (5.16) of the corresponding S-matrix.

Now, to treat all T_j at once and, most importantly, to prove their commutativity, we use the spin chain representation to introduce the concept of transfer matrix. Consider the following object

$$T(p) = \pm S_{1a}(p_1, p) S_{2a}(p_2, p) \ldots S_{Na}(p_N, p), \tag{5.28}$$

where the sign on the right hand side correlates with that in (5.27), *i.e.* it is determined by the behaviour of the S-matrix at coincident momenta. Here "a" stands for an extra copy of V, called "auxiliary space" and p is an associated auxiliary momentum variable. The quantity $T(p)$ is called *monodromy matrix* or simply *monodromy*. The monodromy acts on the space $V^{\otimes N} \otimes V_a$.

Taking the trace of $T(p)$ with respect to the auxiliary space, we obtain an operator

$$\tau(p) = \text{Tr}_a T(p) = \pm \text{Tr}_a \Big[S_{1a}(p_1, p) S_{2a}(p_2, p) \ldots S_{Na}(p_N, p) \Big], \tag{5.29}$$

called *transfer matrix*. The transfer matrix is an operator on the configuration space of the spin chain

$$\tau(p) : \ V^{\otimes N} \to V^{\otimes N}.$$

The fundamental property of the transfer matrix is that it is a generating function for the commuting operators T_j. According to (4.119), the two-body S-matrix for coincident momenta degenerates into a permutation $S_{ij}(p, p) = \pi_{ij}$, which is, of course, compatible with the condition (4.124). We then evaluate the transfer matrix at $p = p_j$

$$\tau(p_j) = \pm \text{Tr}_a \Big[S_{1a}(p_1, p_j) S_{2a}(p_2, p_j) \ldots S_{ja}(p_j, p_j) \ldots S_{Na}(p_N, p_j) \Big]$$
$$= \text{Tr}_a \Big[S_{1a}(p_1, p_j) S_{2a}(p_2, p_j) \ldots \pi_{ja} \ldots S_{Na}(p_N, p_j) \Big].$$

Then, using the brading property (4.120), we pull π_{ja} to the left

$$\tau(p_j) = \text{Tr}_a \Big[\pi_{ja} S_{1j}(p_1, p_j) \ldots S_{j-1j}(p_{j-1}, p_j) \cdot S_{j+1a}(p_{j+1}, p_j) \ldots S_{Na}(p_N, p_j) \Big].$$

Here we separated by · two strings of S-matrices. Since the indices of any S-matrix from one string are different from those for any S-matrix from other string, these two strings commute and we can interchange their position under the trace

$$\tau(p_j) = \text{Tr}_a\Big[\pi_{ja}S_{j+1a}(p_{j+1}, p_j)\ldots S_{Na}(p_N, p_j) \cdot S_{1j}(p_1, p_j)\ldots S_{j-1j}(p_{j-1}, p_j)\Big].$$

Next, we move π_{ja} into the position between two strings to get

$$\tau(p_j) = S_{j+1j}(p_{j+1}, p_j)\ldots S_{Nj}(p_N, p_j) \cdot \text{Tr}_a(\pi_{ja}) \cdot S_{1j}(p_1, p_j)\ldots S_{j-1j}(p_{j-1}, p_j).$$

It remains to note that due to $\text{Tr}_0(\pi_{ja}) = \mathbb{1}$, we get $\tau(p_j) = T_j$, where T_j is given by (5.11).

Working with the spin chain representation and the transfer matrix has an advantage that it allows for an easy proof of commutativity of T_j. It follows from the commutation relation between two monodromy matrices, $T_a(p)$ and $T_b(q)$, each of which is defined with the help of its own independent auxiliary spaces, V_a and V_b, respectively. This commutation relation is derived as follows. Let us consider

$$S_{ab}(p, q)T_b(q)T_a(p),\tag{5.30}$$

where S_{ab} is the S-matrix which acts on auxiliary spaces only. Using the definition of the monodromies, we write

$$S_{ab}(p, q)T_b(q)T_a(p) = S_{ab}(p, q)S_{1b}(p_1, q)\ldots S_{Nb}(p_N, q)S_{1a}(p_1, p)\ldots S_{Na}(p_N, p)$$
$$= S_{ab}(p, q)S_{1b}(p_1, q)S_{1a}(p_1, p)\Big(S_{2b}(p_1, q)\ldots S_{Nb}(p_N, q)S_{2a}(p_1, p)\ldots S_{Na}(p_N, p)\Big),$$

Here we freely moved the matrix S_{1a} next to S_{1b} because it commutes with all the matrices on its way until it meets S_{1b}. Now we can use the Yang-Baxter equation

$$S_{ab}(p, q)S_{1b}(p_1, q)S_{1a}(p_1, p) = S_{1a}(p_1, p)S_{1b}(p_1, q)S_{ab}(p, q)$$

that yields at this stage the following answer

$$S_{ab}(p, q)T_b(q)T_a(p) =$$
$$S_{1a}(p_1, p)S_{1b}(p_1, q)S_{ab}(p, q)\Big(S_{2b}(p_1, q)\ldots S_{Nb}(p_N, q)S_{2a}(p_1, p)\ldots S_{Na}(p_N, p)\Big).$$

Clearly, the matrices S_{1a} and S_{1b} interchanged their initial order and S_{ab} stands again in front of monodromies, the latter being reduced by the elements S_{1a} and S_{1b}. Clearly, we can now repeat the same manipulation for S_{2b} and S_{2a}, and so on until we commute with the help of repeated application of the Yang-Baxter equation the matrix S_{ab} to the right of all the matrices. As a result of these manipulations, we obtain the the following commutation relation between the components of the monodromy matrix

$$S_{ab}(p, q)T_b(q)T_a(p) = T_a(p)T_b(q)S_{ab}(p, q) \,. \tag{5.31}$$

This relation is of fundamental importance, it provides a starting point for the algebraic Bethe Ansatz approach, which will be discussed later. Here we note that (5.31) immediately implies commutativity of T_j. Indeed, we rewrite it as

$$T_a(p)T_b(q) = S_{ab}(p, q)T_b(q)T_a(p)S_{ab}^{-1}(p, q)$$

and then take the trace with respect to each of the two auxiliary spaces. This gives

$$\tau(p)\tau(q) = \mathrm{Tr}_{a,b}\Big(S_{ab}(p, q)T_b(q)T_a(p)S_{ab}^{-1}(p, q)\Big) = \tau(q)\tau(p) \,.$$

Thus, the values of the transfer matrix at different values of momenta commute

$$\tau(p)\tau(q) = \tau(q)\tau(p) \,. \tag{5.32}$$

Taking $p = p_j$ and $q = p_k$ completes the argument.

Considering the system (5.12) in the context of the spin chain representation, we can replace the problem of diagonalising the set $\{T_j\}$ by an equivalent problem of diagonalising the transfer matrix $\tau(p)$ for all values of p. If we denote by $\Lambda(p)$ an eigenvalue of the transfer matrix, then (5.12) results into a set of *Bethe equation*

$$e^{ip_j L} = \Lambda(p_j|\{p_k\}) \,, \quad j = 1, \ldots, N \,, \tag{5.33}$$

where by $\{p_k\}$ we have indicated an implicit dependence of the eigenvalue on all the other momenta than p_j. Equation (5.33) are implications of the periodicity condition for the real space wave function and they can be thought of as the quantisation conditions for asymptotic momenta. The range of their applicability is the same as of the asymptotic wave function.

5.1.4 Yang's Spin-$\frac{1}{2}$ Problem

The simplest physically relevant case of the delta-interaction model for particles with internal degrees of freedom corresponds to spin-$\frac{1}{2}$ fermions (electrons). In this case the S-matrix (5.18) is non-abelian and diagonalisation of the operators T_j becomes a non-trivial task. Here we discuss some group-theoretic aspects of this problem and explain the usefulness of the spin chain representation introduced earlier.

According to our treatment in Sect. 4.3.3, the spin wave function of electrons must transform in the representation $[N - M, M]$ of $\mathfrak{S}_N$, while the coordinate wave function must furnish the corresponding conjugate representation $[2^M 1^{N-2M}]$. Since the relationship between the original representation matrix $\pi(\sigma)$, $\sigma \in \mathfrak{S}_N$, and the matrix of the corresponding conjugate representation consists in replacing $\pi(\sigma) \rightarrow$

sign$(\sigma)\pi(\sigma)$, the S-matrix in $[2^M 1^{N-2M}]$ is simply given by the expression (5.18) evaluated in the representation $\lambda = [N - M, M]$ but with the opposite sign in front of the transposition $\pi_\lambda(\alpha_{ij})$. Working directly with the representation $[N - M, M]$ has the advantage that the latter admits a simple realisation in terms of a closed chain of spin $\frac{1}{2}$-particles, which we now describe in more detail.

Consider a discrete circle which is a collection of ordered points labelled by the index n with the identification $n \equiv n + N$ reflecting periodic boundary conditions. Here N is a positive integer which plays the role of the length (volume) of the space. The numbers $n = 1, \ldots, N$ form the fundamental domain. To each integer n along the chain we associate a two-dimensional vector space $V = \mathbb{C}^2$ with a basis

$$|\uparrow\rangle \equiv \begin{pmatrix} 1 \\ 0 \end{pmatrix}, \qquad |\downarrow\rangle \equiv \begin{pmatrix} 0 \\ 1 \end{pmatrix}. \qquad (5.34)$$

As this notation suggests, the basis elements are identified as "spin up" and "spin down", see Fig. 5.1. Next, we introduce a local spin algebra generated by the spin variables S_n^α, where $\alpha = 1, 2, 3$, with commutation relations

$$[S_m^\alpha, S_n^\beta] = i\hbar\epsilon^{\alpha\beta\gamma}S_n^\gamma\delta_{mn}. \qquad (5.35)$$

The spin operators have the following realization in terms of the standard Pauli matrices: $S_n^\alpha = \frac{\hbar}{2}\sigma^\alpha$. Spin variables are subject to the periodic boundary condition $S_n^\alpha \equiv S_{n+N}^\alpha$. In what follows we choose $\hbar = 1$ and introduce the raising and lowering operators $S_n^\pm = S_n^1 \pm i S_n^2$. They are realised as

$$S_n^+ = \begin{pmatrix} 0 & 1 \\ 0 & 0 \end{pmatrix}, \qquad S_n^- = \begin{pmatrix} 0 & 0 \\ 1 & 0 \end{pmatrix}.$$

The spin operators $S_n^\pm$, S_n^3 acts non-trivially only on the n'th site of the chain, where this action reads as

$$S_n^+|\uparrow_n\rangle = 0, \qquad S_n^+|\downarrow_n\rangle = |\uparrow_n\rangle, \qquad S_n^3|\uparrow_n\rangle = \frac{1}{2}|\uparrow_n\rangle,$$
$$S_n^-|\downarrow_n\rangle = 0, \qquad S_n^-|\uparrow_n\rangle = |\downarrow_n\rangle, \qquad S_n^3|\downarrow_n\rangle = -\frac{1}{2}|\downarrow_n\rangle.$$

The Hilbert space of the spin chain carries a tensor product representation of the Lie algebra $\mathfrak{sl}(2)$; the corresponding generators are realised as

Fig. 5.1 Spin chain. A state of the spin chain can be represented as $|\psi\rangle = |\uparrow\downarrow\uparrow\downarrow\downarrow \cdots \uparrow\downarrow\rangle$

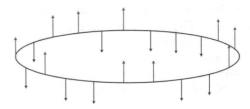

$$S^\alpha = \sum_{n=1}^{N} \mathbb{1} \otimes \cdots \otimes S_n^\alpha \otimes \cdots \otimes \mathbb{1} . \tag{5.36}$$

In particular, $S^\pm = S^1 \pm i S^2$ are the raising and lowering operators.

Finally, we allow the S-matrix to act in the Hilbert space of the spin chain by means of the following formula

$$S_{ij}(p_i, p_j) = \frac{(p_i - p_j)\mathbb{1} + i \varkappa \pi_{ij}}{p_i - p_j + i \varkappa} . \tag{5.37}$$

In comparison to (5.18), we have changed the sign in front of the permutation part and realised the transposition α_{ij} by means of (5.26). In particular, the permutation π can be represented via the Pauli matrices as

$$\pi = \frac{1}{2}(\mathbb{1} \otimes \mathbb{1} + \sigma^\alpha \otimes \sigma^\alpha) . \tag{5.38}$$

At coinciding momenta $S_{ij}(p, p) = \pi_{ij}$, which realises the first of two possibilities in (5.27).

The irreducible representation $[N - M, M]$ we are interested in has the dimension[3]

$$\dim[N - M, M] = \frac{N!(N - 2M + 1)}{(N - M + 1)!M!} . \tag{5.39}$$

To identify the corresponding invariant subspace in the 2^N-dimensional Hilbert space of the spin chain, consider all states of the chain which have $N - M$ spins up and the remaining M spins down. Obviously, there are

$$C_N^M = \frac{N!}{(N - M)!M!} \tag{5.40}$$

such states. They can be dissected in the following way. Let us start from a unique state with all spins up. This state spans a representation space of the 1-dimensional symmetric representation $[N]$ of $\mathfrak{S}_N$. Let us apply to this state the lowering operator S^-. This will produce an $\mathfrak{sl}(2)$ descendent which is a certain linear combination of states with one spin down. On the other hand, there are N independent states with one spin down and, as we see, one of these states is the $\mathfrak{sl}(2)$ descendent, while the remaining states must be $\mathfrak{sl}(2)$ primaries. Thus, the N states with one spin down can be schematically decomposed as

$$N = N - 1 + S^-(1) ,$$

---

[3]This can be easily derived from the determinant formula (4.89).

Fig. 5.2 The $\mathfrak{sl}(2)$-multiplet structure of the $M = 0, 1, 2$ subspaces. Arrows indicate the application of the lowering operator S^-

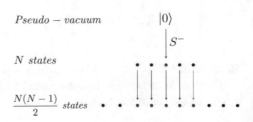

where $S^-(1)$ is the descendent and the other $N - 1$ states give rise to new independent multiplets of $\mathfrak{sl}(2)$. Note that the number $N - 1$ coincides with the dimension of the representation $[N - 1, 1]$ of $\mathfrak{S}_N$. Now, applying S^- to all states with one spin down, we obtain N descendants among all $N(N - 1)/2$ two-spin-down states, see Fig. 5.2. Schematically,

$$\frac{N(N-1)}{2} = \frac{N(N-3)}{2} + S^-(N-1) + S^-S^-(1).$$

Here again, the number $N(N - 3)/2$ is, on the one hand, is a number of new multiplets of $\mathfrak{sl}(2)$ and, on the other hand, is the dimension of the irreducible representation $[N - 2, 2]$, cf. (5.39). Then we repeat this procedure of descending again and again until we reach the representation $[N - M, M]$. Thus, the space of states with M spin downs has the following $\mathfrak{sl}(2) \times \mathfrak{S}_N$ structure[4]

$$C_N^M \text{ states} \simeq \sum_{k=0}^{M} (S^-)^{M-k} [N - k, k]. \tag{5.41}$$

From the point of view of just $\mathfrak{S}_N$ the space of states with M spins down form a reducible representation π_χ that decomposes as

$$\pi_\chi = [N] + [N - 1, 1] + [N - 2, 2] \ldots + [N - M, M]. \tag{5.42}$$

In this decomposition the irreducible component $[N - M, M]$ is characterised by the fact that it comprises the highest weight states of the $\mathfrak{sl}(2)$-multiplets. The formulae (5.41) and (5.42) correctly reproduce the counting of the dimensions

$$\dim \pi_\chi = \sum_{k=0}^{k=M} \dim[N - k, k] = \sum_{k=0}^{k=M} \frac{N!(N - 2k + 1)}{(N - k + 1)!k!} = \frac{N!}{M!(N - M)!}.$$

[4]This structure is an implementation of the Schur-Weyl duality.

5.2 Algebraic Bethe Ansatz

After having understood the combinatorial and group-theoretical properties of the spin chain representation, we are facing the problem of diagonalising the transfer matrix (5.29) in this representation. This will be done with a special technique known under the name *algebraic Bethe Ansatz* [5], which represents a far-reaching embodiment of the general Bethe Ansatz idea. Introducing the R-matrix

$$R_{ab}(p_a, p_b) = S_{ab}^{-1}(p_a, p_b) = \frac{(p_a - p_b)\mathbb{1} - i\varkappa\pi_{ab}}{p_a - p_b - i\varkappa}, \tag{5.43}$$

we rewrite the commutation relations (5.31) between the entries of the monodromy matrix in the form

$$R_{ab}(p_a, p_b)T_a(p_a)T_b(p_b) = T_b(p_b)T_a(p_a)R_{ab}(p_a, p_b), \tag{5.44}$$

where we recall that the indices a and b stand for two auxiliary spaces. One can undoubtedly notice a formal resemblance of (5.44) with the definition (3.10) of the formal quantum group, except that now we are dealing with the R-matrix and T that continuously depend on the parameter p. The relation (5.44) is an essential starting point in the algebraic Bethe Ansatz approach to diagonalisation of the transfer matrix. According to the terminology of [19], (5.44) comprise the *fundamental commutation relations*.

To proceed, we first point out that the monodromy (5.28) and the corresponding transfer matrix (5.29) are *inhomogeneous*, with $p_1, \ldots, p_N$ playing the role of inhomogeneities. Second, using the explicit form (5.37) of the S-matrix, we obtain the following expression for the monodromy

$$T_a(p) = \Omega(p) \prod_{j=1}^{N} \Big((p - p_j)\mathbb{1} - i\varkappa\pi_{ja} \Big), \tag{5.45}$$

where

$$\Omega(p) = \prod_{j=1}^{N} \frac{1}{p - p_j - i\varkappa} \tag{5.46}$$

is a scalar prefactor which was singled out for later convenience. Every term in the product (5.45) acts as a 2×2 matrix in the auxiliary space, which we can write down explicitly with the help of the local spin operators S_j^3 and $S_j^{\pm}$ as

$$L_{ja}(p) \equiv (p - p_j)\mathbb{1} - i\varkappa\pi_{ja} = \begin{pmatrix} p - p_j - \frac{i\varkappa}{2} - i\varkappa S_j^3 & -i\varkappa S_j^- \\ -i\varkappa S_j^+ & p - p_j - \frac{i\varkappa}{2} + i\varkappa S_j^3 \end{pmatrix}. \tag{5.47}$$

In the context of the Inverse Scattering Method, which is another name for the Algebraic Bethe Ansatz, $L_{ja}(p)$ is called Lax operator. The definition of the Lax operator involves the local "quantum" space $V_j \simeq \mathbb{C}^2$. The Lax operator L_{ja} acts in $V_j \otimes V_a$:

$$L_{ja}(p): \quad V_j \otimes V_a \to V_j \otimes V_a. \tag{5.48}$$

We will have to say more about this concept when we come to the discussion of the Heisenberg magnet.

Taking the ordered product (5.45) over all sites of the chain, we obtain a realisation of the monodromy as the 2×2 matrix acting in the auxiliary space and we parametrise its entries as

$$T(p) = \begin{pmatrix} A(p) & B(p) \\ C(p) & D(p) \end{pmatrix}. \tag{5.49}$$

Here $A(p), \ldots, D(p)$ are operators that act on the Hilbert space of the spin chain; they implicitly depend on the inhomogeneities $p_1, \ldots, p_N$. The relations between these operators follow from the fundamental commutation relations (5.44) and those which we need here are

$$B(p)B(q) = B(q)B(p),$$

$$A(p)B(q) = \frac{p-q+i\varkappa}{p-q} B(q)A(p) - \frac{i\varkappa}{p-q} B(p)A(q),$$

$$D(p)B(q) = \frac{p-q-i\varkappa}{p-q} B(q)D(p) + \frac{i\varkappa}{p-q} B(p)D(q). \tag{5.50}$$

The transfer matrix which we aim to diagonalise is given by the following operator

$$\tau(p) = \mathrm{Tr}_a(T_a) = A(p) + D(p). \tag{5.51}$$

The main idea of the algebraic Bethe Ansatz relies on the existence of a reference state $|0\rangle$, also called pseudo-vacuum, such that $C(p)|0\rangle = 0$ for any p and the eigenvectors of $\tau(p)$ with M spins down have the form

$$|\lambda_1, \lambda_2, \cdots, \lambda_M\rangle = B(\lambda_1)B(\lambda_2) \cdots B(\lambda_M)|0\rangle, \tag{5.52}$$

where $\{\lambda_i\}$ are unequal numbers called *Bethe roots*. In our present case the pseudo-vacuum can be naturally identified with the state

$$|0\rangle = \overset{N}{\underset{n=1}{\otimes}} |\uparrow_n\rangle, \tag{5.53}$$

that is the unique state with all spins up. Indeed, since (5.47) acts on this state as

$$L_{ja}(p)|\uparrow_n\rangle = \begin{pmatrix} (p - p_j - i\varkappa)|\uparrow_n\rangle & i\varkappa|\downarrow_n\rangle \\ 0 & (p - p_j)|\uparrow_n\rangle \end{pmatrix},$$

we find that

$$T(p)|0\rangle = \Omega(p) \begin{pmatrix} \prod\limits_{j=1}^{N}(p - p_j - i\varkappa)|0\rangle & \bigstar \\ 0 & \prod\limits_{j=1}^{N}(p - p_j)|0\rangle \end{pmatrix},$$

where $\bigstar$ stands for terms whose explicit form is irrelevant for our further treatment. Taking into account (5.46), we thus have

$$C(p)|0\rangle = 0, \qquad A(p)|0\rangle = |0\rangle, \qquad D(p)|0\rangle = \prod_{j=1}^{N} \frac{p - p_j}{p - p_j - i\varkappa}|0\rangle, \quad (5.54)$$

where the first relation confirms the status of $|0\rangle$ as the pseudo-vacuum. The second two relations show that $|0\rangle$ is an eigenstate of the transfer matrix

$$\tau(p)|0\rangle = \left[1 + \prod_{j=1}^{N} \frac{p - p_j}{p - p_j - i\varkappa}\right]|0\rangle.$$

Excited states are then obtained by multiple application of the "raising operator" B to the vacuum, in accordance with the formula (5.52). These states will form the eigenstates of the transfer matrix, provided the Bethe roots $\{\lambda_i\}$ satisfy certain restrictions, which we are going to determine.

An explicit computation done for small M indicates that the result of acting with $A(p)$ on the state (5.52) should have the following structure

$$A(p)B(\lambda_1)B(\lambda_2)\cdots B(\lambda_M)|0\rangle = \left(\prod_{n=1}^{M} \frac{p - \lambda_n + i\varkappa}{p - \lambda_n}\right)B(\lambda_1)B(\lambda_2)\cdots B(\lambda_M)|0\rangle$$

$$+ \sum_{n=1}^{M} W_n^A(p, \{\lambda_i\})B(p) \prod_{\substack{j=1 \\ j\neq n}}^{M} B(\lambda_j)|0\rangle.$$

Here the coefficients $W_n^A(p, \{\lambda_i\})$ depend on p and the set of Bethe roots $\{\lambda_i\}_{i=1}^{M}$. To determine these coefficients we note that since the operators $B(\lambda)$ commute with each other at different values of λ, we can write

$$|\lambda_1, \lambda_2, \cdots, \lambda_M\rangle = B(\lambda_n) \prod_{\substack{j=1 \\ j\neq n}}^{M} B(\lambda_j)|0\rangle.$$

Thus,

$$A(p)|\lambda_1, \lambda_2, \cdots, \lambda_M\rangle = \frac{p - \lambda_n + i\varkappa}{p - \lambda_n} B(\lambda_n) A(p) \prod_{\substack{j=1 \\ j \neq n}}^{M} B(\lambda_j)|0\rangle$$

$$- \frac{i\varkappa}{p - \lambda_n} B(p) A(\lambda_n) \prod_{\substack{j=1 \\ j \neq n}}^{M} B(\lambda_j)|0\rangle .$$

It is clear from this equation that only the second term on its right hand side will contribute to W_n^A since this term does not contain $B(\lambda_n)$. On the other hand, moving in this term $A(\lambda_n)$ past the string of $B(\lambda_j)$, we see that the only way to avoid the appearance of $B(\lambda_n)$ is to restrict an application of the commutation relation (5.50) to the first term on its right hand side. With this restricted application, we pull the operator $A(\lambda_n)$ through all $B(\lambda_j)$ close to the pseudo-vacuum and, taking into account the second equation in (5.54), find the following contribution

$$- \frac{i\varkappa}{p - \lambda_n} \prod_{\substack{i=1 \\ i \neq n}}^{M} \frac{\lambda_n - \lambda_i + i\varkappa}{\lambda_n - \lambda_i} B(p) \prod_{\substack{j=1 \\ j \neq n}}^{M} B(\lambda_j)|0\rangle ,$$

from which we read off the coefficient W_n^A

$$W_n^A(p, \{\lambda_i\}) = - \frac{i\varkappa}{p - \lambda_n} \prod_{\substack{j=1 \\ j \neq n}}^{M} \frac{\lambda_n - \lambda_j + i\varkappa}{\lambda_n - \lambda_j} .$$

We should point out that this expression for W_n^A is a non-trivial result that comes from cancelling many individual terms arising upon the use of the full commutation relation (5.50). In the same way we obtain

$$D(p) B(\lambda_1) B(\lambda_2) \cdots B(\lambda_M)|0\rangle =$$

$$= \left(\prod_{j=1}^{N} \frac{p - p_j}{p - p_j - i\varkappa} \right) \left(\prod_{n=1}^{M} \frac{p - \lambda_n - i\varkappa}{p - \lambda_n} \right) B(\lambda_1) B(\lambda_2) \cdots B(\lambda_M)|0\rangle$$

$$+ \sum_{n=1}^{M} W_n^D(p, \{\lambda_i\}) B(p) \prod_{\substack{j=1 \\ j \neq n}}^{M} B(\lambda_j)|0\rangle ,$$

where

$$W_n^D(p, \{\lambda_i\}) = \frac{i\varkappa}{p - \lambda_n} \prod_{j=1}^{N} \frac{\lambda_n - p_j}{\lambda_n - p_j - i\varkappa} \prod_{\substack{j=1 \\ j \neq n}}^{M} \frac{\lambda_n - \lambda_j - i\varkappa}{\lambda_n - \lambda_j}.$$

Thus, we will solve the eigenvalue problem

$$\tau(p)|\lambda_1, \cdots, \lambda_M\rangle = \Lambda(p, \{\lambda_n\})|\lambda_1, \cdots, \lambda_M\rangle$$

with

$$\Lambda(p, \{\lambda_n\}) = \prod_{n=1}^{M} \frac{p - \lambda_n + i\varkappa}{p - \lambda_n} + \prod_{j=1}^{N} \frac{p - p_j}{p - p_j - i\varkappa} \prod_{n=1}^{M} \frac{p - \lambda_n - i\varkappa}{p - \lambda_n},$$

provided $W_n^A + W_n^D = 0$ for all n, which means that

$$\prod_{\substack{j=1 \\ j \neq n}}^{M} \frac{\lambda_n - \lambda_j + i\varkappa}{\lambda_n - \lambda_j} = \prod_{j=1}^{N} \frac{\lambda_n - p_j}{\lambda_n - p_j - i\varkappa} \prod_{\substack{j=1 \\ j \neq n}}^{M} \frac{\lambda_n - \lambda_j - i\varkappa}{\lambda_n - \lambda_j}. \tag{5.55}$$

Making a uniform shift of all Bethe roots $\lambda_n \to \lambda_n + \frac{i\varkappa}{2}$, we rewrite the above equations as

$$\prod_{j=1}^{N} \frac{p_j - \lambda_n + \frac{i\varkappa}{2}}{p_j - \lambda_n - \frac{i\varkappa}{2}} = \prod_{j \neq n}^{M} \frac{\lambda_n - \lambda_j - i\varkappa}{\lambda_n - \lambda_j + i\varkappa}. \tag{5.56}$$

These are the *Bethe equations*. Their solutions for the set $\{\lambda_j\}_{j=1}^{M}$ enumerate the eigenstates of the transfer matrix.

Solution of Yang's spin-$\frac{1}{2}$ problem. Thus, Yang's fermion spin-$\frac{1}{2}$ problem reduces to the following set of equations [2]

$$e^{ip_jL} = \prod_{n=1}^{M} \frac{p_j - \lambda_n + \frac{i\varkappa}{2}}{p_j - \lambda_n - \frac{i\varkappa}{2}}, \qquad j = 1, \dots, N, \tag{5.57}$$

$$\prod_{j=1}^{N} \frac{p_j - \lambda_n + \frac{i\varkappa}{2}}{p_j - \lambda_n - \frac{i\varkappa}{2}} = \prod_{j \neq n}^{M} \frac{\lambda_n - \lambda_j - i\varkappa}{\lambda_n - \lambda_j + i\varkappa}. \qquad n = 1, \dots, M. \tag{5.58}$$

Here (5.57) are the Eq. (5.33) for the eigenvalues of the transfer matrix and they express the periodicity condition for the coordinate Bethe wave function. In the present context the variables p_j are called *momentum carrying roots*, while the variables λ_n are usually referred to as *auxiliary roots*. We see that Eq. (5.58) for auxiliary roots are algebraic and they involve momentum carrying roots as parameters. Once the momenta p_j are found the energy of the state described by the corresponding

Bethe wave function is determined as

$$E = \frac{1}{2} \sum_{j=1}^{N} p_j^2 . \tag{5.59}$$

The system of Eqs. (5.57), (5.33) is an example of the so-called *nested Bethe Ansatz*, with one level of nesting given by equations for auxiliary roots. The term *nesting* originates from the hierarchical way of applying the Bethe Ansatz technique for diagonalising an auxiliary spin chain with spins transforming in an arbitrary irreducible representation of the symmetric group $\mathfrak{S}_N$ [20, 21].

To make a connection of our findings with the representation theory of $\mathfrak{sl}(2) \times \mathfrak{S}_N$, let us rewrite the fundamental commutation relations (5.44) in the form

$$R_{ab}(\lambda - \mu) T_a(\lambda) T_b(\mu) = T_b(\mu) T_a(\lambda) R_{ab}(\lambda - \mu) , \tag{5.60}$$

where the momentum variables p_a, p_b were replaced by λ and μ which play the role of *spectral parameters*. We study the behaviour of (5.60) in the limit $\mu \to \infty$. From (5.47) we obtain that in this limit the monodromy expands as

$$T_b(\mu) = \mathbb{1} + \frac{i\varkappa}{\mu} \left(\frac{N}{2} - \sum_{j=1}^{N} S_j^\alpha \otimes \sigma_b^\alpha \right) + \dots , \tag{5.61}$$

where we recognised in the sum $\sum_{j=1}^{N} S_j^\alpha$ the generator (5.36) of the global $\mathfrak{sl}(2)$ algebra and σ_b^α denotes the corresponding Pauli matrix acting in the auxiliary space. Then the relation (5.60) expands as

$$\left((\lambda - \mu)\mathbb{1} - i\varkappa \pi_{ab} \right) T_a(\lambda) \left(\mathbb{1} + \frac{i\varkappa}{\mu} \left(\frac{N}{2} - S^\alpha \otimes \sigma^\alpha \right) + \dots \right) =$$
$$= \left(\mathbb{1} + \frac{i\varkappa}{\mu} \left(\frac{N}{2} - S^\alpha \otimes \sigma^\alpha \right) + \dots \right) T_a(\lambda) \left((\lambda - \mu)\mathbb{1} - i\varkappa \pi_{ab} \right) .$$

Here the leading term in the large μ expansion cancels out and the subleading contribution yields the relation

$$[\pi_{ab}, T_a(\lambda)] + [S^\alpha, T_a(\lambda)] \otimes \sigma^\alpha = 0 . \tag{5.62}$$

Writing π_{ab} via Pauli matrices (5.38), we conclude that (5.62) implies the fulfilment of the following relation

$$[S^\alpha, T_a(\lambda)] = [T_a(\lambda), \tfrac{1}{2}\sigma_a^\alpha] . \tag{5.63}$$

The spin operator S^α acts on the Hilbert space $\mathscr{H}$ of the spin chain. On the left hand side of (5.63) one has the commutator of this operator with each entry of the 2×2

monodromy matrix (5.49), the latter being also operators on $\mathscr{H}$. On the right hand side, one finds a matrix commutator of the monodromy matrix with the corresponding Pauli matrix in the auxiliary space. Thus, Eq. (5.63) is equivalent to three distinct equations

$$[S^3, T_a(\lambda)] = \frac{1}{2}[T_a(\lambda), \sigma_a^3] = \begin{pmatrix} 0 & -B(\lambda) \\ C(\lambda) & 0 \end{pmatrix},$$

$$[S^+, T_a(\lambda)] = [T_a(\lambda), \sigma_a^+] = \begin{pmatrix} -C(\lambda) & A(\lambda) - D(\lambda) \\ 0 & C(\lambda) \end{pmatrix}$$

and

$$[S^-, T_a(\lambda)] = [T_a(\lambda), \sigma_a^-] = \begin{pmatrix} B(\lambda) & 0 \\ D(\lambda) - A(\lambda) & -B(\lambda) \end{pmatrix}.$$

Essentially, we need the following commutation relations

$$[S^3, B] = -B, \qquad [S^+, B] = A - D. \tag{5.64}$$

The action of the symmetry generators on the pseudo-vacuum is

$$S^+|0\rangle = 0, \qquad S^3|0\rangle = \frac{N}{2}|0\rangle.$$

Therefore, the pseudo-vacuum is the highest weight state of the $\mathfrak{sl}(2)$ algebra. With the help of (5.64) we then compute

$$S^3|\lambda_1, \cdots, \lambda_M\rangle = \left(\frac{N}{2} - M\right)|\lambda_1, \cdots, \lambda_M\rangle$$

and

$$S^+|\lambda_1, \cdots, \lambda_M\rangle = \sum_j B(\lambda_1) \ldots B(\lambda_{j-1})(A(\lambda_j) - D(\lambda_j))B(\lambda_{j+1}) \ldots B(\lambda_M)|0\rangle.$$

An attentive look at the last expression reveals that it can be re-expanded as

$$S^+|\lambda_1, \cdots, \lambda_M\rangle = \sum_{n=1}^{M} O_n B(\lambda_1) \ldots B(\lambda_{n-1})\cancel{B(\lambda_n)}B(\lambda_{n+1}) \ldots B(\lambda_M)|0\rangle,$$

where the crossed out term does not appear in the sum. The coefficients O_n are unknown but they can be calculated by invoking the arguments similar to those used for computing W_n^A and W_n^D. The only contributions to O_n will come from

$$B(\lambda_1) \ldots B(\lambda_{k-1})(A(\lambda_k) - D(\lambda_k))B(\lambda_{k+1}) \ldots B(\lambda_M)|0\rangle \qquad \text{with} \quad k \leqslant n.$$

If $k = n$ this contribution will be

$$\prod_{j=n+1}^{M} \frac{\lambda_n - \lambda_j + i\varkappa}{\lambda_n - \lambda_j} - \prod_{j=n+1}^{M} \frac{\lambda_n - \lambda_j - i\varkappa}{\lambda_n - \lambda_j} \prod_{j=1}^{N} \frac{\lambda_n - p_j}{\lambda_n - p_j - i\varkappa}$$

and if $k < n$ the contribution will be

$$W_n^A(\lambda_k, \{\lambda\}_{k+1}^M) - W_n^D(\lambda_k, \{\lambda\}_{k+1}^M),$$

where it is convenient to represent W_n^A and W_n^D in the following split form

$$W_n^A(\lambda_k, \{\lambda\}_{k+1}^M) = \frac{i\varkappa}{\lambda_n - \lambda_k} \prod_{j=n+1}^{M} \frac{\lambda_n - \lambda_j + i\varkappa}{\lambda_n - \lambda_j} \prod_{j=k+1}^{n-1} \frac{\lambda_n - \lambda_j + i\varkappa}{\lambda_n - \lambda_j},$$

$$W_n^D(\lambda_k, \{\lambda\}_{k+1}^M) = -\frac{i\varkappa}{\lambda_n - \lambda_k} \prod_{j=n+1}^{M} \frac{\lambda_n - \lambda_j - i\varkappa}{\lambda_n - \lambda_j} \prod_{j=k+1}^{n-1} \frac{\lambda_n - \lambda_j - i\varkappa}{\lambda_n - \lambda_j} \prod_{j=1}^{N} \frac{\lambda_n - p_j}{\lambda_n - p_j - i\varkappa}.$$

Thus, adding up, we obtain

$$
\begin{aligned}
O_n &= \prod_{j=n+1}^{M} \frac{\lambda_n - \lambda_j + i\varkappa}{\lambda_n - \lambda_j} + \sum_{k=1}^{n-1} W_n^A(\lambda_k, \{\lambda\}_{k+1}^M) \\
&\quad - \prod_{j=n+1}^{M} \frac{\lambda_n - \lambda_j - i\varkappa}{\lambda_n - \lambda_j} \prod_{j=1}^{N} \frac{\lambda_n - p_j}{\lambda_n - p_j - i\varkappa} - \sum_{k=1}^{n-1} W_n^D(\lambda_k, \{\lambda\}_{k+1}^M) = \\
&= \prod_{j=n+1}^{M} \frac{\lambda_n - \lambda_j + i\varkappa}{\lambda_n - \lambda_j} \left(1 + \sum_{k=1}^{n-1} \frac{i\varkappa}{\lambda_n - \lambda_k} \prod_{j=k+1}^{n-1} \frac{\lambda_n - \lambda_j + i\varkappa}{\lambda_n - \lambda_j} \right) \\
&\quad - \prod_{j=n+1}^{M} \frac{\lambda_n - \lambda_j - i\varkappa}{\lambda_n - \lambda_j} \prod_{j=1}^{N} \frac{\lambda_n - p_j}{\lambda_n - p_j - i\varkappa} \left(1 - \sum_{k=1}^{n-1} \frac{i\varkappa}{\lambda_n - \lambda_j} \prod_{j=k+1}^{n-1} \frac{\lambda_n - \lambda_j - i\varkappa}{\lambda_n - \lambda_j} \right).
\end{aligned}
$$

To proceed, we note the following useful identity

$$t_m \equiv 1 + \sum_{k=m}^{n-1} \frac{i\varkappa}{\lambda_n - \lambda_k} \prod_{j=k+1}^{n-1} \frac{\lambda_n - \lambda_j + i\varkappa}{\lambda_n - \lambda_j} = \prod_{j=m}^{n-1} \frac{\lambda_n - \lambda_j + i\varkappa}{\lambda_n - \lambda_j}. \quad (5.65)$$

We will prove this identity by induction over m. For $m = n - 1$ and $m = n - 2$ we have

$$t_{n-1} = 1 + \frac{i\varkappa}{\lambda_n - \lambda_{n-1}} = \frac{\lambda_n - \lambda_{n-1} + i\varkappa}{\lambda_n - \lambda_{n-1}},$$

$$t_{n-2} = 1 + \frac{i\varkappa}{\lambda_n - \lambda_{n-1}} + \frac{i\varkappa}{\lambda_n - \lambda_{n-2}} \frac{\lambda_n - \lambda_{n-1} + i\varkappa}{\lambda_n - \lambda_{n-1}} = \frac{\lambda_n - \lambda_{n-1} + i\varkappa}{\lambda_n - \lambda_{n-1}} \frac{\lambda_n - \lambda_{n-2} + i\varkappa}{\lambda_n - \lambda_{n-2}}.$$

Now we suppose that the formula holds for $m = l$, then we have

$$t_{l-1} = t_l + \frac{i\varkappa}{\lambda_n - \lambda_{l-1}} \prod_{j=l}^{n-1} \frac{\lambda_n - \lambda_j + i\varkappa}{\lambda_n - \lambda_j} = \prod_{j=l-1}^{n-1} \frac{\lambda_n - \lambda_j + i\varkappa}{\lambda_n - \lambda_j},$$

which proves the identity. With formula (5.65) at hand we get

$$1 + \sum_{k=1}^{n-1} \frac{i\varkappa}{\lambda_n - \lambda_k} \prod_{j=k+1}^{n-1} \frac{\lambda_n - \lambda_j + i\varkappa}{\lambda_n - \lambda_j} = \prod_{j=1}^{n-1} \frac{\lambda_n - \lambda_j + i\varkappa}{\lambda_n - \lambda_j}.$$

In the same way one can show that

$$1 - \sum_{k=1}^{n-1} \frac{i\varkappa}{\lambda_n - \lambda_j} \prod_{j=k+1}^{n-1} \frac{\lambda_n - \lambda_j - i\varkappa}{\lambda_n - \lambda_j} = \prod_{j=1}^{n-1} \frac{\lambda_n - \lambda_j - i\varkappa}{\lambda_n - \lambda_j}.$$

This, we found for O_n the following answer

$$O_n = \prod_{\substack{j=1 \\ j \neq n}}^{M} \frac{\lambda_n - \lambda_j + i\varkappa}{\lambda_n - \lambda_j} - \prod_{\substack{j=1 \\ j \neq n}}^{M} \frac{\lambda_n - \lambda_j - i\varkappa}{\lambda_n - \lambda_j} \prod_{j=1}^{N} \frac{\lambda_n - p_j}{\lambda_n - p_j - i\varkappa}.$$

It is quite remarkable that this expression is nothing else but the Bethe equation (5.55) for the root λ_n, and, therefore, if the Bethe equations are satisfied all the coefficients O_n vanish. This proves that the eigenstates of the transfer matrix are annihilated by the raising operator S^+, i.e. they are the highest weight vectors of the spin algebra and, for this reason, belong to the representation $[N - M, M]$ of $\mathfrak{S}_N$. For a given M, the number of distinct solutions[5] $\{\lambda_i\}$ of (5.55) with no two λ coincident is equal to $\dim[N - M, M]$, see (5.39). Showing this is, however, non-trivial and represents a variant of the so-called *completeness problem* for the Bethe Ansatz.

5.3 Heisenberg Model

Historically, one of the first many-body problems for which an exact solution has been given was the ground state of the one-dimensional anti-ferromagnetic Heisenberg model. This solution is founded on Bethe's method, originating from his 1931 paper [1], of finding and enumerating the eigenstates of the corresponding hamiltonian. The use of the Bethe Ansatz approach for solving the delta-interaction model, which has been discussed in this chapter, was the first impressive application of Bethe's method outside the theory of magnetism. Further developments in theory

---

[5]That is the solutions which are not related to each other by permutations of some Bethe roots.

of integrable many-body systems turned the Bethe Ansatz into one of the important integrability tools. In this section we, therefore, briefly describe the solution of the Heisenberg model. We will do this in two different ways: by means of the coordinate and algebraic Bethe Ansätze, although, after our treatment of the more complicated Yang's spin-$\frac{1}{2}$ fermionic problem, consideration of the Heisenberg model becomes a plain endeavour. Our main goal is to cast further light on collective excitations and the notion of quasi-particles.

5.3.1   *Exchange Interactions and the Heisenberg Hamiltonian*

To motivate spin interactions and the Heisenberg hamiltonian accounting for them, consider a simple system consisting of just two electrons that interact only electrostatically and are allowed to move in one dimension. Let $\psi(q_1, q_2)$ be the coordinate (orbital) wave function, where q_1 and q_2 are coordinates of electrons. In accordance with the Fermi-Dirac statistics, the total wave function, see (4.126),

$$\Psi(x_1, x_2) = \Psi(q_1, q_2)\chi(s_1, s_2)$$

must be anti-symmetric. Here $\chi(s_1, s_2)$ is the spin wave function which describes a spin state of electrons. For two electrons there are four possible spin states which lead to either the anti-symmetric spin wave function with the total spin $s = 0$:

$$\Longrightarrow \quad \uparrow\downarrow - \downarrow\uparrow$$

or the symmetric one with $s = 1$:

$$\Longrightarrow \quad \begin{array}{ll} S^3 = 1 & \uparrow\uparrow \\ S^3 = 0 & \uparrow\downarrow + \downarrow\uparrow \\ S^3 = -1 & \downarrow\downarrow \end{array}.$$

Here S^3 is the projection of spin on z-axis. Denoting by $\vec{S} = \{S^\alpha\}$ the spin of an individual electron, for the system of two electrons we have

$$\vec{S} = \vec{S}_1 + \vec{S}_2$$

and, taking the square (quantum mechanically), we obtain

$$s(s + 1) = s_1(s_1 + 1) + s_2(s_2 + 1) + 2\vec{S}_1 \cdot \vec{S}_2,$$

so that

$$\vec{S}_1 \cdot \vec{S}_2 = \frac{1}{2}(s(s+1) - s_1(s_1+1) - s_2(s_2+1)).$$

Since $s_1 = s_2 = \frac{1}{2}$, from this formula we therefore find that

$$\vec{S}_1 \cdot \vec{S}_2 = \begin{cases} -\frac{3}{4} \text{ for } s = 0, \\ \frac{1}{4} \text{ for } s = 1 \end{cases}.$$

For the coordinate wave function these results imply

$$\text{for } s = 0 \quad \Psi(q_1, q_2) = \Psi_{\square} \quad \leftarrow \quad \text{symmetric function},$$
$$\text{for } s = 1 \quad \Psi(q_1, q_2) = \Psi_{\square} \quad \leftarrow \quad \text{anti-symmetric function}.$$

Since the symmetric and anti-symmetric functions describe different orbital motion of electrons, they correspond to different values of energies, E_s and E_a, respectively,

$$E_s \leftrightarrow s = 0,$$
$$E_a \leftrightarrow s = 1.$$

Which energy is realised depends on the problem at hand. For instance, for a molecule of hydrogen, H_2, the minimal energy corresponds to the symmetric wave function and for which, therefore, the electron spin S is equal to zero. Thus, we observe that in spite of the fact that electron spins are considered as non-interacting in any direct manner, due to the Pauli exclusion principle there is a correlation between the energy and the total spin, which can be re-interpreted a posteriori as some effective spin interaction. In our present case the corresponding spin hamiltonian can be defined as

$$H = \frac{1}{4}(E_s + 3E_a) + (E_a - E_s)\vec{S}_1 \cdot \vec{S}_2.$$

Here the first term $\frac{1}{4}(E_s + 3E_a) \equiv \bar{E}$ does not depend on spin and represents the energy averaged over all spin states (three states for $s = 1$ and one state for $s = 0$). The second term depends on spins of electrons. Introducing $J = E_a - E_s$, we can write

$$H = \bar{E} - J\vec{S}_1 \cdot \vec{S}_2 \tag{5.66}$$

This allows to relate energetic preference of states with $s = 0$ and $s = 1$ with the sign of J. For $J < 0$ the anti-parallel configuration of spins is preferred, while for $J > 0$ the parallel one. The parameter J is called an *exchange integral*. The hamiltonian H describes the so-called exchange interaction.

In the general case we consider a closed spin chain of spin-$\frac{1}{2}$ particles, see Fig. 5.1. At each site of the chain we have a two-dimensional vector space with the basis (5.34). The local spin algebra is (5.35). The effective spin interactions are described by the Heisenberg hamiltonian that generalises (5.66) to the multi-particle case, namely,

$$H = -J \sum_{n=1}^{L} S_n^\alpha S_{n+1}^\alpha \,, \tag{5.67}$$

where J is the coupling constant and L is a number of sites. The hamiltonian is local in the sense that it couples only the nearest-neighbouring spins. The model defined by (5.67) is known as the XXX *model*. A more general hamiltonian of the form

$$H = -J \sum_{n=1}^{L} J^\alpha S_n^\alpha S_{n+1}^\alpha \,,$$

where all three constants J^α are different defines the so-called XYZ model. In what follows we consider the XXX model only. Our basic problem now is to find the spectrum of the hamiltonian (5.67).

5.3.2   Coordinate Bethe Ansatz

We start with the observation that the hamiltonian (5.67) commutes with the global spin operators. Indeed,

$$[H, S^\alpha] = -J \sum_{n,m=1}^{L} [S_n^\beta S_{n+1}^\beta, S_m^\alpha] = -J \sum_{n,m=1}^{L} [S_n^\beta, S_m^\alpha] S_{n+1}^\beta + S_n^\beta [S_{n+1}^\beta, S_m^\alpha]$$

$$= -i \sum_{n,m=1}^{L} \left(\delta_{nm} \epsilon^{\alpha\beta\gamma} S_n^\beta S_{n+1}^\gamma - \delta_{n+1,m} \epsilon^{\alpha\beta\gamma} S_n^\beta S_{n+1}^\gamma \right) = 0 \,.$$

In other words, the hamiltonian is central with respect to all $\mathfrak{sl}(2)$ generators. Thus, the spectrum of the model will be degenerate, *i.e.* all states within one $\mathfrak{sl}(2)$ multiplet have the same energy. On a state $|\psi\rangle$ with M spins down we have

$$S^3 |\psi\rangle = \left(\frac{1}{2}(L - M) - \frac{1}{2}M \right) |\psi\rangle = \left(\frac{1}{2}L - M \right) |\psi\rangle \,. \tag{5.68}$$

Since $[H, S^3] = 0$ the hamiltonian can be diagonalised within each subspace of the full Hilbert space with a given total spin (which is uniquely characterised by the number of spins down).

In terms of raising and lowering operators the hamiltonian (5.67) takes the form

$$H = -J \sum_{n=1}^{L} \left[\tfrac{1}{2}(S_n^+ S_{n+1}^- + S_n^- S_{n+1}^+) + S_n^3 S_{n+1}^3 \right].$$

The fact that the spin is closed is reflected in the boundary conditions we impose on spin operators: $S_{L+1}^\alpha = S_1^\alpha$.

Example. As an explicit example, consider a closed spin chain with $L = 2$ sites. The corresponding hamiltonian is

$$H = -J \left(S^+ \otimes S^- + S^- \otimes S^+ + 2S^3 \otimes S^3 \right) = -J \begin{pmatrix} \tfrac{1}{2} & 0 & 0 & 0 \\ 0 & -\tfrac{1}{2} & 1 & 0 \\ 0 & 1 & -\tfrac{1}{2} & 0 \\ 0 & 0 & 0 & \tfrac{1}{2} \end{pmatrix}.$$

This matrix has three eigenvalues equal to $-\tfrac{1}{2}J$ and one equal to $\tfrac{3}{2}J$. The three states

$$v_{s=1}^{\text{hw}} = \begin{pmatrix} 1 \\ 0 \\ 0 \\ 0 \end{pmatrix}, \quad \begin{pmatrix} 0 \\ 1 \\ 1 \\ 0 \end{pmatrix}, \quad \begin{pmatrix} 0 \\ 0 \\ 0 \\ 1 \end{pmatrix},$$

all correspond to the eigenvalue $-\tfrac{1}{2}J$ and form a representation of $\mathfrak{sl}(2)$ with spin $s = 1$, while the state

$$v_{s=0}^{\text{hw}} = \begin{pmatrix} 0 \\ -1 \\ 1 \\ 0 \end{pmatrix},$$

corresponding to $\tfrac{3}{2}J$ is a singlet of $\mathfrak{sl}(2)$. Indeed, the generators of the global $\mathfrak{sl}(2)$ are realised as

$$S^+ = \begin{pmatrix} 0 & 1 & 1 & 0 \\ 0 & 0 & 0 & 1 \\ 0 & 0 & 0 & 1 \\ 0 & 0 & 0 & 0 \end{pmatrix}, \quad S^- = \begin{pmatrix} 0 & 0 & 0 & 0 \\ 1 & 0 & 0 & 0 \\ 1 & 0 & 0 & 0 \\ 0 & 1 & 1 & 0 \end{pmatrix}, \quad S^3 = \begin{pmatrix} 1 & 0 & 0 & 0 \\ 0 & 0 & 0 & 0 \\ 0 & 0 & 0 & 0 \\ 0 & 0 & 0 & -1 \end{pmatrix}.$$

The vectors $v_{s=1}^{\text{hw}}$ and $v_{s=0}^{\text{hw}}$ are the highest-weight vectors of the $s = 1$ and $s = 0$ representations respectively; they are annihilated by S^+ and are the eigenstates of S^3. In fact, $v_{s=0}^{\text{hw}}$ is also annihilated by S^- which shows that this state has zero spin. This completely describes the structure of the Hilbert space for $L = 2$.

Bethe Ansatz. In general, the hamiltonian can be realised as a $2^L \times 2^L$ symmetric matrix meaning that it has a complete orthogonal system of eigenvectors. The Hilbert space splits into a sum of irreducible representations of $\mathfrak{sl}(2)$. Thus, for L being finite the problem of finding the eigenvalues of H reduces to the problem of diagonalising a symmetric $2^L \times 2^L$ matrix. Provided L is sufficiently small, this can be easily achieved on a computer. However, for the physically important regime $L \to \infty$ that corresponds to taking the thermodynamic limit, a development of new analytic methods becomes indispensable and this is where the Bethe Ansatz construction makes its essential contribution.

Let $M < L$ be the number of spins down. If $M = 0$ we have a unique state, which was called pseudo-vacuum in Sect. 5.2:

$$|0\rangle = |\uparrow \cdots \uparrow\rangle .$$

This state is an eigenstate of the hamiltonian with eigenvalue $E_0 = -\frac{JL}{4}$:

$$H|0\rangle = -J \sum_{n=1}^{L} S_n^3 S_{n+1}^3 |\uparrow \cdots \uparrow\rangle = -\frac{JL}{4}|\uparrow \cdots \uparrow\rangle .$$

Let M be arbitrary. Since the M'th space has dimension $\frac{L!}{(L-M)!M!}$ one should find the same number of eigenvectors of H in this subspace. So let us write the eigenvectors of H in the form

$$|\psi\rangle = \sum_{1 \leqslant n_1 < \cdots < n_M \leqslant L} a(n_1, \ldots, n_M)|n_1, \ldots, n_M\rangle$$

with some unknown coefficients $a(n_1, \ldots, n_M)$. Here

$$|n_1, \ldots, n_M\rangle = S_{n_1}^- S_{n_2}^- \ldots S_{n_M}^- |0\rangle$$

and non-coincident integers describe the positions of the overturned spins. Obviously, the coefficients $a(n_1, \ldots, n_M)$ must satisfy the following requirement of periodicity:

$$a(n_2, \ldots, n_M, n_1 + L) = a(n_1, \ldots, n_M) . \tag{5.69}$$

The coordinate Bethe Ansatz postulates the form of these coefficients [1][6]

$$a(n_1, \ldots, n_M) = \sum_{\tau \in \mathfrak{S}_M} \mathcal{A}(\tau) \exp\left(i \sum_{j=1}^{M} p_{\tau(j)} n_j\right) . \tag{5.70}$$

Here for each of the M overturned spins we introduced the variable p_j which is called *quasi-momentum* and, as usual, $\mathfrak{S}_M$ denotes the symmetric group acting on the labels $\{1, \ldots, M\}$. To determine the coefficients $\mathcal{A}(\tau)$, as well as the set of quasi-momenta

---

[6]The reader undoubtedly noticed a striking similarity of this expression to the Bethe wave function (4.100) in the fundamental sector. The "particle" coordinates are now positive integers that form an ordered set. The formula (5.70) constitute the Bethe hypothesis which was the starting point to obtain the solution of the delta-interaction Bose gas.

$\{p_j\}$, we have to use the eigenvalue equation for H and the periodicity condition for $a(n_1, \ldots, n_M)$. First, it is instructive to work in detail the cases $M = 1$ and $M = 2$.

Case $M = 1$. For the case $M = 1$ we have

$$|\psi\rangle = \sum_{n=1}^{L} a(n)|n\rangle, \qquad a(n) = \mathcal{A} e^{ipn},$$

where $\mathcal{A}$ is a normalisation constant Thus, in this case

$$|\psi\rangle = \mathcal{A} \sum_{n=1}^{L} e^{ipn}|n\rangle$$

is nothing else but the Fourier transform of the "coordinate" wave function $|n\rangle$. The periodicity condition leads to determination of the quasi-momenta

$$a(n + L) = a(n) \quad \Longrightarrow \quad e^{ipL} = 1,$$

that is the $\frac{L!}{(L-1)!1!} = L$ allowed values of the pseudo-momenta are

$$p = \frac{2\pi k}{L} \quad \text{with} \quad k = 0, \cdots, L - 1.$$

Further, we have the eigenvalue equation

$$H|\psi\rangle = -\frac{J\mathcal{A}}{2} \sum_{m,n=1}^{L} e^{ipm}\left[S_n^+ S_{n+1}^- + S_n^- S_{n+1}^+ + 2S_n^3 S_{n+1}^3 \right] |m\rangle = E(p)|\psi\rangle.$$

To work out the left hand side, we have to use the formulae

$$S_n^+ S_{n+1}^-|m\rangle = \delta_{nm}|m + 1\rangle, \qquad S_n^- S_{n+1}^+|m\rangle = \delta_{n+1,m}|m - 1\rangle,$$

as well as

$$2S_n^3 S_{n+1}^3|m\rangle = \frac{1}{2}|m\rangle, \qquad \text{for} \quad m \neq n, n + 1,$$

$$2S_n^3 S_{n+1}^3|m\rangle = -\frac{1}{2}|m\rangle, \qquad \text{for} \quad m = n, \text{ or } m = n + 1.$$

Taking this into account, we obtain

$$H|\psi\rangle = -\frac{J\mathcal{A}}{2}\Big[\sum_{n=1}^{L}\Big(e^{ipn}|n+1\rangle + e^{ip(n+1)}|n\rangle\Big) + \frac{1}{2}\sum_{m=1}^{L}\sum_{\substack{n=1\\n\neq m,m-1}}^{L} e^{ipm}|m\rangle$$

$$-\frac{1}{2}\sum_{n=1}^{L}e^{ipn}|n\rangle - \frac{1}{2}\sum_{n=1}^{L}e^{ip(n+1)}|n+1\rangle\Big].$$

Using the periodicity condition, we finally get

$$H|\psi\rangle = -\frac{J\mathcal{A}}{2}\sum_{n=1}^{L}\Big(e^{ip(n-1)} + e^{ip(n+1)} + \frac{L-4}{2}e^{ipn}\Big)|n\rangle = -\frac{J}{2}\Big(e^{-ip} + e^{ip} + \frac{L-4}{2}\Big)|\psi\rangle.$$

From here we read off the eigenvalue

$$E - E_0 = J(1 - \cos p) = 2J\sin^2\frac{p}{2},$$

where $E_0 = -\frac{JL}{4}$. An elementary excitation of the spin chain around the pseudo-vacuum $|0\rangle$ carrying the quasi-momentum p is called a *magnon*.

Ferromagnetism and anti-ferromagnetism. Magnons. The concept of a magnon is due to F. Bloch who introduced it to explain a decrease of spontaneous magnetisation in a ferromagnet under temperature increase [22]. Many crystalline materials possess an ordered magnetic structure. This means that in absence of an external magnetic field the averaged quantum-mechanical magnetic moment in each elementary crystal cell is different from zero. In the ferromagnetic crystals (Fe, Ni, Co) the averaged values of magnetic moments of all the atoms have the same orientation unless the temperature does not exceed a certain critical value known as the Curie temperature. At zero temperature a ferromagnet is in the lowest energy state, in which all the magnetic moments (atomic spins) point in the same direction. With temperature increasing, due to thermal fluctuations, more and more spins deviate from the common direction, thus increasing the internal energy and diminishing the net magnetisation. Considering the fully magnetised state at zero temperature as the vacuum state, a low-temperature state of a ferromagnet with a few spins out of a perfect alignment can be thought of as a gas of quasi-particles, called magnons. By using the formalism of second quantisation, one can show that magnons behave as bosons. Thus, magnons are bosonic quasi-particles with momenta $p = \frac{2\pi k}{L}$, $k = 0, \ldots, L-1$ and the energy

$$E = 2J\sin^2\frac{p}{2}. \tag{5.71}$$

The last expression is the dispersion relation for one-magnon states.

It is important to stress that different signs of the coupling constant J correspond to rather different physical situations. If $J > 0$ then the energy (5.71) of the one-magnon states is positive and $|0\rangle$ is indeed a state with the lowest energy (the ground state).

This is a phenomenon of ferromagnetism mentioned above, it can be described as a tendency of the nearest-neighbouring spins to be parallel. The ferromagnetic ground state corresponds to $M = 0$ and, therefore, carries maximal spin $S^3 = \frac{1}{2}L$, see (5.68).

In contrast, if $J < 0$, then the energy (5.71) is negative and $|0\rangle$ is not the ground state, *i.e.* a state with the lowest energy. In other words, $|0\rangle$ is not a true vacuum, but rather a pseudo-vacuum, or "false" vacuum. The true ground state is non-trivial and needs some work to be identified. The case $J < 0$ is called anti-ferromagnetic, as it is related to anti-ferromagnetism, *i.e.* to the tendency of the nearest-neighbouring spins to be antiparallel. Later on we will see that the anti-ferromagnetic ground state corresponds to $M = \frac{1}{2}L$ and, therefore, it is spinless. Physically, anti-ferromagnetism is observed in more complicated crystals (carbons, sulfates, oxides), in which the averaged values of magnetic moments of individual atoms compensate each other within every elementary crystal cell.

Case $M = 2$. After these comments, we turn to the case $M = 2$. Here we have

$$|\psi\rangle = \sum_{1 \leqslant n_1 < n_2 \leqslant L} a(n_1, n_2)|n_1, n_2\rangle \,,$$

where the general ansatz (5.70) specifies to

$$a(n_1, n_2) = \mathcal{A}(12)e^{i(p_1 n_1 + p_2 n_2)} + \mathcal{A}(21)e^{i(p_2 n_1 + p_1 n_2)} \,. \tag{5.72}$$

The eigenvalue equation for H imposes conditions on $a(n_1, n_2)$ analogous to the $M = 1$ case. However, special care is needed, when two down-spins are sitting next to each other. We need to consider

$$H|\psi\rangle = -\frac{J}{2} \sum_{1 \leqslant n_1 < n_2 \leqslant L} a(n_1, n_2) \sum_{m=1}^{L} \left[S_m^+ S_{m+1}^- + S_m^- S_{m+1}^+ + 2S_m^3 S_{m+1}^3 \right]|n_1, n_2\rangle$$

$$= \left\{ -\frac{J}{2}\Bigg[\sum_{\substack{1 \leqslant n_1 < n_2 \leqslant L \\ n_2 > n_1 + 1}} a(n_1, n_2)\Big(|n_1 + 1, n_2\rangle + |n_1, n_2 + 1\rangle + |n_1 - 1, n_2\rangle + |n_1, n_2 - 1\rangle\Big) \right.$$

$$\left. + \frac{L-4}{2} \sum_{\substack{1 \leqslant n_1 < n_2 \leqslant L \\ n_2 > n_1 + 1}} a(n_1, n_2)|n_1, n_2\rangle - \frac{1}{2}4 \sum_{\substack{1 \leqslant n_1 < n_2 \leqslant L \\ n_2 > n_1 + 1}} a(n_1, n_2)|n_1, n_2\rangle \Bigg] \right\} +$$

$$+ \left\{ -\frac{J}{2} \sum_{1 \leqslant n_1 \leqslant L} a(n_1, n_1 + 1)\Big[|n_1, n_1 + 2\rangle + |n_1 - 1, n_1 + 1\rangle + \Big(\frac{L-2}{2} - 1\Big)|n_1, n_1 + 1\rangle\Big] \right\} \,.$$

Here in the first curly bracket we collect the terms with $n_2 > n_1 + 1$, while the second bracket represents the result of the action of H on terms with $n_2 = n_1 + 1$. Using the periodicity condition (5.69), we can make shifts of the summation variables n_1, n_2 in the first bracket to bring all the states to the uniform expression $|n_1, n_2\rangle$. We therefore get

$$H|\psi\rangle = -\frac{J}{2}\left\{\sum_{n_2 > n_1} a(n_1 - 1, n_2)|n_1, n_2\rangle + \sum_{n_2 > n_1 + 2} a(n_1, n_2 - 1)|n_1, n_2\rangle\right.$$

$$+ \sum_{n_2 > n_1 + 2} a(n_1 + 1, n_2)|n_1, n_2\rangle + \sum_{n_2 > n_1} a(n_1, n_2 + 1)|n_1, n_2\rangle$$

$$\left. + \frac{L - 8}{2}\sum_{n_2 > n_1 + 1} a(n_1, n_2)|n_1, n_2\rangle\right\}$$

$$-\frac{J}{2}\left\{\sum_{1 \leqslant n_1 \leqslant L} a(n_1, n_1 + 1)\left[|n_1, n_1 + 2\rangle + |n_1 - 1, n_1 + 1\rangle + \frac{L - 4}{2}|n_1, n_1 + 1\rangle\right]\right\}.$$

Now we complete the sums in the first bracket for their summation indices to run over the range $n_2 > n_1$. This is achieved by adding and subtracting the missing terms. As a result, after reducing similar terms, we get

$$H|\psi\rangle =$$

$$-\frac{J}{2}\left\{\sum_{n_2 > n_1}\left(a(n_1 - 1, n_2) + a(n_1, n_2 - 1)\right.\right.$$

$$\left.\left. + a(n_1 + 1, n_2) + a(n_1, n_2 + 1) + \frac{L - 8}{2}a(n_1, n_2)\right)|n_1, n_2\rangle\right\}$$

$$+\frac{J}{2}\left\{\sum_{1 \leqslant n_1 \leqslant L}\left(a(n_1, n_1) + a(n_1 + 1, n_1 + 1) - 2a(n_1, n_1 + 1)\right)|n_1, n_1 + 1\rangle\right\}.$$

It is now clear that if we require the coefficients $a(n_1, n_2)$ to satisfy the condition

$$a(n_1, n_1) + a(n_1 + 1, n_1 + 1) - 2a(n_1, n_1 + 1) = 0, \tag{5.73}$$

then the second bracket in the eigenvalue equation above vanishes and the eigenvalue problem reduces to the following equation

$$(E - E_0)a(n_1, n_2) = J\left[2a(n_1, n_2) - \frac{1}{2}\sum_{\sigma = \pm 1}\left(a(n_1 + \sigma, n_2) + a(n_1, n_2 + \sigma)\right)\right]. \tag{5.74}$$

We first consider Eq. (5.74). Substituting in this equation the expression (5.72) for $a(n_1, n_2)$, we observe that (5.74) factorises in such a manner that the dependence on the coefficients $\mathscr{A}(12)$ and $\mathscr{A}(21)$ drop out and we are left with the following expression for the energy of two-magnon states

$$E - E_0 = J\left(2 - \cos p_1 - \cos p_2\right) = 2J \sum_{k=1}^{2} \sin^2 \frac{p_k}{2} \,.$$

Miraculously, the energy appears to be additive, *i.e.* for two magnons it looks like the sum of energies of individual one-magnon states. This shows, indeed, that in spite of interactions being present, magnons essentially behave as non-interacting quasi-particles. In this respect it is pertinent to recall that the additive nature of the spectrum is one of the facets of integrability that we have already observed a few times throughout the book.

Now we turn our attention to (5.73). Substituting here the ansatz (5.72), we get

$$\mathcal{A}(12)e^{(p_1+p_2)n} + \mathcal{A}(21)e^{i(p_1+p_2)n} + \mathcal{A}(12)e^{(p_1+p_2)(n+1)} + \mathcal{A}(21)e^{i(p_1+p_2)(n+1)}$$
$$- 2\big(\mathcal{A}(12)e^{i(p_1 n + p_2(n+1))} + \mathcal{A}(21)e^{i(p_2 n + p_1(n+1))}\big) = 0\,.$$

The last equation allows one to find the ratio

$$\frac{\mathcal{A}(21)}{\mathcal{A}(12)} = -\frac{e^{i(p_1+p_2)} + 1 - 2e^{ip_2}}{e^{i(p_1+p_2)} + 1 - 2e^{ip_1}} \tag{5.75}$$

It is not hard to see that the above expression is the pure phase, which we parametrise as

$$\frac{\mathcal{A}(21)}{\mathcal{A}(12)} = e^{-i\theta(p_1, p_2)} \equiv S(p_1, p_2)\,, \tag{5.76}$$

where S is the scattering matrix.[7] We further note that it obeys the following relation

$$S(p_1, p_2)S(p_2, p_1) = 1\,,$$

implying that the phase is skew-symmetric, $\theta(p_1, p_2) = -\theta(p_2, p_1)$. Another important fact about the two-body S-matrix is that by simple algebraic manipulation the expression (5.75) can be brought to the form

$$S(p_1, p_2) = \frac{\frac{1}{2} \cot \frac{p_1}{2} - \frac{1}{2} \cot \frac{p_2}{2} - i}{\frac{1}{2} \cot \frac{p_1}{2} - \frac{1}{2} \cot \frac{p_2}{2} + i}\,. \tag{5.77}$$

This expression motivates to introduce the variable $\lambda = \frac{1}{2} \cot \frac{p}{2}$, called *rapidity*, so that in terms of this variable the S-matrix reads

$$S(\lambda_1, \lambda_2) = \frac{\lambda_1 - \lambda_2 - i}{\lambda_1 - \lambda_2 + i}$$

---

[7] From the point of view of our construction in Sect. 4.3.2 the S-matrix (5.75) coincides with Yang's operator $Y_1(p_1, p_2)$.

and it appears to be a function that depends on the difference of rapidities, i.e. it is, in fact, a function of a single variable, sharing this property with the S-matrix (5.18) of the Lieb-Liniger-Yang model. For later use we note also the identity

$$e^{ip} = \frac{\lambda + i/2}{\lambda - i/2}. \tag{5.78}$$

Thus, with the formula (5.76) the two-magnon Bethe Ansatz is further specified to have the form

$$a(n_1, n_2) = e^{i(p_1 n_1 + p_2 n_2)} + S(p_1, p_2)e^{i(p_2 n_1 + p_1 n_2)},$$

where we factored out the unessential normalisation coefficient $\mathscr{A}(12)$. Finally, we have to impose the periodicity condition $a(n_2, n_1 + L) = a(n_1, n_2)$. This results into

$$e^{i(p_1 n_2 + p_2 n_1)}e^{ip_2 L} + S(p_1, p_2)e^{ip_1 L}e^{i(p_2 n_2 + p_1 n_1)} = e^{i(p_1 n_1 + p_2 n_2)} + S(p_1, p_2)e^{i(p_2 n_1 + p_1 n_2)}$$

which implies

$$e^{ip_1 L} = S(p_2, p_1), \qquad e^{ip_2 L} = S(p_1, p_2).$$

These are the two-particle Bethe equations for the Heisenberg model.

Multi-magnon case. Considerations of the general multi-magnon case can be done following the same steps as for the $M = 1, 2$ cases. One finds the following analogues of Eqs. (5.73) and (5.74) for M magnons

$$a(n_1, \ldots, n_k, n_k, \ldots, n_M) + a(n_1, \ldots, n_k + 1, n_k + 1, \ldots, n_M) =$$
$$= 2a(n_1, \ldots, n_k, n_k + 1, \ldots, n_M), \tag{5.79}$$

and

$$(E - E_0)a(n_1, \ldots, n_M) =$$
$$= \frac{J}{2} \sum_{j=1}^{M} \sum_{\sigma=\pm1} \left[a(n_1, \ldots, n_M) - a(n_1, \ldots, n_j + \sigma, \ldots, n_M) \right]. \tag{5.80}$$

Substitution of the Bethe Ansatz (5.70) into the second equation yields the additive expression for the energy that generalises the two-magnon case

$$E - E_0 = 2J \sum_{k=1}^{M} \sin^2 \frac{p_k}{2}, \tag{5.81}$$

while the first equation gives, up to an unessential overall normalisation, the following solution for the amplitudes

$$\mathcal{A}(\tau) = \exp\left(\frac{i}{2}\sum_{i<j}\theta(p_{\tau(i)}, p_{\tau(j)})\right), \quad \tau \in \mathfrak{S}_M, \tag{5.82}$$

where θ is the same as in (5.76). This expression for the amplitudes shows that all of them are determined by the two-body phase shift (S-matrix), precisely as in our earlier discussion of Factorised Scattering Theory.

Finally, the periodicity condition (5.69) gives

$$\sum_{\tau \in \mathfrak{S}_M} e^{\frac{i}{2}\sum_{i<j}\theta(p_{\tau(i)}, p_{\tau(j)})} e^{i\sum_{i=1}^{M} p_{\tau(i)} n_i} =$$

$$= \sum_{\sigma \in \mathfrak{S}_M} e^{\frac{i}{2}\sum_{i<j}\theta(p_{\sigma(i)}, p_{\sigma(j)})} e^{i\sum_{i=2}^{M} p_{\sigma(i-1)} n_i + i p_{\sigma(M)} n_1} e^{i p_{\sigma(M)} L}.$$

This equation must hold for any $n_1, \ldots, n_M$ and, therefore, the terms on the left and the right hand sides of this equation that have the same dependence on n_k must be equal. Let us pick up on the left hand side a term corresponding to a fixed permutation τ and on the right hand side a term corresponding a permutation σ that is related to τ as

$$\sigma(i-1) = \tau(i), \quad i = 2, \ldots, M,$$
$$\sigma(M) = \tau(1).$$

These terms have the same dependence on n_k and their equality implies the following relation

$$e^{i p_{\tau(1)} L} = e^{\frac{i}{2}\sum_{i<j}\theta(p_{\tau(i)}, p_{\tau(j)}) - \frac{i}{2}\sum_{i<j}\theta(p_{\sigma(i)}, p_{\sigma(j)})}.$$

We then consider the difference of the phases

$$\sum_{i<j}\theta(p_{\tau(i)}, p_{\tau(j)}) - \sum_{i<j}\theta(p_{\sigma(i)}, p_{\sigma(j)}) = \sum_{2\leqslant i<j\leqslant M}\theta(p_{\tau(i)}, p_{\tau(j)}) + \sum_{j=2}^{M}\theta(p_{\tau(1)}, p_{\tau(j)})$$

$$- \sum_{1\leqslant i<j\leqslant M-1}\theta(p_{\tau(i+1)}, p_{\tau(j+1)}) - \sum_{i=1}^{M-1}\theta(p_{\tau(i+1)}, p_{\tau(1)}).$$

Here on the right hand side the first terms in the first and the second line cancel, while the remaining two terms are equal and add up because the phase is skew-symmetric. In this way we obtain

$$e^{i p_{\tau(1)} L} = e^{i\sum_{j\neq1}^{M}\theta(p_{\tau(1)}, p_{\tau(j)})}.$$

Lastly, the requirement that this equation must be satisfied for any $\tau \in \mathfrak{S}_M$ yields the M-magnon Bethe equations

$$e^{ip_n L} = \prod_{j \neq n}^{M} S(p_j, p_n) \,, \tag{5.83}$$

where $n = 1, \ldots, M$. Substituting the explicit form of the S-matrix, we get

$$e^{ip_n L} = \prod_{j \neq n}^{M} \frac{\frac{1}{2} \cot \frac{p_n}{2} - \frac{1}{2} \cot \frac{p_j}{2} + i}{\frac{1}{2} \cot \frac{p_n}{2} - \frac{1}{2} \cot \frac{p_j}{2} - i} \,. \tag{5.84}$$

In terms of rapidities λ_n these equations take the form

$$\left(\frac{\lambda_n + \frac{i}{2}}{\lambda_n - \frac{i}{2}} \right)^L = \prod_{j \neq n}^{M} \frac{\lambda_n - \lambda_j + i}{\lambda_n - \lambda_j - i} \,, \tag{5.85}$$

while the energy of the corresponding M-magnon state reads

$$E - E_0 = \frac{J}{2} \sum_{n=1}^{L} \frac{1}{\lambda_n^2 + \frac{1}{4}} \,. \tag{5.86}$$

The reader might notice that the Bethe equations (5.85) coincide with the auxiliary Bethe equations (5.56) upon setting in the latter equations all $p_i = 0$, $\varkappa = 1$ and identifying the spin chain length L with N. The model with $p_i = 0$ is called the *homogeneous* Heisenberg model, in contrast to the inhomogeneous model with $p_i \neq 0$, which we have already considered in the context of Yang's spin-$\frac{1}{2}$ model.

Solutions of the two-particle Bethe equations. To give the reader a flavour of the types of solutions of the Bethe equations (5.83) or (5.85), here we will look at the two-particle case.

Taking the logarithm of the Bethe equations, we obtain

$$p_1 = \frac{2\pi m_1}{L} + \frac{1}{L} \theta(p_1, p_2) \,, \qquad p_2 = \frac{2\pi m_2}{L} + \frac{1}{L} \theta(p_2, p_1) \,, \tag{5.87}$$

where the integers $m_i \in \{0, 1, \ldots, L - 1\}$ are called *Bethe quantum numbers*. The Bethe quantum numbers provide a useful counting label to distinguish eigenstates with different physical properties. It is clear that the magnon interactions are encoded in the phase shift θ, which is responsible for the deviation of the momenta p_1, p_2 from the free one-magnon values. Note that Eq. (5.87) yield the following value of the total momentum

$$P = p_1 + p_2 = \frac{2\pi}{L} (m_1 + m_2) \,,$$

which does not involve the phase shift. Different types of states arise due to the fact that magnons can either scatter off each other or form the bound states.

The first problem in the analysis of (5.87) is to find all possible quantum numbers (m_1, m_2) for which Bethe equations have solutions. It turns out that the allowed pairs (m_1, m_2) are restricted to

$$0 \leqslant m_1 \leqslant m_2 \leqslant L - 1.$$

This is because switching m_1 and m_2 simply interchanges p_1 and p_2 and produces the same solution. There are $\frac{1}{2}L(L + 1)$ pairs which meet this restriction but only $\frac{1}{2}L(L - 1)$ of them yield a solution of the Bethe equations. Some of these solutions have real p_1 and p_2, the others yield the complex conjugate momenta $p_2 = p_1^*$.

The simplest solutions are the pairs for which one of the Bethe numbers is zero, e.g. $m_1 = 0$, $m = m_2 = 0, 1, \ldots, L - 1$. For such a pair we have

$$Lp_1 = \theta(p_1, p_2), \qquad Lp_2 = 2\pi m + \theta(p_2, p_1),$$

which is solved by $p_1 = 0$ and $p_2 = \frac{2\pi m}{L}$. Indeed, for $p_1 = 0$ the phase shift vanishes: $\theta(0, p_2) = 0$. These solutions have the dispersion relation

$$E - E_0 = 2J \sin^2 \frac{p}{2}, \qquad p = p_2$$

which is the same as the dispersion for the one-magnon states. These solutions are nothing else but $\mathfrak{sl}(2)$-descendants of the solutions with $M = 1$.

One can show that for $M = 2$ all solutions are divided into three distinct classes with a different number of states in each class

$$\underbrace{\text{Descendents}}_{L}, \qquad \underbrace{\text{Scattering States}}_{\frac{L(L-5)}{2}+3}, \qquad \underbrace{\text{Bound States}}_{L-3},$$

so that

$$L + \frac{L(L - 5)}{2} + 3 + L - 3 = \frac{1}{2}L(L - 1)$$

states yield a complete solution space of the two-magnon problem. We refer the reader to [23] for the thorough discussion of this classification. Here we only point out that the most simple description of the bound states is obtained in the limit when $L \to \infty$. If p_n has a non-trivial negative imaginary part, $\text{Im } p_n < 0$, then $e^{ip_n L}$ tends in this limit to infinity and this means that the bound states are determined by poles of the right hand side of the Bethe equations (5.84). In particular, for the case two-magnon case the bound states correspond to poles in the two-body S-matrix. The pole condition reads in this case as

$$\frac{1}{2} \cot \frac{p_1}{2} - \frac{1}{2} \cot \frac{p_2}{2} = i. \tag{5.88}$$

The corresponding state has the total momentum $P = p_1 + p_2$ that must be real. This suggests the following parametrisation of the particle momenta

$$p_1 = \frac{P}{2} - iv, \qquad p_2 = \frac{P}{2} + iv, \quad v > 0, \quad p \in \mathbb{R}.$$

Substitution of this parametrisation into (5.88) yields

$$\cos \tfrac{1}{2}\left(\tfrac{P}{2} - iv\right) \sin \tfrac{1}{2}\left(\tfrac{P}{2} + iv\right) - \cos \tfrac{1}{2}\left(\tfrac{P}{2} + iv\right) \sin \tfrac{1}{2}\left(\tfrac{P}{2} - iv\right)$$
$$= 2i \sin \tfrac{1}{2}\left(\tfrac{P}{2} + iv\right) \sin \tfrac{1}{2}\left(\tfrac{P}{2} - iv\right),$$

from where it follows that

$$\cos \frac{P}{2} = e^{-v}. \tag{5.89}$$

The energy of such a state is

$$E = 2J\left(\sin^2 \frac{p_1}{2} + \sin^2 \frac{p_2}{2} \right) = 2J\left(\sin^2 \left(\frac{P}{4} - i\frac{v}{2}\right) + \sin^2 \left(\frac{P}{4} + i\frac{v}{2}\right) \right).$$

Simplifying this expression and substituting (5.89), we obtain

$$E = 2J\left(1 - \cos \frac{P}{2} \cosh v\right) = 2J\left(1 - \cos \frac{P}{2} \frac{\cos^2 \frac{P}{2} + 1}{2\cos \frac{P}{2}}\right) = J \sin^2 \frac{P}{2}.$$

This expression can be regarded as the dispersion relation of the bound states.

5.3.3 Algebraic Bethe Ansatz

The coordinate Bethe Ansatz provides a solution of the Heisenberg model in the sense that it reduces it to a set of algebraic equations for a set of rapidity variables. In spite of some very suggestive features such as additivity of the spectrum and a distinguished role of the two-body phase shift, the relation of this approach to integrability remains, however, rather obscure. It is the algebraic Bethe Ansatz that makes the integrable structure of the model manifest. We discuss this method applied to the homogeneous Heisenberg model rather briefly, as it essentially follows the same steps as in the case of its inhomogeneous counterpart.

Lax operator, monodromy and transfer matrices. The starting point of the algebraic Bethe Ansatz approach is the Lax operator (5.48), which for the present model is given by

$$L_{ia}(\lambda) = \lambda \mathbb{1}_i \otimes \mathbb{1}_a + i \sum_\alpha S_i^\alpha \otimes \sigma_a^\alpha,$$

where $\mathbb{1}_i$ and S_i^α acts in the local "quantum" space $V_i \simeq \mathbb{C}^2$, while $\mathbb{1}_a$ and the Pauli matrices σ_a^α operate on an auxiliary space which is another copy of $\mathbb{C}^2$. The complex number λ is the spectral parameter. In terms of the permutation π_{ia}, see (5.38), this Lax operator is

$$L_{ia}(\lambda) = \left(\lambda - \frac{i}{2}\right)\mathbb{1}_{ia} + i\pi_{ia}.$$

Finally, as a 2×2 matrix in the auxiliary space with operator entries acting on the quantum space, the Lax operator looks as

$$L_{ia}(\lambda) = \begin{pmatrix} \lambda + iS_i^3 & iS_i^- \\ iS_i^+ & \lambda - iS_i^3 \end{pmatrix}.$$

The Lax operators acting in two different auxiliary spaces are intertwined with the help of the quantum R-matrix as $R_{ab}(\lambda_1, \lambda_2) = R_{ab}(\lambda_1 - \lambda_2)$ such that the following relation is true

$$R_{ab}(\lambda_1 - \lambda_2)L_{ia}(\lambda_1)L_{ib}(\lambda_2) = L_{ib}(\lambda_2)L_{ia}(\lambda_1)R_{ab}(\lambda_1 - \lambda_2), \qquad (5.90)$$

where $R_{ab}(\lambda)$ is

$$R_{ab} = \lambda\mathbb{1}_{ab} + i\pi_{ab}.$$

It is convenient to suppress the index of the quantum space and write the fundamental commutation relations (5.90) as

$$R_{ab}(\lambda_1 - \lambda_2)L_a(\lambda_1)L_b(\lambda_2) = L_b(\lambda_2)L_a(\lambda_1)R_{ab}(\lambda_1 - \lambda_2).$$

The consistency condition for these relations is the quantum Yang-Baxter equation. Further, the monodromy is the following ordered product of L-operators along the chain[8]

$$T_a(\lambda) = L_{L,a}(\lambda)\ldots L_{1,a}(\lambda).$$

The relations between the entries of the monodromy matrix are given by (5.60) from where we deduce that the corresponding transfer matrix $\tau(\lambda) = \mathrm{Tr}_a T_a(\lambda)$ evaluated at different values of the spectral parameter yields operators which commute with each other. We represent that transfer matrix as a 2×2 matrix in the auxiliary space

$$T(\lambda) = \begin{pmatrix} A(\lambda) & B(\lambda) \\ C(\lambda) & D(\lambda) \end{pmatrix},$$

where the entries are operators acting in the space $\overset{L}{\underset{i=1}{\otimes}} V_i$. From the definition of the monodromy and the L-operator it is clear that T is a polynomial in λ and

---

[8]Compared to (5.28) we took the product of the Lax operators in an opposite order which is more convenient for our present treatment.

$$T(\lambda) = \lambda^L + i\lambda^{L-1} \sum_{n=1}^{L} S_n^\alpha \otimes \sigma^\alpha + \cdots$$

Thus, the transfer matrix is also polynomial of degree L:

$$\tau(\lambda) = \mathrm{tr}_a T_a(\lambda) = A(\lambda) + D(\lambda) = 2\lambda^L + \sum_{j=0}^{L-2} I_j \lambda^j \,.$$

Note that the subleading term of order λ^{L-1} is absent because Pauli matrices are traceless. It follows from our discussion that the coefficients I_j mutually commute

$$[I_i, I_j] = 0 \,.$$

The hamiltonian and the total momentum. To show that the model is quantum integrable, we need to find the Heisenberg hamiltonian among the commutative family of operators generated by the transfer matrix. This can be done as follows. First, we notice that the L-operator has two special points on the spectral parameter plane, namely,

(1) $\lambda = \frac{i}{2}$, where $L_{i,a}(i/2) = i\pi_{ia}$.
(2) $\lambda = \infty$.

Concerning the second point, we have

$$\frac{1}{i}\mathrm{Res}\frac{T(\lambda)}{\lambda^L} = \sum_{n=1}^{L} S^\alpha \otimes \sigma^\alpha = S^\alpha \otimes \sigma^\alpha \,.$$

Thus, this point is related to the realization of the global $\mathfrak{su}(2)$ symmetry of the model.

Let us investigate the first point. At $\lambda = i/2$ the monodromy is

$$T_a(i/2) = i^L \pi_{L,a}\pi_{L-1,a} \cdots \pi_{1,a} = i^L \pi_{L-1,L}\pi_{L-2,L} \cdots \pi_{1,L}\pi_{L,a} =$$
$$= i^L \pi_{L-2,L-1}\pi_{L-3,L-1} \cdots \pi_{1,L-1}\pi_{L-1,L}\pi_{L,a} = \cdots = i^L \pi_{12}\pi_{23} \cdots \pi_{L-1,L}\pi_{L,a} \,.$$

Thus, we have managed to isolate a single permutation carrying the index of the auxiliary subspace. Taking the trace and recalling that $\mathrm{Tr}_a \pi_{L,a} = \mathbb{1}_L$, we obtain the corresponding value of the transfer matrix at this point

$$\tau(i/2) = i^L \pi_{12}\pi_{23} \cdots \pi_{L-1,L} \equiv \mathcal{U} \,.$$

The operator $\mathcal{U}$ is unitary $\mathcal{U}^\dagger \mathcal{U} = \mathcal{U}\mathcal{U}^\dagger = \mathbb{I}$ and it generates a shift along the chain

$$\mathcal{U}^{-1} X_n \mathcal{U} = X_{n-1} \,.$$

If we denote by P the operator of momentum, then it is related to the shift operator $\mathcal{U}$ on the lattice as

$$\mathcal{U} = e^{iP}.$$

Second, we differentiate the logarithm of the transfer matrix and further evaluate it at $\lambda = i/2$. We have

$$\frac{d\tau(\lambda)}{d\lambda}\bigg|_{\lambda=i/2} = i^{L-1} \sum_n \pi_{L,a} \cdots \overline{\pi_{n,a}} \cdots \pi_{1,a} = i^{L-1} \sum_n \pi_{12}\pi_{23} \cdots \pi_{n-1,n+1} \cdots \pi_{L-1,L},$$

where the crossed out term is absent. This further gives

$$\frac{d\tau(\lambda)}{d\lambda}\tau(\lambda)^{-1}\bigg|_{\lambda=i/2} =$$

$$= i^{-1}\left(\sum_n \pi_{12}\pi_{23} \cdots \pi_{n-1,n+1} \cdots \pi_{L-1,L}\right)\left(\pi_{L,L-1}\pi_{L-1,L-2} \cdots \pi_{2,1}\right) = \frac{1}{i}\sum_n^L \pi_{n,n+1},$$

On the other hand, the Heisenberg hamiltonian is

$$H = -J \sum_{n=1}^L S_n^\alpha S_{n+1}^\alpha = -\frac{J}{4} \sum_{n=1}^L \sigma_n^\alpha \sigma_{n+1}^\alpha = -J\left(\frac{1}{2}\sum_{n=1}^L \pi_{n,n+1} - \frac{L}{4}\right).$$

Hence,

$$H = -J\left(\frac{i}{2}\frac{d\tau(\lambda)}{d\lambda}\tau(\lambda)^{-1} - \frac{L}{4}\right)\bigg|_{\lambda=i/2},$$

i.e. the hamiltonian does belong to the family of $L - 1$ commuting integrals. To obtain L commuting integrals, we can add to this family the operator S^3.

The spectrum of the Heisenberg model. The relevant commutation relations which follow from (5.60) are

$$[B(\lambda), B(\mu)] = 0,$$

$$A(\lambda)B(\mu) = \frac{\lambda - \mu - i}{\lambda - \mu}B(\mu)A(\lambda) + \frac{i}{\lambda - \mu}B(\lambda)A(\mu), \qquad (5.91)$$

$$D(\lambda)B(\mu) = \frac{\lambda - \mu + i}{\lambda - \mu}B(\mu)D(\lambda) - \frac{i}{\lambda - \mu}B(\lambda)D(\mu).$$

The pseudo-vacuum (5.53) is annihilated by all $C(\lambda)$ and the the eigenvectors of $\tau(\lambda)$ with M spins down are searched in the form (5.52). We find

$$T(\lambda)|0\rangle = \begin{pmatrix} (\lambda + \frac{i}{2})^L|0\rangle & \bigstar \\ 0 & (\lambda - \frac{i}{2})^L|0\rangle \end{pmatrix},$$

where $\bigstar$ stands for irrelevant terms. Thus,

$$C(\lambda)|0\rangle = 0, \qquad A(\lambda)|0\rangle = \left(\lambda + \frac{i}{2}\right)^L |0\rangle, \qquad D(\lambda)|0\rangle = \left(\lambda - \frac{i}{2}\right)^L |0\rangle.$$

Acting on states (5.52) with $\tau(\lambda)$, we find that they are the eigenstates of the transfer matrix provided the Bethe roots $\lambda_n, n = 1, \ldots, M$, satisfy the same Bethe equations (5.85) as coming from the coordinate Bethe Ansatz considerations. The solution of the eigenvalue problem

$$\tau(\lambda)|\lambda_1, \cdots, \lambda_M\rangle = \Lambda(\lambda, \{\lambda_n\})|\lambda_1, \cdots, \lambda_M\rangle$$

reads as

$$\Lambda(\lambda, \{\lambda_n\}) = \left(\lambda + \frac{i}{2}\right)^L \prod_{n=1}^{M} \frac{\lambda - \lambda_n - i}{\lambda - \lambda_n} + \left(\lambda - \frac{i}{2}\right)^L \prod_{n=1}^{M} \frac{\lambda - \lambda_n + i}{\lambda - \lambda_n}.$$

One can further show, just as has been done for the case of Yang's model, that the states with $\lambda_n = \frac{1}{2} \cot \frac{p_n}{2} = \infty$, *i.e.* with $p_n = 0$, represent descendants of the $\mathfrak{su}(2)$ symmetry algebra. Finally, we can compute the eigenvalues of the transfer matrix on the corresponding Bethe eigenvectors. The result is the same as expected from the coordinate Bethe Ansatz considerations, namely,

$$E = -J\left(\frac{i}{2}\frac{d\tau(\lambda)}{d\lambda}\tau(\lambda)^{-1}\Big|_{\lambda=i/2} - \frac{L}{4}\right) = E_0 + \frac{J}{2}\sum_{j=1}^{L}\frac{1}{\lambda_j^2 + \frac{1}{4}} = 2\sum_{j=1}^{L}\sin^2\frac{p_j}{2}.$$

Let us summarise the most important observations about the Bethe Ansatz construction. First, the Heisenberg model has $\mathfrak{su}(2)$ symmetry that splits the eigenvectors of the transfer matrix calculated by using the Bethe Ansatz procedure into irreducible representations of $\mathfrak{su}(2)$. For finite values of the Bethe roots λ_j the eigenvectors of the algebraic Bethe Ansatz are the always the highest weight states of $\mathfrak{su}(2)$. Descendants of the highest weight vectors correspond to Bethe roots λ_j at infinity, correspondingly at $p_j = 0$. Second, the algebraic Bethe Ansatz enables us to prove integrability of the model and it gives an explicit construction of the Hilbert space of states in terms of simultaneous eigenvectors of commuting integrals of motion. Comparing to the action-angle variables arising in the classical inverse scattering method [24], one may notice that the transfer matrix $\tau(\lambda)$ resembles the action variable, while $B(\lambda)$ plays the role of the angle variable.

References

1. Bethe, H.: On the theory of metals. 1. Eigenvalues and eigenfunctions for the linear atomic chain. Z. Phys. **71**, 205–226 (1931)
2. Yang, C.N.: Some exact results for the many body problems in one dimension with repulsive delta function interaction. Phys. Rev. Lett. **19**, 1312–1314 (1967)

3. Gaudin, M.: The Bethe Wavefunction (Translated from French original "La fonction d'onde de Bethe" (1983) by Caux, J.-S.). Cambridge University Press (2014)
4. Sutherland, B.: Beautiful Models: 70 Years of Exactly Solved Quantum Many-body Problems, p. 381. World Scientific (2004)
5. Faddeev, L.D.: Quantum completely integral models of field theory. Sov. Sci. Rev. C **1**, 107–155 (1980)
6. Takhtajan, L.A., Faddeev, L.D.: The quantum method of the inverse problem and the Heisenberg XYZ model. Russ. Math. Surveys **34**(5), 11–68 (1979). [Usp. Mat. Nauk 34, no.5, 13(1979)]
7. Kulish, P.P., Sklyanin, E.K.: Quantum spectral transform method. Recent developments. Lect. Notes Phys. **151**, 61–119 (1982)
8. Gutzwiller, M.C.: The Quantum Mechanical Toda Lattice. II. Annals Phys. **133**, 304–331 (1981)
9. Sklyanin, E.K.: Goryachev-Chaplygin top and the inverse scattering method. J. Sov. Math. **31**, 3417–3431 (1985). Zap. Nauchn. Semin. 133, 236 (1984)
10. Sklyanin, E.K.: The Quantum Toda chain. Lect. Notes Phys. **226**, 196–233 (1985)
11. Baxter, R.J.: Generalized ferroelectric model on a square lattice. Stud. Appl. Math. **50**, 51–69 (1971)
12. Baxter, R.J.: Partition function of the eight vertex lattice model. Annals Phys. **70**, 193–228 (1972). Annals Phys. 281, 187 (2000)
13. Yang, C.N., Yang, C.P.: Thermodynamics of one-dimensional system of bosons with repulsive delta function interaction. J. Math. Phys. **10**, 1115–1122 (1969)
14. Korepin, V.E., Bogoliubov, N.M., Izergin, A.G.: Quantum Inverse Scattering Method and Correlation Functions. Cambridge University Press, Cambridge Monographs on Mathematical Physics (1997)
15. Essler, F.H.L., Frahm, H., Göhmann, F., Klümper, A., Korepin, V.E.: The One-Dimensional Hubbard Model. Cambridge University Press (2005)
16. Šamaj, L., Bajnok, Z.: Introduction to the Statistical Physics of Integrable Many-body Systems. Cambridge University Press (2013)
17. Lieb, E.H., Liniger, W.: Exact analysis of an interacting Bose gas. 1. The General solution and the ground state. Phys. Rev. **130**, 1605–1616 (1963)
18. Lieb, E.H., Liniger, W.: Exact analysis of an interacting bose gas. II.The excitation spectrum. Phys. Rev. **130**, 1616 (1963)
19. Faddeev, L.D.: How algebraic Bethe ansatz works for integrable model. In: Proceedings of Relativistic gravitation and gravitational radiation: Les Houches School of Physics, France, 26 Sept.–6 Oct., 1995, pp. 149–219 (1996)
20. Sutherland, B.: Further results for the many-body problem in one dimension. Phys. Rev. Lett. **20**, 98–100 (1968)
21. Kulish, P.P., Reshetikhin, N.Yu.: Diagonalization of $GL(N)$ invariant transfer matrices and quantum n wave system (Lee model). J. Phys. A **16**, L591–L596 (1983)
22. Bloch, F.Z.: Zur Theorie des Ferromagnetismus. Z. Physik **61**, 206 (1930)
23. Karbach, M., Muller, G.: Introduction to the Bethe ansatz I. Comput. Phys. **11**, 36–43 (1997)
24. Faddeev, L.D., Takhtajan, L.A.: Hamiltonian Methods in the Theory of Solitons. Springer-Verlag, Berlin Heidelberg (1987)

Chapter 6
Integrable Thermodynamics

This is a very controversial point of the thermodynamic Bethe-ansatz equations for soluble models, except for the repulsive boson case, which has no string solutions. But equations obtained using the string hypothesis seem to give the correct free energy and other thermodynamic quantities.

Minoru Takahashi
Thermodynamics of one-dimensional solvable models

In this chapter we consider the thermodynamics of integrable models. We assume that these models have a phase shift that can be computed exactly or approximately. For models with no bound states, by applying the thermodynamic limit to the corresponding Bethe equations, we obtain equations for the ground state. We then consider the case of finite temperature and derive the Yang-Yang equation describing the state of thermodynamic equilibrium. Further, using the example of a one-dimensional electron gas with delta-function interaction, we formulate the so-called string hypothesis and derive the corresponding Thermodynamic Bethe Ansatz equations, both canonical and simplified. We also exhibit solutions of these equations at weak and strong coupling.

6.1 Zero-Temperature Thermodynamics

At zero temperature a system is in its ground state and the main problem consists in evaluating the ground state energy as a function of particle density and the coupling constant(s). The central role in solving this problem as well as in developing the

© Springer Nature Switzerland AG 2019
G. Arutyunov, *Elements of Classical and Quantum Integrable Systems*,
UNITEXT for Physics, https://doi.org/10.1007/978-3-030-24198-8_6

finite-temperature thermodynamics is played by the Bethe equations that encode the asymptotic spectrum of the model in a one-dimensional "box" of a finite but large size L.

6.1.1 General Treatment

We therefore start with the Bethe equations written in the following universal form

$$e^{ip_j L} = (-1)^{\iota(N)} \prod_{k \neq j}^{N} e^{i\theta(p_j, p_k)} \,. \tag{6.1}$$

Here the N-dependent prefactor $(-1)^{\iota(N)}$ is determined by the model and by the boundary conditions. Note that in writing (6.1) we followed the conventions accepted in condensed matter literature, in the literature on relativistic quantum field theories one takes [1]

$$e^{ip_j L} \prod_{k \neq j}^{N} e^{i\delta(p_j, p_k)} = (-1)^{\iota(N)} \,, \tag{6.2}$$

so that $\delta(p_j, p_k) = -\theta(p_j, p_k)$.

Taking the logarithm of (6.1), one gets

$$p_j L = 2\pi I_j + \sum_{k \neq j}^{N} \theta(p_j, p_k) \,. \tag{6.3}$$

Here the numbers $I_j = \frac{1}{2}\iota(N) + n_j, n_j \in \mathbb{Z}$, are integers for $\iota(N)$ even and half-odd integers for $\iota(N)$ odd. A skew-symmetric phase $\theta(p, q) = -\theta(q, p)$ is defined in such a way that it vanishes for the limiting cases of either free fermions or impenetrable bosons. For these cases $p_j = 2\pi I_j/L$, where all I_j are all distinct. Due to the skew-symmetry of θ, the summation in (6.3) can be extended to include $k = j$.

The treatment of thermodynamics below is applicable for models whose Bethe equations (6.1) admit *real solutions only* and for which all the solutions $\{p_j\}$ are in one to one correspondence with sets of permitted *quantum numbers* $\{I_j\}$. The Lieb-Liniger model in the repulsive regime, the rational CMS and RS models are of this type but not, for instance, the Heisenberg model which has complex solutions. Analysis of models admitting complex configurations of Bethe roots is more subtle and can typically be done in the thermodynamic limit by invoking the so-called *string hypothesis*.

By summing up Eq. (6.3) and taking into account that the phase is skew-symmetric, one obtains an expression for the total momentum P

$$P = \sum_{j=1}^{N} p_j = \frac{2\pi}{L} \sum_{j=1}^{N} I_j,$$

which does not depend on the coupling constant. To treat bosons and fermions on equal footing, we assume without loss of generality that N is odd.[1]

The quantum numbers of the ground state must be the same as for the corresponding non-interacting system, they are symmetrically distributed around the origin

$$I_j = -\frac{N+1}{2} + j, \quad j = 1, \dots, N,$$ (6.4)

to achieve the minimum energy and vanishing of the total momentum.

Introduce the so-called *counting function* $\eta = \eta(p)$

$$\eta(p) = \frac{p}{2\pi} - \frac{1}{2\pi L} \sum_{k=1}^{N} \theta(p, p_k),$$ (6.5)

where p_k are Bethe roots corresponding to a given set $\{I_j\}$ of quantum numbers. To proceed, we assume that $-\theta(p, p_j)$ considered as a function of p is a *monotonically increasing* function.[2] With this assumption, it is evident from (6.5) that $\eta(p)$ is a monotonically increasing function of p. The value of $\eta(p)$ on a Bethe root p_j is

$$\eta(p_j) = I_j/L.$$ (6.6)

The inverse $p \equiv p(\eta)$ of the counting function is defined by the equation

$$\eta = \frac{1}{2\pi} p(\eta) - \frac{1}{2\pi L} \sum_{k=1}^{N} \theta(p(\eta), p_k).$$ (6.7)

The single-valued function $p(\eta)$ is also monotonically increasing[3] and at $\eta = I_j/L$ its value is

$$p(I_j/L) = p_j, \quad I_j \in \{I_j\}.$$ (6.8)

---

[1] For N even the ground state of a fermion system is doubly degenerate, see footnote 2 on Sect. 5.1.2.

[2] As follows from the variational argument based on the Yang-Yang functional [2], this implies that solutions of (6.1) are uniquely parametrised by a set of quantum numbers I_j, in accordance with our initial assumption.

[3] These important properties of $p(\eta)$ also follow from the variational principle, see [3].

It follows from monotonicity of $p(\eta)$ that if $I_i < I_j$ then $p_i < p_j$. Therefore, any solution of the Bethe equations with a set $\{I_j\}$ different from the one for the ground state will have larger energy.

At this point it is convenient to switch on to the standard terminology of condensed matter physics by introducing the concept of particles and holes, which use will require us to adopt different labelling of momenta in the Bethe Ansatz solutions.

Consider the function $p(\eta)$ at points m/L, where $m \in \mathbb{Z}$ runs through the set of all integers and denote

$$p(m/L) \equiv p_m. \tag{6.9}$$

Given the solution of the Bethe equations specified by a chosen set of quantum numbers $\{I_j\}$, p_m will coincide with some of the momenta present in this solution, provided $m \in \{I_j\}$. In this situation p_m is referred to as the momentum of a *particle* with a quantum number m. If $m \notin \{I_j\}$, then p_m is said to be the momentum of a *hole* with a quantum number m. Thus, each integer defines a vacancy which is either a particle or a hole. A vacancy taken by a particle is called filled or occupied, otherwise it is unoccupied. For instance, for the ground state all vacancies in the interval $[-(N-1)/2, (N+1)/2]$ are occupied and for an excited state some of the particles will move outside this interval leaving the holes inside, see Fig. 6.1.

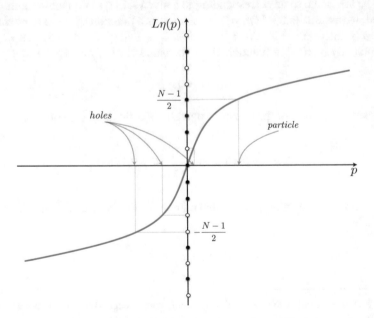

Fig. 6.1 Graph of a contingent counting function for an excited state. Black dots are integers from a given set $\{I_j\}$ of $N = 9$ quantum numbers. Three particles have their quantum numbers outside the ground state interval giving rise to three holes inside

Adopting the numbering of momenta of particles and holes as in (6.9), consider the value of the counting function at two neighbouring integers j and $j + 1$

$$\eta(p_{j+1}) = \frac{j+1}{L}, \quad \eta(p_j) = \frac{j}{L},$$

where p_j and p_{j+1} are the momenta of the corresponding vacancies, each of them corresponds to either a particle or a hole. Then

$$\eta(p_{j+1}) - \eta(p_j) = \frac{1}{L}.$$

When L becomes large the momentum tends to a continuous distribution $p_m \to p(m)$ and we can approximate

$$\eta'(p_j)(p_{j+1} - p_j) \approx \frac{1}{L}. \tag{6.10}$$

Define now the distribution density of vacancies ρ_v as[4]

$$\rho_v(p_j) = \lim_{L \to \infty} \frac{1}{L(p_{j+1} - p_j)}. \tag{6.11}$$

Thus, owing to (6.10), in the limit $L \to \infty$ one has

$$\rho_v(p) = \frac{d\eta(p)}{dp} > 0. \tag{6.12}$$

Taking into account that $\mathbb{Z} = \mathbb{Z}_p \cup \mathbb{Z}_h$, where $\mathbb{Z}_p$ and $\mathbb{Z}_h$ denote the sets of quantum numbers of particles and holes, respectively, we introduce the densities of particles and holes (these are introduced for finite L and N)

$$\rho(p) = \frac{1}{L} \sum_{k \in \mathbb{Z}_p} \delta(p - p_k), \quad \bar{\rho}(p) = \frac{1}{L} \sum_{k \in \mathbb{Z}_h} \delta(p - p_k). \tag{6.13}$$

In particular, the cardinality of $\mathbb{Z}_p = N$. Both densities are non-negative and their sum gives the density of vacancies

$$\rho_v(p) = \rho(p) + \bar{\rho}(p).$$

---

[4]In physical units this density has the dimension $1/[\hbar]$.

Differentiating (6.5), one gets

$$\rho_v(p) = \frac{1}{2\pi} + \frac{1}{L} \sum_{k \in \mathbb{Z}_p} K(p, p_k), \qquad (6.14)$$

where we denoted the derivative of the phase as

$$K(p, p_k) = -\frac{1}{2\pi} \frac{d\theta(p, p_k)}{dp}. \qquad (6.15)$$

In the thermodynamic limit, where both L and N tend to infinity, any thermodynamic state will be characterised by a certain distribution of particles and holes to which we associate the densities $\rho(p)$ and $\bar{\rho}(p)$, respectively. The energy of a state is then given by

$$E = \sum_{k \in \mathbb{Z}_p} e(p_k) = L \int dp\, e(p)\rho(p), \qquad (6.16)$$

where $e(p)$ is the dispersion relation specific to the model.

At zero temperature the system is in its ground state. The N quantum numbers of the ground state fill the interval $[-(N-1)/2, (N+1)/2]$ according to (6.4), with no holes. In the limit $N \to \infty$ there is a maximal possible value of the momentum $a = \lim_{N \to \infty} p_{(N-1)/2}$ and the roots p_j condense and fill the interval $[-a, a]$ with the density $\rho(p) = \rho_v(p)$. Equation (6.14) in the thermodynamic limit yields

$$\rho(p) = \frac{1}{2\pi} + \int_{-a}^{a} dq\, K(p, q)\rho(q), \qquad (6.17)$$

for $|p| \leqslant a$ and $\rho(p) = 0$ for $|p| > a$. From (6.13) it follows that the density is normalised as

$$\int_{-a}^{a} dp\, \rho(p) = \lim_{N,L \to \infty} \frac{N}{L} \equiv \mathcal{D}, \qquad (6.18)$$

where $\mathcal{D}$ is the density of particles in the thermodynamic limit. The energy density of the ground state is then

$$\mathcal{E}_0 = \lim_{N,L \to \infty} \frac{E_0}{L} = \int_{-a}^{a} dp\, e(p)\rho(p). \qquad (6.19)$$

Formulae (6.17), (6.18) and (6.19) fully characterise the ground state in the thermodynamic limit.

Generalised rapidity. In many important cases but not always the asymptotic momentum p can be considered as a single-valued function of a generalised rapidity variable u: $p = p(u)$. This function is chosen such that the original phase shift evaluated in terms of the asymptotic momenta, $\theta(p_j, p_k)$, turns into a function that depends on the difference of rapidities. For all the models we discussed such a generalised rapidity variable exists and the corresponding Bethe equations (6.3) can be cast in the form

$$p(u_j)L = 2\pi I_j + \sum_{k=1}^{N} \theta(u_j - u_k). \tag{6.20}$$

Indeed, for the Lieb-Liniger, CMS and RS models, $p(u) = u$, for the Heisenberg model $p(u) = 2\text{arccot}(2u)$. Moreover, (6.1) and (6.20) also encompass the Bethe Ansätze arising in two-dimensional integrable relativistic quantum field theories, where $p(u) = \sinh u$ so that u is the standard relativistic rapidity.

Introducing the u-dependent phase and the density

$$\theta(u) \equiv \theta(p(u), q(u)), \quad \rho(u) \equiv \frac{d\eta}{dp}\frac{dp}{du} = \rho(p(u))\frac{dp(u)}{du}, \tag{6.21}$$

where we assumed that $dp(u)/du > 0$, Eq. (6.17) turns into

$$\rho(u) = \frac{1}{2\pi}\frac{dp}{du} + \int_{-z}^{z} dv\, K(u - v)\rho(v), \quad K(u) = -\frac{1}{2\pi}\frac{d\theta(u)}{du} \tag{6.22}$$

for $|u| \leqslant z$ and $\rho(u) = 0$ for $|u| > z$, where z is related to a as $p(\pm z) = \pm a$. The normalisation condition (6.18) becomes

$$\int_{-z}^{z} du\, \rho(u) = \mathcal{D}. \tag{6.23}$$

An advantage of working with (6.22) instead of (6.17) is that the kernel in the first equation is a symmetric function of a single variable. Since a finite z breaks translational invariance, one cannot immediately profit from this observation by passing to the Fourier transform of the density. However, in the singular limit $z \to \infty$, Eq. (6.22) can be solved by the Fourier transform. As an intermediate step, one needs to evaluate the Fourier transform $\widehat{K}(\omega)$ of the kernel $K(u)$

$$\widehat{K}(\omega) = \int_{-\infty}^{\infty} du\, e^{i\omega u} K(u). \tag{6.24}$$

In particular, when $K(u)$ is not positive, the knowledge of the spectrum of $\widehat{K}(s)$ can be used to prove the existence of a solution for $\rho(u)$. The Fourier transform and the Fourier transformed kernels find their further applications in considering the finite temperature thermodynamics.

6.1.2 Examples

Let us look at (6.17) for two concrete models.

RS models. For the rational RS model the corresponding phase is given by (4.72) and for the kernel (6.15) we get

$$K(p) = (1 - \beta)\delta(p)\,, \tag{6.25}$$

where we recall that $\beta = \gamma/\hbar = \gamma$, as $\hbar = 1$. Clearly, in the attractive regime corresponding to $0 < \beta < 1$ the kernel is positive. Equation (6.17) for the density of the ground state turns into

$$\rho(p) = \frac{1}{2\pi} + (1 - \beta)\rho(p)\,, \quad |p| < a\,,$$
$$\rho(p) = 0\,, \quad |p| > a \tag{6.26}$$

and yields

$$\rho(p) = \begin{cases} \frac{1}{2\pi\beta}\,, & |p| < a\,, \\ 0\,, & |p| > a\,. \end{cases} \tag{6.27}$$

From the normalisation condition (6.18), one finds $a = \pi\gamma\mathcal{D}$. The energy density of the ground state is

$$\mathcal{E}_0 = \int_{-a}^{a} \mathrm{d}p\,\rho(p)\cosh p = \frac{\sinh\pi\gamma\mathcal{D}}{\pi\gamma}\,, \tag{6.28}$$

which is nothing else but (3.222).

Except for the rational case, the equation for the density of the ground state cannot be solved exactly for any other CMS or RS model. We point out that for the hyperbolic model the kernel K, up to the sign, is given by (4.80), and its Fourier transform (6.24) is[5]

$$\widehat{K}(\omega) = \frac{\sinh\frac{\omega}{2\ell}(1 - \beta)}{2\sinh\frac{\omega}{2\ell}}\frac{\sinh\frac{\pi\mu}{\hbar}\omega(1 - \frac{\gamma}{2\pi\mu\ell})}{\sinh\frac{\pi\mu}{\hbar}\omega}\,. \tag{6.29}$$

For possible applications we point out the degenerate cases of this kernel:

(1) Hyperbolic CMS model is obtained in the limit $\mu \to \infty$ and in this limit the kernel (6.29) reduces to

---

[5] All physical parameters are in there, including the mass parameter $\mu = mc$ (c is the speed of light), the length parameter ℓ and the Planck constant $\hbar$.

$$\widehat{K}(\omega) = \frac{\sinh \frac{s}{2\ell}(\beta - 1)}{2 \sinh \frac{\omega}{2\ell}} e^{-\frac{\gamma}{2h\ell}|\omega|} . \tag{6.30}$$

(2) Rational RS model arises in the limit $\ell \to \infty$ and for the corresponding kernel we get the constant

$$\widehat{K}(\omega) = \frac{1}{2}(\beta - 1) . \tag{6.31}$$

This is also the kernel for the rational CMS model.

Lieb-Liniger model. For this model the phase is, see (4.59),

$$\theta(p) = -2 \arctan \frac{p}{\varkappa} ,$$

so that for the kernel one gets

$$K(p) = \frac{1}{\pi} \frac{\varkappa}{\varkappa^2 + p^2} > 0 \quad \text{for} \quad \varkappa > 0 ,$$

and the equation for the ground state takes the form

$$\rho(p) = \frac{1}{2\pi} + \frac{\varkappa}{\pi} \int_{-a}^{a} dq \frac{\rho(q)}{\varkappa^2 + (p - q)^2} .$$

Although this equation cannot be solved analytically, it can be solved perturbatively around the Tonks-Girardeau limit $\varkappa \to \infty$. Expanding the kernel into power series in inverse powers of $\varkappa$,

$$\rho(p) = \frac{1}{2\pi} + \frac{1}{\pi} \sum_{k=0}^{\infty} \int_{-a}^{a} dq \, (-1)^k (p - q)^{2k} \varkappa^{-2k-1} \rho(q) ,$$

one looks for a solution in the form

$$\rho(p) = \sum_{k=0}^{\infty} \rho_k(p) \varkappa^{-k} , \quad \rho_0(p) = \frac{1}{2\pi} ,$$

yielding a perturbative series

$$\rho(p) = \frac{1}{2\pi} \left[1 + 2\left(\frac{a}{\pi\varkappa}\right) + 4\left(\frac{a}{\pi\varkappa}\right)^2 + \left(8 - \frac{2\pi^2}{3} - \frac{2p^2\pi^2}{a^2}\right)\left(\frac{a}{\pi\varkappa}\right)^3 \dots \right] .$$

Computing (6.18) and (6.19), one finds a perturbative dependence of the energy density on the density of particles

$$\mathscr{E}_0 = \frac{\pi^2 \mathcal{D}^3}{3}\left[1 - 4\frac{\mathcal{D}}{\varkappa} + 12\left(\frac{\mathcal{D}}{\varkappa}\right)^2 + \cdots\right].$$

One can show that this series is convergent for $\mathcal{D}/\varkappa < 1/2$.

An expansion around $\varkappa \to 0$ corresponding to the case of free bosons is also possible, although this limit is singular and, therefore, requires special care [4]. One can show that in this limit the density $\rho(p)$ is given by the Wigner semi-circle law

$$\rho(p) = \frac{a}{2\pi\varkappa}\sqrt{1 - \left(\frac{p}{a}\right)^2}, \qquad (6.32)$$

while for the energy density one has

$$\mathscr{E}_0 = \mathcal{D}^3\left[\left(\frac{\varkappa}{\mathcal{D}}\right) - \frac{4}{3\pi}\left(\frac{\varkappa}{\mathcal{D}}\right)^{3/2} + \cdots\right].$$

6.2 Finite-Temperature Thermodynamics and TBA

At finite temperature we are interested in the state describing thermodynamic equilibrium. The main object of study is the free energy F that is determined from the canonical partition function

$$Z = \text{Tr}\, e^{-H/T}, \qquad (6.33)$$

where H is the hamiltonian and T is the temperature, the Boltzmann constant is set to one. The free energy is $F = -T \log Z$. We consider a system consisting of N particles in the one-dimensional volume $V = L$ and we will be interested in finding the value of F in the thermodynamic limit where $N, L \to \infty$ with the ratio $\mathcal{D} = N/L$ kept fixed.

6.2.1 Yang-Yang Equation

At finite temperature there appear holes which in the thermodynamic limit will be distributed in the momentum space with some density $\bar{\rho}(p)$. For an excited state quantum numbers of some of its particles are to be found outside the ground state interval, meaning that in the thermodynamic limit the integration boundaries in (6.17) will not be respected and both distribution densities, $\rho(p)$ and $\bar{\rho}(p)$, will acquire support on the whole real line. In the thermodynamic limit Eq. (6.14) will turn into

$$\rho(p) + \bar{\rho}(p) = \frac{1}{2\pi} + \int_{-\infty}^{\infty} dq\, K(p,q)\rho(q), \qquad (6.34)$$

The equilibrium state we are looking for is not an eigenstate of the hamiltonian, rather it is a mixture of excited states with different quantum numbers $\{I_j\}$, all giving the same macroscopic densities ρ and $\bar{\rho}$. The existence of many different microscopic configurations which lead to the same macroscopic probability distributions gives rise to the entropy factor. To compute the latter, let us recall the physical meaning of the quantities

$L\rho(p)dp$ — # of particles in the interval $[p, p + dp]$,

$L\bar{\rho}(p)dp$ — # of holes in the interval $[p, p + dp]$,

$L\rho_v(p)dp$ — # of vacancies in the interval $[p, p + dp]$.

Thus, the number of ways to put $L\rho(p)dp$ particles into $L\rho_v(p)dp$ vacant places is

$$\frac{[L\rho_v(p)dp]!}{[L\rho(p)dp]![L\bar{\rho}(p)dp]!} = e^{dS}, \tag{6.35}$$

where dS is the contribution to the entropy from the interval $[p, p + dp]$. Formula (6.35) corresponds to the Fermi-Dirac statistical distribution based on the assumption that no two particles can occupy the same position in the momentum space. This is the case for the models we consider in the book. More generally, one can have models with Bethe roots obeying Bose statistics which requires to proper modification of the entropy formula [1].

In the thermodynamic limit the number of particles and holes becomes large which justifies the use of Stirling's approximation

$$\ln n! \approx n \ln n - n, \quad n \to \infty,$$

applied to the logarithm of (6.35). In this way we find

$$dS = L[(\rho + \bar{\rho}) \ln(\rho + \bar{\rho}) - \rho \ln \rho - \bar{\rho} \ln \bar{\rho}]dp, \tag{6.36}$$

where we also made use of $\rho_v(p) = \rho(p) + \bar{\rho}(p)$. Thus, the entropy per unit length is the following functional of ρ and $\bar{\rho}$

$$\frac{S}{L} = \int_{-\infty}^{\infty} dp \left[(\rho + \bar{\rho}) \ln(\rho + \bar{\rho}) - \rho \ln \rho - \bar{\rho} \ln \bar{\rho}\right].$$

Consider now the free energy of the canonical ensemble

$$F = E - TS - \mu(N - \mathcal{D}L).$$

Here the last term containing the lagrangian multiplier μ was added to keep the number of particles N at a fixed density $\mathcal{D}$; as it will become clear in a moment, μ is nothing else but the chemical potential. In the limit $L \to \infty$ we resort to the free energy per unit length

$$F/L = \mathcal{E} - TS/L - \mu(N/L - \mathcal{D}),$$

for which we have

$$\frac{F}{L} = \int_{-\infty}^{\infty} dp\, e(p)\rho(p) \tag{6.37}$$

$$-T \int_{-\infty}^{\infty} dp\left[(\rho + \bar{\rho})\ln(\rho + \bar{\rho}) - \rho \ln \rho - \bar{\rho}\ln \bar{\rho}\right] - \mu\left[\int_{-\infty}^{+\infty} dp\rho(p) - \mathcal{D}\right].$$

The state of thermodynamic equilibrium is obtained by extremising the free energy over the densities of particles and holes. The densities of particles and holes are not independent, rather they are related through (6.34) that also yields a relation between their variations

$$\delta\rho(p) + \delta\bar{\rho}(p) = \int_{-\infty}^{\infty} dq\, K(p, q)\delta\rho(q). \tag{6.38}$$

Extremising the free energy modulo the relation (6.38) and introducing

$$\frac{\rho}{\bar{\rho}} = e^{-\epsilon/T}, \tag{6.39}$$

where the quantity $\epsilon = \epsilon(p)$ is called *pseudo-energy*, we find the following equation

$$\epsilon(p) = e(p) - \mu - T \int_{-\infty}^{\infty} dq\, \ln\left(1 + e^{-\epsilon(q)/T}\right) K(q, p). \tag{6.40}$$

This equation is of the fundamental importance and is known as the *Yang-Yang equation* [2]. One can show that under the assumptions of the positivity and monotonicity of the kernel K a solution $\epsilon = \epsilon(p, \mu, T)$ to this equation exists and it is unique. Once (6.40) is solved, ρ can be determined from the equation

$$\rho(p)\left(1 + e^{\epsilon(p)/T}\right) = \frac{1}{2\pi} + \int_{-\infty}^{\infty} dq\, K(p, q)\rho(q), \tag{6.41}$$

which under the same assumptions gives a unique solution for $\rho = \rho(p, \mu, T)$. Note that the ratio

$$\frac{\rho}{\rho_v} = \frac{\rho}{\rho + \bar{\rho}} = \frac{1}{e^{\epsilon/T} + 1}$$

looks like the standard Fermi distribution, allowing to think of the pseudo-energy $\epsilon(p)$ as the energy of an elementary excitation over the state of thermodynamic equilibrium.

The solutions of (6.40) and (6.41) provide the main contribution to the exact partition in the limit $L \to \infty$, as can be rigorously shown by evaluation the partition function of the canonical ensemble by the method of steepest descent [3].

On the solutions of (6.40) and (6.34) describing the state of thermodynamic equilibrium the free energy F is

$$F = \mu N - T L \int_{-\infty}^{\infty} \frac{dp}{2\pi} \ln\left(1 + e^{-\epsilon(p)/T}\right). \tag{6.42}$$

If one wishes the chemical potential can be replaced by the density through solving the constraint

$$\frac{\partial F}{\partial \mu} = 0 \quad \to \quad \mathcal{D} = \frac{N}{L} = \int_{-\infty}^{\infty} dp \, \rho(p, \mu, T). \tag{6.43}$$

As is well known from elementary thermodynamics, the extensive nature of thermodynamic potentials is captured, in the case of F, by the following formula

$$F = \mu N - V P, \tag{6.44}$$

where P is pressure and V is the volume which in our case is $V = L$. Comparing this formula with (6.42) gives an expression for pressure[6]

$$P = T \int_{-\infty}^{\infty} \frac{dp}{2\pi} \ln\left(1 + e^{-\epsilon(p)/T}\right), \tag{6.45}$$

where P appears as a function of T and μ: $P = P(T, \mu)$. On the other hand, compatibility of the fundamental equation of equilibrium thermodynamics

$$dF = -SdT - PdV + \mu dN,$$

with (6.44) implies the fulfilment of the Gibbs-Duhem equation

$$dP = \frac{S}{V} dT + \frac{N}{V} d\mu = \frac{S}{L} dT + \mathcal{D} d\mu. \tag{6.46}$$

Let us show that P given by (6.45) does satisfy this equation so that the interpretation of P as pressure and μ as chemical potential is correct. Taking the differential of (6.45), one gets

---

[6]We warn the reader that we use for pressure the same notation as for the total momentum of a microscopic system.

$$dP = \left[\int_{-\infty}^{\infty} \frac{dp}{2\pi} \ln\left(1 + e^{-\epsilon/T}\right) + \int_{-\infty}^{\infty} \frac{dp}{2\pi} \frac{\left(\frac{\epsilon}{T} - \frac{\partial\epsilon}{\partial T}\right)}{1 + e^{\epsilon/T}} \right] dT + \left[\int_{-\infty}^{\infty} \frac{dp}{2\pi} \frac{\left(-\frac{\partial\epsilon}{\partial\mu}\right)}{1 + e^{\epsilon/T}} \right] d\mu .$$

Differentiating (6.40) over T and μ and using (6.40), one obtains

$$
\begin{aligned}
\frac{\epsilon(p)}{T} - \frac{\partial\epsilon(p)}{\partial T} &= \frac{e(p) - \mu}{T} + \int_{-\infty}^{+\infty} dq \frac{\left(\frac{\epsilon(q)}{T} - \frac{\partial\epsilon(q)}{\partial T}\right)}{1 + e^{\epsilon(q)/T}} K(q, p), \\
-\frac{\partial\epsilon(p)}{\partial\mu} &= 1 + \int_{-\infty}^{+\infty} dq \frac{\left(-\frac{\partial\epsilon(q)}{\partial\mu}\right)}{1 + e^{\epsilon(q)/T}} K(q, p) .
\end{aligned}
$$

(6.47)

The first equation is then multiplied by $\rho(p)$ and integrated over p. The usage of (6.41) inside the resulting expression gives

$$\int_{-\infty}^{\infty} \frac{dp}{2\pi} \frac{1}{1 + e^{\epsilon/T}} \left(\frac{\epsilon}{T} - \frac{\partial\epsilon}{\partial T} \right) = \frac{1}{T} \int_{-\infty}^{\infty} dp \, \rho(p)(e(p) - \mu) .$$

Next, using the symmetry of the kernel $K(p, q) = K(q, p)$, we compare the second equation in (6.47) with (6.41), and relying on the uniqueness of the solution of the latter equation, we conclude that

$$2\pi\rho = -\frac{1}{1 + e^{\epsilon/T}} \frac{\partial\epsilon}{\partial\mu} .$$

(6.48)

Thus, for the differential dP we get

$$dP = \left[\int_{-\infty}^{\infty} \frac{dp}{2\pi} \ln\left(1 + e^{-\epsilon(p)/T}\right) + \frac{1}{T} \int_{-\infty}^{\infty} dp \, \rho(p)(e(p) - \mu) \right] dT + \mathcal{D} \, d\mu .$$

It remains to note that the expression in the brackets is S/L, where S is the entropy of the equilibrium state, so that the Gibbs-Duhem equation (6.46) holds confirming the consistency of the thermodynamic description.

The above-described method of treating thermodynamics of an integrable model based on the Yang-Yang equation is called *Thermodynamic Bethe Ansatz* (TBA). Note that we could arrive to the same results concerning equilibrium thermodynamics by starting from the grand canonical partition function $\Xi \equiv \Xi(\mu, L)$ and the Gibbs free energy $\Omega = -T \ln \Xi$. The number of particles and the pressure at equilibrium are then found as

$$N = -\frac{\partial\Omega}{\partial\mu}, \quad P = -\frac{\partial\Omega}{\partial L} .$$

The Gibbs free energy is related to F in the standard way $\Omega = F - \mu N = -PL$.

6.2.2 TBA for the Rational RS Model

Here we consider the application of the TBA approach to the rational RS model. The corresponding Yang-Yang equation is

$$\epsilon(p) = \cosh p - \mu + T(\beta - 1) \int_{-\infty}^{\infty} \delta(p - q) \ln\left(1 + e^{-\epsilon(q)/T}\right), \quad (6.49)$$

where we made use of the kernel (6.25) and the dispersion relation $e(p) = \cosh p$. Integrating the delta-function, we get the following transcendental equation

$$\epsilon(p) = \cosh p - \mu + T(\beta - 1) \ln\left(1 + e^{-\epsilon(p)/T}\right). \quad (6.50)$$

that is equivalent to

$$\left(e^{\epsilon/T}\right)^{\beta} = e^{(\cosh p - \mu)/T}\left(1 + e^{\epsilon/T}\right)^{\beta-1}. \quad (6.51)$$

It is clear that $\epsilon(p)$ is an even function: $\epsilon(p) = \epsilon(-p)$. Differentiating (6.50) over the chemical potential, we obtain

$$\frac{\partial \epsilon}{\partial \mu} = -\frac{1 + e^{\epsilon/T}}{\beta + e^{\epsilon/T}}, \quad (6.52)$$

so that from (6.48) we find the density of particles at the equilibrium

$$\rho(p) = \frac{1}{2\pi} \frac{1}{\beta + e^{\epsilon(p)/T}}. \quad (6.53)$$

Further, we rewrite (6.51) as

$$\begin{aligned}
e^{(\mu - \cosh p)/T} &= e^{-\epsilon/T}\left(1 + e^{-\epsilon/T}\right)^{\beta-1} \\
&= \left((1 + e^{-\epsilon/T}) - 1\right)\left(1 + e^{-\epsilon/T}\right)^{\beta-1} = \zeta^{\beta} - \zeta^{\beta-1},
\end{aligned}$$

where we have introduced $\zeta = 1 + e^{-\epsilon/T}$. Setting

$$\zeta = e^{\omega}, \quad \alpha = e^{(\mu - \cosh p)/T},$$

the last equation takes the form

$$\alpha = e^{\beta\omega} - e^{(\beta-1)\omega} \quad (6.54)$$

and can be perturbatively inverted around $\alpha \approx 0 \approx \omega$, in the same way as done in [5] for the rational CMS model. One finds

$$\omega = \sum_{n=1}^{\infty} c_n \frac{\alpha^n}{n!} \, , \qquad c_n = \frac{(-1)^{n+1}}{n(\beta-1)} \frac{\Gamma(n\beta)}{\Gamma(n(\beta-1))} \tag{6.55}$$

For the pressure (6.45) one therefore finds

$$P = T \int_{-\infty}^{\infty} \frac{dp}{2\pi} \, \omega(p) = T \sum_{n=1}^{\infty} \frac{c_n}{n!} e^{n\mu/T} \int_{-\infty}^{\infty} \frac{dp}{2\pi} \exp(-n \cosh(p)/T)$$

$$= \frac{T}{\pi} \sum_{n=1}^{\infty} \frac{c_n K_0(n/T)}{n!} e^{n\mu/T} \, , \tag{6.56}$$

where the last integral was computed in terms of the modified Bessel function of the second kind $K_0(x)$. Formula (6.56) represents an example of the high-temperature expansion, $T \to \infty$. It is organised in powers of the fugacity $x = e^{\mu/T}$. The series (6.56) has a finite radius R of convergency

$$R = e^{1/T} \left| \frac{(\beta-1)^{\beta-1}}{\beta^{\beta}} \right| \, . \tag{6.57}$$

For $|x| > R$ the function P is given by an analytic continuation of the power series (6.56).

Consider now the low-temperature expansion $T \to 0$. First we have to reproduce the results concerning the ground state at $T = 0$. The perturbative analysis of (6.50) shows that for positive μ and non-zero density $\mathcal{D}$, the function $\epsilon(p)$ must have two real zeros $\epsilon(\pm a) = 0$, $a > 0$, and exhibit the following behaviour

$$\epsilon(p) < 0 \quad \text{for} \quad |p| < a \, ,$$
$$\epsilon(p) > 0 \quad \text{for} \quad |p| > a \, .$$

Taking these general properties into account, for the ground state solution $\epsilon_0(p)$ in the interval $|p| < a$, we get from (6.50) the following equation

$$\epsilon_0(p) = \cosh p - \mu + (\beta-1) \lim_{T \to 0} T \ln(e^{-\frac{\epsilon_0(p)}{T}}) = \cosh p - \mu - (\beta-1)\epsilon_0(p), \quad |p| < a \, ,$$

which yields

$$\epsilon_0(p) = \frac{1}{\beta}(\cosh p - \mu), \quad |p| < a \, . \tag{6.58}$$

For $|p| > a$ Eq. (6.50) together with positivity of $\epsilon_0(p)$ implies in the limit $T \to 0$ that

$$\epsilon_0(p) = \cosh p - \mu, \quad |p| > a \, . \tag{6.59}$$

Fig. 6.2 Pseudo-energy of
the ground state

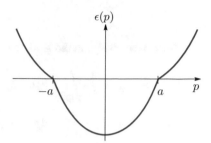

The formulae (6.58) and (6.59) give the pseudo-energy of the ground state, see the
plot of a hypothetical $\epsilon(p)$ on Fig. 6.2. It is clear that $\epsilon(\pm a) = 0$ relates a to the
chemical potential as $\mu = \cosh a$. The pseudo-energy is not differentiable at $\pm a$
unless $\beta = 1$ (free fermions). With this $\epsilon(p)$, formula (6.53) yields the expected
particle density (6.27), so that $a = \pi \gamma \mathcal{D}$.
    The free energy of the ground state is

$$F = N\mu - L \lim_{T \to 0} T \int_{-\infty}^{+\infty} \frac{dp}{2\pi} \ln\left(1 + e^{-\epsilon_0(p)/T}\right) = N\mu + \frac{L}{2\pi} \int_{-a}^{a} dp\, \epsilon_0(p) \quad (6.60)$$

Substituting here (6.58), we find the following answer

$$\frac{F}{L} = \mathcal{D}\mu + \frac{1}{2\pi\gamma} \int_{-a}^{a} dp(\cosh p - \mu) = \left(\mathcal{D} - \frac{a}{\pi\gamma}\right)\mu + \frac{\sinh a}{\pi\gamma} = \frac{\sinh \pi\gamma\mathcal{D}}{\pi\gamma}.$$

Evidently, the free energy of the ground state coincides with its energy, see (6.28).
    One can also determine the leading temperature correction to the ground state.
For $|p| < a$ we rewrite (6.50) as

$$\epsilon(p) = \frac{1}{\beta}(\cosh p - \mu) + T \frac{\beta - 1}{\beta} \ln\left(1 + e^{\epsilon(p)/T}\right).$$

Assuming an expansion

$$\epsilon = \epsilon_0 + T\epsilon_1(T) + \dots,$$

for the leading correction we obtain

$$\epsilon_1 = \frac{\beta - 1}{\beta} \ln\left(1 + e^{\epsilon_0/T} e^{\epsilon_1/T}\right) \approx \frac{\beta - 1}{\beta} e^{\epsilon_0/T}. \quad (6.61)$$

On the other hand, for $|p| > a$ one gets

$$\epsilon_0 + T\epsilon_1 = \cosh p - \mu + T(\beta - 1)e^{-\epsilon_0/T}, \quad (6.62)$$

yielding

$$\epsilon_1 = (\beta - 1)e^{-\epsilon_0/T} .$$

Thus, the free energy at the leading order in T is

$$F(T) = N\mu - \frac{L}{2\pi} \int_{|p|<a} dp(-\epsilon_0 - T\epsilon_1 + Te^{\epsilon_0/T}) - \frac{TL}{2\pi} \int_{|p|>a} e^{-\epsilon_0/T}$$

$$= N\mu + \frac{L}{2\pi} \int_{|p|<a} dp\epsilon_0 - \frac{TL}{2\pi\beta} \int_{|p|<a} dpe^{\epsilon_0/T} - \frac{TL}{2\pi} \int_{|p|>a} dpe^{-\epsilon_0/T} ,$$

so that

$$\delta F = F(T) - F(0) = -\frac{TL}{\pi} \left[\frac{1}{\beta} \int_0^a dp\, e^{(\cosh p - \mu)/(\beta T)} + \int_a^\infty dp\, e^{(\mu - \cosh p)/T} \right].$$

The leading asymptotic of the integrals above can be obtained by applying the Laplace method and we find

$$\delta F \approx \frac{2LT^2}{\pi\sqrt{\mu^2 - 1}} .$$

Correspondingly, the free energy of the ground state at low temperatures receives its first correction of order T^2,

$$\frac{F}{L} = \left(\mathcal{D} - \frac{a}{\pi\gamma}\right)\mu + \frac{\sinh a}{\pi\gamma} - \frac{2T^2}{\pi\sqrt{\mu^2 - 1}} + \cdots .$$

where a is found from $\cosh a = \mu$. Excluding the chemical potential in favour of the density, one finally gets

$$\frac{F}{L} = \frac{\sinh \pi\gamma\mathcal{D}}{\pi\gamma} - \frac{2T^2}{\pi \sinh \pi\gamma\mathcal{D}} + \cdots .$$

6.3 Thermodynamics of Yang's Model

Here we apply the TBA approach to Yang's model of a one-dimensional electron gas with delta-function interaction [6–8]. An essential new feature of this model is that it admits complex solutions so that the previous formalism should be extended to account for this fact. We start from the Bethe Ansatz equations (5.57) for Yang's model

$$e^{ip_j L} = \prod_{n=1}^{M} \frac{p_j - \lambda_n + \frac{i\varkappa}{2}}{p_j - \lambda_n - \frac{i\varkappa}{2}}, \qquad j = 1, \dots, N, \quad (6.63)$$

$$\prod_{j=1}^{N} \frac{p_j - \lambda_n + \frac{i\varkappa}{2}}{p_j - \lambda_n - \frac{i\varkappa}{2}} = - \prod_{j=1}^{M} \frac{\lambda_n - \lambda_j - i\varkappa}{\lambda_n - \lambda_j + i\varkappa}, \qquad n = 1, \dots, M. \quad (6.64)$$

The equations depend on the length L and two numbers N and M, where N is the number of momentum carrying particles and M is the number of auxiliary roots. Solutions which differ only by a reordering of the roots are considered as equivalent. The thermodynamic limit corresponds to taking $L, N, M \to \infty$ with the ratios N/L and M/L kept fixed.

6.3.1 String Hypothesis

In order to derive the thermodynamics of this system, we first need to understand the structure of solutions of the Bethe equations (6.63) and (6.64) in the thermodynamic limit. Some of the solutions contributing to this limit will involve complex Bethe roots and will be naturally interpreted as bound states.

Suppose we have a set $\{p_1, \dots, p_N; \lambda_1, \dots, \lambda_M\}$ that solves these equations. One can start from the following assumption the consistency of which can be justified a posteriori. Namely, one can assume that if a complex p (or λ) is from this set, then its complex conjugate $\bar{p}$ (or $\bar{\lambda}$) also belongs to this set. In other words, p's (λ's) are either real or come in complex conjugate pairs and, therefore, they are symmetrically distributed with respect to the real axis. Such an arrangement of p's guarantees that the energy $E = \sum_{j=1}^{N} p_j^2$ is automatically real.

To proceed further, we need the following properties

$$\left| e^{ixL} \right| = \begin{cases} \leqslant 1 & \text{if } \operatorname{Im} x \geqslant 0, \\ \geqslant 1 & \text{if } \operatorname{Im} x \leqslant 0 \end{cases}$$

and

$$\left| \frac{x + ic}{x - ic} \right| = \begin{cases} \geqslant 1 & \text{if } \operatorname{Im} x \geqslant 0, \\ \leqslant 1 & \text{if } \operatorname{Im} x \leqslant 0 \end{cases} \quad (6.65)$$

for any constant $c > 0$. These properties immediately show that for $\varkappa > 0$ a solution with all λ's real has all p's real as well. Indeed, take some p_j and assume that $\operatorname{Im} p_j > 0$, then the absolute value of the left hand side of (6.63) is less than unity, while the absolute value of the right hand side is bigger than unity and, therefore, (6.63) cannot be satisfied. If $\operatorname{Im} p_j < 0$, then the absolute values of the left and right hand sides are respectively bigger and smaller than unity, so they mismatch again. This leaves room for real p_j only.

For the attractive case $\varkappa < 0$ the situation is different and one can have solutions with complex momenta. Indeed, consider the simplest configuration of just two particles with momenta p_1 and p_2 and one auxiliary root λ. Equations (6.63) and (6.64) boil down to

$$e^{ip_1 L} = \frac{p_1 - \lambda + \frac{i\varkappa}{2}}{p_1 - \lambda - \frac{i\varkappa}{2}}, \qquad e^{ip_2 L} = \frac{p_2 - \lambda + \frac{i\varkappa}{2}}{p_2 - \lambda - \frac{i\varkappa}{2}}, \tag{6.66}$$

$$\frac{p_1 - \lambda + \frac{i\varkappa}{2}}{p_1 - \lambda - \frac{i\varkappa}{2}} \frac{p_2 - \lambda + \frac{i\varkappa}{2}}{p_2 - \lambda - \frac{i\varkappa}{2}} = 1. \tag{6.67}$$

Assuming $\mathrm{Im}\, p_1 > 0$, in the limit $L \to \infty$ the left hand side of the first equation in (6.66) is exponentially suppressed and the only way to balance this behaviour on the right hand side is to assume that

$$p_1 = \lambda - \frac{i\varkappa}{2} + \mathcal{O}(e^{-\delta L}), \tag{6.68}$$

where $\delta > 0$. Evidently, Eq. (6.67) will be then satisfied if

$$p_2 = \lambda + \frac{i\varkappa}{2} + \mathcal{O}(e^{-\delta L}). \tag{6.69}$$

Thus Eq. (6.66) can be replaced with exponential accuracy by (6.68) and (6.69).

For the energy to be real, $\bar{p}_1 = p_2$, which implies that λ is real. The solution for p_1 we found is then compatible with our assumption $\mathrm{Im}\, p_1 > 0$ only for $\varkappa < 0$. Thus, in the attractive regime we will have complex solutions for p_j as, for instance, the two-particle configuration[7] found above, but for the opposite case of $\varkappa > 0$ this is not possible and only real p-s are allowed, at least in the thermodynamic limit. Thus, the repulsive case is technically less challenging and in the following we consider it only. For the treatment of the attractive case we refer the reader to the book [9].

So far we have been looking at the Eq. (6.63) in the limit $L \to \infty$, keeping N and M finite. Now we would like to identify the relevant configurations for $N \to \infty$ with M finite. Let us pick some $\lambda_n \equiv \lambda$. Since all p_j are real, in the limit $N \to \infty$ from (6.65) we observe the following behaviour

$$\prod_{j=1}^{N} \frac{p_j - \lambda + \frac{i\varkappa}{2}}{p_j - \lambda - \frac{i\varkappa}{2}} \to 0 \ \text{ if }\ \mathrm{Im}\, \lambda > 0,$$

$$\prod_{j=1}^{N} \frac{p_j - \lambda + \frac{i\varkappa}{2}}{p_j - \lambda - \frac{i\varkappa}{2}} \to \infty \ \text{ if }\ \mathrm{Im}\, \lambda < 0.$$

---

[7]This configuration is called p-λ *string*.

Let us assume for definiteness that Im $\lambda_1 > 0$. The only way to satisfy (6.64) for λ_1, is to assume that there exist λ_2 such that

$$\lambda_1 - \lambda_2 - i\varkappa = 0 \quad \rightarrow \quad \lambda_2 = \lambda_1 - i\varkappa = 0 . \tag{6.70}$$

Now there are two possibilities.

The first possibility corresponds to the situation when Im $\lambda_2 < 0$. In this case, considering Eq. (6.64) for λ_2 shows that it will be automatically satisfied, up to exponentially suppressed terms, because

$$\prod_{j=1}^{N} \frac{p_j - \lambda_2 + \frac{i\varkappa}{2}}{p_j - \lambda_2 - \frac{i\varkappa}{2}} \rightarrow \infty \quad \text{and} \quad \frac{\lambda_2 - \lambda_1 - i\varkappa}{\lambda_2 - \lambda_1 + i\varkappa} \prod_{j \neq 1,2}^{M} \frac{\lambda_2 - \lambda_j - i\varkappa}{\lambda_2 - \lambda_j + i\varkappa} \rightarrow \infty ,$$

as $\lambda_2 \rightarrow \lambda_1 - i\varkappa$. According to our assumption on the distribution of the Bethe roots, $\overline{\lambda_2} = \lambda_1$. Combining this with Eq. (6.70) yields the so-called $2|\lambda$-string configuration

$$\lambda_1 = \lambda + \tfrac{i}{2}\varkappa , \quad \lambda_2 = \lambda - \tfrac{i}{2}\varkappa ,$$

where λ is an arbitrary real number called the *string's center*.

The second possibility corresponds to Im $\lambda_2 > 0$. In this case to have a solution, there must exist $\lambda_3 = \lambda_2 - i\varkappa$ which, in its turn, gives rise to two possibilities for the sign of Im λ_3. Having Im $\lambda_3 < 0$ will leave us with a $3|\lambda$-string, otherwise we have to continue and assume the existence of $\lambda_4 = \lambda_3 - i\varkappa$, and so on. A generic configuration of this type is the $m|\lambda$-string which represents the following complex of m Bethe roots λ_j

$$\lambda_j = \lambda + \tfrac{i\varkappa}{2}(m + 1 - 2j), \quad j = 1, \ldots, m . \tag{6.71}$$

The roots in this complex are distributed equidistantly along the imaginary axis and symmetrically around the real center $\lambda \in \mathbb{R}$, justifying the name *Bethe string* for this configuration of Bethe roots. This classification also includes the $1|\lambda$-string which consists of a single real root λ. According to the *string hypothesis* [9, 10], in the thermodynamic limit all solutions of the Bethe equations for $\varkappa > 0$ are given by Bethe strings (6.71) with corrections of order $\mathcal{O}(\exp(-\delta N))$. Strings belong to the category of bound states owing to the fact that the corresponding wave function shows an exponential decay with the distance between Bethe roots forming a string configuration.

Let us fix a state which for any positive integer m contains M_m $m|\lambda$-strings, so that the total number M of λ roots is

$$M = \sum_{m=1}^{\infty} m M_m , \tag{6.72}$$

where M_m is called an *occupation number*. For finite M only a finite number of M_m is non-zero. Individual Bethe roots of a $m|\lambda$-string will be conveniently parametrised as

$$\lambda_{\alpha,m}^{(s)} = \lambda_{\alpha,m} + \tfrac{i\varkappa}{2}(m + 1 - 2s) \quad s = 1, \ldots, m, \quad \alpha = 1, \ldots, M_n, \quad (6.73)$$

where $\lambda_{\alpha,m} \in \mathbb{R}$ is the center of this (α, m)-string.

The next step consists in expressing the original Bethe equations (6.63), (6.64) via the real centers of λ-strings. Assuming that M λ-roots are organised into Bethe strings according to the pattern (6.72), equation (6.63) takes the form

$$e^{ip_j L} = \prod_{m=1}^{\infty} \prod_{\alpha=1}^{M_m} \prod_{s=1}^{m} \frac{p_j - \lambda_{\alpha,m} - \tfrac{i\varkappa}{2}(m + 1 - 2s) + \tfrac{i\varkappa}{2}}{p_j - \lambda_{\alpha,m} - \tfrac{i\varkappa}{2}(m + 1 - 2s) - \tfrac{i\varkappa}{2}}. \qquad (6.74)$$

Here the product over the variable s can be easily computed by writing it explicitly

$$\frac{p_j - \lambda_{\alpha,m} - \tfrac{i\varkappa}{2}(m - 2)}{p_j - \lambda_{\alpha,m} - \tfrac{i\varkappa}{2}m} \; \frac{p_j - \lambda_{\alpha,m} - \tfrac{i\varkappa}{2}(m - 4)}{p_j - \lambda_{\alpha,m} - \tfrac{i\varkappa}{2}(m - 2)} \; \cdots \; \frac{p_j - \lambda_{\alpha,m} + \tfrac{i\varkappa}{2}m}{p_j - \lambda_{\alpha,m} + \tfrac{i\varkappa}{2}(m - 2)},$$

and noting the cancellation pattern where the numerator of the first fraction cancels against the denominator of the second one, the numerator of the second fraction cancels against the denominator of the third one and so on. Thus, individual components of each $m|\lambda$-string have been "fused" and we obtain the following set of equations

$$e^{ip_j L} = \prod_{m=1}^{\infty} \prod_{\alpha=1}^{M_m} \frac{p_j - \lambda_{\alpha,m} + \tfrac{i\varkappa}{2}m}{p_j - \lambda_{\alpha,m} - \tfrac{i\varkappa}{2}m}, \quad j = 1, \ldots, N. \qquad (6.75)$$

Next, we take the product of equations (6.64) for individual components of the (α, m)-string and, upon fusing the left hand side of this product as above, obtain

$$\prod_{j=1}^{N} \frac{p_j - \lambda_{\alpha,m} + \tfrac{i\varkappa}{2}m}{p_j - \lambda_{\alpha,m} - \tfrac{i\varkappa}{2}m} = \left(\prod_{s \neq r}^{m} \frac{\lambda_{\alpha,m}^{(s)} - \lambda_{\alpha,m}^{(r)} - i\varkappa}{\lambda_{\alpha,m}^{(s)} - \lambda_{\alpha,m}^{(r)} + i\varkappa} \right) \prod_{s=1}^{m} \prod_{(\beta,n) \neq (\alpha,m)} \prod_{r=1}^{n} \frac{\lambda_{\alpha,m}^{(s)} - \lambda_{\beta,n}^{(r)} - i\varkappa}{\lambda_{\alpha,m}^{(s)} - \lambda_{\beta,n}^{(r)} + i\varkappa}.$$

Here on the the right hand side we singled out the self-interaction terms of the $\lambda_{\alpha,m}$ string which contribute trivially because

$$\prod_{s \neq r}^{m} \frac{\lambda_{\alpha,m}^{(s)} - \lambda_{\alpha,m}^{(r)} - i\varkappa}{\lambda_{\alpha,m}^{(s)} - \lambda_{\alpha,m}^{(r)} + i\varkappa} = 1.$$

The rest can be expressed as

$$\prod_{(\beta,n)\neq(\alpha,m)} \prod_{s=1}^{m} \prod_{r=1}^{n} \frac{\lambda_{\alpha,m}^{(s)} - \lambda_{\beta,n}^{(r)} - i\varkappa}{\lambda_{\alpha,m}^{(s)} - \lambda_{\beta,n}^{(r)} + i\varkappa} = \prod_{(\beta,n)\neq(\alpha,m)} S_{mn}(\lambda_{\beta,n} - \lambda_{\alpha,m}),$$

where we have introduced the function

$$S_{mn}(u) = \frac{u + \frac{i\varkappa}{2}|m-n|}{u - \frac{i\varkappa}{2}|m-n|} \frac{u + \frac{i\varkappa}{2}(m+n)}{u - \frac{i\varkappa}{2}(m+n)} \prod_{j=1}^{\min(m,n)-1} \left(\frac{u + \frac{i\varkappa}{2}(|m-n|+2j)}{u - \frac{i\varkappa}{2}(|m-n|+2j)} \right)^2 .$$

It has the properties

$$S_{mn}(u) = S_{nm}(u), \quad S_{mn}(-u)S_{mn}(u) = 1,$$

and also $S_{mm}(0) = -1$, $S_{mn}(0) = 1$ if $m \neq n$. Using these properties, the equation for $\lambda_{\alpha,m}$ can be brought to the form

$$-1 = \prod_{j=1}^{N} \frac{p_j - \lambda_{\alpha,m} + \frac{i\varkappa}{2}m}{p_j - \lambda_{\alpha,m} - \frac{i\varkappa}{2}m} \prod_{n=1}^{\infty} \prod_{\beta=1}^{M_n} S_{mn}(\lambda_{\alpha,m} - \lambda_{\beta,n}) . \tag{6.76}$$

Introducing the function

$$S_m(u) = \frac{u + \frac{i\varkappa}{2}m}{u - \frac{i\varkappa}{2}m} , \tag{6.77}$$

Equations (6.75) and (6.76) can be compactly written as

$$e^{ip_j L} = \prod_{m=1}^{\infty} \prod_{\alpha=1}^{M_m} S_m(p_j - \lambda_{\alpha,m}), \quad j = 1, \ldots, N, \tag{6.78}$$

$$\prod_{j=1}^{N} S_m(\lambda_{\alpha,m} - p_j) = -\prod_{n=1}^{\infty} \prod_{\beta=1}^{M_n} S_{mn}(\lambda_{\alpha,m} - \lambda_{\beta,n}), \quad \alpha = 1, \ldots, M_m. \tag{6.79}$$

These formulae provide an implementation of the string hypothesis at the level of the Bethe equations. Equations (6.78), (6.79) are equivalent to the original, "microscopic" Bethe equations up to terms exponentially suppressed in the length L and the number of particles N. From the physical point of view, they look like the usual Bethe equations for the full asymptotic spectrum of the model that now in addition to the fundamental particles with real momenta p_j also involves particles with real rapidities $\lambda_{\alpha,m}$. In particular, $S_m(u)$ can be interpreted as the scattering matrix of a fundamental particle and an $m|\lambda$-string and $S_{mn}(u)$ has a similar interpretation as the S-matrix which describes scattering of two bound states corresponding to $m|\lambda$- and $n|\lambda$-strings.

6.3.2 TBA Equations

To proceed with derivation of the corresponding TBA equations, we note that for real u one has

$$\frac{1}{i} \log S_m(u) = \pi + \theta_m(u), \quad \theta_m(u) = \theta\left(\frac{2u}{m}\right), \tag{6.80}$$

where

$$\theta(u) = -2\arctan\frac{u}{\varkappa} \tag{6.81}$$

is the phase shift of the Lieb-Liniger model. The phase $\theta(u)$ is a monotonically decreasing function of a real variable u and it takes values in the interval $-\pi \leq \theta(u) \leq \pi$. It reaches the maximum at $u \to -\infty$ and the minimum at $u \to +\infty$. Analogously, introducing

$$\Theta_{mn}(u) = (1 - \delta_{mn})\theta_{|m-n|}(u) + \theta_{m+n}(u) + 2\sum_{j=1}^{\min(m,n)-1} \theta_{|m-n|+2j}(u), \tag{6.82}$$

we relate it to the logarithm of the corresponding S-matrix

$$\frac{1}{i} \log S_{mn}(u) = \pi\left[2\min(m,n) - \delta_{mn}\right] + \Theta_{mn}(u) \quad (\text{mod } 2\pi). \tag{6.83}$$

The logarithm of (6.75) and (6.76) yields

$$\begin{aligned}
p_j L &= 2\pi I_j + \sum_{m=1}^{\infty}\sum_{\alpha=1}^{M_m} \theta_m\left(p_j - \lambda_{\alpha,m}\right), \\
\sum_{j=1}^{N} \theta_m\left(\lambda_{\alpha,m} - p_j\right) + 2\pi \mathcal{I}_{\alpha,m} &= \sum_{n=1}^{\infty}\sum_{\beta=1}^{M_n} \Theta_{mn}\left(\lambda_{\alpha,m} - \lambda_{\beta,n}\right).
\end{aligned} \tag{6.84}$$

Here for I_j are distinct integers or half-odd integers depending on whether the sum $\sum_{n=1}^{\infty} M_n$ is even or odd, respectively. Taking into account that

$$S_m(u) = e^{i\pi} e^{i\theta_m(u)}, \quad S_{mn}(u) = e^{-i\pi\delta_{mn}} e^{i\Theta_{mn}(u)},$$

the overall accumulation of the π-phase in (6.79) is $e^{i\pi(N-M_m+1)}$. This shows that the distinct numbers $\mathcal{I}_{\alpha,m}$ are integers if $N - M_m$ is odd and are half-odd integers if $N - M_m$ is even. The numbers $\mathcal{I}_{\alpha,m}$ are bounded; the bounds are determined by taking the limits $u \to \pm\infty$ in the second set of equations in (6.84). Since $\theta(u) \to -\mp\infty$ as $u \to \pm\infty$ and taking into account that $\mathcal{I}_{\alpha,m}$ are integers or half-odd integers depending on whether $N - M_m$ is odd or even, we find

$$|\mathcal{I}_{\alpha,m}| \leq \frac{1}{2}\left(N - 1 - \sum_{n=1}^{\infty}(2\min(m,n) - \delta_{m,n})M_n\right).$$

According to the string hypothesis every set of allowed (half-)integers $\{I_j, \mathcal{I}_{\alpha,m}\}$ yields a unique solution $\{p_j, \lambda_{\alpha,m}\}$. Mathematically, this relationship between solutions and their numbers is captured by the counting functions $\eta(p)$ and $\eta_m(\lambda), m \geq 1$, that are introduced as

$$
\begin{aligned}
\eta(p) &= \frac{p}{2\pi} - \frac{1}{2\pi L}\sum_{m=1}^{\infty}\sum_{\alpha=1}^{M_m}\theta_n\big(p - \lambda_{\alpha,m}\big), \\
\eta_m(\lambda) &= \frac{1}{2\pi L}\sum_{n=1}^{\infty}\sum_{\beta=1}^{M_n}\Theta_{mn}\big(\lambda - \lambda_{\beta,n}\big) - \frac{1}{2\pi L}\sum_{j=1}^{N}\theta_m\big(\lambda - p_j\big).
\end{aligned}
\tag{6.85}
$$

The counting functions are monotonically increasing functions of their arguments and they map solutions of the Bethe equations (6.84) into the corresponding quantum numbers

$$\eta(p_j) = \frac{I_j}{L}, \quad \eta_m(\lambda_{\alpha,m}) = \frac{\mathcal{I}_{\alpha,m}}{L}.$$

According to our general definition of particles and holes in Sect. 6.1, (half)-integer numbers that do not belong to a solution set $\{I_j, \mathcal{I}_{\alpha,m}\}$ are mapped to holes in the p- and λ-axes. In the thermodynamic limit, which involves taking $M \to \infty$, each occupation number M_m becomes large and $\lambda_{\alpha,m}$ form a dense distribution with the density $\sigma_m(\lambda)$. Also holes condense and form a distribution with the density $\bar{\sigma}_m(\lambda)$. The densities of particles and holes corresponding to fundamental particles with microscopic momenta p_j will be denoted, as before, $\rho(p)$ and $\bar{\rho}(p)$, respectively. It then follows from the definition of the densities that they are related to the derivatives of the counting functions as

$$\rho(p) + \bar{\rho}(p) = \frac{d\eta}{dp}, \quad \sigma_m(\lambda) + \bar{\sigma}_m(\lambda) = \frac{d\eta_m}{d\lambda}.$$

The macroscopic particle densities are

$$\frac{N}{L} = \int_{-\infty}^{\infty} dp\,\rho(p), \quad \frac{M}{L} = \sum_{m=1}^{\infty} m\int_{-\infty}^{\infty} d\lambda\,\sigma_m(\lambda).\tag{6.86}$$

Differentiating (6.85), we obtain in the thermodynamic limit the following integral equations

$$\rho(p) + \bar{\rho}(p) = \frac{1}{2\pi} + \sum_{m=1}^{\infty} \int_{-\infty}^{\infty} dq \, K_m(p-q)\sigma_n(q) \,,$$

$$\sigma_m(p) + \bar{\sigma}_m(p) = \int_{-\infty}^{\infty} dq \, K_m(p-q)\rho(q) - \sum_{n=1}^{\infty} \int_{-\infty}^{\infty} dq \, K_{mn}(p-q)\sigma_n(q) \,,$$
(6.87)

where we uniformised the notation for the arguments of the functions ρ and σ_m. Here we introduced the kernels

$$K_m(u) = -\frac{1}{2\pi}\frac{d\theta_m(u)}{du} = \frac{1}{\pi}\frac{\frac{\varkappa}{2}m}{u^2 + (\frac{\varkappa}{2}m)^2} \,,$$
(6.88)

$$K_{mn}(u) = -\frac{1}{2\pi}\frac{d\Theta_{mn}(u)}{du}$$
(6.89)

$$= (1 - \delta_{mn})K_{|m-n|}(u) + K_{m+n}(u) + 2\sum_{j=1}^{\min(m,n)-1} K_{|m-n|+2j}(u) \,.$$

Both kernels are obviously positive for real values of their arguments.

The entropy receives contributions from all kind of particles and is given by

$$\frac{S}{L} = \int_{-\infty}^{\infty} dp \left[(\rho + \bar{\rho}) \ln(\rho + \bar{\rho}) - \rho \ln \rho - \bar{\rho} \ln \bar{\rho} \right]$$
(6.90)

$$+ \sum_{m=1}^{\infty} \int_{-\infty}^{\infty} d\lambda \left[(\sigma_m + \bar{\sigma}_m) \ln(\sigma_m + \bar{\sigma}_m) - \sigma_m \ln \sigma_m - \bar{\sigma}_m \ln \bar{\sigma}_m \right].$$

The free energy is

$$F = E - TS - \mu(N - \mathcal{D}L) - h(N - 2M - \mathcal{M}L) \,.$$
(6.91)

Here $\mathcal{M}$ is the magnetisation per unit length and we included the new term with the lagrangian multiplier h to keep magnetisation at a fixed value. Since the number of particles is controlled by the density $\mathcal{D}$, fixing $\mathcal{M}$ is equivalent to fixing the number M of electrons with spin up. Physically, the parameter h can be identified with an external magnetic field.

Taking into account (6.86) and (6.90), we write (6.91) as the following functional of the densities $\{\rho, \bar{\rho}, \sigma_m, \bar{\sigma}_m\}$

$$\frac{F}{L} = \int_{-\infty}^{\infty} dp \, (e(p) - \mu - h)\rho(p) + 2h \sum_{m=1}^{\infty} m \int dp \, \sigma_m(p) + \mu \mathcal{D} + h \mathcal{M}$$

$$-T \int_{-\infty}^{\infty} dp \left[(\rho + \bar{\rho}) \ln(\rho + \bar{\rho}) - \rho \ln \rho - \bar{\rho} \ln \bar{\rho} \right] \qquad (6.92)$$

$$-T \sum_{m=1}^{\infty} \int_{-\infty}^{\infty} dp \left[(\sigma_m + \bar{\sigma}_m) \ln(\sigma_m + \bar{\sigma}_m) - \sigma_m \ln \sigma_m - \bar{\sigma}_m \ln \bar{\sigma}_m \right].$$

Extremising the free energy modulo the constraints imposed on the variations of densities of particles and holes by equations (6.87), we find the conditions for the thermodynamic equilibrium. To present them in the most convenient and compact way, we introduce the left and right convolutions of a function g with any kernel K

$$(K \star g)(p) = \int_{-\infty}^{\infty} dq \, K(p, q) g(q), \quad (g \star K)(p) = \int_{-\infty}^{\infty} dq \, g(q) K(q, p).$$

If the kernel is symmetric (as the kernels in our present case) these convolutions are the same and we need not distinguish between them.

Introducing the pseudo-energies

$$\frac{\rho(p)}{\bar{\rho}(p)} = e^{-\epsilon(p)/T}, \quad \frac{\sigma_m(p)}{\bar{\sigma}_m(p)} = e^{-\epsilon_m(p)/T}, \qquad (6.93)$$

the conditions of the thermodynamic equilibrium are then given by the following set of coupled non-linear integral equations

$$\epsilon = \frac{p^2 - \mu - h}{T} - \sum_{m=1}^{\infty} \ln\left(1 + e^{-\epsilon_m/T}\right) \star K_m,$$

$$\epsilon_m = \frac{2hm}{T} - \ln\left(1 + e^{-\epsilon/T}\right) \star K_m + \sum_{n=1}^{\infty} \ln\left(1 + e^{-\epsilon_n/T}\right) \star K_{nm}. \qquad (6.94)$$

These are the so-called *canonical TBA equations* because they follow directly from the above-described canonical procedure [10], the latter can be applied to derive the TBA equations in any integrable model.[8] For finite T and $\varkappa$ the TBA equations cannot be solved analytically, but an analytic solution is possible for certain limits on these parameters.

Finally, for the canonical ensemble of N electrons with M of those having spin down, the free energy at the equilibrium is

$$F = \mu N + h(N - 2M) - TL \int_{-\infty}^{+\infty} \frac{dp}{2\pi} \ln\left(1 + e^{-\epsilon(p)}\right). \qquad (6.95)$$

---

[8]Including, for instance, integrable models arising in the context of the AdS/CFT correspondence [11–13], see [14, 15] for the reviews.

so that for the pressure one gets

$$P = T \int_{-\infty}^{+\infty} \frac{\mathrm{d}p}{2\pi} \ln\left(1 + e^{-\epsilon(p)}\right). \tag{6.96}$$

It is the same expression as found for the Lieb-Liniger model, confer (6.45).

TBA kernels. To study the TBA equations, we need some properties of the TBA kernels K_m and K_{mn}. Most easily these properties can be revealed in the Fourier space, owing to the fact that convolution of functions is mapped under the Fourier transform to the product of their Fourier images. Performing the Fourier transform (6.24) of the kernels, we find

$$\widehat{K}_m(\omega) = e^{-\frac{\varkappa}{2}m|\omega|} \,,$$

$$\widehat{K}_{mn}(\omega) = \coth\left(\tfrac{\varkappa}{2}|\omega|\right)\left(e^{-\frac{\varkappa}{2}|m-n||\omega|} - e^{-\frac{\varkappa}{2}(m+n)|\omega|}\right) - \delta_{mn} \,.$$

Consider the matrix $(K+1)_{mn}(u) \equiv \delta_{mn}\delta(u) + K_{mn}(u)$. The Fourier image of its inverse is defined as

$$\sum_{k=1}^{\infty} \widehat{(K+1)}_{mk}^{-1}\left(\delta_{kn} + \widehat{K}_{kn}(\omega)\right) = \delta_{mn}$$

and one finds

$$\widehat{(K+1)}_{mn}^{-1} = \delta_{mn} - \widehat{s}(\omega)(\delta_{m+1,n} + \delta_{m-1,n})\,, \quad \widehat{s}(\omega) = \frac{1}{2\cosh\frac{\varkappa}{2}\omega}\,. \tag{6.97}$$

Therefore, the inverse of $K + 1$ in the sense of the convolution is

$$(K+1)_{mn}^{-1}(u) = \delta_{mn}\delta(u) - s(u)(\delta_{m+1,n} + \delta_{m-1,n})\,, \tag{6.98}$$

where we defined

$$s(u) = \frac{1}{2\pi} \int_{-\infty}^{\infty} \mathrm{d}\omega\, e^{-iu\omega}\,\widehat{s}(\omega) = \frac{1}{2\varkappa\cosh(\frac{\pi u}{\varkappa})}\,. \tag{6.99}$$

Further, in the Fourier space the following relation is true

$$\sum_{m=1}^{\infty} \widehat{K}_m \widehat{(K+1)}_{mn}^{-1} = \widehat{s}(\omega)\,\delta_{n,1}\,.$$

Fourier transforming this relation, we obtain

$$\sum_{m=1}^{\infty} K_m \star (K+1)_{mn}^{-1} = s(u)\delta_{n,1} \,. \tag{6.100}$$

The last relation can alternatively be proved by noting the following convolutions

$$\begin{aligned} s \star (K_{m-1} + K_{m+1}) &= K_m \,, \\ s \star K_2 + s &= K_1 \,. \end{aligned} \tag{6.101}$$

Finally, by going to the Fourier space, it is elementary to see that

$$\sum_{m=1}^{\infty} m \star (K+1)_{mn}^{-1} = 0 \,. \tag{6.102}$$

Simplified TBA equations. The TBA equations we obtained can be further simplified and brought to a quasi-local form (to be discussed below) by exploiting the properties of the TBA kernels established above. First we introduce Y-functions as the following exponentials of the pseudo-energies

$$Y(p) = e^{-\epsilon(p)/T} \,, \quad Y_m(p) = e^{\epsilon_m(p)/T} \,, \quad m \geq 1 \,. \tag{6.103}$$

The function Y is called the main Y-function and Y_m are the auxiliary ones. In terms of these function the canonical TBA equations acquire the form

$$-\ln Y = \frac{p^2 - \mu - h}{T} - \sum_{m=1}^{\infty} \ln\left(1 + Y_m^{-1}\right) \star K_m \,, \tag{6.104}$$

$$\ln Y_m = \frac{2hm}{T} - \ln(1 + Y) \star K_m + \sum_{n=1}^{\infty} \ln\left(1 + Y_n^{-1}\right) \star K_{nm} \,. \tag{6.105}$$

We start from (6.105) by convoluting it with $(K+1)_{mn}^{-1}$ and summing over m

$$\sum_{m=1}^{\infty} \ln Y_m \star (K+1)_{mn}^{-1} = \frac{2h}{T} \sum_{m=1}^{\infty} m \star (K+1)_{mn}^{-1} - \ln(1+Y) \star \sum_{m=1}^{\infty} K_m \star (K+1)_{mn}^{-1}$$

$$+ \sum_{k,m=1}^{\infty} \ln(1 + Y_k^{-1}) \star K_{km} \star (K+1)_{mn}^{-1} \,.$$

By using the relations (6.98), (6.100) and (6.102), we obtain the *simplified TBA equations* for the Y_m-functions

$$\ln Y_1 = \ln \frac{1 + Y_2}{1 + Y} \star s \,, \tag{6.106}$$

$$\ln Y_m = \ln(1 + Y_{m-1})(1 + Y_{m+1}) \star s \,, \quad m \geq 2 \,. \tag{6.107}$$

Further, we consider (6.105) for $m = 1$ and convolute it with s

$$\ln Y_1 \star s = \frac{2h}{T} \star s - \ln(1 + Y) \star K_1 \star s + \sum_{m=1}^{\infty} \ln(1 + Y_m^{-1}) \star K_{m1} \star s$$

From the definition (6.89) of K_{mn} one sees that

$$K_{11} = K_2 \,,$$
$$K_{m1} = K_{m-1} + K_{m+1} \,, \quad m \geq 2 \,,$$

so that

$$K_{11} \star s = K_2 \star s = K_1 - s \,,$$
$$K_{m1} \star s = (K_{m-1} + K_{m+1}) \star s = K_m \,, \quad m \geq 2 \,.$$

Using these formulae together with $1 \star s = \frac{1}{2}$, we arrive at

$$\ln(1 + Y_1) \star s = \frac{h}{T} - \ln(1 + Y) \star K_1 \star s + \sum_{m=1}^{\infty} \ln(1 + Y_m^{-1}) \star K_m \,.$$

The infinite sum in this expression is then substituted from (6.104) and we obtain the following simplified TBA equation for the main Y-function

$$\ln Y = -\frac{p^2 - \mu}{T} + \ln(1 + Y) \star K_1 \star s + \ln(1 + Y_1) \star s \,. \tag{6.108}$$

Here the kernel $K_1 \star s$ is

$$(K_1 \star s)(u) = \frac{1}{4\pi} \int_{-\infty}^{\infty} d\omega \, e^{-i\omega u} \frac{e^{-\frac{\varkappa}{2}|\omega|}}{\cosh \frac{\varkappa}{2}\omega} \,.$$

The system of the simplified TBA equations (6.106), (6.107) and (6.108) is quasi-local in the sense that it does not involve infinite sums of Y-functions. On the other hand, this system is not completely equivalent to the original TBA equations. For instance, the magnetic field h, while present in the canonical TBA equations, completely drops from the simplified ones. The reason for this is that the operator $(K + 1)_{mn}^{-1}$, by which we acted on the original TBA equations, has a non-trivial kernel so that an application of this operator erases some information on solutions. This information can be reinstated though by specifying asymptotic or/and analytic behaviour of solutions of the simplified equations. In our present case the magnetic field can be restored by requiring Y_m to respect the limit

$$\lim_{m \to \infty} \ln Y_m / m = 2h/T \,. \tag{6.109}$$

Another concern is related to the fact that in the intermediate steps of the simplification procedure one should use the canonical TBA equations only rather than the already obtained simplified equations for some of the Y-functions. To illustrate this point, we will attempt to derive the equation for Y in a different way, by relying on the already obtained (6.106) and (6.107). We start with evaluating the infinite sum in (6.110) directly

$$\sum_{m=1}^{\infty} \ln(1 + Y_m^{-1}) \star K_m = \ln(1 + Y_1^{-1}) \star K_1 + \sum_{m=2}^{\infty} \ln(1 + Y_m^{-1}) \star s \star (K_{m-1} + K_{m+1})$$

$$= \ln(1 + Y_1^{-1}) \star s + \ln(1 + Y_2^{-1}) \star s \star K_1 + \sum_{m=2}^{\infty} \ln(1 + Y_{m-1}^{-1})(1 + Y_{m+1}^{-1}) \star s \star K_m \,.$$

By using the simplified TBA equations for Y_m, one then finds

$$\sum_{m=2}^{\infty} \ln(1 + Y_{m-1}^{-1})(1 + Y_{m+1}^{-1}) \star s \star K_m = \sum_{m=2}^{\infty} (\ln Y_m - \ln Y_{m+1} Y_{m-1} \star s) \star K_m$$

$$= \sum_{m=3}^{\infty} \ln Y_m \star \left(K_m - s \star (K_{m-1} + K_{m+1})\right) + \ln Y_2 \star K_2 - \ln Y_1 \star s \star K_2 - \ln Y_2 \star s \star K_3$$

$$= \ln Y_2 \star s \star K_1 - \ln Y_1 \star (K_1 - s) \,,$$

where the relations (6.101) were used. Thus,

$$\sum_{m=1}^{\infty} \ln(1 + Y_m^{-1}) \star K_m = \ln(1 + Y_1) \star s + \ln(1 + Y_2) \star s \star K_1 - \ln Y_1 \star K_1$$

$$= \ln(1 + Y_1) \star s + \ln(1 + Y_2) \star s \star K_1 - \ln \frac{1 + Y_2}{1 + Y} \star s \star K_1$$

$$= \ln(1 + Y_1) \star s + \ln(1 + Y) \star s \star K_1 \,,$$

where in the middle line equation (6.106) was applied. In this way we were able to replace the infinite sum in (6.110) with only two terms involving Y and Y_1. The equation for the main Y-function is then

$$\ln Y = -\frac{p^2 - \mu - h}{T} + \ln(1 + Y_1) \star s + \ln(1 + Y) \star s \star K_1 \,.$$

Since $s \star K_1 = K_1 \star s$, the only difference between this equation and (6.108) is the presence of the magnetic field. This difference occurs because of the use of the simplified equations for Y_m, the latter already miss the information about the so-called driving term with the magnetic field.

Ground state. The equations for the ground state are obtained in the limit $T \to 0$. Since $Y_m = e^{\epsilon_m/T} > 0$, equation (6.107) implies that $\epsilon_m = \ln Y_m$ are all positive for

$m \geq 2$ and, therefore, Y_m^{-1} for these values of m are exponentially suppressed in the limit $T \to 0$. The canonical TBA equations reduce in this limit to

$$\epsilon = p^2 - \mu - h - T \ln(1 + Y_1^{-1}) \star K_1 , \tag{6.110}$$

$$\epsilon_1 = 2h - T \ln(1 + Y) \star K_1 + T \ln(1 + Y_1^{-1}) \star K_2 , \tag{6.111}$$

where we used that $K_{11} = K_2$. The functions ϵ and ϵ_1 are both increasing functions of p^2 and they are negative in the regions $[-a, a]$ and $[-b, b]$. With these findings, equations (6.87) boil down to

$$\rho(p) = \frac{1}{2\pi} + \int_{-a}^{a} dq \, K_1(p - q)\sigma_1(q) ,$$

$$\sigma_1(p) = \int_{-a}^{a} dq \, K_1(p - q)\rho(q) - \int_{-b}^{b} dq \, K_2(p - q)\sigma_1(q) . \tag{6.112}$$

These are the equations for the ground state [16]. The integration bounds a and b are found from the fixing the values of a macroscopic particle density and magnetisation of the canonical ensemble

$$\frac{N}{L} = \int_{-a}^{a} dp \, \rho(p) , \qquad \frac{M}{L} = \int_{-b}^{b} dp \, \sigma_1(p) .$$

Assuming without loss of generality that N is even and $M \leq N/2$ is odd, for the ground state solution the half-odd integers I_α and the integers $J_{\alpha,1}$ will uniformly fill the intervals $[-(N-1)/2, (N-1)/2]$ and $[-(M-1)/2, (M-1)/2]$, respectively.

Weak and strong coupling limits. The strong and weak coupling limits of the TBA equations for the delta-interaction model of repulsive electrons were worked out in [8, 9], see also the books [9, 17]. Here for the reader convenience, we reproduce the corresponding results in our notation.

In the weak coupling limit $\varkappa \to 0$ both kernels s and $s \star K_1$ tend to $\frac{1}{2}\delta$, so that Eqs. (6.106), (6.107) and (6.108) reduce to

$$Y_1^2 = \frac{1 + Y_2}{1 + Y} , \tag{6.113}$$

$$Y_m^2 = (1 + Y_{m-1})(1 + Y_{m+1}) , \quad m \geq 2 \tag{6.114}$$

and

$$Y^2 = \exp\left(-2\frac{p^2 - \mu}{T}\right)(1 + Y)(1 + Y_1) . \tag{6.115}$$

The system (6.114) has a general solution

$$Y_m = \left(\frac{z^m w - z^{-m} w^{-1}}{z - z^{-1}}\right)^2 - 1, \tag{6.116}$$

where z, w are arbitrary. Matching this expression with the asymptotics (6.109) gives $z = e^{h/T}$. The parameter w is then determined from (6.113) via Y. Finally, equation (6.115) is solved for Y. One finds

$$Y_m = -1 + \frac{\left(\frac{\sinh \frac{h}{T} m}{\sinh \frac{h}{T}} + \frac{\sinh \frac{h}{T}(m+1)}{\sinh \frac{h}{T}} e^{-\frac{p^2 - \mu}{T}}\right)^2}{1 + Y},$$

and

$$Y = -1 + \left(1 + e^{-(p^2 - \mu + h)/T}\right)\left(1 + e^{-(p^2 - \mu - h)/T}\right),$$

so that

$$P = T \int_{-\infty}^{+\infty} \frac{dp}{2\pi} \ln\left[\left(1 + e^{-(p^2 - \mu + h)/T}\right)\left(1 + e^{-(p^2 - \mu - h)/T}\right)\right],$$

which is the well-known expression for the pressure of a free fermion gas in an external magnetic field [18].

In the strong coupling limit $\varkappa \to \infty$ (Tonks-Girardeau limit corresponding to impenetrable bosons), the kernels s and $K_1 \star s$ tend to flat distributions of height $1/\varkappa$, so that the normalisation conditions hold

$$\int_{-\infty}^{+\infty} du \, s(u) = \frac{1}{2}, \qquad \int_{-\infty}^{+\infty} du \, (s \star K_1)(u) = \frac{1}{2}.$$

This behaviour implies that $(g \star s)(u) = 1/2g$ if g is a constant and $(g \star s)(u) \to 0$ as $\varkappa \to \infty$ uniformly in u, if f vanishes at infinity. Hence, it follows from (6.106) and (6.107) that all Y_m are all constant in the strong coupling limit, satisfying

$$\begin{aligned} Y_1^2 &= 1 + Y_2, \\ Y_m^2 &= (1 + Y_{m-1})(1 + Y_{m+1}), \quad m \geq 2. \end{aligned} \tag{6.117}$$

In particular, the term $\ln(1 + Y) \star s$ decouples from (6.106) because Y is not a constant function, rather it obeys (6.108) which in the strong coupling limit takes the form

$$Y^2 = \exp\left(-2\frac{p^2 - \mu}{T}\right)(1 + Y_1). \tag{6.118}$$

Combing the general solution (6.116) with the first equation in (6.117) and the asymptotics (6.109), one finds $z = w = e^{h/T}$ and

$$Y_m = -1 + \frac{\sinh^2(m+1)\frac{h}{T}}{\sinh^2 \frac{h}{T}}. \tag{6.119}$$

It is then follows from (6.118) that

$$Y = 2\cosh\frac{h}{T}e^{-\frac{p^2-\mu}{T}}, \tag{6.120}$$

so that

$$P = T \int_{-\infty}^{\infty} \frac{dp}{2\pi} \ln\left(1 + 2\cosh\frac{h}{T}e^{-\frac{p^2-\mu}{T}}\right). \tag{6.121}$$

This completes our presentation of thermodynamics of integrable models by means of the TBA approach. Introduction into other related and important topics, such as the T-, Y- and Q-systems, Hirota equations, non-linear integral equations of the Klümper-Pearce-Destri-de Vega type, the TBA approach for computing the spectra of integrable field theories in finite volume can be found, for instance, in the reviews [19, 20] and the references therein.

References

1. Zamolodchikov, A.B.: Thermodynamic Bethe Ansatz in relativistic models. Scaling three state potts and Lee-yang models. Nucl. Phys. B **342**, 695–720 (1990)
2. Yang, C.N., Yang, C.P.: Thermodynamics of one-dimensional system of bosons with repulsive delta function interaction. J. Math. Phys. **10**, 1115–1122 (1969)
3. Korepin, V.E., Bogoliubov, N.M., Izergin, A.G.: Quantum Inverse Scattering Method and Correlation Functions. Cambridge Monographs on Mathematical Physics, Cambridge University Press (1997)
4. Gaudin, M.: Boundary energy of a Bose gas in one dimension. Phys. Rev. A **4**, 386–394 (1971)
5. Sutherland, B.: Quantum many body problem in one-dimension. J. Math. Phys. **12**, 251–256 (1971)
6. Takahashi, M.: One-dimensional electron gas with delta-function interaction at finite temperature. Prog. Theor. Phys. **46**, 1388–1406 (1971)
7. Lai, C.K.: Thermodynamics of fermions in one dimension with a δ-function interaction. Phys. Rev. Lett. **26**, 1472–1475 (1971)
8. Lai, C.K.: Thermodynamics of a one-dimensions system of fermions with a repulsive δ-function interaction. Phys. Rev. A **8**, 2567–2573 (1973)
9. Takahashi, M.: Thermodynamics of One-Dimensional Solvable Models. Cambridge University Press (1999)
10. Essler, F.H.L., Frahm, H., Göhmann, F., Klümper, A., Korepin, V.E.: The One-Dimensional Hubbard Model. Cambridge University Press (2005)
11. Arutyunov, G., Frolov, S.: String hypothesis for the AdS$_5$ × S^5 mirror. JHEP **03**, 152 (2009)
12. Arutyunov, G., Frolov, S.: Thermodynamic Bethe Ansatz for the AdS$_5$ × S^5 mirror model. JHEP **05**, 068 (2009)

13. Arutyunov, G., de Leeuw, M., van Tongeren, S.J.: The quantum deformed mirror TBA I. JHEP **10**, 090 (2012)

14. Arutyunov, G., Frolov, S.: Foundations of the AdS$_5 \times$ S^5 superstring. Part I. J. Phys. A **42**, 254003 (2009)

15. Beisert, N., et al.: Review of AdS/CFT integrability: an overview. Lett. Math. Phys. **99**, 3–32 (2012)

16. Yang, C.N.: Some exact results for the many body problems in one dimension with repulsive delta function interaction. Phys. Rev. Lett. **19**, 1312–1314 (1967)

17. Šamaj, L., Bajnok, Z.: Introduction to the statistical physics of integrable many-body systems. Cambridge University Press (2013)

18. Landau, L.D., Lifshitz, E.M., Pitaevskij, L.P.: Statistical Physics, (Course of Theoretical Physics), 2 edn., vol. 5. Oxford (1980)

19. Kuniba, A., Nakanishi, T., Suzuki, J.: T-systems and Y-systems in integrable systems. J. Phys. A **44**, 103001 (2011)

20. van Tongeren, S.J.: Introduction to the thermodynamic Bethe ansatz. J. Phys. A **49**, 323005 (2016)

Appendix A
Frobenius Theorem

In this appendix belonging to Chap. 1 we recall the proof of the Frobenius theorem.

A vector field ξ on a manifold $\mathscr{M}$ gives rise to an integral curve through each point of $\mathscr{M}$, such that ξ is tangent to this curve everywhere. The Frobenius theorem deals with the more general case of determining *integral submanifolds* of distributions generated by a set of vector fields on $\mathscr{M}$.

Distributions and integral submanifolds. Let $\mathscr{M}$ be a real smooth N-dimensional manifold without boundary and let $T\mathscr{M}$ be its tangent bundle. A family $V = \{V_x\}$ of linear m-dimensional subspaces $V_x \subset T_x\mathscr{M}$ which depend smoothly on the point $x \in \mathscr{M}$, i.e. a subbundle $V \subset T\mathscr{M}$, is called a m-dimensional *distribution* on $\mathscr{M}$.

For a m-dimensional distribution V passing through a point $x \in \mathscr{M}$ there exist m smooth vector fields ξ_i, $i = 1, \ldots, m$ such that in the neighbourhood of x

$$V = \mathrm{span}\{\xi_1, \ldots, \xi_m\}.$$

Note that V is a $C^\infty(\mathscr{M})$-submodule in the space $\mathfrak{X}(\mathscr{M})$. We say that a vector field ξ belongs to V if $\xi_x \in V_x$ for all $x \in \mathscr{M}$.

In many cases one has to deal with the more general concept of a singular distribution, where the dimension of $V_x \subset T_x\mathscr{M}$ may vary with x. This happens, for instance, for distributions associated to the hamiltonian vector fields on a Poisson manifold. Thus, more generally, we define a distribution V as the linear subspace of vector fields on $\mathscr{M}$ which is a $C^\infty(\mathscr{M})$-module.

A connected immersed submanifold $\mathscr{N}$ of $\mathscr{M}$ is called an integral submanifold of the distribution V if $T_x\mathscr{N} \subset V_x$ for any point $x \in \mathscr{N}$. In the case when $T_x\mathscr{N} = V_x$, $x \in \mathscr{N}$ such an integral submanifold is called *maximal*. If a maximal integral submanifold of V exists through each point of $\mathscr{N}$, the distribution V is said to be integrable or holonomic. An integrable distribution gives a foliation of $\mathscr{M}$, that is the decomposition of $\mathscr{M}$ into maximal integral submanifolds called leaves. In contrast to the case of integral curves, integral manifolds may not exist in general, even locally. In fact, the integrability of a distribution and, as a consequence, the existence of a

© Springer Nature Switzerland AG 2019
G. Arutyunov, *Elements of Classical and Quantum Integrable Systems*,
UNITEXT for Physics, https://doi.org/10.1007/978-3-030-24198-8

foliation, is a rather rare situation. To formulate a criterion for the integrability of a distribution, we need the notion of vector fields in involution.

A set of vector fields $\{\xi_1, \ldots, \xi_m\}$ on $\mathcal{M}$ is *in involution* if there exist smooth real-valued functions $c_{ij}^k(x)$, $x \in \mathcal{M}$, $i, j, k = 1, \ldots, m$, such that

$$[\xi_i, \xi_j] = c_{ij}^k(x)\xi_k .$$

A distribution V generated by vector fields ξ_i is called involutive if the set $\{\xi_1, \ldots, \xi_m\}$ is in involution.

Frobenius theorem. Let V be a m-distribution on a smooth manifold $\mathcal{M}$ generated by vector fields ξ_i, $i = 1, \ldots m$. Then

(1) V is integrable if and only if it is involutive, i.e. the vector fields ξ_i are in involution,
(2) an integrable distribution V is generated by commuting vector fields.

The proof goes as follows. See e.g. [1–3]. First, let us assume that V is integrable and that ξ, η are two vector fields belonging to V. We will show that $[\xi, \eta]$ also belongs to V. Indeed, since by our assumption V is integrable, there exists a maximal integrable manifold $\mathcal{N}$ of V passing through a point $x \in \mathcal{M}$. Let $m = \dim \mathcal{N}$. In a neighbourhood U of the point x one can choose a coordinate system $(q_1, \ldots, q_N)$, such that $q_1(x) = \ldots = q_N(x) = 0$ and that the intersection $\mathcal{N} \cap U$ defines a local chart on $\mathcal{N}$ given by $(q_1, \ldots, q_m, q_{m+1} = \ldots = q_N = 0)$. In this coordinate chart the expressions for the vector fields ξ and η that are tangent to $\mathcal{N}$ are

$$\xi = \sum_{i=1}^{m} u_j \frac{\partial}{\partial q_j} , \quad \eta = \sum_{i=1}^{m} v_j \frac{\partial}{\partial q_j} ,$$

where $u_j = u_j(q_1, \ldots, q_m, 0, \ldots, 0)$ and $v_j = v_j(q_1, \ldots, q_m, 0, \ldots, 0)$. For the commutator we then find

$$[\xi, \eta] = \sum_{j=1}^{m} w_i \frac{\partial}{\partial q_i} , \quad w_i = \sum_{j=1}^{m} \left(u_j \frac{\partial v_i}{\partial q_j} - v_j \frac{\partial u_i}{\partial q_j} \right) ,$$

that is $[\xi, \eta]_x \in V_x$. In simple words, the commutator of two vector fields tangent to a manifold is also tangent to this manifold.

Second, let us assume that a distribution V is involutive and show that there is an integral submanifold $\mathcal{N}$ passing through an arbitrary point $x \in \mathcal{M}$. In the neighbourhood U of x we choose a coordinate system $(q_1, \ldots q_N)$ and N independent vector fields

$$\xi_i = \sum_{j=1}^{N} u_{ij} \frac{\partial}{\partial q_j} , \quad i = 1, \ldots, N ,$$

such that ξ_i for $i = 1, \ldots, m$ generate V. Since ξ_i are independent, the submatrix $(u_{ij})_{i,j=1}^m$ is non-degenerate, and, therefore, is invertible in the domain U. We denote the inverse of this submatrix as $(t_{ij})_{i,j=1}^m$: $\sum_{k=1}^m t_{ik} u_{kj} = \delta_{ij}$ for $i, j = 1, \ldots m$. Further, consider the following vector fields

$$\eta_i = \sum_{k=1}^m t_{ik} \xi_k , \quad i = 1, \ldots, m . \tag{A.1}$$

Explicitly, we have

$$\eta_i = \sum_{k=1}^m \sum_{j=1}^N t_{ik} u_{kj} \frac{\partial}{\partial q_j} = \sum_{k=1}^m \sum_{j=1}^m t_{ik} u_{kj} \frac{\partial}{\partial q_j} + \sum_{k=1}^m \sum_{j=m+1}^N t_{ik} u_{kj} \frac{\partial}{\partial q_j}$$
$$= \frac{\partial}{\partial q_i} + \sum_{j=m+1}^N v_{ij} \frac{\partial}{\partial q_j} , \quad i = 1, \ldots, m , \tag{A.2}$$

where we introduced

$$v_{ij} = \sum_{k=1}^m t_{ik} u_{kj} , \quad i = 1, \ldots, m , \quad j = m+1, \ldots, N . \tag{A.3}$$

Evidently, the vector fields η_i are linearly independent and, since V is a $C^\infty(\mathcal{M})$-module, η_i are generators of V. For their commutator we get

$$[\eta_i, \eta_j] = \sum_{k=m+1}^N \left(\frac{\partial v_{jk}}{\partial q_i} - \frac{\partial v_{ik}}{\partial q_j} \right) \frac{\partial}{\partial q_k} + \sum_{k=m+1}^N \sum_{k'=m+1}^N \left(v_{ik'} \frac{\partial v_{jk}}{\partial q_{k'}} - v_{jk'} \frac{\partial v_{ik}}{\partial q_{k'}} \right) \frac{\partial}{\partial q_k} .$$

In other words, this commutator has the structure

$$[\eta_i, \eta_j] = \sum_{k=m+1}^N f_{ij}^k \frac{\partial}{\partial q_k} , \quad i, j = 1, \ldots, m . \tag{A.4}$$

On the other hand, by our assumption the distribution V is involutive, which means that there exists functions c_{ij}^k, $i, j, k = 1, \ldots, m$, such that

$$[\eta_i, \eta_j] = \sum_{k=1}^m c_{ij}^k \eta_k . \tag{A.5}$$

Substituting in the right hand side of this formula the expression (A.2), we obtain

$$[\eta_i, \eta_j] = \sum_{k=1}^{m} c_{ij}^{k} \frac{\partial}{\partial q_k} + \sum_{k=m+1}^{N} \sum_{s=1}^{m} c_{ij}^{s} v_{sk} \frac{\partial}{\partial q_k}.$$

This structure of the commutator should be compared to that of (A.4), from where we deduce that $c_{ij}^{k} = 0$. It then follows from (A.5) that $[\eta_i, \eta_j] = 0$ proving thereby the statement (2) of the Frobenius theorem. To rephrase the result obtained, by the proper change of basis in the space of vector fields, a system of non-commutative but involutive vector fields were brought to the commutative form. Finally, in the neighbourhood U of x the original distribution V is generated by

$$V = \text{span} \left\{ \frac{\partial}{\partial q_1}, \ldots, \frac{\partial}{\partial q_m} \right\}.$$

In terms of these local coordinates the maximal integral submanifold $\mathcal{N}$ through x is given by a connected component of

$$\{ z \in U \mid q_i(z) = q_i(x), \text{ for } i = m+1, \ldots, N \}.$$

This means, that maximal integral submanifolds of an integrable distribution V are leaves of a foliation on $\mathcal{M}$.

Note that for singular involutive distributions the corresponding set of vector fields may have integral submanifolds of different dimensions.

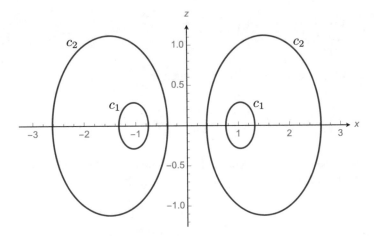

Fig. A.1 Sections $y = 0$ of the foliation of $\mathbb{R}^3$ by integral submanifolds $f(z, y, z) = c, c > 2$. The plotted sections correspond to the tori $c_1 = 2.08$ and $c_2 = 3$

Example [3]. Consider the vector fields

$$\xi = -y\frac{\partial}{\partial x} + x\frac{\partial}{\partial y}, \quad \eta = 2xz\frac{\partial}{\partial x} + 2yz\frac{\partial}{\partial y} + (z^2 + 1 - x^2 - y^2)\frac{\partial}{\partial z}$$

on $\mathbb{R}^3$. One finds that $[\xi, \eta] = 0$, i.e. the fields commute and the Frobenius theorem applies. The subspace of $T\mathbb{R}^3$ spanned by ξ and η at a point (x, y, z) is two-dimensional, except on the z-axis $\{x = y = 0\}$ and the circle $\{x^2 + y^2 = 1, z = 0\}$, where it is one -dimensional. Indeed, on the z-axis the field ξ vanishes, while η vanishes on the circle. Both the z-axis and the circle are one-dimensional integral submanifolds. All the other integral submanifolds are two-dimensional tori described by an equation

$$f(x, y, z) = (x^2 + y^2)^{-1/2}(x^2 + y^2 + z^2 + 1) = c, \tag{A.6}$$

where $c > 2$, see Fig. A.1. Clearly, $f(x, y, x)$ is annihilated by both ξ and η, i.e. ξ and η are tangent to the manifolds (A.6).

References

1. Treves, F.: *Introduction to Pseudodifferential and Fourier Integral Operators*, vol. 2. Springer Science+Business Media, New York (1981)
2. Morita, S.: *Geometry of Differential Forms*. Translations of Mathematical Monographs, vol. 201. American Mathematical Society (2001)
3. Olver, P.J.: *Applications of Lie Groups to Differential Equations*. Graduate Texts in Mathematics, vol. 107. Springer Verlag, New York (2000)

Appendix B
Jacobi Identity for the Dirac Bracket

In Chap. 2 we introduced the Dirac bracket

$$\{f, h\}_D = \{f, h\} - \{f, \mu_{\bar{\alpha}}\} \Psi^{-1}_{\bar{\alpha}\bar{\beta}} \{\mu_{\bar{\beta}}, h\}, \quad \Psi_{\bar{\alpha}\bar{\beta}} = \{\mu_{\bar{\alpha}}, \mu_{\bar{\beta}}\}.$$

Below we present a proof that this bracket satisfies the Jacobi identity.
Consider

$$(fgh) \equiv \{\{f, g\}_D, h\}_D = \{\{f, g\}_D, h\} - \{\{f, g\}_D, \mu_{\bar{\alpha}}\}\{\mu_{\bar{\alpha}}, \mu_{\bar{\beta}}\}^{-1}\{\mu_{\bar{\beta}}, h\} =$$

$$= \left\{\{f, g\} - \{f, \mu_{\bar{\alpha}}\}\{\mu_{\bar{\alpha}}, \mu_{\bar{\beta}}\}^{-1}\{\mu_{\bar{\beta}}, g\}, h\right\}$$

$$- \left\{\{f, g\} - \{f, \mu_{\bar{\gamma}}\}\{\mu_{\bar{\gamma}}, \mu_{\bar{\mu}}\}^{-1}\{\mu_{\bar{\mu}}, g\}, \mu_{\bar{\alpha}}\right\}\{\mu_{\bar{\alpha}}, \mu_{\bar{\beta}}\}^{-1}\{\mu_{\bar{\beta}}, h\}.$$

Next we have

$$(fgh) = \{\{f, g\}, h\} - \{\{f, \mu_{\bar{\alpha}}\}, h\}\Psi^{-1}_{\bar{\alpha}\bar{\beta}}\{\mu_{\bar{\beta}}, g\} - \{f, \mu_{\bar{\alpha}}\}\Psi^{-1}_{\bar{\alpha}\bar{\beta}}\{\{\mu_{\bar{\beta}}, g\}, h\} +$$

$$+ \{f, \mu_{\bar{\alpha}}\}\Psi^{-1}_{\bar{\alpha}\bar{\gamma}}\{\{\mu_{\bar{\gamma}}, \mu_{\bar{\mu}}\}, h\}\Psi^{-1}_{\bar{\mu}\bar{\beta}}\{\mu_{\bar{\beta}}, g\}$$

$$- \{\{f, g\}, \mu_{\bar{\alpha}}\}\Psi^{-1}_{\bar{\alpha}\bar{\beta}}\{\mu_{\bar{\beta}}, h\} + \{\{f, \mu_{\bar{\gamma}}\}, \mu_{\bar{\alpha}}\}\Psi^{-1}_{\bar{\alpha}\bar{\beta}}\{\mu_{\bar{\beta}}, h\}\Psi^{-1}_{\bar{\gamma}\bar{\mu}}\{\mu_{\bar{\mu}}, g\}$$

$$+ \{f, \mu_{\bar{\gamma}}\}\Psi^{-1}_{\bar{\gamma}\bar{\mu}}\{\{\mu_{\bar{\mu}}, g\}, \mu_{\bar{\alpha}}\}\Psi^{-1}_{\bar{\alpha}\bar{\beta}}\{\mu_{\bar{\beta}}, h\}$$

$$- \{f, \mu_{\bar{\gamma}}\}\Psi^{-1}_{\bar{\gamma}\bar{\rho}}\{\{\mu_{\bar{\rho}}, \mu_{\bar{\nu}}\}, \mu_{\bar{\alpha}}\}\Psi^{-1}_{\bar{\nu}\bar{\mu}}\{\mu_{\bar{\mu}}, g\}\Psi^{-1}_{\bar{\alpha}\bar{\beta}}\{\mu_{\bar{\beta}}, h\}. \tag{B.1}$$

To proceed, we introduce a concise notation

$$f_{\bar{\alpha}} \equiv \{f, \mu_{\bar{\gamma}}\}\Psi^{-1}_{\bar{\gamma}\bar{\alpha}}, \qquad g_{\bar{\alpha}} \equiv \{g, \mu_{\bar{\gamma}}, \}\Psi^{-1}_{\bar{\gamma}\bar{\alpha}}, \qquad h_{\bar{\alpha}} \equiv \{h, \mu_{\bar{\gamma}}\}\Psi^{-1}_{\bar{\gamma}\bar{\alpha}}.$$

Then, up to the minus sign, the last line of Eq. (B.1) is $\{\{\mu_{\bar{\rho}}, \mu_{\bar{\nu}}\}, \mu_{\bar{\alpha}}\}f_{\bar{\rho}}g_{\bar{\nu}}h_{\bar{\alpha}}$ and upon taking sum of cycling permutations of (fgh), we get

© Springer Nature Switzerland AG 2019
G. Arutyunov, *Elements of Classical and Quantum Integrable Systems*,
UNITEXT for Physics, https://doi.org/10.1007/978-3-030-24198-8

$$\{\{\mu_{\bar\rho}, \mu_{\bar\nu}\}, \mu_{\bar\alpha}\} f_{\bar\rho} g_{\bar\nu} h_{\bar\alpha} + \{\{\mu_{\bar\rho}, \mu_{\bar\nu}\}, \mu_{\bar\alpha}\} h_{\bar\rho} f_{\bar\nu} g_{\bar\alpha} + \{\{\mu_{\bar\rho}, \mu_{\bar\nu}\}, \mu_{\bar\alpha}\} g_{\bar\rho} h_{\bar\nu} f_{\bar\alpha}$$

$$= \Big(\{\{\mu_{\bar\rho}, \mu_{\bar\nu}\}, \mu_{\bar\alpha}\} + \{\{\mu_{\bar\alpha}, \mu_{\bar\rho}\}, \mu_{\bar\nu}\} + \{\{\mu_{\bar\nu}, \mu_{\bar\alpha}\}, \mu_{\bar\rho}\}\Big) f_{\bar\rho} g_{\bar\nu} h_{\bar\alpha} = 0$$

due to the Jacobi identity. The first term in the first line of (B.1) also vanishes upon adding cycling permutations. Thus, the only non-trivial terms left are

$$(fgh) = -\{\{f, \mu_{\bar\alpha}\}, h\} \Psi^{-1}_{\bar\alpha\bar\beta} \{\mu_{\bar\beta}, g\} - \{f, \mu_{\bar\alpha}\} \Psi^{-1}_{\bar\alpha\bar\beta} \{\{\mu_{\bar\beta}, g\}, h\} +$$

$$+ \{f, \mu_{\bar\alpha}\} \Psi^{-1}_{\bar\alpha\bar\gamma} \{\{\mu_{\bar\gamma}, \mu_{\bar\mu}\}, h\} \Psi^{-1}_{\bar\mu\bar\beta} \{\mu_{\bar\beta}, g\}$$

$$- \{\{f, g\}, \mu_{\bar\alpha}\} \Psi^{-1}_{\bar\alpha\bar\beta} \{\mu_{\bar\beta}, h\} + \{\{f, \mu_{\bar\gamma}\}, \mu_{\bar\alpha}\} \Psi^{-1}_{\bar\alpha\bar\beta} \{\mu_{\bar\beta}, h\} \Psi^{-1}_{\bar\gamma\bar\mu} \{\mu_{\bar\mu}, g\}$$

$$+ \{f, \mu_{\bar\gamma}\} \Psi^{-1}_{\bar\gamma\bar\mu} \{\{\mu_{\bar\mu}, g\}, \mu_{\bar\alpha}\} \Psi^{-1}_{\bar\alpha\bar\beta} \{\mu_{\bar\beta}, h\} \,.$$

Now, using our previously introduced notation, we group the three underlined terms together and add to them their cyclic permutations

$$g_{\bar\alpha}\{\{\mu_{\bar\alpha}, f\}, h\} - f_{\bar\alpha}\{\{\mu_{\bar\alpha}, g\}, h\} - h_{\bar\alpha}\{\{f, g\}, \mu_{\bar\alpha}\} +$$
$$h_{\bar\alpha}\{\{\mu_{\bar\alpha}, g\}, f\} - h_{\bar\alpha}\{\{\mu_{\bar\alpha}, f\}, g\} - g_{\bar\alpha}\{\{h, f\}, \mu_{\bar\alpha}\} +$$
$$f_{\bar\alpha}\{\{\mu_{\bar\alpha}, h\}, g\} - g_{\bar\alpha}\{\{\mu_{\bar\alpha}, h\}, f\} - f_{\bar\alpha}\{\{g, h\}, \mu_{\bar\alpha}\} =$$

$$= f_{\bar\alpha}\Big(- \{\{\mu_{\bar\alpha}, g\}, h\} + \{\{\mu_{\bar\alpha}, h\}, g\} - \{\{g, h\}, \mu_{\bar\alpha}\}\Big) +$$

$$+ g_{\bar\alpha}\Big(\{\{\mu_{\bar\alpha}, f\}, h\} - \{\{\mu_{\bar\alpha}, h\}, f\} - \{\{h, f\}, \mu_{\bar\alpha}\}\Big) +$$

$$+ h_{\bar\alpha}\Big(- \{\{f, g\}, \mu_{\bar\alpha}\} + \{\{\mu_{\bar\alpha}, g\}, f\} - \{\{\mu_{\bar\alpha}, f\}, g\}\Big) \,.$$

Each of the three lines in the right hand of the last formula vanishes due to the Jacobi identity for the original Poisson bracket. The remaining terms to analyse are therefore

$$(fgh) + \text{cycl. perm.} = f_{\bar\gamma} g_{\bar\mu} \{\{\mu_{\bar\gamma}, \mu_{\bar\mu}\}, h\} + g_{\bar\gamma} h_{\bar\mu} \{\{f, \mu_{\bar\gamma}\}, \mu_{\bar\mu}\} + f_{\bar\mu} h_{\bar\gamma} \{\{\mu_{\bar\mu}, g\}, \mu_{\bar\gamma}\} +$$

$$+ h_{\bar\gamma} f_{\bar\mu} \{\{\mu_{\bar\gamma}, \mu_{\bar\mu}\}, g\} + f_{\bar\gamma} g_{\bar\mu} \{\{h, \mu_{\bar\gamma}\}, \mu_{\bar\mu}\} + h_{\bar\mu} g_{\bar\gamma} \{\{\mu_{\bar\mu}, f\}, \mu_{\bar\gamma}\} +$$

$$+ g_{\bar\gamma} h_{\bar\mu} \{\{\mu_{\bar\gamma}, \mu_{\bar\mu}\}, f\} + h_{\bar\gamma} f_{\bar\mu} \{\{g, \mu_{\bar\gamma}\}, \mu_{\bar\mu}\} + g_{\bar\mu} f_{\bar\gamma} \{\{\mu_{\bar\mu}, h\}, \mu_{\bar\gamma}\} =$$

$$= f_{\bar\gamma} g_{\bar\mu} \Big(\{\{\mu_{\bar\gamma}, \mu_{\bar\mu}\}, h\} + \{\{h, \mu_{\bar\gamma}\}, \mu_{\bar\mu}\} + \{\{\mu_{\bar\mu}, h\}, \mu_{\bar\gamma}\}\Big) +$$

$$+ g_{\bar\gamma} h_{\bar\mu} \Big(\{\{f, \mu_{\bar\gamma}\}, \mu_{\bar\mu}\} + \{\{\mu_{\bar\mu}, f\}, \mu_{\bar\gamma}\} + \{\{\mu_{\bar\gamma}, \mu_{\bar\mu}\}, f\}\Big) +$$

$$+ f_{\bar\mu} h_{\bar\gamma} \Big(\{\{\mu_{\bar\mu}, g\}, \mu_{\bar\gamma}\} + \{\{\mu_{\bar\gamma}, \mu_{\bar\mu}\}, g\} + \{\{g, \mu_{\bar\gamma}\}, \mu_{\bar\mu}\}\Big) = 0 \,,$$

which again sum up to zero due to the Jacobi identity.

Appendix C
Details on the Double

In this appendix we collect a number of technical details concerning the double of a Lie algebra and related Poisson structures introduced in Chap. 2.

C.1 Lie Bracket of $\mathscr{D}$

First, we verify the Jacobi identity for the Lie bracket that endows $\mathscr{D}$ with the Lie algebra structure. To demonstrate the rigidity of this bracket, we put in the definition of the latter two constants α and β

$$[(X_1, \ell_1), (X_2, \ell_2)] = \left([X_1, X_2] + \alpha\big(\mathrm{ad}^*_{\ell_2} X_1 - \mathrm{ad}^*_{\ell_1} X_2\big), [\ell_1, \ell_2]_* + \beta\big(\mathrm{ad}^*_{X_1} \ell_2 - \mathrm{ad}^*_{X_2} \ell_1\big)\right),$$

and show that the Jacobi identity fixes these constants uniquely. To this end, we compute

$$[[(X_1, \ell_1), (X_2, \ell_2)], (X_3, \ell_3)] + \text{c.p.} \equiv (X, \ell), \tag{C.1}$$

where

$$X = [[X_1, X_2], X_3] + \alpha[\mathrm{ad}^*_{\ell_2} X_1 - \mathrm{ad}^*_{\ell_1} X_2, X_3] + \alpha\,\mathrm{ad}^*_{\ell_3}\left([X_1, X_2] + \alpha\big(\mathrm{ad}^*_{\ell_2} X_1 - \mathrm{ad}^*_{\ell_1} X_2\big)\right)$$
$$- \alpha\,\mathrm{ad}^*_{[\ell_1, \ell_2]_* + \beta(\mathrm{ad}^*_{X_1}\ell_2 - \mathrm{ad}^*_{X_2}\ell_1)} X_3 + \text{c.p.}$$

and

$$\ell = [[\ell_1, \ell_2]_*, \ell_3]_* + \beta[\mathrm{ad}^*_{X_1}\ell_2 - \mathrm{ad}^*_{X_2}\ell_1, \ell_3]_* + \beta\,\mathrm{ad}^*_{[X_1, X_2]+\alpha(\mathrm{ad}^*_{\ell_2}X_1 - \mathrm{ad}^*_{\ell_1}X_2)}\ell_3$$
$$- \beta\mathrm{ad}^*_{X_3}\left([\ell_1, \ell_2]_* + \beta(\mathrm{ad}^*_{X_1}\ell_2 - \mathrm{ad}^*_{X_2}\ell_1)\right) + \text{c.p.},$$

© Springer Nature Switzerland AG 2019
G. Arutyunov, *Elements of Classical and Quantum Integrable Systems*,
UNITEXT for Physics, https://doi.org/10.1007/978-3-030-24198-8

and c.p. stands for the cyclic permutations. We start with analysis of the expression for X. First, the contribution of $[[X_1, X_2], X_3]$ taken together with its cyclic permutations vanishes because $\mathfrak{g}$ is a Lie algebra. To prove the vanishing of the remainder of X, we pair X with an arbitrary element $n \in \mathfrak{g}^*$. The following term is quadratic in X and its cyclic permutations enter the expression for $\langle n, X \rangle$

$$
\begin{aligned}
\langle n, [\mathrm{ad}^*_{\ell_2} X_1 - \mathrm{ad}^*_{\ell_1} X_2, X_3] \rangle + \text{c.p.} &= \langle \mathrm{ad}^*_{X_3} n, \mathrm{ad}^*_{\ell_2} X_1 - \mathrm{ad}^*_{\ell_1} X_2 \rangle + \text{c.p.} = \\
&- \langle [\ell_2, \mathrm{ad}^*_{X_3} n]_*, X_1 \rangle + \langle [\ell_1, \mathrm{ad}^*_{X_3} n]_*, X_2 \rangle + \text{c.p.} = \\
- \langle [\ell_2, \mathrm{ad}^*_{X_3} n]_*, X_1 \rangle + \langle [\ell_1, \mathrm{ad}^*_{X_3} n]_*, X_2 \rangle &- \langle [\ell_1, \mathrm{ad}^*_{X_2} n]_*, X_3 \rangle + \langle [\ell_3, \mathrm{ad}^*_{X_2} n]_*, X_1 \rangle \\
- \langle [\ell_3, \mathrm{ad}^*_{X_1} n]_*, X_2 \rangle &+ \langle [\ell_2, \mathrm{ad}^*_{X_1} n]_*, X_3 \rangle .
\end{aligned}
\tag{C.2}
$$

In a similar manner, we proceed with another quadratic in X term and its cyclic permutations

$$
\begin{aligned}
\mathrm{ad}^*_{\mathrm{ad}^*_{X_1} \ell_2 - \mathrm{ad}^*_{X_2} \ell_1} X_3(n) + \text{c.p.} &= - \langle [\mathrm{ad}^*_{X_1} \ell_2 - \mathrm{ad}^*_{X_2} \ell_1, n]_*, X_3 \rangle \\
&- \langle [\mathrm{ad}^*_{X_3} \ell_1 - \mathrm{ad}^*_{X_1} \ell_3, n]_*, X_2 \rangle - \langle [\mathrm{ad}^*_{X_2} \ell_3 - \mathrm{ad}^*_{X_3} \ell_2, n]_*, X_1 \rangle .
\end{aligned}
\tag{C.3}
$$

From here we see that if $\beta = 1$, the relevant combination of (C.2) and (C.3) yields

$$
\begin{aligned}
\langle n, [\mathrm{ad}^*_{\ell_2} X_1 - \mathrm{ad}^*_{\ell_1} X_2, X_3] \rangle &- \beta \langle n, [\mathrm{ad}^*_{\ell_2} X_1 - \mathrm{ad}^*_{\ell_1} X_2, X_3] \rangle + \text{c.p.} = \\
&= \langle [\mathrm{ad}^*_{X_1} \ell_2, n]_* + [\ell_2, \mathrm{ad}^*_{X_1} n]_*, X_3 \rangle - \langle [\mathrm{ad}^*_{X_3} \ell_2, n]_* + [\ell_2, \mathrm{ad}^*_{X_3} n]_*, X_1 \rangle \\
&+ \langle [\mathrm{ad}^*_{X_2} \ell_3, n]_* + [\ell_3, \mathrm{ad}^*_{X_2} n]_*, X_1 \rangle - \langle [\mathrm{ad}^*_{X_1} \ell_3, n]_* + [\ell_3, \mathrm{ad}^*_{X_1} n]_*, X_2 \rangle \\
&+ \langle [\mathrm{ad}^*_{X_3} \ell_1, n]_* + [\ell_1, \mathrm{ad}^*_{X_3} n]_*, X_2 \rangle - \langle [\mathrm{ad}^*_{X_2} \ell_1, n]_* + [\ell_1, \mathrm{ad}^*_{X_2} n]_*, X_3 \rangle \\
&= \langle \mathrm{ad}^*_{X_1} [\ell_2, n]_*, X_3 \rangle + \langle \mathrm{ad}^*_{X_2} [\ell_3, n]_*, X_1 \rangle + \langle \mathrm{ad}^*_{X_3} [\ell_1, n]_*, X_2 \rangle \\
&= - \langle [\ell_2, n]_*, [X_1, X_3] \rangle - \langle [\ell_3, n]_*, [X_2, X_1] \rangle - \langle [\ell_1, n]_*, [X_3, X_2] \rangle \\
&= \langle n, \mathrm{ad}^*_{\ell_2} [X_1, X_3] + \mathrm{ad}^*_{\ell_3} [X_2, X_1] + \mathrm{ad}^*_{\ell_1} [X_3, X_2] \rangle ,
\end{aligned}
$$

where Eq. (2.96) was used. Upon substituting in $\langle n, X \rangle$ this result cancels exactly against $\alpha \langle n, \mathrm{ad}^*_{\ell_3} [X_1, X_2] \rangle$ and cyclic permutations thereof leaving no restrictions for α. Finally, the only remaining terms are linear in X and they cancel out due to the Jacobi identity in $\mathfrak{g}^*$ provided we choose $\alpha = -1$.

Now we turn our attention to the quantity ℓ. The contribution of $[[\ell_1, \ell_2]_*, \ell_3]_*$ taken together with its cyclic permutations disappears because $\mathfrak{g}^*$ is a Lie algebra. To prove the vanishing of ℓ, we pair it with an arbitrary $Y \in \mathfrak{g}$. Consider then a term

$$
\begin{aligned}
\mathrm{ad}^*_{\mathrm{ad}^*_{\ell_2} X_1 - \mathrm{ad}^*_{\ell_1} X_2} \ell_3(Y) &= - \ell_3([\mathrm{ad}^*_{\ell_2} X_1 - \mathrm{ad}^*_{\ell_1} X_2, Y]) \\
&= \ell_3([Y, \mathrm{ad}^*_{\ell_2} X_1 - \mathrm{ad}^*_{\ell_1} X_2]) = - \langle \mathrm{ad}^*_Y \ell_3, \mathrm{ad}^*_{\ell_2} X_1 - \mathrm{ad}^*_{\ell_1} X_2 \rangle \\
&= \langle [\ell_2, \mathrm{ad}^*_Y \ell_3]_*, X_1 \rangle - \langle [\ell_1, \mathrm{ad}^*_Y \ell_3]_*, X_2 \rangle .
\end{aligned}
$$

All terms in $\ell(Y)$ which are linear in X must cancel by themselves. Combining these terms we get

$$\beta \langle [\mathrm{ad}^*_{X_1}\ell_2 - \mathrm{ad}^*_{X_2}\ell_1, \ell_3]_*, Y\rangle + \alpha\beta\Big(\langle [\ell_2, \mathrm{ad}^*_Y\ell_3]_*, X_1\rangle - \langle [\ell_1, \mathrm{ad}^*_Y\ell_3]_*, X_2\rangle\Big)$$
$$-\beta \langle \mathrm{ad}^*_{X_3}[\ell_1, \ell_2]_*, Y\rangle + \text{c.p.}$$
$$= + \beta\Big(\langle [\mathrm{ad}^*_{X_1}\ell_2, \ell_3]_* + [\ell_2, \mathrm{ad}^*_{X_1}\ell_3]_* + [\mathrm{ad}^*_{X_2}\ell_3, \ell_1]_* + [\ell_3, \mathrm{ad}^*_{X_2}\ell_1]_*$$
$$+[\mathrm{ad}^*_{X_3}\ell_1, \ell_2]_* + [\ell_1, \mathrm{ad}^*_{X_3}\ell_2]_*, Y\rangle\Big)$$
$$+ \alpha\beta\Big(\langle [\mathrm{ad}^*_Y\ell_2, \ell_3]_* + [\ell_2, \mathrm{ad}^*_Y\ell_3]_*, X_1\rangle + \langle [\mathrm{ad}^*_Y\ell_3, \ell_1]_* + [\ell_3, \mathrm{ad}^*_Y\ell_1]_*, X_2\rangle$$
$$+\langle [\mathrm{ad}^*_Y\ell_1, \ell_2]_* + [\ell_2, \mathrm{ad}^*_Y\ell_1]_*, X_3\rangle\Big)$$
$$- \beta \langle \mathrm{ad}^*_{X_1}[\ell_2, \ell_3]_* + \mathrm{ad}^*_{X_2}[\ell_3, \ell_1]_* + \mathrm{ad}^*_{X_3}[\ell_1, \ell_2]_*, Y\rangle.$$

Now we replace terms in the middle line of the last expression by using (2.96), for instance,

$$\langle [\mathrm{ad}^*_Y\ell_2, \ell_3]_* + [\ell_2, \mathrm{ad}^*_Y\ell_3]_*, X_1\rangle = \langle [\mathrm{ad}^*_{X_1}\ell_2, \ell_3]_* + [\ell_2, \mathrm{ad}^*_{X_1}\ell_3]_*, Y\rangle - \langle \mathrm{ad}^*_{X_1}[\ell_2, \ell_3]_*, Y\rangle.$$

All the terms then cancel provided $\alpha = -1$. The remaining terms are quadratic in X

$$\beta\mathrm{ad}^*_{[X_1, X_2]}\ell_3 - \beta^2 (\mathrm{ad}^*_{X_1}\ell_2 - \mathrm{ad}^*_{X_2}\ell_1) + \text{c.p.} \qquad (C.4)$$

and they all cancel due to the Jacobi identity for the commutator in $\mathfrak{g}$ provided $\beta = 1$. Thus, from the condition of vanishing of X and ℓ in (C.1) we fix the values of α and β.

Concerning the ad-invariance of the form (2.121) on $\mathscr{D}$, we compute

$$\langle [(X_1, \ell_1), (X_2, \ell_2)], (X_3, \ell_3)\rangle_{\mathscr{D}} =$$
$$= \langle [\ell_1, \ell_2]_* + \mathrm{ad}^*_{X_1}\ell_2 - \mathrm{ad}^*_{X_2}\ell_1, X_3\rangle + \langle \ell_3, [X_1, X_2] + \mathrm{ad}^*_{\ell_1}X_2 - \mathrm{ad}^*_{\ell_2}X_1\rangle =$$
$$= \langle [\ell_2, \ell_3]_*, X_1\rangle + \langle [\ell_3, \ell_1]_*, X_2\rangle + \langle [\ell_1, \ell_2]_*, X_3\rangle$$
$$+ \langle \ell_1, [X_2, X_3]\rangle + \langle \ell_2, [X_3, X_1]\rangle + \langle \ell_3, [X_1, X_2]\rangle.$$

The right hand side here is invariant under the cyclic permutations of the arguments. Thus, taking into account that the bilinear form is symmetric, we get

$$\langle [(X_1, \ell_1), (X_2, \ell_2)], (X_3, \ell_3)\rangle_{\mathscr{D}} = \langle [(X_3, \ell_3), (X_1, \ell_1)], (X_2, \ell_2)\rangle_{\mathscr{D}} =$$
$$= -\langle (X_2, \ell_2), [(X_1, \ell_1), (X_3, \ell_3)]\rangle_{\mathscr{D}},$$

which means the ad-invariance of this form.

C.2   Yang-Baxter Equation for $\mathscr{R}_{\pm}$

Let us show that the matrices $\mathscr{R}_{\pm}$ defined by (2.124) are solutions of the CYBE. For $\mathscr{R}_{+}$ we have

$$
\begin{aligned}
[[\mathscr{R}_{+}, \mathscr{R}_{+}]] &= [\mathscr{R}_{+12}, \mathscr{R}_{+13}] + [\mathscr{R}_{+12}, \mathscr{R}_{+23}] + [\mathscr{R}_{+13}, \mathscr{R}_{+23}] \\
&= [(e_i, 0), (e_j, 0)] \otimes (0, e^i) \otimes (0, e^j) + (e_i, 0) \otimes [(0, e^i), (e_j, 0)] \otimes (0, e^j) \\
&\quad + (e_i, 0) \otimes (e_j, 0) \otimes [(0, e^i), (0, e^j)] \,.
\end{aligned}
$$

Using the commutation relations for $\mathscr{D}$, we get

$$
\begin{aligned}
[[\mathscr{R}_{+}, \mathscr{R}_{+}]] &= ([e_i, e_j], 0) \otimes (0, e^i) \otimes (0, e^j) + (e_i, 0)_1 \otimes (\mathrm{ad}^*_{e^i} e_j, -\mathrm{ad}^*_{e_j} e^i) \otimes (0, e^j) \\
&\quad + (e_i, 0) \otimes (e_j, 0) \otimes (0, [e^i, e^j]_*) \,.
\end{aligned}
$$

Since

$$
\mathrm{ad}^*_{e_j} e^i = -f^i_{jk} e^k \,, \qquad \mathrm{ad}^*_{e^i} e_j = -\lambda^{ik}_j e_k \,,
$$

we have

$$
(\mathrm{ad}^*_{e^i} e_j, -\mathrm{ad}^*_{e_j} e^i) = -\lambda^{ik}_j (e_k, 0) + f^i_{jk} (0, e^k) \,.
$$

Hence,

$$
\begin{aligned}
[[\mathscr{R}_{+}, \mathscr{R}_{+}]] &= ([e_i, e_j], 0) \otimes (0, e^i) \otimes (0, e^j) \\
&\quad - (e_i, 0) \otimes (e_k, 0) \otimes (0, \lambda^{ik}_j e^j) - (f^i_{kj} e_i, 0) \otimes (0, e^k) \otimes (0, e^j) \\
&\quad + (e_i, 0) \otimes (e_j, 0) \otimes (0, [e^i, e^j]_*) = 0 \,.
\end{aligned}
$$

The demonstration for $\mathscr{R}_{-}$ is analogous.

C.3   Lie Algebra Homomorphism $\rho : \mathscr{D} \to \mathscr{D}^{\mathbb{R}}$

Let us show that the map ρ given by (2.143) is the Lie algebra homomorphism of the double $\mathscr{D}$ into $\mathscr{D}^{\mathbb{R}} = \mathfrak{g} \oplus \mathfrak{g}$ equipped with the structure of the direct sum of two copies of the Lie algebra $\mathfrak{g}$.

To achieve to more clarify, we adopt the notation ℓ for an element of the dual space $\mathfrak{g}^*$, which in (2.143) is the same as $Y \in \mathfrak{g}$ under the identification $\mathfrak{g} \simeq \mathfrak{g}^*$. Consider

$$
\begin{aligned}
\rho\Big([(X_1, \ell_1), (X_2, \ell_2)]_{\mathscr{D}}\Big) &= \\
&= \Big([X_1, X_2] + \mathrm{ad}^*_{\ell_1} X_2 - \mathrm{ad}^*_{\ell_2} X_1 + \boldsymbol{r}_{+}\big([\ell_1, \ell_2]_* + \mathrm{ad}^*_{X_1} \ell_2 - \mathrm{ad}^*_{X_2} \ell_1\big), \\
&\qquad [X_1, X_2] + \mathrm{ad}^*_{\ell_1} X_2 - \mathrm{ad}^*_{\ell_2} X_1 + \boldsymbol{r}_{-}\big([\ell_1, \ell_2]_* + \mathrm{ad}^*_{X_1} \ell_2 - \mathrm{ad}^*_{X_2} \ell_1\big)\Big) \,.
\end{aligned}
$$

On the other hand, we have

$$[\rho(X_1, \ell_1), \rho(X_2, \ell_2)] = [(X_1 + r_+\ell_1, X_1 + r_-\ell_1), (X_2 + r_+\ell_2, X_2 + r_-\ell_2)] =$$
$$= \Big([X_1 + r_+\ell_1, X_2 + r_+\ell_2], [X_1 + r_-\ell_1, X_2 + r_-\ell_2]\Big) =$$
$$= \Big([X_1, X_2] + [X_1, r_+\ell_2] + [r_+\ell_1, X_2] + r_+[\ell_1, \ell_2]_*,$$
$$[X_1, X_2] + [X_1, r_-\ell_2] + [r_-\ell_1, X_2] + r_-[\ell_1, \ell_2]_*\Big),$$

where we have taken into account that $r_\pm$ are homomorphic embeddings of $\mathfrak{g}^*$ into $\mathscr{D}^{\mathbb{R}}$. Hence showing that ρ is a homomorphism reduces to proving the following identities

$$\mathrm{ad}^*_{\ell_1} X_2 - \mathrm{ad}^*_{\ell_2} X_1 + r_\pm\big(\mathrm{ad}^*_{X_1}\ell_2 - \mathrm{ad}^*_{X_2}\ell_1\big) \overset{?}{=} [X_1, r_\pm\ell_2] + [r_\pm\ell_1, X_2]. \quad \text{(C.5)}$$

Upon identification $\mathfrak{g}^* \simeq \mathfrak{g}$, the pairing $\langle \cdot, \cdot \rangle$ in the formulae (2.93), (2.94) is by means of the bilinear symmetric ad-invariant form. In this setting, we find that

$$\mathrm{ad}^*_X \ell = [X, \ell], \quad \text{(C.6)}$$

which, together with the fact that X_1 and X_2 are independent, shows that the identities above should originate from a single relation

$$\mathrm{ad}^*_\ell X - \tfrac{1}{2} r(\mathrm{ad}^*_X \ell) \overset{?}{=} \tfrac{1}{2}[r(\ell), X]. \quad \text{(C.7)}$$

From (2.94) and (2.103) we have[1]

$$\langle n, \mathrm{ad}^*_\ell X \rangle = -\langle [\ell, n]_*, X \rangle = -\tfrac{1}{2}\langle \ell, [r(n), X] \rangle + \tfrac{1}{2}\langle n, [r(\ell), X] \rangle$$
$$= \tfrac{1}{2}\langle [\ell, X], r(n) \rangle + \tfrac{1}{2}\langle n, [r(\ell), X] \rangle = -\tfrac{1}{2}\langle n, r([\ell, X]) \rangle + \tfrac{1}{2}\langle n, [r(\ell), X] \rangle,$$

where we have used the ad-invariance of the bilinear form and the skew-symmetric property of r. Thus, we find

$$\mathrm{ad}^*_\ell X = \tfrac{1}{2}[r(\ell), X] - \tfrac{1}{2} r([\ell, X]). \quad \text{(C.8)}$$

The formulae (C.6) and (C.8) show that (C.7) holds and, therefore, the claim is proved.

---

[1] Recall the factor $1/2$ in (2.140) and (2.141).

C.4   Poisson Brackets of Coordinate Functions on $D_{\pm}$

Evaluation of the Poisson brackets (2.209) and (2.210) for various coordinate functions relies on the knowledge of the corresponding left and right differentials taking values in $\mathscr{D}^{\mathbb{R}} = \mathfrak{g} \oplus \mathfrak{g}$. For any $f \in \mathscr{F}(G \times G)$ we define these differentials in the standard fashion, namely,

$$\langle\!\langle \mathscr{D}^{\ell} f, (X, Y)\rangle\!\rangle = \frac{d}{dt} f\left(e^{t(X,Y)}(x, y)\right)\Big|_{t=0}, \quad \langle\!\langle \mathscr{D}^r f, (X, Y)\rangle\!\rangle = \frac{d}{dt} f\left((x, y)e^{t(X,Y)}\right)\Big|_{t=0}.$$

where $(X, Y) \in \mathscr{D}^{\mathbb{R}}$ and $(x, y) \in D \simeq G \times G$. For the variables x and y themselves one immediately gets

$$\mathscr{D}^{\ell} x = (e_i x)(e_i, 0), \qquad \mathscr{D}^r x = (x e_i)(e_i, 0),$$
$$\mathscr{D}^{\ell} y = (e_i y)(0, -e_i), \quad \mathscr{D}^r y = (y e_i)(0, -e_i).$$

Here the term in the first brackets in the right hand side of each expression carries the same representation label (matrix indices) as the corresponding coordinate function on the left hand side. Written explicitly, we will have, for instance, $\mathscr{D}^{\ell} x_{\alpha\beta} = (e_i x)_{\alpha\beta}(e_i, 0)$. In order not to overload our notation, we will not specify these representation labels unless it is really necessary. Recalling the action of $\widehat{\mathscr{R}}_{\mathbb{R}}$

$$\widehat{\mathscr{R}}_{\mathbb{R}}(X, Y) = \left(2r_+ Y - rX, \; rY - 2r_- X\right). \tag{C.9}$$

from (2.209) and (2.210) we obtain, for example,

$$\{x_1, y_2\}_{\pm} =$$
$$= -\tfrac{1}{2}(e_i x)_1 (e_j y)_2 \langle\!\langle (e_i, 0), (2r_+(e_j), r(e_j))\rangle\!\rangle \mp \tfrac{1}{2}(x e_i)_1 (y e_i)_2 \langle\!\langle (e_i, 0), (2r_+(e_j), r(e_j))\rangle\!\rangle$$
$$= -r_+^{ij} (e_i x)_1 (e_j y)_2 \mp (x e_i)_1 (y e_i)_2 r_+^{ij} = -(r_+ x_1 y_2 \pm x_1 y_2 r_+).$$

The remaining brackets between generators x and y are obtained in the same fashion with the result (2.211).

Consider now a system of generators $(\mathscr{L}_{\pm}, g)$ related to the factorisation problem (2.212). For left and right shifts of $(\mathscr{L}_+ g^{-1}, \mathscr{L}_- g^{-1})$ by elements of one-parametric subgroups one has

$$e^{tX} \mathscr{L}_+ g^{-1} = \mathscr{L}_+(t) g^{-1}(t), \quad \mathscr{L}_+ g^{-1} e^{tX} = \mathscr{L}_+(t) g^{-1}(t),$$
$$e^{tY} \mathscr{L}_- g^{-1} = \mathscr{L}_-(t) g^{-1}(t), \quad \mathscr{L}_- g^{-1} e^{tY} = \mathscr{L}_-(t) g^{-1}(t).$$

First, we concentrate on the equations corresponding to the left shifts. Upon differentiating over t and setting $t = 0$, these equations can be cast in the form

$$\mathscr{L}_+^{-1} X \mathscr{L}_+ = \mathscr{L}_+^{-1} \mathscr{L}_+' - g^{-1} g',$$
$$\mathscr{L}_-^{-1} Y \mathscr{L}_- = \mathscr{L}_-^{-1} \mathscr{L}_-' - g^{-1} g'. \tag{C.10}$$

Here $\mathcal{L}'_\pm$ and g' are derivatives of the corresponding functions evaluated at $t = 0$. Obviously, $\mathcal{L}_\pm^{-1}\mathcal{L}'_\pm$ are the images of a unique element $Q \in \mathfrak{g}^*$ under the maps $r_\pm$, that is,

$$r_\pm Q = \mathcal{L}_\pm^{-1}\mathcal{L}'_\pm . \tag{C.11}$$

Also $g^{-1}g' \in \mathfrak{g}$ and, therefore, Eq. (C.10) fit the Lie algebra factorisation problem (2.147). From (C.10) and (C.11) and with the help of identity $r_+ - r_- = \mathbb{1}$ we find

$$Q = \mathcal{L}_+^{-1}X\mathcal{L}_+ - \mathcal{L}_-^{-1}Y\mathcal{L}_- .$$

Thus, for $\mathcal{L}'_\pm$ and g' we find

$$\mathcal{L}'_\pm = (\mathcal{L}_\pm e_i)\left(r_\pm^{ij}\,\langle e_j, \mathcal{L}_+^{-1}X\mathcal{L}_+\rangle - r_\pm^{ij}\,\langle e_j, \mathcal{L}_-^{-1}Y\mathcal{L}_-\rangle\right),$$

$$g' = (ge_i)\left(r_-^{ij}\,\langle e_j, \mathcal{L}_+^{-1}X\mathcal{L}_+\rangle - r_+^{ij}\,\langle e_j, \mathcal{L}_-^{-1}Y\mathcal{L}_-\rangle\right).$$

These expressions can be further rewritten as follows

$$\mathcal{L}'_\pm = (\mathcal{L}_\pm e_i)\,\langle\!\langle(X, Y),\,(r_\pm^{ij}\mathcal{L}_+e_j\mathcal{L}_+^{-1}, r_\pm^{ij}\mathcal{L}_-e_j\mathcal{L}_-^{-1})\rangle\!\rangle,$$

$$g' = (ge_i)\,\langle\!\langle(X, Y),\,(r_-^{ij}\mathcal{L}_+e_j\mathcal{L}_+^{-1}, r_+^{ij}\mathcal{L}_-e_j\mathcal{L}_-^{-1})\rangle\!\rangle,$$

which allows one to read off the form of the left differentials

$$\mathscr{D}^\ell\mathcal{L}_\pm = (\mathcal{L}_\pm e_i)\left(r_\pm^{ij}\,\mathcal{L}_+e_j\mathcal{L}_+^{-1}, r_\pm^{ij}\,\mathcal{L}_-e_j\mathcal{L}_-^{-1}\right),$$

$$\mathscr{D}^\ell g = (ge_i)\left(r_-^{ij}\,\mathcal{L}_+e_j\mathcal{L}_+^{-1}, r_+^{ij}\,\mathcal{L}_-e_j\mathcal{L}_-^{-1}\right). \tag{C.12}$$

Analogously, for the right shifts the corresponding equations are

$$g^{-1}Xg = \mathcal{L}_+^{-1}\mathcal{L}'_+ - g^{-1}g',$$

$$g^{-1}Yg = \mathcal{L}_-^{-1}\mathcal{L}'_- - g^{-1}g'.$$

Therefore, there should exist Q such that

$$r_\pm Q = \mathcal{L}_\pm^{-1}\mathcal{L}'_\pm .$$

This time we find for Q an expression

$$Q = g^{-1}Xg - g^{-1}Yg .$$

As a result,

$$\mathcal{L}'_\pm = (\mathcal{L}_\pm e_i)\left(r_\pm^{ij}\langle e_j, g^{-1}Xg\rangle - r_\pm^{ij}\langle e_j, g^{-1}Yg\rangle\right),$$
$$g' = (ge_i)\left(r_-^{ij}\langle e_j, g^{-1}Xg\rangle - r_+^{ij}\langle e_j, g^{-1}Yg\rangle\right).$$

From the last formulae the we read off the right differentials

$$\mathcal{D}^r\mathcal{L}_\pm = (\mathcal{L}_\pm e_i)\left(r_\pm^{ij}\,ge_jg^{-1},\ r_\pm^{ij}\,ge_jg^{-1}\right),$$
$$\mathcal{D}^r g = (ge_i)\left(r_-^{ij}\,ge_jg^{-1},\ r_+^{ij}\,ge_jg^{-1}\right). \tag{C.13}$$

Now we point out a very simple but important identity. Namely, as a consequence of $r_\pm^t = -r_\mp$, one has

$$r_\pm^{ij}e_j = -r_\mp(e_i) = -r_\pm(e_i) \pm e_i. \tag{C.14}$$

With the help of this identity, one can see that $\mathcal{D}^l\mathcal{L}_\pm$ and $\mathcal{D}^l g$ can be represented in the form

$$\mathcal{D}^l\mathcal{L}_+ = (\mathcal{L}_+ e_i)\left[-\mathscr{X}_i + \left(\mathcal{L}_+ e_i\mathcal{L}_+^{-1}, 0\right)\right],$$
$$\mathcal{D}^l\mathcal{L}_- = (\mathcal{L}_- e_i)\left[-\mathscr{X}_i + \left(0, -\mathcal{L}_- e_i\mathcal{L}_-^{-1}\right)\right],$$
$$\mathcal{D}^l g = -(ge_i)\mathscr{X}_i.$$

Here $\mathscr{X}_i \in \mathfrak{g}_+ \oplus \mathfrak{g}_- \in \mathscr{D}^\vee$ is the following element

$$\mathscr{X}_i = \left(\mathcal{L}_+ r_+(e_i)\mathcal{L}_+^{-1},\ \mathcal{L}_- r_-(e_i)\mathcal{L}_-^{-1}\right).$$

Taking into account definition (2.148) of $\widehat{\mathscr{R}}_\mathbb{R}$ as a difference of projectors, it is now easy to find the action of $\widehat{\mathscr{R}}_\mathbb{R}$ on the left differentials

$$\widehat{\mathscr{R}}_\mathbb{R}(\mathcal{D}^l\mathcal{L}_+) = (\mathcal{L}_+ e_i)\left[\mathscr{X}_i + \widehat{\mathscr{R}}_\mathbb{R}\left(\mathcal{L}_+ e_i\mathcal{L}_+^{-1}, 0\right)\right],$$
$$\widehat{\mathscr{R}}_\mathbb{R}(\mathcal{D}^l\mathcal{L}_-) = (\mathcal{L}_- e_i)\left[\mathscr{X}_i + \widehat{\mathscr{R}}_\mathbb{R}\left(0, -\mathcal{L}_- e_i\mathcal{L}_-^{-1}\right)\right],$$
$$\widehat{\mathscr{R}}_\mathbb{R}(\mathcal{D}^l g) = (ge_i)\mathscr{X}_i.$$

Concerning the right differentials of $\mathcal{L}_\pm$, we have

$$\mathcal{D}^r\mathcal{L}_\pm = -(\mathcal{L}_\pm e_i)\mathscr{Y}_{\mp i},$$
$$\widehat{\mathscr{R}}_\mathbb{R}(\mathcal{D}^r\mathcal{L}_\pm) = -(\mathcal{L}_\pm e_i)\mathscr{Y}_{\mp i}.$$

where we have introduced the elements $\mathscr{Y}_{\pm i}$ taking values in the diagonal subalgebra

$$\mathscr{Y}_{\pm i} = \left(g\boldsymbol{r}_{\pm}(e_i)g^{-1},\ g\boldsymbol{r}_{\pm}(e_i)g^{-1}\right) \in \mathfrak{g}^{\delta} .$$

Finally, we have

$$\mathscr{D}^r g = (ge_i)\left[-\mathscr{Y}_{+i} + (0, ge_i g^{-1})\right],$$

$$\widehat{\mathscr{R}}_{\mathbb{R}}(\mathscr{D}^r g) = (ge_i)\left[-\mathscr{Y}_{+i} + \widehat{\mathscr{R}}_{\mathbb{R}}(0, ge_i g^{-1})\right].$$

Now we are ready to compute the brackets between the generators $(\mathscr{L}_{\pm}, g)$. First, we note that since $\mathfrak{g}^{\delta}$ is stable under $\widehat{\mathscr{R}}_{\mathbb{R}}$ and isotropic, the contribution to the brackets $\{\mathscr{L}_{\pm}, \mathscr{L}_{\pm}\}_{\pm}$ from the right differentials vanishes. Therefore, for any choice of the sign we will have

$$\{\mathscr{L}_{+1}, \mathscr{L}_{+2}\}_{\pm} = \tfrac{1}{2}\langle\!\langle \mathscr{D}^{\ell}\mathscr{L}_{+1}, \widehat{\mathscr{R}}_{\mathbb{R}}(\mathscr{D}^{\ell}\mathscr{L}_{+2})\rangle\!\rangle$$
$$= \tfrac{1}{2}(\mathscr{L}_{+}e_i)_1(\mathscr{L}_{+}e_j)_2\,\langle\!\langle -\mathscr{X}_i + (\mathscr{L}_{+}e_i\mathscr{L}_{+}^{-1}, 0),\ \mathscr{X}_j + \widehat{\mathscr{R}}_{\mathbb{R}}\left(\mathscr{L}_{+}e_j\mathscr{L}_{+}^{-1}, 0\right)\rangle\!\rangle ,$$

$$\{\mathscr{L}_{+1}, \mathscr{L}_{-2}\}_{\pm} = \tfrac{1}{2}\langle\!\langle \mathscr{D}^{\ell}\mathscr{L}_{+1}, \widehat{\mathscr{R}}_{\mathbb{R}}(\mathscr{D}^{\ell}\mathscr{L}_{-2})\rangle\!\rangle$$
$$= \tfrac{1}{2}(\mathscr{L}_{+}e_i)_1(\mathscr{L}_{-}e_j)_2\,\langle\!\langle -\mathscr{X}_i + (\mathscr{L}_{+}e_i\mathscr{L}_{+}^{-1}, 0),\ \mathscr{X}_j + \widehat{\mathscr{R}}_{\mathbb{R}}\left(0, -\mathscr{L}_{-}e_j\mathscr{L}_{-}^{-1}\right)\rangle\!\rangle ,$$

$$\{\mathscr{L}_{-1}, \mathscr{L}_{-2}\}_{\pm} = \tfrac{1}{2}\langle\!\langle \mathscr{D}^{\ell}\mathscr{L}_{-1}, \widehat{\mathscr{R}}_{\mathbb{R}}(\mathscr{D}^{\ell}\mathscr{L}_{-2})\rangle\!\rangle$$
$$= \tfrac{1}{2}(\mathscr{L}_{-}e_i)_1(\mathscr{L}_{-}e_j)_2\,\langle\!\langle -\mathscr{X}_i + (0, -\mathscr{L}_{-}e_i\mathscr{L}_{-}^{-1}),\ \mathscr{X}_j + \widehat{\mathscr{R}}_{\mathbb{R}}\left(0, -\mathscr{L}_{-}e_j\mathscr{L}_{-}^{-1}\right)\rangle\!\rangle .$$

Second, by using skew-symmetry of $\widehat{\mathscr{R}}_{\mathbb{R}}$ together with $\widehat{\mathscr{R}}_{\mathbb{R}}(\mathscr{X}_i) = -\mathscr{X}_i$ and the fact that $\mathfrak{g}_{+} \oplus \mathfrak{g}_{-}$ is isotropic with respect to the scalar product (2.145), we obtain

$$\{\mathscr{L}_{+1}, \mathscr{L}_{+2}\}_{\pm} = -\tfrac{1}{2}(\mathscr{L}_{+}e_i)_1(\mathscr{L}_{+}e_j)_2\left(\langle\mathscr{L}_{+}e_i\mathscr{L}_{+}^{-1}, \boldsymbol{r}(\mathscr{L}_{+}e_j\mathscr{L}_{+}^{-1})\rangle - \langle e_i, \boldsymbol{r}(e_j)\rangle\right),$$

$$\{\mathscr{L}_{+1}, \mathscr{L}_{-2}\}_{\pm} = -(\mathscr{L}_{+}e_i)_1(\mathscr{L}_{-}e_j)_2\left(\langle\mathscr{L}_{+}e_i\mathscr{L}_{+}^{-1}, \boldsymbol{r}_{+}(\mathscr{L}_{-}e_j\mathscr{L}_{-}^{-1})\rangle - \langle e_i, \boldsymbol{r}_{+}(e_j)\rangle\right),$$

$$\{\mathscr{L}_{-1}, \mathscr{L}_{-2}\}_{\pm} = -\tfrac{1}{2}(\mathscr{L}_{-}e_i)_1(\mathscr{L}_{-}e_j)_2\left(\langle\mathscr{L}_{-}e_i\mathscr{L}_{-}^{-1}, \boldsymbol{r}(\mathscr{L}_{-}e_j\mathscr{L}_{-}^{-1})\rangle - \langle e_i, \boldsymbol{r}(e_j)\rangle\right).$$

Finally, by invoking the Ad-invariance of the element $C = e_i \otimes e_i$, conform (2.67), these formulae can be simplified and written by using the r-matrices

$$\{\mathscr{L}_{+1}, \mathscr{L}_{+2}\}_{\pm} = -\tfrac{1}{2}[r, \mathscr{L}_{+1}\mathscr{L}_{+2}],$$

$$\{\mathscr{L}_{+1}, \mathscr{L}_{-2}\}_{\pm} = -[r_{+}, \mathscr{L}_{+1}\mathscr{L}_{-2}],$$

$$\{\mathscr{L}_{-1}, \mathscr{L}_{-2}\}_{\pm} = -\tfrac{1}{2}[r, \mathscr{L}_{-1}\mathscr{L}_{-2}].$$

The bracket $\{\mathscr{L}_{1-}, \mathscr{L}_{2+}\}_\pm$ can be easily obtained as follows

$$\{\mathscr{L}_{-1}, \mathscr{L}_{+2}\}_\pm = -\{\mathscr{L}_{+2}, \mathscr{L}_{-1}\}_\pm$$
$$= -C_{12}\{\mathscr{L}_{+1}, \mathscr{L}_{-2}\}_\pm C_{12} = [Cr_+C, \mathscr{L}_{-1}\mathscr{L}_{+2}] = -[r_-, \mathscr{L}_{-1}\mathscr{L}_{+2}].$$

The rest of the brackets is computed in a similar fashion. We have

$$\{\mathscr{L}_{+1}, g_2\}_\pm = \tfrac{1}{2}\langle\!\langle\mathcal{D}^\ell\mathscr{L}_{+1}, \widehat{\mathscr{R}}_\mathbb{R}(\mathcal{D}^\ell g_2)\rangle\!\rangle \pm \tfrac{1}{2}\langle\!\langle\mathcal{D}^r\mathscr{L}_{+1}, \widehat{\mathscr{R}}_\mathbb{R}(\mathcal{D}^r g_2)\rangle\!\rangle$$
$$= \tfrac{1}{2}(\mathscr{L}_{+}e_i)_1(ge_j)_2\Big[\langle\!\langle -\mathscr{X}_i + (\mathscr{L}_{+}e_i\mathscr{L}_{+}^{-1}, 0), \mathscr{X}_j\rangle\!\rangle \pm \langle\!\langle -\mathscr{Y}_{-i}, -\mathscr{Y}_{+j} + \widehat{\mathscr{R}}_\mathbb{R}(0, ge_j g^{-1})\rangle\!\rangle\Big],$$

$$\{\mathscr{L}_{-1}, g_2\}_\pm = \tfrac{1}{2}\langle\!\langle\mathcal{D}^\ell\mathscr{L}_{-1}, \widehat{\mathscr{R}}_\mathbb{R}(\mathcal{D}^\ell g_2)\rangle\!\rangle \pm \tfrac{1}{2}\langle\!\langle\mathcal{D}^r\mathscr{L}_{-1}, \widehat{\mathscr{R}}_\mathbb{R}(\mathcal{D}^r g_2)\rangle\!\rangle$$
$$= \tfrac{1}{2}(\mathscr{L}_{+}e_i)_1(ge_j)_2\Big[\langle\!\langle -\mathscr{X}_i + (0, -\mathscr{L}_{-}e_i\mathscr{L}_{-}^{-1}), \mathscr{X}_j\rangle\!\rangle \pm \langle\!\langle -\mathscr{Y}_{+i}, -\mathscr{Y}_{+j} + \widehat{\mathscr{R}}_\mathbb{R}(0, ge_j g^{-1})\rangle\!\rangle\Big],$$

$$\{g_1, g_2\}_\pm = \pm\tfrac{1}{2}(ge_i)_1(ge_j)_2\langle\!\langle -\mathscr{Y}_{+i} + (0, ge_i g^{-1}), -\mathscr{Y}_{+j} + \widehat{\mathscr{R}}_\mathbb{R}(0, ge_j g^{-1})\rangle\!\rangle.$$

Computing further, we find

$$\{\mathscr{L}_{+1}, g_2\}_+ = \mathscr{L}_{+1}g_2 r_+, \qquad \{\mathscr{L}_{+1}, g_2\}_- = 0,$$
$$\{\mathscr{L}_{-1}, g_2\}_+ = \mathscr{L}_{-1}g_2 r_-, \qquad \{\mathscr{L}_{-1}, g_2\}_- = 0$$

and

$$\{g_1, g_2\}_\pm = \mp\tfrac{1}{2}[r, g_1 g_2].$$

For reader's convenience we also present the formulae for the differentials of the coordinate functions corresponding to the opposite factorisation (2.215). For the left differentials one has

$$\mathcal{D}^\ell\mathscr{L}_{\pm}' = (\mathscr{L}_{\pm}'e_i)\Big(g'r_{\mp}(e_i)g'^{-1}, g'r_{\mp}(e_i)g'^{-1}\Big) = (\mathscr{L}_{\pm}'e_i)\mathscr{Y}_{\mp i}',$$
$$\mathcal{D}^\ell g' = (g'e_i)\Big(g'r_+(e_i)g'^{-1}, g'r_-(e_i)g'^{-1}\Big) = (g'e_i)\Big[\mathscr{Y}_{+i}' + (0, -g'e_i g'^{-1})\Big].$$

Analogously, for the right differentials one gets

$$\mathcal{D}^r\mathscr{L}_{\pm}' = (\mathscr{L}_{\pm}'e_i)\Big(\mathscr{L}_{+}'r_{\mp}(e_i)\mathscr{L}_{+}'^{-1}, \mathscr{L}_{-}'r_{\mp}(e_i)\mathscr{L}_{-}'^{-1}\Big),$$
$$\mathcal{D}^r g' = (g'e_i)\Big(\mathscr{L}_{+}'r_+(e_i)\mathscr{L}_{+}'^{-1}, \mathscr{L}_{-}'r_-(e_i)\mathscr{L}_{-}'^{-1}\Big) = (g'e_i)\mathscr{X}_i',$$

so that

$$\mathcal{D}^r\mathscr{L}_{+}' = (\mathscr{L}_{+}'e_i)\Big[\mathscr{X}_i' - (\mathscr{L}_{+}'e_i\mathscr{L}_{+}'^{-1}, 0)\Big],$$
$$\mathcal{D}^r\mathscr{L}_{-}' = (\mathscr{L}_{-}'e_i)\Big[\mathscr{X}_i' + (0, \mathscr{L}_{-}'e_i\mathscr{L}_{-}'^{-1})\Big].$$

Appendix D
Details on RS Models

In this appendix we collect some further information and technical details that should help the reader to discern the content of Sect. 2.2 of Chap. 2.

D.1 Decompositions and Projectors

Here we discuss a decomposition of $\mathfrak{g} = \mathrm{Mat}_N(\mathbb{C})$ into a direct sum of vector spaces one of them being the isotropy subalgebra of the moment map.

The isotropy subalgebra $\mathfrak{f}_e \subset \mathfrak{g}$ of the element (2.283) is spanned by the following basis elements

$$(\mathfrak{f}_e)_{ij} = E_{11} - E_{i1} - E_{1j} + E_{ij}, \quad i, j = 2, \ldots, N, \tag{D.1}$$

and has the dimension $(N-1)^2$. The Lie algebra $\mathfrak{g}$ decomposes into a direct sum of vector spaces

$$\mathfrak{g} = \mathfrak{f}_e \oplus \mathfrak{c} \oplus \mathfrak{a} \oplus \mathfrak{d}. \tag{D.2}$$

Here $\mathfrak{a}$ and $\mathfrak{c}$ are two abelian $N-1$-dimensional Lie subalgebras for which we can choose the following bases

$$\mathfrak{c}_j = \sum_{i=1}^{N}(E_{i1} - E_{ij}), \quad j = 2, \ldots, N, \tag{D.3}$$

$$\mathfrak{a}_j = \sum_{i=1}^{N}(E_{1i} - E_{ji}), \quad j = 2, \ldots, N. \tag{D.4}$$

© Springer Nature Switzerland AG 2019
G. Arutyunov, *Elements of Classical and Quantum Integrable Systems*,
UNITEXT for Physics, https://doi.org/10.1007/978-3-030-24198-8

Finally, $\mathfrak{d}$ is generated by $\mathfrak{d} = \sum_{ij} E_{ij}$. Decomposition (D.2) is *not* orthogonal with respect to the bilinear form on $\mathfrak{g}$ given by $(A, B) = \mathrm{Tr}(A, B)$, although the subspace $\mathfrak{f}_e \oplus \mathfrak{d}$ is orthogonal to the rest with respect to this form.

Denote by $\pi^{(i)}$ with $i = 0, 1, 2, 3$ projectors on $\mathfrak{f}_e$, $\mathfrak{c}$, $\mathfrak{a}$ and $\mathfrak{d}$, respectively. For any $X \in \mathfrak{g}$ introduce the following quantities

$$t^{(0)}(X)_{ij} = X_{ij} - \frac{1}{N} \sum_a X_{aj} - \frac{1}{N} \sum_a X_{ia} + \frac{1}{N^2} \sum_{ab} X_{ab}, \quad i, j = 2, \ldots, N,$$

$$t^{(1)}(X)_j = \frac{1}{N^2} \sum_{ab} X_{ab} - \frac{1}{N} \sum_a X_{aj}, \quad j = 2, \ldots, N,$$

$$t^{(2)}(X)_j = \frac{1}{N^2} \sum_{ab} X_{ab} - \frac{1}{N} \sum_a X_{ja}, \quad j = 2, \ldots, N, \tag{D.5}$$

$$t^{(3)}(X) = \frac{1}{N^2} \sum_{ab} X_{ab}.$$

Then, projectors $\pi^{(i)}$ have the following action on X

$$\pi^{(0)}(X) = \sum_{i,j=2}^N (\mathfrak{f}_e)_{ij}\, t^{(0)}(X)_{ij}, \qquad \pi^{(1)}(X) = \sum_{j=2}^N \mathfrak{c}_j\, t^{(1)}(X)_j,$$

$$\pi^{(2)}(X) = \sum_{j=2}^N \mathfrak{a}_j\, t^{(2)}(X)_j, \qquad \pi^{(3)}(X) = \sum_{i,j=1}^N E_{ij}\, t^{(3)}(X). \tag{D.6}$$

In terms of projectors decomposition (D.2) reads

$$X = \sum_{k=0}^3 \pi^{(k)}(X).$$

Computing the Poisson bracket of the moment map $\mu = \ell - g\ell g^{-1}$ with an element T from (2.265), we get

$$\{\mu_{ij}, T_{kl}\} = -\delta_{kj} T_{il} + T_{kl} T_{lj}^{-1}. \tag{D.7}$$

From here we deduce that

$$\{\mu_{ii} - \mu_{ji}, T_{kl}\} = \delta_{ki}(T_{jl} - T_{il}) = (E_{ij} - E_{ii})_{km} T_{ml} = -(f_{ij}T)_{kl},$$

where we recall that $f_{ij} = E_{ii} - E_{ij}$ is the basis in $\mathfrak{f}$. Thus, $\mu_{ii} - \mu_{ji}$ with $i \neq j$ are the Poisson generators of the Frobenius subalgebra $\mathfrak{f}$.

D.2 Poisson Bracket of the Lax Matrix on D_+

Consider the following matrix function on the Heisenberg double

$$L = T^{-1}BT , \qquad (D.8)$$

where T is the Frobenius solution of the factorisation problem (2.356). As was discussed in (2.2.2), on the reduced phase space L turns into the Lax matrix of the hyperbolic RS model. For this reason we continue to call (D.8) the Lax matrix and below we compute the Poisson brackets between the entries of L considered as functions on the Heisenberg double. This will constitute the first step towards evaluation of the corresponding Dirac bracket.

The standard manipulations give

$$\{L_1, L_2\} = \mathbb{T}_{12}L_1L_2 - L_1\mathbb{T}_{12}L_2 - L_2\mathbb{T}_{12}L_1 + L_1L_2\mathbb{T}_{12} \qquad (D.9)$$
$$+ T_1^{-1}T_2^{-1}\{B_1 B_2\}T_1 T_2 + \mathbb{B}_{21}L_2 - L_2\mathbb{B}_{21} - \mathbb{B}_{12}L_1 + L_1\mathbb{B}_{12} ,$$

where we defined the following quantities

$$\mathbb{T}_{12} = T_1^{-1}T_2^{-1}\{T_1, T_2\} ,$$
$$\mathbb{B}_{12} = T_1^{-1}T_2^{-1}\{T_1, B_2\}T_2 .$$

By using (2.260), we get

$$T_1^{-1}T_2^{-1}\{B_1, B_2\}T_1 T_2 = -\check{r}_- L_1 L_2 - L_1 L_2 \check{r}_+ + L_1 \check{r}_- L_2 + L_2 \check{r}_+ L_1 .$$

Here we introduced the *dressed r*-matrices

$$\check{r}_\pm = T_1^{-1}T_2^{-1}r_\pm T_1 T_2 , \qquad (D.10)$$

which are proved to be a useful tool for the present calculation. The dressed r-matrices have essentially the same properties as their undressed counterparts, most importantly,

$$\check{r}_+ - \check{r}_- = C_{12} , \qquad (D.11)$$

because C_{12} is an invariant element. Thus, for (D.9) we get

$$\{L_1, L_2\} = (\mathbb{T}_{12} - \check{r}_-)L_1L_2 + L_1L_2(\mathbb{T}_{12} - \check{r}_+) + L_1(\check{r}_- - \mathbb{T}_{12})L_2 + L_2(\check{r}_+ - \mathbb{T}_{12})L_1$$
$$+ \mathbb{B}_{21}L_2 - L_2\mathbb{B}_{21} - \mathbb{B}_{12}L_1 + L_1\mathbb{B}_{12} . \qquad (D.12)$$

Now we proceed with evaluation of $\mathbb{T}$. Taking onto account that T satisfies the condition (2.266), see (2.271), in components we have

$$\mathbb{T}_{ij,kl} = T_{ip}^{-1}T_{kq}^{-1}\frac{\delta T_{pj}}{\delta A_{mn}}\frac{\delta T_{ql}}{\delta A_{rs}}\{A_{mn}, A_{rs}\}$$

$$= \sum_{a\neq j}\sum_{b\neq l}\frac{1}{Q_{ja}Q_{lb}}(\delta_{ia}T_{nj}T_{am}^{-1} + \delta_{ij}T_{na}T_{jm}^{-1})(\delta_{kb}T_{sl}T_{br}^{-1} + \delta_{kl}T_{sb}T_{lr}^{-1})\{A_{mn}, A_{rs}\}$$

$$= \sum_{a\neq j}\sum_{b\neq l}\frac{1}{Q_{ja}Q_{lb}}(\delta_{ia}\delta_{kb}\,\zeta_{aj,bl} + \delta_{ij}\delta_{kb}\,\zeta_{ja,bl} + \delta_{ia}\delta_{kl}\,\zeta_{aj,lb} + \delta_{ij}\delta_{kl}\,\zeta_{ja,lb})\,. \quad\text{(D.13)}$$

Here $Q_{ij} = Q_i - Q_j$ and we introduced a concise notation

$$\zeta_{12} = T_1^{-1}T_2^{-1}\{A_1, A_2\}T_1 T_2\,.$$

Using (2.260) and the fact that $A = TQT^{-1}$, we find that

$$\zeta_{12} = -\check{r}_- \, Q_1 Q_2 - Q_1 Q_2 \check{r}_+ + Q_1 \check{r}_- Q_2 + Q_2 \check{r}_+ Q_1\,.$$

With the help of (D.11) we find in components

$$\zeta_{ij,kl} = -Q_{ij}(\check{r}_{-ij,kl}Q_{kl} + C_{ij,kl}Q_k)\,,$$

where $C_{ij,kl} = \delta_{jk}\delta_{il}$ are the entries of C_{12}. Substitution of this tensor into (D.13) yields the following expression

$$\mathbb{T}_{ij,kl} = \sum_{a\neq j}\sum_{b\neq l}\left(-\delta_{ia}\delta_{kb}\check{r}_{-aj,bl} + \delta_{ij}\delta_{kb}\check{r}_{-ja,bl} + \delta_{ia}\delta_{kl}\check{r}_{-aj,lb} - \delta_{ij}\delta_{kl}\check{r}_{-ja,lb}\right)$$

$$+ \sum_{a\neq j}\sum_{b\neq l}\frac{1}{Q_{lb}}\left(\delta_{ia}\delta_{kb}C_{aj,bl}Q_b - \delta_{ij}\delta_{kb}C_{ja,bl}Q_b + \delta_{ia}\delta_{kl}C_{aj,lb}Q_l - \delta_{ij}\delta_{kl}C_{ja,lb}Q_l\right)\,.$$

In the first line the summation can be extended to all values of a and b, because the expression which is summed vanishes for $a = j$ and independently for $b = l$. For the same reason, we can extended a summation over a in the second line, where we also substitute the explicit value for $C_{ij,kl} = \delta_{jk}\delta_{il}$. In this way we find

$$\mathbb{T}_{ij,kl} = \sum_{ab}\left(-\delta_{ia}\delta_{kb}\check{r}_{-aj,bl} + \delta_{ij}\delta_{kb}\check{r}_{-ja,bl} + \delta_{ia}\delta_{kl}\check{r}_{-aj,lb} - \delta_{ij}\delta_{kl}\check{r}_{-ja,lb}\right)$$

$$+ \sum_{a}\sum_{b\neq l}\frac{1}{Q_{lb}}\left(\delta_{ia}\delta_{kb}\delta_{al}\delta_{jb}Q_b - \delta_{ij}\delta_{jl}\delta_{kb}\delta_{ab}Q_b + \delta_{ia}\delta_{kl}\delta_{ab}\delta_{jl}Q_l - \delta_{ij}\delta_{kl}\delta_{al}\delta_{jb}Q_l\right)\,.$$

This further yields the following expression

$$\mathbb{T}_{ij,kl} = -\check{r}_{-ij,kl} + \delta_{ij}\sum_a \check{r}_{-ia,kl} + \delta_{kl}\sum_a \check{r}_{-ij,ka} - \delta_{ij}\delta_{kl}\sum_{ab}\check{r}_{-ia,kb}$$

$$+ \sum_{b\neq l}\frac{1}{Q_{lb}}\left(\delta_{kb}\delta_{il}\delta_{jb}Q_b - \delta_{ij}\delta_{jl}\delta_{kb}Q_b + \delta_{kl}\delta_{ib}\delta_{jl}Q_l - \delta_{ij}\delta_{kl}\delta_{jb}Q_l\right)\,.$$

Here the second line can be written in the concise form as the matrix element $r_{Q\,ij,kl}$ of the following matrix

$$r_Q = \sum_{a \neq b} \frac{Q_b}{Q_{ab}} (E_{aa} - E_{ab}) \otimes (E_{bb} - E_{ba}) \qquad (D.14)$$

that is nothing else but a deformed analogue of (2.273). Therefore,

$$\mathbb{T}_{ij,kl} = r_{Q\,ij,kl} - \check{r}_{-ij,kl} + \delta_{ij} \sum_a \check{r}_{-ia,kl} + \delta_{kl} \sum_a \check{r}_{-ij,ka} - \delta_{ij}\delta_{kl} \sum_{ab} \check{r}_{-ia,kb} \,.$$

Hence,

$$\mathbb{T}_{12} = r_{Q\,12} - \check{r}_{-12} + a_{12} + b_{12} - c_{12} \,. \qquad (D.15)$$

where we introduced three r-matrices, a, b and c with entries

$$a_{ij,kl} = \delta_{ij} \sum_a \check{r}_{-ia,kl} \,, \quad b_{ij,kl} = \delta_{kl} \sum_a \check{r}_{-ij,ka} \,, \quad c_{ij,kl} = \delta_{ij}\delta_{kl} \sum_{ab} \check{r}_{-ia,kb} \,. \quad (D.16)$$

Needless to say, that the bracket thus obtained is compatible with the Frobenius condition (2.266) which means that

$$\sum_a \mathbb{T}_{ia,kl} = 0 \,, \quad \sum_a \mathbb{T}_{ij,ka} = 0 \,,$$

for any values of free indices.

Now we turn out attention to $\mathbb{B}_{12}$, which in components reads as

$$\mathbb{B}_{ij,kl} = \sum_{a \neq j} \frac{1}{Q_{ja}} (\delta_{ia}\eta_{aj,kl} + \delta_{ij}\eta_{ja,kl}) \,,$$

where we introduced a notation

$$\eta_{12} = T_1^{-1} T_2^{-1} \{A_1, B_2\} T_1 T_2 \,.$$

With the help of (2.260) we get

$$\eta_{12} = -\check{r}_- Q_1 L_2 - Q_1 L_2 \check{r}_- + Q_1 \check{r}_- L_2 + L_2 \check{r}_+ Q_1 \,,$$

and by using (D.11) obtain for components the following expression

$$\eta_{aj,kl} = Q_{ja} (L_{ks}\check{r}_{-aj,sl} - \check{r}_{-aj,ks} L_{sl}) + L_{ks} C_{aj,sl} Q_j \,.$$

With this expression at hand, we get

$$\mathbb{B}_{ij,kl} = \sum_{a\neq j} \left(\delta_{ia}(L_{ks}\check{r}_{-aj,sl} - \check{r}_{-aj,ks}L_{sl}) - \delta_{ij}(L_{ks}\check{r}_{-ja,sl} - \check{r}_{-ja,ks}L_{sl}) \right)$$
$$+ L_{ks} \sum_{a\neq j} \frac{1}{Q_{ja}} \left(\delta_{ia}\delta_{al}\delta_{js}Q_j + \delta_{ij}\delta_{jl}\delta_{as}Q_a \right).$$

Here the summation in the first line can be extended to include the term with $a = j$ because the latter vanishes. The second line can be conveniently written as a matrix element of some r-matrix. Namely,

$$\mathbb{B}_{ij,kl} = L_{ks}\left(\check{r}_{-ij,sl} - \delta_{ij}\sum_a \check{r}_{-ja,sl} \right) - \left(\check{r}_{-ij,ks} - \delta_{ij}\sum_a \check{r}_{-ja,ks} \right)L_{sl}$$
$$+ L_{ks} \sum_{a\neq b} \frac{Q_b}{Q_{ab}}(E_{aa} - E_{ab})_{ij} \otimes (E_{ba})_{sl}.$$

In the matrix form

$$\mathbb{B}_{12} = L_2(\check{r}_{-12} - a_{12}) - (\check{r}_{-12} - a_{12})L_2 + L_2 d_{12}, \qquad (D.17)$$

where a_{12} is the same matrix as in (D.15) and we introduced

$$d_{12} = \sum_{a\neq b} \frac{Q_b}{Q_{ab}}(E_{aa} - E_{ab}) \otimes E_{ba}. \qquad (D.18)$$

We also need

$$\mathbb{B}_{21} = L_1(\check{r}_{-21} - a_{21}) - (\check{r}_{-21} - a_{21})L_1 + L_1 d_{21},$$

Since $\check{r}_{-21} = -\check{r}_{+12}$, we have

$$\mathbb{B}_{21} = -L_1(\check{r}_{+12} + a_{21}) + (\check{r}_{+12} + a_{21})L_1 + L_1 d_{21}. \qquad (D.19)$$

Now everything is ready to obtain the bracket (D.12). Substituting in (D.12) expressions (D.15), (D.17) and (D.19), we conclude that (D.12) has a structure

$$\{L_1, L_2\} = k_{12}^+ L_1 L_2 + L_1 L_2 k_{12}^- + L_1 s_{12}^- L_2 + L_2 s_{12}^+ L_1, \qquad (D.20)$$

where the coefficients are

$$\begin{aligned}
k_{12}^+ &= r_{Q\,12} + C_{12} + (a_{21} + b_{12} - c_{12}), \\
k_{12}^- &= r_{Q\,12} + d_{12} - d_{21} + (a_{21} + b_{12} - c_{12}), \\
s_{12}^+ &= -r_{Q\,12} - d_{12} - (a_{21} + b_{12} - c_{12}), \\
s_{12}^- &= -r_{Q\,12} - C_{12} + d_{21} - (a_{21} + b_{12} - c_{12}).
\end{aligned} \qquad (D.21)$$

First, we note that these coefficients satisfy

$$k^+ + k^- + s^+ + s^- = 0 \,, \tag{D.22}$$

which guarantees that spectral invariants of L are in involution on the Heisenberg double. Second, in (D.21) an apparent dependence on the variable T occurs in a single combination $a_{21} + b_{12} - c_{12}$. To make further progress, consider

$$a_{21} = C_{12} a_{12} C_{12} \,,$$

as C_{12} acts as the permutation. We have for components

$$(a_{21})_{ij,kl} = C_{im,kn} (a_{12})_{mr,ns} C_{rj,sl} = \delta_{mk} \delta_{in} \left(\delta_{mr} \sum_a \check{r}_{-ma,ns} \right) \delta_{js} \delta_{rl}$$

$$= \delta_{kl} \sum_a \check{r}_{-ka,ij} = -\delta_{kl} \sum_a \check{r}_{+ij,ka} \,.$$

Therefore,

$$(a_{21} + b_{12})_{ij,kl} = -\delta_{kl} \sum_a \check{r}_{+ij,ka} + \delta_{kl} \sum_a \check{r}_{-ij,ka} = -\delta_{kl} \sum_a C_{ij,ka}$$

$$= -\sum_a \delta_{kl} \delta_{jk} \delta_{ia} = -\sum_{ab} (E_{ab})_{ij} \otimes (E_{bb})_{kl} \,.$$

The dependence on T disappears and we find a simple answer

$$a_{21} + b_{12} = -\sum_{ab} E_{ab} \otimes E_{bb} \,. \tag{D.23}$$

The only T-dependence is in the coefficient c_{12}. This coefficient cannot be simplified or cancelled, so we leave it in the present form. Substituting in (D.21) the matrices (D.14), (D.18) and (D.23) and, performing necessary simplifications, we obtain our final result for the coefficients of the bracket (D.20)

$$k_{12}^+ = \sum_{a \neq b} \left(\frac{Q_b}{Q_{ab}} E_{aa} - \frac{Q_a}{Q_{ab}} E_{ab} \right) \otimes (E_{bb} - E_{ba}) - c_{12} \,,$$

$$k_{12}^- = \sum_{a \neq b} \frac{Q_a}{Q_{ab}} E_{aa} \otimes E_{bb} - \sum_{a \neq b} \frac{Q_a}{Q_{ab}} E_{ab} \otimes E_{ba} - \mathbb{1} \otimes \mathbb{1} - c_{12} \,,$$

$$s_{12}^+ = -\sum_{a \neq b} \frac{Q_a}{Q_{ab}} (E_{aa} - E_{ab}) \otimes E_{bb} + \mathbb{1} \otimes \mathbb{1} + c_{12} \,,$$

$$s_{12}^- = -\sum_{a \neq b} \frac{Q_b}{Q_{ab}} E_{aa} \otimes (E_{bb} - E_{ba}) + c_{12} \,. \tag{D.24}$$

In fact, the identity matrix $\mathbb{1} \otimes \mathbb{1}$ appearing in k^- and s^+ can be omitted as it cancels out in the expression (D.20). As was already mentioned, the only T-dependence left over is in the term c_{12}, namely,

$$(c_{12})_{ij,kl} = \delta_{ij}\delta_{kl} \sum_{ab} \check{r}_{-ia,kb} = \delta_{ij}\delta_{kl} T_{im}^{-1} T_{kn}^{-1} \sum_{ab} r_{-ma,nb} \,. \tag{D.25}$$

It is this term which violates the invariance of the bracket (D.20) under transformations from the Frobenius group.

To complete our discussion, we consider

$$(c_{21})_{ij,kl} = \delta_{ij}\delta_{kl} \sum_{ab} \check{r}_{-ka,ib} = -\delta_{ij}\delta_{kl} \sum_{ab} \check{r}_{+ia,kb} \,.$$

This gives

$$(c_{21} + c_{12})_{ij,kl} = -\delta_{ij}\delta_{kl} \sum_{ab} C_{ia,kb} = -\delta_{ij}\delta_{kl} \sum_{ab} \delta_{ib}\delta_{ka} = -\delta_{ij}\delta_{kl} = -(\mathbb{1} \otimes \mathbb{1})_{ij,kl} \,,$$

or in other words,

$$c_{21} + c_{12} = -\mathbb{1} \otimes \mathbb{1} \,. \tag{D.26}$$

Equation (D.26) leads to the following relations between the coefficients

$$k_{12}^+ + k_{21}^+ = C_{12} - 2\,(\mathbb{1} \otimes \mathbb{1})\,, \quad k_{12}^- + k_{21}^- = -C_{12}\,, \quad s_{12}^- = -s_{21}^+\,. \tag{D.27}$$

Notice that the fact that the right hand side of the first two expressions is an invariant tensor. Relations (D.27) guarantee that the bracket (D.20) is skew symmetric.

Following similar steps, we can derive the Poisson brackets involving other Frobenius invariants on the Heisenberg double, for instance, the coordinates P_i. Introducing the notations

$$r_{\pm}^{hg} = h_1^{-1} g_2^{-1} r_{\pm} h_1 g_2\,, \quad (c_{12}^{hg})_{ijkl} = \delta_{ij}\delta_{kl} \sum_{\alpha,\beta} (r_-^{hg})_{i\alpha k\beta}\,,$$

which, for Frobenius elements g, h satisfies

$$c_{21}^{hg} + c_{12}^{gh} = -\mathbb{1} \otimes \mathbb{1}\,,$$

we can write

$$\{P_1, P_2\} = P_1 P_2 \,(c_{12}^{UT} + c_{12}^{TU} - c_{12}^{TT} - c_{12}^{UU})\,, \tag{D.28}$$

where matrices r_{12} and $\bar{r}_{12}$ are defined in (2.370). The c^{hg}-like terms in the brackets (D.28) are not Frobenius invariants, despite the arguments of the brackets are so.

These terms disappear after imposing Dirac constraints in the reduced phase space, as it will explicitly shown for the $\{L_1, L_2\}$-bracket below.

D.3 Dirac Bracket of the Lax Matrix

Here we outline the construction of the Dirac bracket between the entries of the Lax matrix (D.8). We argue that the contribution to the Dirac bracket coming from the second class constraints has the same matrix structure as (D.20) and that this contribution precisely cancels all the terms c_{12} in (D.24), so that the resulting coefficients describing the Dirac bracket on the constraint surface are given by expressions (2.370) in the main text.

We start with the Poisson algebra (2.249) of the non-abelian moment map (2.352)

$$\{m_1, m_2\} = -r_+ m_1 m_2 - m_1 m_2 r_- + m_1 r_- m_2 + m_2 r_+ m_1 . \quad \text{(D.29)}$$

This is the Semenov-Tian-Shansky type bracket; it has N Casimir functions $\text{Tr}(m^k)$ with $k = 1, \ldots, N$. On the constraint surface $\mathcal{S}$ the moment map is fixed to the following value

$$m = t\mathbb{1} + \beta e \otimes e^t , \quad \text{(D.30)}$$

see (2.354) and (2.355). Substituting this expression into the right hand side of (D.29) yields the following answer

$$m_{ij,kl} \equiv \{m_{ij}, m_{kl}\}\Big|_{\mathcal{S}} = \beta\Big[(t^{1-N} - \beta(i - \tfrac{1}{2}))\delta_{il} - \tfrac{\beta}{2}\delta_{jl} + \beta\,\Theta(l - j)$$
$$- (t^{1-N} - \beta(j - \tfrac{1}{2}))\delta_{jk} + \tfrac{\beta}{2}\delta_{ik} - \beta\,\Theta(k - i)\Big], \quad \text{(D.31)}$$

where Θ is the Heaviside step function

$$\Theta(j) = \begin{cases} 1, & j \geq 0, \\ 0, & j < 0 \end{cases} . \quad \text{(D.32)}$$

From this explicit formula it is readily seen that

$$\{t^{(3)}(m), m_{kl}\} = \frac{1}{N^2} \sum_{ab} \{m_{ab}, m_{kl}\} = 0 .$$

Analogously, we find

$$\{t^{(0)}(M)_{ij}, m_{kl}\} = \{m_{ij} - \tfrac{1}{N}\sum_a m_{aj} - \tfrac{1}{N}\sum_a m_{ia}, m_{kl}\} = 0, \quad i, j = 2, \ldots, N .$$

Here $t^{(i)}(M)$ refer to the coefficients (D.5). Thus, projections $\pi^{(0)}(m)$ and $\pi^{(3)}(m)$ constitute $(N - 1)^2 + 1 = N^2 - 2N + 2$ constraints of the first class. Projections

$\pi^{(1)}$ and $\pi^{(2)}$ yield a non-degenerate matrix of Poisson brackets and, therefore, represent $2(N-1)$ constraints of the second class. This matrix should be inverted and used to define the corresponding Dirac bracket. Even simpler, the matrix (D.31) has rank $2(N-1)$ and we can use its any non-degenerate submatrix of this rank to define the corresponding Dirac bracket.

Now we investigate the Poisson relations between the moment map m and the Lax matrix given by (D.8). First, we compute

$$\{m_{ij}, T_{kl}\} = \frac{\delta T_{kl}}{\delta A_{rs}} \{m_{ij}, A_{rs}\} =$$
$$= -((r_+ m_1 - m_1 r_-)T_2)_{ij,kl} + T_{kl} \sum_a (T_2^{-1}(r_+ m_1 - m_1 r_-))_{ij,la} \quad \text{(D.33)}$$

This formula is a deformed analogue of (D.7); deriving it we have used (2.263) as well as the fact that $T \in F$. Next, we obtain

$$\{m_{ij}, L_{kl}\} = L_{kl} \sum_{sp} (T_{ls}^{-1} - T_{ks}^{-1})(r_+ m_1 - m_1 r_-)_{ij,sp}. \quad \text{(D.34)}$$

It is clear that the diagonal entries from this expression of L commute with all the constraints: $\{m_{ij}, L_{kk}\} = 0$, even without restricting to the constrained surface.

On the constrained surface where m is given by (D.30), we have

$$\{m_{ij}, L_{kl}\}\big|_\mathcal{S} = t^{1-N} L_{kl}(T_{lj}^{-1} - T_{kj}^{-1}) + \beta L_{kl} \sum_{sp}(T_{ls}^{-1} - T_{ks}^{-1})\Omega_{ijs}.$$

Here

$$\Omega_{ijs} \equiv \sum_p [r_+, (e \otimes e^t)_1]_{ij,sp} = -\tfrac{1}{2}\delta_{is} - (j - \tfrac{1}{2})\delta_{js} + \Theta(s - i). \quad \text{(D.35)}$$

From the explicit expression (D.35) and the fact that T is an element of the Frobenius group, we further deduce that

$$\{t^{(0)}(m)_{ij}, L_{kl}\}\big|_\mathcal{S} = 0, \quad \{t^{(3)}(m)_{ij}, L_{kl}\}\big|_\mathcal{S} = 0.$$

In other words, L commutes on the constraint surface with all constraints of the first class, independently on the value of T.

With the help of (D.35) we obtain

$$\{m_{ij}, L_{kl}\}\big|_\mathcal{S} = L_{kl}\Big[(t^{1-N} - \beta(j - \tfrac{1}{2}))(T_{lj}^{-1} - T_{kj}^{-1})$$
$$+ \tfrac{\beta}{2}(T_{li}^{-1} - T_{ki}^{-1}) + \beta \sum_{s>i}(T_{ls}^{-1} - T_{ks}^{-1})\Big].$$

Taking into account that

$$\sum_{s>i}^{N}(T_{ls}^{-1} - T_{ks}^{-1}) + \sum_{s<i}^{N}(T_{ls}^{-1} - T_{ks}^{-1}) + (T_{li}^{-1} - T_{ki}^{-1}) = 0\,,$$

we can write

$$\{m_{ij}, L_{kl}\}\Big|_{\mathscr{S}} = L_{kl}\Big[(t^{1-N} - \beta(j - \tfrac{1}{2}))(T_{lj}^{-1} - T_{kj}^{-1})$$

$$+ \tfrac{\beta}{2}\Big(\sum_{s>i}^{N}(T_{ls}^{-1} - T_{ks}^{-1}) - \sum_{s<i}^{N}(T_{ls}^{-1} - T_{ks}^{-1})\Big)\Big].$$

Now we come to the Dirac bracket construction. By picking a non-degenerate sub-matrix Ψ of the matrix $m_{ij,kl}$, we invert it and define the corresponding Dirac bracket

$$\{L_1, L_2\}_{\mathrm{D}} = \{L_1, L_2\} - \sum_{I,J=1}^{2N-2}\{L_1, m_I\}\Psi_{IJ}^{-1}\{m_J, L_2\}\,. \tag{D.36}$$

Here $I = (ij)$ is a generalised index which we use to label matrix elements of $m_{ij,kl}$ that comprise the non-degenerate matrix Ψ_{IJ}. To give an example, for $N = 3$ we can take as Ψ the following matrix

$$\Psi = \begin{pmatrix} m_{11,11} & m_{11,12} & m_{11,13} & m_{11,21} \\ m_{12,11} & m_{12,12} & m_{12,13} & m_{12,21} \\ m_{13,11} & m_{12,12} & m_{13,13} & m_{13,21} \\ m_{21,11} & m_{21,12} & m_{21,13} & m_{21,21} \end{pmatrix} = \beta \begin{pmatrix} 0 & -t-2\beta & -t-2\beta & -t-2\beta \\ t+2\beta & 0 & \tfrac{\beta}{2} & 0 \\ t+2\beta & -\tfrac{\beta}{2} & 0 & t+\tfrac{3}{2}\beta \\ t+2\beta & 0 & -t-\tfrac{3}{2}\beta & 0 \end{pmatrix}.$$

In particular $\det \Psi = \beta^4(t + \beta)^2(t + 2\beta)^2$. Inverting Ψ, we find that

$$\sum_{I,J=1}^{2N-2}\{L_1, m_I\}\Psi_{IJ}^{-1}\{m_J, L_2\} = k_{\mathrm{D}12}^{+}L_1L_2 + L_1L_2k_{\mathrm{D}12}^{-} + L_1s_{\mathrm{D}12}^{-}L_2 + L_2s_{\mathrm{D}12}^{+}L_1\,,$$

that is, the contribution of the second class constraints has precisely the same structure as (D.20). Moreover, the corresponding coefficients are

$$k_{\mathrm{D}12}^{+} = -c_{12}\,, \quad k_{\mathrm{D}12}^{-} = -c_{12}\,, \quad s_{\mathrm{D}12}^{+} = c_{12}\,, \quad s_{\mathrm{D}12}^{-} = c_{12}\,, \tag{D.37}$$

where c_{12} is given by (D.25). Thus, in (D.36) all the terms c_{12} cancel out. Similar computations can be explicitly done for $N = 4, 5, 6$ with the same result. We leave it to the reader to verify this result analytically for arbitrary N.

In summary, on the reduced phase space the Dirac bracket between the components of the Lax matrix has the form (D.20) with the following coefficients

$$k_{12}^+ = \sum_{a \neq b} \left(\frac{Q_b}{Q_{ab}} E_{aa} - \frac{Q_a}{Q_{ab}} E_{ab} \right) \otimes (E_{bb} - E_{ba}),$$

$$k_{12}^- = \sum_{a \neq b} \frac{Q_a}{Q_{ab}} E_{aa} \otimes E_{bb} - \sum_{a \neq b} \frac{Q_a}{Q_{ab}} E_{ab} \otimes E_{ba},$$

$$s_{12}^+ = -\sum_{a \neq b} \frac{Q_a}{Q_{ab}} (E_{aa} - E_{ab}) \otimes E_{bb},$$ (D.38)

$$s_{12}^- = -\sum_{a \neq b} \frac{Q_b}{Q_{ab}} E_{aa} \otimes (E_{bb} - E_{ba}).$$

The coefficients have the following properties

$$k_{12}^\pm + k_{21}^\pm = \pm (C_{12} - \mathbb{1} \otimes \mathbb{1}), \quad s_{12}^- = -s_{21}^+,$$ (D.39)

which guarantee, in particular, skew-symmetry of (D.20). In addition, they satisfy the relation (D.22). In the main text we present the formula (D.20) in the r-matrix form (2.369) with the following identifications

$$k^+ = r(Q), \quad s^+ = -\bar{r}(Q), \quad k^- = -\underline{r}(Q).$$

Lax matrix and Poisson structure. One way to present the Lax matrix and the Poisson structure of the hyperbolic RS model is the following. Write the Lax matrix (2.363) in the form

$$L = (1 - e^{-\gamma}) \sum_{i,j=1}^N \frac{Q_i}{Q_i - e^{-\gamma} Q_j} f_j E_{ij}, \quad f_j \equiv \prod_{a \neq j} \frac{Q_a - e^{-\gamma} Q_j}{Q_a - Q_j} e^{-p_j},$$

where we recall that $Q_i = e^{q_i}$ and γ is the coupling constant. Then the Poisson structure (2.369) results into the following expressions

$$\{f_i, f_j\} = (1 - e^{-\gamma})^2 \frac{e^{q_i} + e^{q_j}}{e^{q_i} - e^{q_j}} \frac{e^{q_i}}{e^{q_i} - e^{-\gamma + q_j}} \frac{e^{q_j}}{e^{q_j} - e^{-\gamma + q_i}} f_i f_j, \quad i \neq j,$$

$$\{q_i, f_j\} = \delta_{ij} f_j.$$

Here the first bracket can also be written as

$$\{f_i, f_j\} = \frac{e^{q_i} + e^{q_j}}{e^{q_i} - e^{q_j}} f_i f_j + \frac{1}{1 - e^{-\gamma}} (f_j L_{ji} - L_{ij} f_i).$$ (D.40)

The variables f_i are convenient for constructing generalisations of the RS models that include spin variables [1].

D.4 Derivation of the Spectral-Dependent r-Matrices

To determine the r-matrices governing the structure (2.395), we start with computing the Poisson brackets between the components of $L(\lambda)$ given by (2.392). Applying the Poisson brackets (2.368) and (2.369), we obtain

$$
\begin{aligned}
\{L_1(\lambda), L_2(\mu)\} ={}& r_{12}L_1L_2 - L_1L_2\underline{r}_{12} + L_1\bar{r}_{21}L_2 - L_2\bar{r}_{12}L_1 \\
&- \frac{1}{\lambda}\Big[Q_1^{-1}r_{12}Q_1L_1'L_2 - L_1'L_2(Q_1^{-1}\underline{r}_{12}Q_1 - \overline{C}_{12}) + L_1'Q_1^{-1}\bar{r}_{21}Q_1L_2 - L_2(Q_1^{-1}\bar{r}_{12}Q_1 + \overline{C}_{12})L_1'\Big] \\
&- \frac{1}{\mu}\Big[Q_2^{-1}r_{12}Q_2L_1L_2' - L_1L_2'(Q_2^{-1}\underline{r}_{12}Q_2 + \overline{C}_{12}) + L_1(Q_2^{-1}\bar{r}_{21}Q_2 + \overline{C}_{12})L_2' - L_2'Q_2^{-1}\bar{r}_{12}Q_2L_1\Big] \\
&+ \frac{1}{\lambda\mu}\Big[Q_1^{-1}Q_2^{-1}r_{12}Q_1Q_2L_1'L_2' - L_1'L_2'Q_1^{-1}Q_2^{-1}\underline{r}_{12}Q_1Q_2 \\
&\qquad + L_1'(Q_1^{-1}Q_2^{-1}\bar{r}_{21}Q_1Q_2 + \overline{C}_{12})L_2' - L_2'(Q_1^{-1}Q_2^{-1}\bar{r}_{12}Q_1Q_2 + \overline{C}_{12})L_1'\Big].
\end{aligned}
\tag{D.41}
$$

Further developments are based on the following observation about the properties of the r-matrices rotated by Q's. First, we find that

$$
\begin{aligned}
Q_1^{-1}r_{12}Q_1 &= r_{12} - \sigma_{12} - C_{12} + \mathbb{1}\otimes\mathbb{1}, \\
Q_2^{-1}r_{12}Q_2 &= r_{12} + \sigma_{21} + V_{12} - \mathbb{1}\otimes\mathbb{1}, \\
Q_1^{-1}Q_2^{-1}r_{12}Q_1Q_2 &= r_{12} + \sigma_{21} - \sigma_{12},
\end{aligned}
\tag{D.42}
$$

where σ_{12} is given by (2.396) and we introduced

$$
V_{12} = \sum_{i,j=1}^{N} \frac{Q_i}{Q_j} E_{ij}\otimes E_{ji}.
$$

Second,

$$
\begin{aligned}
Q_1^{-1}\underline{r}_{12}Q_1 - \overline{C}_{12} &= \underline{r}_{12} - C_{12}, \\
Q_2^{-1}\underline{r}_{12}Q_2 + \overline{C}_{12} &= \underline{r}_{12} + V_{12}, \\
Q_1^{-1}Q_2^{-1}\underline{r}_{12}Q_1Q_2 &= \underline{r}_{12}.
\end{aligned}
\tag{D.43}
$$

Finally,

$$
\begin{aligned}
Q_1^{-1}\bar{r}_{12}Q_1 + \overline{C}_{12} &= \bar{r}_{12} - \sigma_{12} + \mathbb{1}\otimes\mathbb{1}, \\
Q_2^{-1}\bar{r}_{12}Q_2 &= \bar{r}_{12}, \\
Q_1^{-1}Q_2^{-1}\bar{r}_{12}Q_1Q_2 + \overline{C}_{12} &= \bar{r}_{12} - \sigma_{12} + \mathbb{1}\otimes\mathbb{1}.
\end{aligned}
\tag{D.44}
$$

With the help of (D.42), (D.43) and (D.44) the bracket (D.41) turns into

$$
\begin{aligned}
\{L_1(\lambda), L_2(\mu)\} = {} & r_{12}L_1L_2 - L_1L_2\underline{r}_{12} + L_1\bar{r}_{21}L_2 - L_2\bar{r}_{12}L_1 \\
& -\frac{1}{\lambda}\Big[(r_{12} - \sigma_{12} - C_{12})L_1'L_2 - L_1'L_2(\underline{r}_{12} - C_{12}) + L_1'\bar{r}_{21}L_2 - L_2(\bar{r}_{12} - \sigma_{12})L_1'\Big] \\
& -\frac{1}{\mu}\Big[(r_{12} + \sigma_{21})L_1L_2' - L_1L_2'\underline{r}_{12} + L_1(\bar{r}_{21} - \sigma_{21})L_2' - L_2'\bar{r}_{12}L_1\Big] \\
& +\frac{1}{\lambda\mu}\Big[(r_{12} - \sigma_{12} + \sigma_{21})L_1'L_2' - L_1'L_2'\underline{r}_{12} \\
& \qquad\qquad +L_1'(\bar{r}_{21} - \sigma_{21})L_2' - L_2'(\bar{r}_{12} - \sigma_{12})L_1'\Big] .
\end{aligned}
\tag{D.45}
$$

Notice that the element V_{12} totally decouples from from the right hand side of (D.45), as it satisfies an identity

$$
V_{12}L_1L_2' = L_1L_2'V_{12} ,
$$

which can be straightforwardly verified by computing its matrix elements,

$$
(V_{12}L_1L_2')_{mn,kl} = t L_{ml} L_{kn} Q_l Q_k^{-1} = (L_1L_2'V_{12})_{mn,kl} .
$$

The next progress relies on the identity (2.388), i.e.,

$$
L' = L - \frac{1 - t^N}{N} e \otimes c^t L ,
\tag{D.46}
$$

and the special (Frobenius) structure of the r-matrices. Indeed, from (D.46) it follows that

$$
(E_{ii} - E_{ij})L' = (E_{ii} - E_{ij})L , \quad \forall i, j = 1, \ldots, N .
$$

This observation immediately shows that

$$
\begin{aligned}
\bar{r}_{12}L_1' = \bar{r}_{12}L_1 , \quad & \bar{r}_{21}L_2' = \bar{r}_{21}L_2 , \\
\sigma_{12}L_1' = \sigma_{12}L_1 , \quad & \sigma_{21}L_2' = \sigma_{21}L_2 .
\end{aligned}
\tag{D.47}
$$

Analogously,

$$
r_{12}L_2' = r_{12}L_2 , \quad r_{21}L_1' = r_{21}L_1 .
\tag{D.48}
$$

Owing to the identity (2.371), we then have

$$
r_{12}(L_1' - L_1) = (-r_{21} + C_{12} - \mathbb{1} \otimes \mathbb{1})(L_1' - L_1) = (C_{12} - \mathbb{1} \otimes \mathbb{1})(L_1' - L_1) ,
$$

or, in other words,

$$r_{12}L'_1 = (C_{12} - \mathbb{1} \otimes \mathbb{1})L'_1 + (r_{12} - C_{12} + \mathbb{1} \otimes \mathbb{1})L_1 . \qquad (D.49)$$

Thus, to obtain an irreducible expression for the bracket (D.45), whenever its is possible we will use the reduction formulae (D.47), (D.48) and (D.49) to replace L' with L on the right hand side of (D.45). This replacement leads to the following result

$$
\begin{aligned}
\{L_1(\lambda), L_2(\mu)\} &= \\
&= \Big(r_{12} - \frac{1}{\lambda}(r_{12} - \sigma_{12} - C_{12} + \mathbb{1} \otimes \mathbb{1}) - \frac{1}{\mu}(r_{12} + \sigma_{21}) + \frac{1}{\lambda\mu}(r_{12} - C_{12} + \mathbb{1} \otimes \mathbb{1})\Big)L_1 L_2 \\
&\quad - L_1 L_2 \underline{r}_{12} + \frac{1}{\lambda}L'_1 L_2 \underline{r}_{12} + \frac{1}{\mu}L_1 L'_2 \underline{r}_{12} - \frac{1}{\lambda\mu}L'_1 L'_2 \underline{r}_{12} \\
&\quad + L_1\Big(\bar{r}_{21} - \frac{1}{\mu}(\bar{r}_{21} - \sigma_{21})\Big)L_2 - L_2\Big(\bar{r}_{12} - \frac{1}{\lambda}(\bar{r}_{12} - \sigma_{12})\Big)L_1 \\
&\quad - \frac{1}{\lambda}L'_1\Big(\bar{r}_{21} - \frac{1}{\mu}(\bar{r}_{21} - \sigma_{21})\Big)L_2 + \frac{1}{\mu}L'_2\Big(\bar{r}_{12} - \frac{1}{\lambda}(\bar{r}_{12} - \sigma_{12})\Big)L_1 \\
&\quad + \frac{1}{\lambda}\Big(\mathbb{1} \otimes \mathbb{1} + \frac{1}{\mu}(C_{12} + \sigma_{21} - \mathbb{1} \otimes \mathbb{1})\Big)L'_1 L_2 - \frac{1}{\lambda}\Big(C_{12} + \frac{1}{\mu}\sigma_{12}\Big)L_1 L'_2 . \qquad (D.50)
\end{aligned}
$$

We will now search for the spectral dependent r-matrices r^s that allow one to present the bracket above in the form

$$
\begin{aligned}
\{L_1(\lambda), L_2(\mu)\} = {}& r^s_{12}L_1(\lambda)L_2(\mu) - L_1(\lambda)L_2(\mu)\underline{r}^{\,s}_{12} \\
& + L_1(\lambda)\bar{r}^s_{21}L_2(\mu) - L_2(\mu)\bar{r}^s_{12}L_1(\lambda) . \qquad (D.51)
\end{aligned}
$$

An examination of this expression shows that it involves the following matrices r_{12}, $\bar{r}_{12}$, $\bar{r}_{21}$, σ_{12}, σ_{21} and C_{12}. There is also the identity matrix $\mathbb{1} \otimes \mathbb{1}$ but we ignore its presence for the moment. Thus, the structure of (D.50) motivates to try for the spectral-dependent r-matrices the following minimal ansatz

$$
\begin{aligned}
r^s_{12} &= r_{12} + \alpha\sigma_{12} + \beta\sigma_{21} + \gamma C_{12} \\
\bar{r}^s_{12} &= \bar{r}_{12} + \delta_{12}\sigma_{12} , \\
\bar{r}^s_{21} &= \bar{r}_{21} + \delta_{21}\sigma_{21} , \\
\underline{r}^{\,s}_{12} &= \underline{r}_{12} + \gamma C_{12} .
\end{aligned}
$$

This ansatz depends on five undermined parameters: $\alpha, \beta, \gamma, \delta_{12}$ and δ_{21}, which should eventually be expressed via λ and μ. We then plug this ansatz together with the expression (2.392) for the spectral-dependent Lax matrix into (D.51) and, by using the reduction formulae (D.47), (D.48) and (D.49), bring the resulting expression to the following irreducible form

$$\{L_1(\lambda), L_2(\mu)\} =$$

$$= \Big[r_{12} + \alpha\sigma_{12} + \beta\sigma_{21} - \frac{1}{\lambda}(r_{12} - C_{12} + \mathbb{1} \otimes \mathbb{1} + \alpha\sigma_{12})$$

$$- \frac{1}{\mu}(r_{12} + \beta\sigma_{21}) + \frac{1}{\lambda\mu}(r_{12} - C_{12} + \mathbb{1} \otimes \mathbb{1}) \Big] L_1 L_2$$

$$- L_1 L_2 \underline{r}_{12} + \frac{1}{\lambda} L_1' L_2 \underline{r}_{12} + \frac{1}{\mu} L_1 L_2' \underline{r}_{12} - \frac{1}{\lambda\mu} L_1' L_2' \underline{r}_{12}$$

$$+ L_1 \Big[(\bar{r}_{21} + \delta_{21}\sigma_{21}) - \frac{1}{\mu}(\bar{r}_{21} + \delta_{21}\sigma_{21}) \Big] L_2 - L_2 \Big[(\bar{r}_{12} + \delta_{12}\sigma_{12}) - \frac{1}{\lambda}(\bar{r}_{12} + \delta_{12}\sigma_{12}) \Big] L_1$$

$$- \frac{1}{\lambda} L_1' \Big[(\bar{r}_{21} + \delta_{21}\sigma_{21}) - \frac{1}{\mu}(\bar{r}_{21} + \delta_{21}\sigma_{21}) \Big] L_2 + \frac{1}{\mu} L_2' \Big[(\bar{r}_{12} + \delta_{12}\sigma_{12}) - \frac{1}{\lambda}(\bar{r}_{12} + \delta_{12}\sigma_{12}) \Big] L_1$$

$$+ \Big[-\frac{1}{\lambda}(C_{12} - \mathbb{1} \otimes \mathbb{1} + \beta\sigma_{21} + \gamma C_{12}) + \frac{1}{\mu}\gamma C_{12} + \frac{1}{\lambda\mu}(C_{12} + \beta\sigma_{21} - \mathbb{1} \otimes \mathbb{1}) \Big] L_1' L_2$$

$$+ \Big[\frac{1}{\lambda}\gamma C_{12} - \frac{1}{\mu}(\alpha\sigma_{12} + \gamma C_{12}) + \frac{1}{\lambda\mu}\alpha\sigma_{12} \Big] L_1 L_2'. \tag{D.52}$$

Comparison of the first lines of (D.50) and (D.52) yields a unique solution for α and β,

$$\alpha = \frac{1}{\lambda - 1}, \qquad \beta = -\frac{1}{\mu - 1}.$$

Comparison of third lines yields

$$\delta_{12} = \frac{1}{\lambda - 1}, \qquad \delta_{21} = \frac{1}{\mu - 1},$$

which automatically makes the fourth lines of (D.50) and (D.52) to be equal. Finally, with α and β already determined, comparison of the terms in front of $L_1' L_2$ or $L_1 L_2'$ gives an unambiguous solution for γ,

$$\gamma = \frac{\mu}{\lambda - \mu}.$$

Thus, we end up with the following expressions for the spectral-dependent r-matrices realising the Poisson algebra (D.51)

$$r_{12}(\lambda, \mu) = \frac{\lambda r_{12} + \mu r_{21}}{\lambda - \mu} + \frac{\sigma_{12}}{\lambda - 1} - \frac{\sigma_{21}}{\mu - 1} + \frac{\mu}{\lambda - \mu} \mathbb{1} \otimes \mathbb{1},$$

$$\bar{r}_{12}(\lambda) = \bar{r}_{12} + \frac{\sigma_{12}}{\lambda - 1}, \tag{D.53}$$

$$\underline{r}_{12}(\lambda, \mu) = r_{12}(\lambda, \mu) + \bar{r}_{21}(\mu) - \bar{r}_{12}(\lambda) = \frac{\lambda \underline{r}_{12} + \mu \underline{r}_{21}}{\lambda - \mu} + \frac{\mu}{\lambda - \mu} \mathbb{1} \otimes \mathbb{1},$$

where we used the relation (2.371) to bring the result to a more symmetric form. Finally, using the shift symmetry (2.403), we can omit in (D.53) the terms proportional to the identity matrix, obtaining a slightly simpler solution (2.397).

Reference

1. Arutyunov, G.E., Frolov, S.A.: On Hamiltonian structure of the spin Ruijsenaars-Schneider model. J. Phys. A **31**, 4203–4216 (1998)

Appendix E
Non-stationary Scattering Theory

In Chap. 3 we introduced the two-body scattering matrix by using the time-independent scattering theory. To recall the physical meaning of the S-matrix (4.42), here we turn to the time-dependent picture where one considers an actual scattering process of (sharp) wave packets. The time-dependent Schrödinder equation yields

$$i\frac{d\varphi}{dt} = H\varphi, \quad \longrightarrow \quad \varphi(t) = e^{-iHt}\varphi(0).$$

We assume that $\varphi(0)$ is expandable over the stationary states from the continuous spectrum. In particular, using the function Ψ_1, see Table 4.2, we construct the following time-dependent solution $\varphi_1(q, t)$ of the Schrödinger equation

$$\varphi_1(q, t) = \frac{1}{\sqrt{2\pi}} \int\limits_{p_1 > p_2} dp_1 dp_2\, \chi(p_1, p_2) e^{-i(p_1^2/2 + p_2^2/2)t} \Psi_1(q_1, q_2)$$

$$= \frac{1}{\sqrt{2\pi}} \int\limits_{p_1 > p_2} dp_1 dp_2\, \chi(p_1, p_2) e^{-i(p_1^2/2 + p_2^2/2)t} e^{iP(q_1 + q_2)/2} \psi_1(k). \quad \text{(E.1)}$$

We assume that $\chi(p_1, p_2)$ is sharply concentrated around p_1^0 and p_2^0, i.e. it is non-zero in a small neighbourhood of the momenta p_1^0 and p_2^0 that satisfy inequality $p_1^0 > p_2^0$,

$$p_1^0 + \tfrac{1}{2}\Delta p_1 \leqslant p_1 \leqslant p_1^0 + \tfrac{1}{2}\Delta p_1,$$
$$p_2^0 + \tfrac{1}{2}\Delta p_2 \leqslant p_2 \leqslant p_2^0 + \tfrac{1}{2}\Delta p_2,$$

and it is normalised as

$$\int\limits_{p_1 > p_2} dp_1 dp_2\, |\chi(p_1, p_2)|^2 = 1.$$

© Springer Nature Switzerland AG 2019
G. Arutyunov, *Elements of Classical and Quantum Integrable Systems*,
UNITEXT for Physics, https://doi.org/10.1007/978-3-030-24198-8

Then we have

$$\int_{-\infty}^{\infty}\int_{-\infty}^{\infty} dq_1 dq_2\, |\varphi_1(q,t)|^2 = \frac{1}{2\pi} \int\limits_{p_1>p_2} dp_1 dp_2 \int\limits_{p_1'>p_2'} dp_1' dp_1'\, \overline{\chi(p_1,p_2)}\chi(p_1',p_2') \times$$

$$\times\, e^{i(p_1^2/2+p_2^2/2-p_1'^2/2-p_2'^2/2)t} \times \underbrace{\frac{1}{2}\int_{-\infty}^{+\infty} dQ\, e^{i(P'-P)Q/2}}_{\delta(P'-P)}\underbrace{\int_{-\infty}^{\infty} dx\, \overline{\psi(k,x)}\psi(k',x)}_{2\pi\delta(k-k')}\,,$$

where we made a change of variables $x = q_2 - q_1$, $Q = q_1 + q_2$. Resolution of the delta-functions gives $p_1' = p_1$ and $p_2' = p_2$, and, therefore,

$$\int_{-\infty}^{\infty}\int_{-\infty}^{\infty} dq_1 dq_2\, |\varphi_1(q,t)|^2 = \int\limits_{p_1>p_2} dp_1 dp_2\, |\chi(p_1,p_2)|^2 = 1\,.$$

Thus, $\varphi_1(q,t)$ is a *square-integrable* solution of the non-stationary Schrödinger equation that represents a combination of two moving wave packets.

By using the asymptotics of Ψ_1, it is straightforward to find the expressions for $\varphi_1(q,t)$ in the asymptotic regions $q_1 \ll q_2$ and $q_1 \gg q_2$. Introducing $k^0 = p_1^0 - p_2^0 > 0$, we have

$$q_1 \ll q_2: \quad \varphi_1(q,t) = \frac{1}{\sqrt{2\pi}} \int\limits_{p_1>p_2} dp_1 dp_2\, \chi(p_1,p_2) \times$$

$$\times \left(e^{ip_1q_1+ip_2q_2} + A(k)e^{ip_2q_1+ip_1q_2} \right)e^{-i(p_1^2/2+p_2^2/2)t}$$

$$\approx \varphi_{\rightarrow}(q,t) + A(k_0)\varphi_{\leftarrow}(q,t); \qquad (E.2)$$

$$q_1 \gg q_2: \quad \varphi_1(q,t) = \frac{1}{\sqrt{2\pi}} \int\limits_{p_1>p_2} dp_1 dp_2\, \chi(p_1,p_2)B(k)e^{ip_1q_1+ip_2q_2}e^{-i(p_1^2/2+p_2^2/2)t}$$

$$\approx B(k_0)\varphi_{\rightarrow}(q,t)\,. \qquad (E.3)$$

Here we used the fact that $\chi(p_1,p_2)$ is sharply concentrated around (p_1^0, p_2^0) and introduced the following functions

$$\varphi_{\rightarrow}(q,t) = \frac{1}{\sqrt{2\pi}} \int\limits_{p_1>p_2} dp_1 dp_2\, \chi(p_1,p_2)e^{ip_1q_1+ip_2q_2-i(p_1^2/2+p_2^2/2)t}\,,$$

$$\varphi_{\leftarrow}(q,t) = \frac{1}{\sqrt{2\pi}} \int\limits_{p_1>p_2} dp_1 dp_2\, \chi(p_1,p_2)e^{ip_1q_2+ip_2q_1-i(p_1^2/2+p_2^2/2)t}\,. \qquad (E.4)$$

These functions can be approximately evaluated. We have

$$\varphi_{\rightarrow}(q,t) \approx \frac{\chi(p_1^0, p_2^0)}{\sqrt{2\pi}} \prod_{j=1}^{2} \int_{p_j^0 - \frac{\Delta p_j}{2}}^{p_j^0 + \frac{\Delta p_j}{2}} dp_j \, e^{ip_j q_j - ip_j^2 t/2}$$

$$\approx \underbrace{\frac{\chi(p_1^0, p_2^0)}{\sqrt{2\pi}} \frac{\sin \frac{\Delta p_1}{2}(q_1 - p_1^0 t)}{\frac{1}{2}(q_1 - p_1^0 t)} \frac{\sin \frac{\Delta p_2}{2}(q_2 - p_2^0 t)}{\frac{1}{2}(q_2 - p_2^0 t)}}_{\text{amplitude}} e^{ip_1^0 q_1 + ip_2^0 q_2 - i((p_1^0)^2/2 + (p_2^0)^2/2)t}.$$

Obviously, the amplitude of $\varphi_{\rightarrow}(x,t)$ is essentially non-zero in the neighbourhood of the points

$$q_1 = p_1^0 t, \quad q_2 = p_2^0 t.$$

Thus, $\varphi_{\rightarrow}$ describes a state of two free particles moving with momenta $p_1^0 > p_2^0$. Analogously, $\varphi_{\leftarrow}$ corresponds to a two-particle state with the probability amplitude concentrated on the trajectories

$$q_1 = p_2^0 t, \quad q_2 = p_1^0 t.$$

Consider now the behaviour of $\varphi_1(q,t)$ for $t \to \pm\infty$. For $t \to -\infty$, the wave $\varphi_{\rightarrow}$ is supported in the region $q_1 \ll q_2$ and vanishes for $q_1 \gg q_2$, while $\varphi_{\leftarrow}$ has the opposite behaviour, so it is present only in the region $q_1 \gg q_2$. From (E.2) and (E.3) we deduce that for $t \to -\infty$

$$q_1 \ll q_2 : \quad \varphi_1(q,t) = \varphi_{\rightarrow}(q,t),$$
$$q_1 \gg q_2 : \quad \varphi_1(q,t) = 0.$$

The faster particle is on the left from the slower one and, as time grows, the particles (wave packets) approach each other.

For $t \to +\infty$, we see that the wave $\varphi_{\rightarrow}$ is supported in the region $q_1 \gg q_2$ and vanishes for $q_1 \ll q_2$, while $\varphi_{\leftarrow}$ has asymptotic support for $q_1 \ll q_2$. Thus, for $t \to +\infty$ one finds

$$q_1 \ll q_2 : \quad \varphi_1(q,t) = A(k_0)\varphi_{\leftarrow}(q,t),$$
$$q_1 \gg q_2 : \quad \varphi_1(q,t) = B(k_0)\varphi_{\rightarrow}(q,t).$$

In the standard terminology $A(k)$ and $B(k)$ are called *reflection and transmission coefficients* the, respectively. The equality $|A|^2 + |B|^2 = 1$ renders conservation of probability. Initially the particles are localised and move towards each other. Entering the interaction region set up by the potential, each of the individual wave packets splits into reflected and transmitted pieces, entangled so they are either both reflected or both transmitted.

In a similar manner we can construct the time-dependent solution $\varphi_2(q, t)$ that corresponds to the second solution Ψ_2 from Table 4.2. It describes the scattering process, where initially the faster particle is on the right of the slower one.

Index

© Springer Nature Switzerland AG 2019
G. Arutyunov, *Elements of Classical and Quantum Integrable Systems*,
UNITEXT for Physics, https://doi.org/10.1007/978-3-030-24198-8

Printed in the United States
By Bookmasters